Fascinating Life Sciences

This interdisciplinary series brings together the most essential and captivating topics in the life sciences. They range from the plant sciences to zoology, from the microbiome to macrobiome, and from basic biology to biotechnology. The series not only highlights fascinating research; it also discusses major challenges associated with the life sciences and related disciplines and outlines future research directions. Individual volumes provide in-depth information, are richly illustrated with photographs, illustrations, and maps, and feature suggestions for further reading or glossaries where appropriate.

Interested researchers in all areas of the life sciences, as well as biology enthusiasts, will find the series' interdisciplinary focus and highly readable volumes especially appealing.

More information about this series at https://link.springer.com/bookseries/15408

Gerald Mayr

Paleogene Fossil Birds

Second Edition

Springer

Gerald Mayr
Senckenberg Research Institute
Frankfurt am Main, Germany

ISSN 2509-6745　　　　　　ISSN 2509-6753　(electronic)
Fascinating Life Sciences
ISBN 978-3-030-87647-0　　ISBN 978-3-030-87645-6　(eBook)
https://doi.org/10.1007/978-3-030-87645-6

© The Editor(s) (if applicable) and The Author(s), under exclusive license to Springer Nature Switzerland AG 2009, 2022

This work is subject to copyright. All rights are solely and exclusively licensed by the Publisher, whether the whole or part of the material is concerned, specifically the rights of translation, reprinting, reuse of illustrations, recitation, broadcasting, reproduction on microfilms or in any other physical way, and transmission or information storage and retrieval, electronic adaptation, computer software, or by similar or dissimilar methodology now known or hereafter developed.

The use of general descriptive names, registered names, trademarks, service marks, etc. in this publication does not imply, even in the absence of a specific statement, that such names are exempt from the relevant protective laws and regulations and therefore free for general use.

The publisher, the authors, and the editors are safe to assume that the advice and information in this book are believed to be true and accurate at the date of publication. Neither the publisher nor the authors or the editors give a warranty, expressed or implied, with respect to the material contained herein or for any errors or omissions that may have been made. The publisher remains neutral with regard to jurisdictional claims in published maps and institutional affiliations.

This Springer imprint is published by the registered company Springer Nature Switzerland AG.
The registered company address is: Gewerbestrasse 11, 6330 Cham, Switzerland

Preface

The first edition of this book was written at a time when comprehensive analyses of molecular sequence data just began to provide a robust framework for the interrelationships of extant birds. Some results that were new and unexpected back then are now widely accepted and supported by multiple independent data sets. This phylogenetic scaffold opened a fresh perspective on the interpretation of unusual character mosaics seen in some fossil taxa. On the other hand, some of these very fossils also bridge the morphological gaps between disparate avian groups now known to be closely related.

In the 13 years that have elapsed since the publication of the first edition, numerous fossils have been described from Paleogene sites. This geological period was critical for the evolution of modern birds, and the data obtained from new specimens not only improved our knowledge of various extinct groups of birds but also yielded insights into avifaunas of previously underrepresented geographic areas and geological intervals. For the first edition, I could draw on an already rich fossil record from Europe and North America, whereas Paleogene avifaunas of the Southern Hemisphere were poorly known. This disparity has now been diminished by new fossils from Africa, Antarctica, and New Zealand, which were reported during the past decade.

The first edition of this book provided a comprehensive overview of the then known Paleogene fossil record of birds, but the large numbers of new fossils and our improved understanding of the interrelationships of avian higher-level taxa necessitated substantial revisions of many systematic sections. To shorten the extensive reference lists, various earlier citations of phylogenetic analyses were substituted with up-to-date references. For improved readability of the text, I also removed the indication of taxonomic authorities after species names. Finally, this new edition benefits from a larger format and an increased number of figures, many of which are now in color.

Fossil birds are only found after intensive collecting efforts, and more than in other areas of vertebrate paleontology, they are often chance discoveries of enthusiastic amateur collectors. Such specimens in private hand tend to be ignored in the scientific literature. However, an account of Paleogene fossil birds would be incomplete without consideration of what is the largest and most significant collection of Paleogene birds assembled by a single person, namely the London Clay specimens of the late Michael Daniels. This extraordinary collection, which encompasses fossils of several hundred bird individuals, has been repeatedly mentioned in the literature, but published images exist of just a few of the specimens. Owing to the courtesy of Michael Daniels, I was able to include photographs of some key specimens in this revised edition, which enables the reader to assess osteological characteristics of early Eocene birds that are not easily discerned in images of compression fossils.

Frankfurt am Main, Germany   Gerald Mayr

Acknowledgments

I am particularly indebted to Michael Daniels for permitting the inclusion of photos of some of his key fossils in the new edition of this book; I also owe Michael many thanks for repeated access to his collection, inspiring discussions on the fossils, and help in their identification. During the past years, I have benefited much from collaborations with other enthusiastic "amateurs" (in the best sense of the word), who discovered many finds mentioned and figured in this book. Most pivotal for the study of large flightless diving birds were James Goedert, who collected the bulk of the plotopterid material from the Olympic Peninsula, and Leigh Love, who found most specimens from the Waipara Greensand. The fossils from the Ikovo site in Ukraine were collected by Evgenij Zvonok, whom I also thank for involving me in the study of this material. Sven Tränkner is thanked for taking most of the photographs in this book. Additional images of Messel birds were provided by Anika Vogel and Wolfgang Fuhrmannek. Further photographs of fossil specimens are courtesy of Herculano Alvarenga, Lance Grande, Peter Houde, Cécile Mourer-Chauviré, and Ilka Weidig. Access to fossil specimens was kindly enabled by Sandra Chapman, Gilles Cuny, Annelise Folie, Eberhard Frey, Norbert Hauschke, Amy Henrici, Carl Mehling, Norbert Micklich, Cécile Mourer-Chauviré, Stephan Schaal, Paul Scofield, Thierry Smith, Basil Thüring, and Nicholas Tourment. Herculano Alvarenga, Zbigniew Bochenski, Vanesa De Pietri, James Goedert, Daniel Ksepka, Albrecht Manegold, Cécile Mourer-Chauviré, Paul Scofield, Thierry Smith, Min Wang, and Nikita Zelenkov are thanked for fruitful and pleasant scientific collaborations on Paleogene birds during the past years. Olaf Vogel was pivotal in the preparation of many key fossils. Lars Koerner of Springer publishing house encouraged me to revise this book and enabled the new format of this work; his impetus initiated this second edition. Again, however, I owe the greatest thanks to my wife, Eun-Joo, for her tolerance and encouragement during the months I spent at the computer after my regular work.

Contents

1	**Introduction**		1
2	**Basic Terminology and the Broader Phylogenetic and Geological Framework**		3
	2.1	Phylogenetic Terminology	3
	2.2	Bird Bones and Often Used Osteological Terms	4
	2.3	The Higher-Level Phylogeny of Extant Birds	7
	2.4	Mesozoic Neornithes	12
	2.5	Stratigraphy and Major Paleogene Fossil Localities	17
		2.5.1 Europe	17
		2.5.2 Asia	20
		2.5.3 North America	20
		2.5.4 Central and South America	21
		2.5.5 Africa	22
		2.5.6 Australia, New Zealand, and Antarctica	22
	References		22
3	**Palaeognathous Birds**		29
	3.1	Lithornithidae	29
	3.2	Palaeotididae, Geranoididae, and Eogruidae—Northern Hemispheric Stem Group Representatives of the Struthioniformes	30
		3.2.1 The European Palaeotididae	32
		3.2.2 The North American Geranoididae	33
		3.2.3 The Asian Eogruidae and Ergilornithidae	35
	3.3	Remiornithidae	37
	3.4	Eremopezidae	37
	3.5	Rheiformes (Rheas)	38
	3.6	Casuariiformes (Emus and Cassowaries)	40
	3.7	Putative Palaeognathous Bird from the Eocene of Antarctica	40
	References		40
4	**Pelagornithidae, Gastornithidae, and Crown Group Galloanseres**		43
	4.1	Pelagornithidae (Bony-Toothed Birds)	43
	4.2	Gastornithidae	50
	4.3	Galliformes (Landfowl)	52
		4.3.1 The Eocene Gallinuloididae—Stem Group Galliformes from North America and Europe	52
		4.3.2 Other Early and Middle Eocene Stem Group Galliforms	54
		4.3.3 The Paraortygidae—A Well-Represented Group of Stem Group Galliformes	56
		4.3.4 The North American Taxa *Procrax*, *Archaealectrornis*, and *Palaeonossax*	57
		4.3.5 Quercymegapodiidae	57
		4.3.6 The Australian Megapodiidae (Megapodes)	58

		4.3.7	Phasianoidea (Guinea Fowl, Quails, Pheasants, and Allies)	58
	4.4	The Australian Dromornithidae		59
	4.5	Anseriformes (Waterfowl)		59
		4.5.1	Anhimidae (Screamers)	60
		4.5.2	Presbyornithidae	62
		4.5.3	Anseranatidae (Magpie Geese)	64
		4.5.4	Anatidae (Ducks, Geese, and Swans)	65
	References			68

5 Mirandornithes (Grebes and Flamingos), Charadriiformes (Shorebirds and Allies), and Gruiformes (Rails, Cranes, and Allies) ... 73

	5.1	Podicipediformes (Grebes) and Phoenicopteriformes (Flamingos)		73
	5.2	Charadriiformes (Shorebirds and Allies)		76
		5.2.1	Lari (Gulls, Auks, and Allies)	78
		5.2.2	Charadrii (Plovers and Allies)	80
		5.2.3	Scolopaci (Sandpipers and Allies)	80
	5.3	Gruiformes (Rails, Cranes, and Allies)		81
		5.3.1	Messelornithidae and *Walbeckornis*	81
		5.3.2	Songziidae	83
		5.3.3	Ralloidea (Finfoots, Flufftails, and Rails)	84
		5.3.4	Parvigruidae	87
		5.3.5	Gruoidea (Trumpeters, Limpkins, and Cranes)	87
	References			89

6 Opisthocomiformes (Hoatzins), "Columbaves" (Doves, Cuckoos, Bustards, and Allies), and Strisores (Nightjars, Swifts, Hummingbirds, and Allies) ... 93

	6.1	Opisthocomiformes (Hoatzin)		93
	6.2	Otididae (Bustards)		94
	6.3	Foratidae		94
	6.4	Musophagiformes (Turacos)		94
	6.5	Cuculiformes (Cuckoos)		96
	6.6	Columbiformes (Doves), Pterocliformes (Sandgrouse), and Mesitornithiformes (Mesites)		96
	6.7	Eopachypterygidae, *Eocuculus*, and *Carpathiavis*		96
	6.8	Strisores (Nightjars, Swifts, Hummingbirds, and Allies)		98
		6.8.1	Archaeotrogonidae	98
		6.8.2	Caprimulgiformes (Nightjars)	100
		6.8.3	Nyctibiiformes (Potoos)	101
		6.8.4	Steatornithiformes (Oilbirds)	102
		6.8.5	Fluvioviridavidae	102
		6.8.6	Parvicuculidae	105
		6.8.7	Podargiformes (Frogmouths)	105
		6.8.8	Protocypselomorphus	105
		6.8.9	Aegotheliformes (Owlet Nightjars) and Apodiformes (Swifts and Hummingbirds)	106
	References			114

7 Phaethontiformes and Aequornithes: The Aquatic and Semi-aquatic Neoavian Taxa ... 117

	7.1	Phaethontiformes (Tropicbirds)		117
	7.2	Gaviiformes (Loons)		119
	7.3	Procellariiformes (Tubenoses)		122
		7.3.1	The Early Oligocene Diomedeoididae	123

		7.3.2	Diomedeidae (Albatrosses)...........................	126
	7.4	Sphenisciformes (Penguins).................................		126
	7.5	Ciconiidae (Storks).......................................		134
	7.6	Suliformes: Fregatidae (Frigatebirds) and Suloidea (Gannets, Boobies, Cormorants, and Anhingas)................................		135
		7.6.1	Protoplotidae.......................................	136
		7.6.2	Fregatidae (Frigatebirds)............................	136
		7.6.3	Sulidae (Gannets and Boobies).......................	137
		7.6.4	Phalacrocoracidae (Cormorants) and Anhingidae (Anhingas)....	138
		7.6.5	Plotopteridae......................................	140
	7.7	Threskiornithidae (Ibises and Spoonbills)....................		144
	7.8	Scopidae (Hamerkop), Balaenicipitidae (Shoebill), and Pelecanidae (Pelicans)..		146
	7.9	Ardeidae (Herons).......................................		146
	7.10	Xenerodiopidae...		146
	References...			147

8 Accipitriformes (New World Vultures, Hawks, and Allies), Falconiformes (Falcons), and Cariamiformes (Seriemas and Allies).................... 153

	8.1	Accipitriformes (New World Vultures, Hawks, and Allies)..........		153
		8.1.1	Teratornithidae and Cathartidae (New Word Vultures).........	153
		8.1.2	Sagittariidae.......................................	154
		8.1.3	Pandionidae (Ospreys), and Accipitridae (Hawks and Allies)....	154
		8.1.4	Horusornithidae....................................	156
	8.2	Falconiformes (Falcons)..................................		157
	8.3	Masillaraptoridae.......................................		157
	8.4	Cariamiformes (Seriemas and Allies).........................		159
		8.4.1	Phorusrhacidae.....................................	159
		8.4.2	Idiornithidae.......................................	161
		8.4.3	Bathornithidae.....................................	164
		8.4.4	*Strigogyps* and the Ameghinornithidae....................	165
	8.5	Cariamiform-like Paleogene Taxa of Uncertain Affinities..........		168
		8.5.1	*Elaphrocnemus* and *Paracrax*...........................	168
		8.5.2	Eleutherornithidae..................................	170
		8.5.3	Salmilidae...	171
		8.5.4	*Gradiornis*..	172
	References...			173

9 Psittacopasseres: Psittaciformes (Parrots) and Passeriformes (Passerines)... 177

	9.1	Halcyornithidae and Messelornithidae: Owl/Parrot Mosaics of Uncertain Affinities...		177
		9.1.1	Halcyornithidae....................................	178
		9.1.2	Messelasturidae....................................	179
	9.2	*Vastanavis*, *Avolatavis*, *Eurofluvioviridavis*: Further Eocene parrot-like Birds of Controversial Affinities.............................		182
	9.3	Cladornithidae...		184
	9.4	Psittaciformes (Parrots)..................................		185
	9.5	Passeriformes (Passerines)................................		185
		9.5.1	Psittacopedidae and Allies............................	185
		9.5.2	Zygodactylidae.....................................	188
		9.5.3	Passeriformes......................................	191
	References...			193

10	**Strigiformes (Owls), Coliiformes (Mousebirds), and Cavitaves (Trogons, Rollers, Woodpeckers, and Allies)**		197
	10.1	Strigiformes (Owls)	197
		10.1.1 *Berruornis* and Sophiornithidae	197
		10.1.2 Ogygoptyngidae	198
		10.1.3 Protostrigidae, *Primoptynx*, and *Eostrix*	198
		10.1.4 Necrobyinae, Palaeoglaucidae, and Selenornithinae	201
	10.2	Coliiformes (Mousebirds)	202
		10.2.1 Sandcoleidae	202
		10.2.2 Coliidae	205
	10.3	Cavitaves: Birds that Nest in Burrows and Tree Cavities	208
		10.3.1 Leptosomiformes (Courols)	208
		10.3.2 Trogoniformes (Trogons)	209
		10.3.3 Bucerotes (Hornbills, Hoopoes, and Woodhoopoes)	212
		10.3.4 Coraciiformes (Rollers, Bee-eaters, Kingfishers, and Allies)	212
		10.3.5 Piciformes (Jacamars, Puffbirds, Woodpeckers, and Allies)	218
	References		222
11	**Paleogene Avifaunas: A Synopsis of General Biogeographic and Paleoecological Aspects**		227
	11.1	Continental avifaunas of the Northern Hemisphere	227
		11.1.1 Biogeography	227
		11.1.2 Climatic Cooling and avifaunal Turnovers	228
	11.2	Continental avifaunas of the Southern Hemisphere	231
		11.2.1 Biogeography	231
		11.2.2 Extant Southern Hemispheric "endemics" in the Paleogene of the Northern Hemisphere	233
	11.3	Ecological Interactions	234
		11.3.1 Mammalian Evolution and Terrestrial avifaunas	234
		11.3.2 The Impact of Passerines on the Diversity of Paleogene Avian Insectivores	236
		11.3.3 Marine avifaunas	236
	References		237

Introduction

With more than 9000 extant species, birds are the most species-rich group of land vertebrates, rivaled only by squamates (snakes, lizards, and allies). As seed dispersers, flower pollinators, predators, prey, and through numerous other interactions they play an important ecological role in today's world. Even though a picture of past ecosystems is therefore likely to be incomplete without consideration of their avifaunas, Cenozoic fossil birds are still significantly underrepresented in many treatises of vertebrate paleontology, which is particularly true for Paleogene taxa.

The Paleogene covers the first half of the Cenozoic, from the mass extinction event at the end of the Mesozoic era, 66 million years ago (Ma), to the beginning of the Miocene, 23 Ma. It has long been recognized that this geological period was pivotal for the early diversification of modern mammals and birds. However, whereas Paleogene mammals are intensely studied and set into a palaeobiogeographic and palaeoecological context, the fossil record of birds has long played a subordinate role in evolutionary considerations.

Until a few decades ago, our knowledge of the early evolution of modern birds was indeed patchy and mainly based on fragmentary bones of often uncertain phylogenetic affinities. As will be evident from the present work, this situation has dramatically changed. At least in the Northern Hemisphere, the Paleogene fossil record of birds is no longer much short of the mammalian one with regard to the number of well-represented higher-level taxa. In some renowned fossil localities, such as the London Clay in England and the Green River Formation in North America, birds are even much more abundant than mammalian remains.

Even though the evolution of modern birds (Neornithes) already commenced in the Late Cretaceous, the fossil record from this period, which is also reviewed in the present book, is still very scant. From Paleogene sites, by contrast, numerous well-preserved avian specimens are known, many of which represent the earliest unambiguously identified records of extant higher-level taxa. These fossils not only afford information on major morphological transformations that occurred in the evolutionary lineages of the extant avian taxa but also provide critical data on the historical biogeography of certain groups. Many fossil taxa furthermore belong to remarkable extinct groups without extant counterparts and give exciting insights into the past diversity of long-vanished avifaunas.

Meanwhile, there is not only a confusing diversity of fossil taxa but significant progress has also been made in unraveling the higher-level phylogeny of extant birds. Well-supported hypotheses unite very different avian groups and allow a better understanding of unusual morphological characteristics found in Paleogene fossil birds. The present book aims at bringing this information together, with much of the following data being based on first-hand examination of fossil specimens.

After a brief introduction of phylogenetic and osteological terminology, current hypotheses on the higher-level phylogeny of birds are outlined in the second chapter, followed by an overview of Mesozoic Neornithes and key fossil localities for Paleogene birds. A survey of the Paleogene avian fossil record forms the main body of the book and is distributed over eight chapters, in which the reader finds data on fundamental morphological characteristics of the various taxa and their temporal and geographic distributions. If possible, the fossils are placed in a phylogenetic and evolutionary context based on current hypotheses on the interrelationships of extant birds. General aspects of their palaeobiogeographic and palaeoecological significance are summarized in a concluding chapter.

Although I aimed at providing a comprehensive overview of the Paleogene fossil record, I did not intend to write a catalog, and very fragmentary specimens of uncertain affinities are usually not accounted for. Fossil remains other than bones (e.g., trackways, feathers, and eggs) are also mentioned just occasionally, because often these cannot be assigned to a particular taxon with confidence.

This book sets fossil taxa in an evolutionary context and addresses the relevance of some of these as "evolutionary mosaics," which complement phylogenetic insights gained from analyses of molecular data. As such, it may not only be consulted by readers expecting a detailed review of the early fossil record of Neornithes but also by those with a general interest in the evolutionary history of modern birds.

Basic Terminology and the Broader Phylogenetic and Geological Framework

Many readers will be acquainted with phylogenetic terminology and avian osteology, and it is beyond the scope of the present work to provide an in-depth overview of these topics, each of which could fill a book on its own. For those less familiar with essential terms and definitions, these are outlined in the present chapter, which also introduces major features of the skull and some of the limb and pectoral girdle bones. Current hypotheses on the interrelationships of extant birds are reviewed, which constitute a phylogenetic framework for the study of fossil taxa. In order to set the following chapters on Paleogene birds into a full context, the Mesozoic fossil record of neornithine birds is furthermore discussed and an overview is given of major Paleogene fossil localities.

2.1 Phylogenetic Terminology

In the following, the term taxon (plural: taxa) is used for supraspecific taxonomic units, the delimitation of which is more arbitrary than that of species. Although taxa are hierarchically classified according to their content and phylogenetic position, a clear definition of higher-ranking taxa, such as "genera," "families," or "orders," is not possible. The latter terms are therefore avoided in the present work.

Phylogenetic systematics, or cladistics, aims to identify monophyletic groups, or clades, which descended from a common ancestor that is not shared with other taxa. A clade is therefore by definition monophyletic, with the often-used term "monophyletic clade" being a pleonasm. Two taxa that descend from an exclusively shared stem species are termed sister groups or sister taxa. In a given phylogeny, sister taxa are the closest relatives of one another. A sister group relationship can be denoted by a spelling in parentheses and a separation of the involved taxa by the plus sign. In this notation, hierarchically ranked taxa are indicated by a sequence of parentheses.

A clade is diagnosed by derived, or apomorphic, characters, which represent evolutionary novelties of a species and its descendants. Characters, which unite two clades or taxa are termed synapomorphies, whereas autapomorphies are derived characters that characterize a single taxon. Plesiomorphic characters, by contrast, represent primitive traits, which cannot be used to characterize clades.

An assessment of the character distribution within certain clades is often not straightforward because the convergent evolution of derived features is a common evolutionary phenomenon. Derived characters can also be secondarily lost, which further complicates the detection of phylogenetic relationships. The overarching term for the independent gain or loss of characters is homoplasy, or homoplastic character evolution.

A taxonomic entity is termed paraphyletic, if it does not include all taxa that descended from its last common ancestor, whereas it is polyphyletic, if it consists of only distantly related taxa. Many traditional avian higher-level taxa have turned out to be para- or polyphyletic assemblages, which in the following are set in quotation marks.

The crown group of a taxon is the clade including the last common ancestor of the extant species of this taxon and all its extant and extinct descendants. Stem group representatives are all taxa outside the crown group; a stem group is therefore by definition paraphyletic and does not represent a clade. Stem group representatives are always extinct, but not all crown group representatives need to be extant taxa. However, a crown group cannot be composed of extinct taxa only, and an extinct clade without living descendants does not have a crown group. The stem lineage of a clade is the evolutionary lineage of stem group representatives, which leads directly to the crown group. Whereas the stem group may include coeval species, the stem lineage represents a temporal sequence of species. Unspecified clade names used in this book refer to the total group, that is, the clade including stem and crown group representatives, which may be denoted by the

© The Author(s), under exclusive license to Springer Nature Switzerland AG 2022
G. Mayr, *Paleogene Fossil Birds*, Fascinating Life Sciences, https://doi.org/10.1007/978-3-030-87645-6_2

prefix "Pan-". The term Neornithes is applied to crown group Aves.

2.2 Bird Bones and Often Used Osteological Terms

Throughout this book, English equivalents of the Latin standard nomenclature of avian anatomy (Baumel and Witmer 1993) are used for osteological features. Anatomical directions of the avian skeleton refer to the standing bird with spread wings, and instead of the designations anterior and posterior, which are commonly applied to non-avian vertebrates, the directional terms cranial and caudal are employed in avian anatomy. The ends of most major limb bones are denoted as proximal (directing toward the center of the body) and distal (directing away from the body center).

Even though the avian skull offers numerous traits of taxonomic and phylogenetic significance, well-preserved skulls are available for just a few Paleogene birds. If complete skulls are known, these often belong to compression fossils, are badly crushed, and allow the recognition of a limited amount of osteological details. The following notes are therefore restricted to a few hallmark features of the avian skull.

The shape of the beak determines many aspects of the avian skull and shows considerable variation across different avian clades. Earlier anatomists paid much attention to different conditions of the nostrils (narial openings), which exhibit two basic types in neornithine birds: In the holorhinal beak, the nostrils are ovate openings with a concave caudal margin; this is the presumably primitive condition found in most extant birds. Another nostril type is termed schizorhinal and is characterized by long and slit-like nostrils, the caudal ends of which reach beyond the nasofrontal hinge (the transition zone between the beak and the neurocranium). Schizorhinal nostrils are associated with a rhynchokinetic type of cranial kinesis, in which the skull exhibits bending zones within the upper beak. This nostril type increases the flexibility of the tip of the upper beak and is often found in species with a long beak, which probe substrate for food.

The main components of the bony palate of birds are the premaxillary and maxillary bones, the vomer, as well as the pterygoids and palatines. Different types of palate morphology separate the two major clades of neornithine birds, the Palaeognathae ("old jaws") and the Neognathae ("new jaws"). In palaeognathous birds, the pterygoid is co-ossified with the palatines and the rigid unit formed by both bones articulates with the skull via well-developed basipterygoid processes. This palatal type is likely to be plesiomorphic for neornithine birds (see, however, Torres et al. 2021). In all neognathous birds, by contrast, there is a movable joint between the pterygoid and the palatine, which increases the mobility of the upper beak; in addition, basipterygoid processes are reduced in many neognathous birds.

The neornithine skull is characterized by an extensive fusion of its constituent bones (Fig. 2.1b). However, in all extant birds, the skull shows cranial kinesis, that is, the beak is movable against the rest of the skull. A critical bone in the caudal portion of the cranium is the quadrate, which articulates with the jugal bar and the pterygoid, and by pushing these elements toward the bill tip (rostrally), raises the upper beak. Its ventral portion articulates with the mandible (lower jaw), so that quadrate, mandible, and upper beak represent a functional unit. In most neornithine birds, the tips of the two mandibular rami are co-ossified and form a mandibular symphysis.

The pectoral girdle consists of two paired elements, the coracoids and scapulae, as well as the furcula and the sternum. The scapulae are attached to the ribcage, and the coracoids articulate with the sternum. Both bones form the glenoid fossa for the articulation with the proximal end of the humerus and, hence, the wing. Scapula, coracoid, and furcula also contribute to the formation of the triosseal canal, which represents a pulley for the tendon of the supracoracoideus muscle; this latter muscle elevates the wing, although it originates ventral of it on the sternum.

The coracoid (Fig. 2.2a–d) anchors the wing to the sternum. Because the bone is oriented in an oblique position relative to the transverse body plane, the terms proximal and distal do not correctly define its "upper" and "lower" ends, which are termed omal ("upper") and sternal ("lower") extremities. In most extant birds, the coracoid is elongated and strut-like. The omal extremity terminates with the acrocoracoid process, which provides the articular surface for the furcula. The omal extremity also bears articular facets for the humerus and scapula, with the latter showing different morphologies that range from a shallow facet to a deeply concave cotyla. Finally, the omal extremity often exhibits a procoracoid process, which directs medially and guides the tendon of the supracoracoideus muscle (in some birds without a well-developed procoracoid process, this function is assumed by the cranial end of the scapula or by the omal extremity of the furcula). In many taxa, the shaft of the coracoid is pierced by a foramen for the supracoracoideus nerve, which appears to be a plesiomorphic neornithine feature and is lost in most birds with slender and strut-like coracoids. The broad sternal end of the coracoid articulates with the sternum and often forms a long lateral process. In some flightless birds, coracoid and scapula are co-ossified and form a scapulocoracoid.

The paired clavicles are fused to form the furcula. The two shanks of this V- or U-shaped bone end with the omal extremities. The sternal end often bears a rod-shaped or blade-like process, the furcular apophysis.

The sternum is a plate-like bone, which exhibits a deep midline keel, or carina, that is one of the prerequisites of powered flapping flight. The cranial end of the sternum bears sulci for the articulation of the coracoid. In some taxa, there is

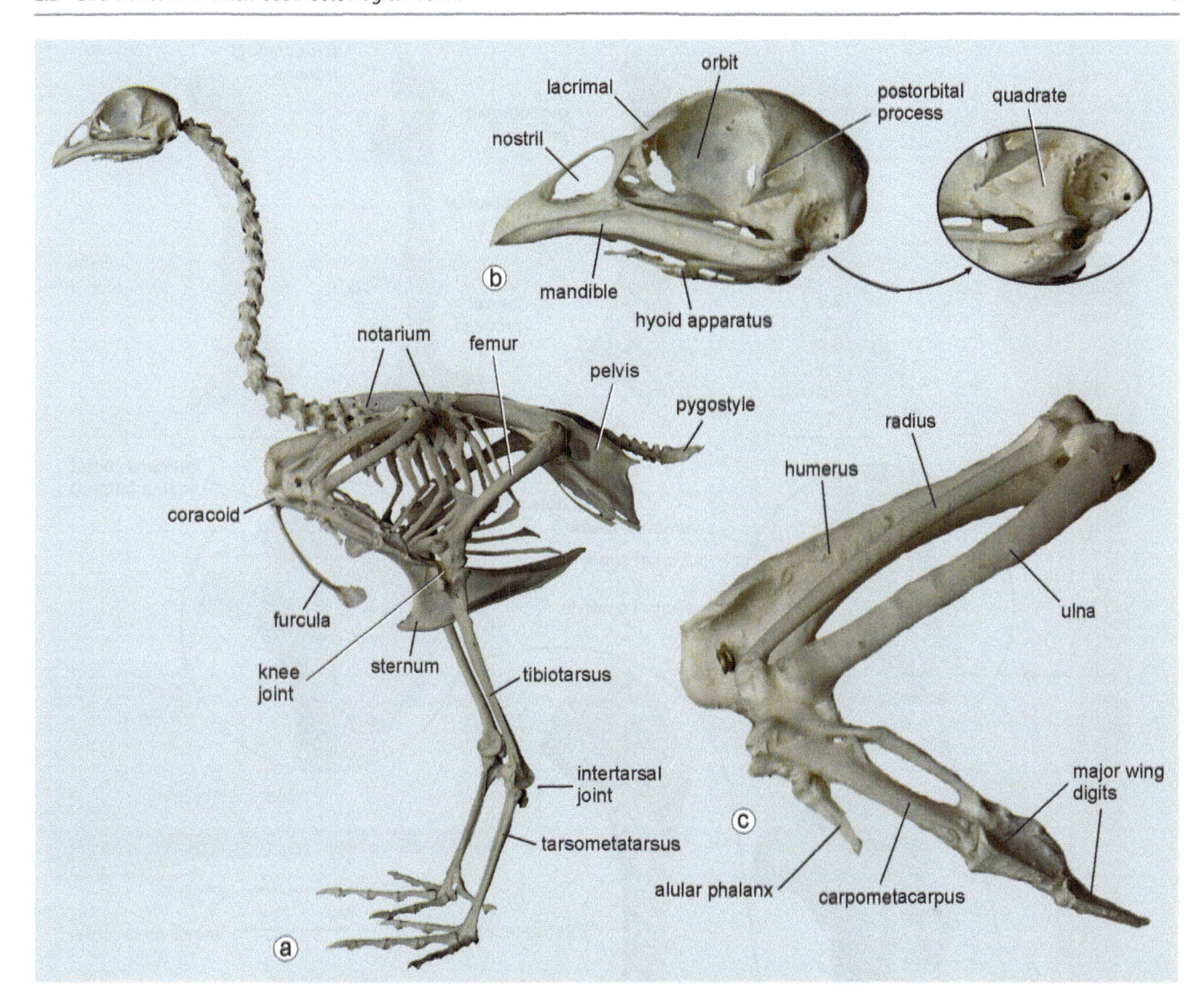

Fig. 2.1 (**a**) Skeleton of a domestic fowl (*Gallus gallus*), with major skeletal elements indicated. (**b**) Skull of *G. gallus* with a detail of the otic region. (**c**) Left wing of *G. gallus*. (Images not to scale, all photos by Sven Tränkner)

furthermore a cranially directed spine, the spina externa, which may have a bifurcated tip and is sometimes accompanied by another, more dorsally located projection, the spina interna. The caudal margin of the sternum usually forms one or two pairs of incisions, which are particularly pronounced in taxa capable of powerful burst take-offs, such as tinamous (Tinamiformes) and landfowl (Galliformes). The sternum is very long in diving birds, such as loons (Gaviiformes), whereas it is short in many soaring birds, such as frigatebirds (Fregatidae).

The wing skeleton is formed by the humerus, a midsection consisting of ulna and radius, and a distal hand section to which the primary feathers are attached (Fig. 2.1c). The proportions of the humerus (Fig. 2.2e–g) are determined by wing length and flight mode, with the bone being very long in soaring birds and shorter and stockier in many broad-winged arboreal taxa. The humerus has proximal and distal ends, and its main surfaces are designated as cranial and caudal; its narrow sides represent its dorsal and ventral surfaces. The pectoral muscles deflecting the wing insert on the deltopectoral crest on the dorsal side of the proximal end of the humerus. The dorsal tubercle (tuberculum dorsale) in the proximodorsal portion of the bone serves as the attachment site for the tendon of the supracoracoideus muscle; this tubercle is very pronounced in birds capable of powerful flapping flight and burst take-offs, whereas it is weakly developed in soaring birds. The caudomedial portion of the proximal end bears another landmark feature of the avian humerus, the pneumotricipital fossa. This fossa derives its name from the fact that its bottom often exhibits pneumatic openings for the entrance of air sac diverticula into the hollow shaft of the humerus; in addition, the pneumotricipital fossa serves as an attachment site for the tricipital pectoral muscles. The distal end of the humerus bears two condyles for the

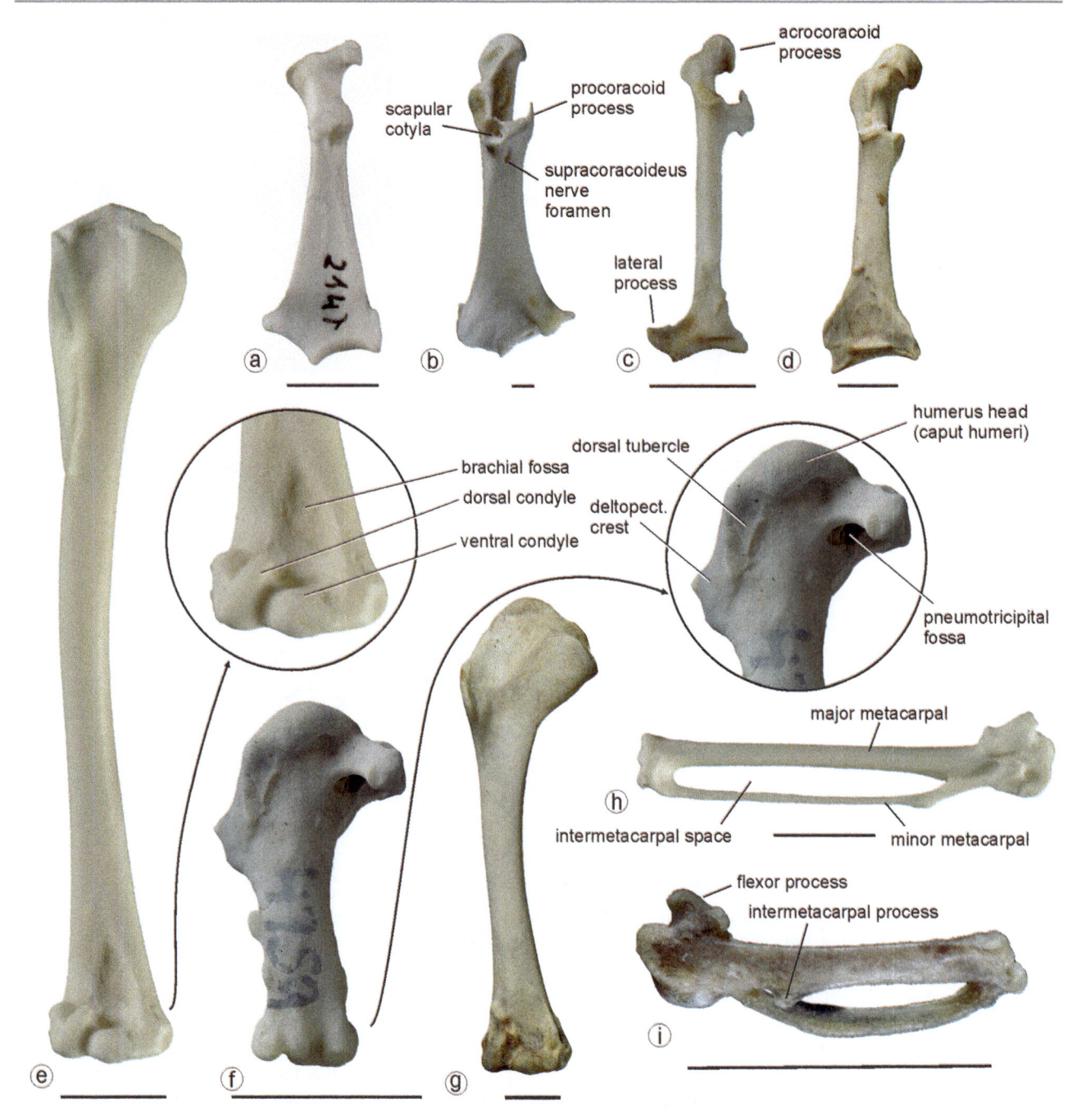

Fig. 2.2 Osteological features of major pectoral girdle and wing bones. (**a**)–(**d**) Left coracoid (dorsal view) of (**a**) *Rhynchotus rufescens* (Tinamidae, Tinamiformes), (**b**) *Pelecanus onocrotalus* (Pelecanidae, Pelecaniformes), (**c**) *Centropus burchellii* (Cuculidae, Cuculiformes), and (**d**) *Nestor notabilis* (Psittacidae, Psittaciformes). (**e**)–(**g**) Humeri of (**e**) *Tyto alba* (Tytonidae, Strigiformes; right side, cranial view), (**f**) *Cypseloides senex* (Apodidae, Apodiformes; left side, caudal view), and (**g**) *N. notabilis* (right side, cranial view); the arrows indicate enlarged views of the proximal and distal ends of the bone. (**h**), (**i**) Carpometacarpi of (**h**) *T. alba* (right side, ventral view) and (**i**) *Colius colius* (Coliidae, Coliiformes; right side, dorsal view). The scale bars equal 1 cm. (All photos by Sven Tränkner)

articulation of the radius (dorsal condyle) and ulna (ventral condyle). Just proximal to these condyles, there is the brachial fossa (fossa musculi brachialis). The distal end of the bone furthermore provides attachment sites for various ligaments and tendons, the development of which may be of taxonomic significance. This is particularly true for the dorsal supracondylar tubercle (tuberculum supracondylare dorsale), which in some taxa, such as the Charadriiformes, Apodiformes, or Passeriformes, forms a pronounced process (dorsal supracondylar process).

The ulna has proximal and distal ends as well as cranial/caudal and dorsal/ventral surfaces. The bone serves for the attachment of the secondary feathers, and sometimes it forms marked tubercles at the insertion sites of these feathers. In the plesiomorphic neornithine condition, the ulna is subequal to the humerus in length. In many long-winged birds, however, it is much longer than the humerus, whereas it is distinctly shorter in some wing-propelled divers. The proximal end exhibits two articular facets for the condyles of the humerus. The distal end forms condyles for articulation with the carpometacarpus.

The carpometacarpus (Fig. 2.2h, i) is the sole compound bone of the wing and developed from co-ossification of two metacarpal bones (major and minor metacarpals) and some of the carpals. The latter form the carpal trochlea, which articulates with ulna and radius. The cranially directed extensor process serves for the attachment of muscles, which extend the hand section of the wing. The carpometacarpus also serves as the main attachment site for the proximal primary feathers. The bone is usually long and narrow in birds with long wings, whereas it is short and with a wide intermetacarpal space in birds with short and broadly rounded wings. In some taxa, the major metacarpal forms a distinct intermetacarpal process, which bridges the intermetacarpal spaces between the major and minor metacarpal and increases the leverage of a muscle that flexes the hand section of the wing.

The avian leg skeleton consists of the femur (Fig. 2.3a), the tibiotarsus (Fig. 2.3b, c), and fibula, as well as the foot, which is formed by the tarsometatarsus (Fig. 2.3d–h) and the toes. These bones form two major joints, the knee (between femur and tibiotarsus) and an intertarsal joint between the tarsal (ankle) bones (Fig. 2.1a).

In avian evolution, the tibiotarsus developed through co-ossification of the tibia with the proximal tarsal bones, which contribute to the articular surfaces of the distal end. The bone has proximal and distal ends as well as cranial and caudal surfaces. Its proximal end articulates with the femur and bears the cnemial crests, which serve as attachment sites of the shank musculature and are proximally projected in many foot-propelled swimming and diving birds. The cnemial crests are cranially prominent in long-legged species, whereas they are reduced in arboreal birds with short legs. On the distal end of the bone, the extensor muscles of the toes run in a sulcus (extensor sulcus), which in most neognathous birds is bridged by an osseous arch, the supratendinal bridge. The two condyles formed by the distal tibiotarsus articulate with the tarsometatarsus.

The tarsometatarsus is another compound bone, which developed through fusion of three metatarsals with the distal tarsal bones. The bone has proximal and distal ends, as well as plantar and dorsal surfaces. Its length and proportions show much variation across Neornithes, with the tarsometatarsi of wading or cursorial birds being long and slender, whereas those of many aerial insectivores or arboreal birds are short. The plantar surface of the proximal end of the bone bears the hypotarsus, a structure of taxonomic significance, which guides the tendons of the flexor muscles of the toes (Fig. 2.3i). In the lateral portion of the distal tarsometatarsus end, there is usually a distal vascular foramen, which serves for the passage of blood vessels. The distal end terminates with the trochleae for the three anterior toes. Their configuration shows much variation in neornithine birds and allows inferences about the locomotion of a certain taxon. In aquatic birds, for example, the trochlea for the second toe is short and plantarly deflected, whereas in most birds with webbed toes the trochlea for the fourth toe is laterally slanted (Fig. 2.3h). In many arboreal perching birds, the trochleae are short and have a similar distal extent (Fig. 2.3f). In birds with reversed toes (see below), the respective tarsometatarsal trochleae are deflected and often form accessory trochleae (Fig. 2.3g).

Most birds have four toes: three anterior ones directing forward and a hind toe, or hallux, which is turned backward. This plesiomorphic toe arrangement is termed anisodactyl. In many terrestrial or cursorial taxa, the hallux is reduced or completely lost, whereas it is elongated in birds, in which the foot has a grasping function (e.g., owls and diurnal birds of prey). In birds with a zygodactyl foot, the fourth toe is permanently reversed, with this toe arrangement showing some variation in different taxa: in facultatively zygodactyl birds the fourth toe can be moved back and forth, whereas it is laterally spread in semi-zygodactyl birds, and permanently directed backward in fully zygodactyl birds. Trogons are the sole taxon, in which the second toed permanently directs backward, forming a heterodactyl foot morphology. In the pamprodactyl foot, all four toes are directed forward.

2.3 The Higher-Level Phylogeny of Extant Birds

Many Paleogene birds exhibit an unusual character mosaic which impedes a straightforward classification. Knowledge of the interrelationships of the extant avian taxa is essential for a phylogenetic placement of these fossils and for an evaluation of their evolutionary significance. The higher-level phylogeny of neornithine birds is still incompletely understood, but some consensus has been reached in recent analyses, which provides a framework for an assessment of fossil taxa (Fig. 2.4; Prum et al. 2015; Kuhl et al. 2021).

The advent of molecular analyses greatly contributed to a better resolved higher-level phylogeny of birds. However, quite a few of the strongly supported clades have already been proposed by earlier systematists based on analyses of anatomical data (Fig. 2.5), and some taxa that have traditionally proven difficult to classify also pose problems in analyses of sequence data. Results of molecular studies are

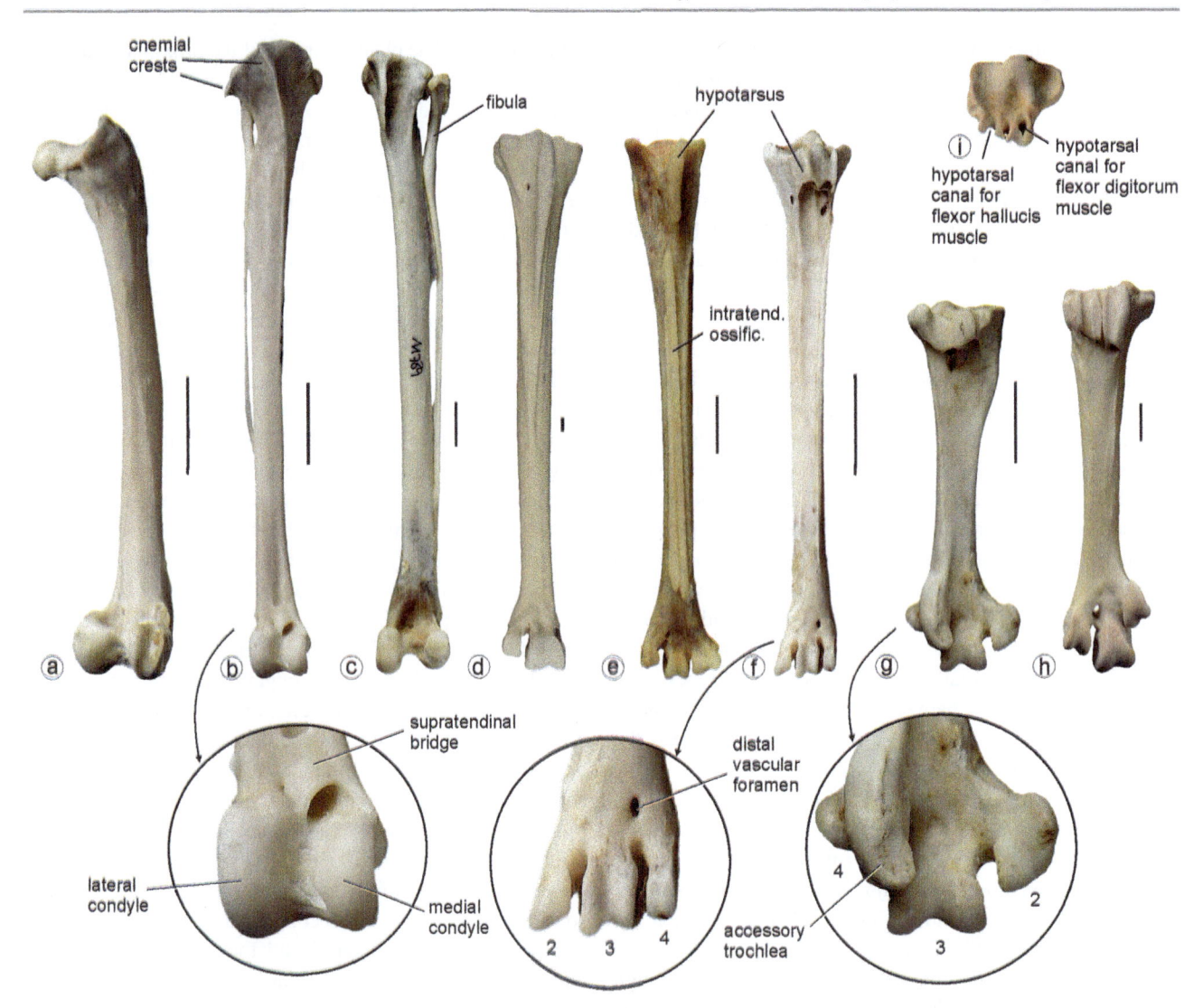

Fig. 2.3 Osteological features of major leg bones. (**a**) Right femur (caudal view) of *Alectoris barbara* (Phasianidae, Galliformes). (**b**) right tibiotarsus (cranial view) of *A. barbara*; the arrow indicates an enlarged view of the distal end of the bone. (**c**) Left tibiotarsus (cranial view) of *Bubo bubo* (Strigidae, Strigidae). (**d**)–(**h**) Morphologically disparate tarsometatarsi of (**d**) *Struthio camelus* (Struthionidae, Struthioniformes, left side, plantar view), (**e**) *Caracara plancus* (Falconidae, Falconiformes; left side, plantar view), (**f**) *Corvus frugilegus* (Corvidae, Passeriformes; right side, plantar view), (**g**) *Nestor notabilis* (Psittacidae, Psittaciformes; left side, plantar view), and (**h**) *Cygnus olor* (Anserinae, Anseriformes; left side, plantar view); the arrows indicates a enlarged views of the distal ends of the bones, in which the trochleae are numbered. (**i**) Tarsometatarsus of *C. olor* in proximal view to show the canals and sulci formed by the hypotarsus. The scale bars equal 1 cm. (All photos by Sven Tränkner)

particularly convincing, if clades are congruently obtained in analyses of independent data, such as nuclear and mitochondrial DNA or gene sequences on different chromosomes. However, although molecular analyses are an important tool for the reconstruction of the higher-level phylogeny of extant birds, only a phylogeny based on morphological characters allows a placement of fossil taxa.

There have been some attempts to analyze the higher-level phylogeny of extant birds with large morphological data sets (e.g., Mayr and Clarke 2003; Livezey and Zusi 2007; Fig. 2.5b). Concerning several major clades, the results of these analyses are, however, not in concordance with well-supported phylogenies obtained in studies of molecular data. As detailed elsewhere (Mayr 2008), such large-scale analyses of equally weighted morphological characters run the risk that many simple homoplastic characters overrule fewer ones of greater phylogenetic significance. Analyses of smaller sets of well-defined characters may therefore be a more appropriate approach in the case of morphological data. The problems of incongruent tree topologies derived from molecular and morphological data may be overcome by the combination of molecular and morphological data or by an enforcement of a molecular scaffold to phylogenetic analyses. However, it remains an open question whether such analyses indeed

2.3 The Higher-Level Phylogeny of Extant Birds

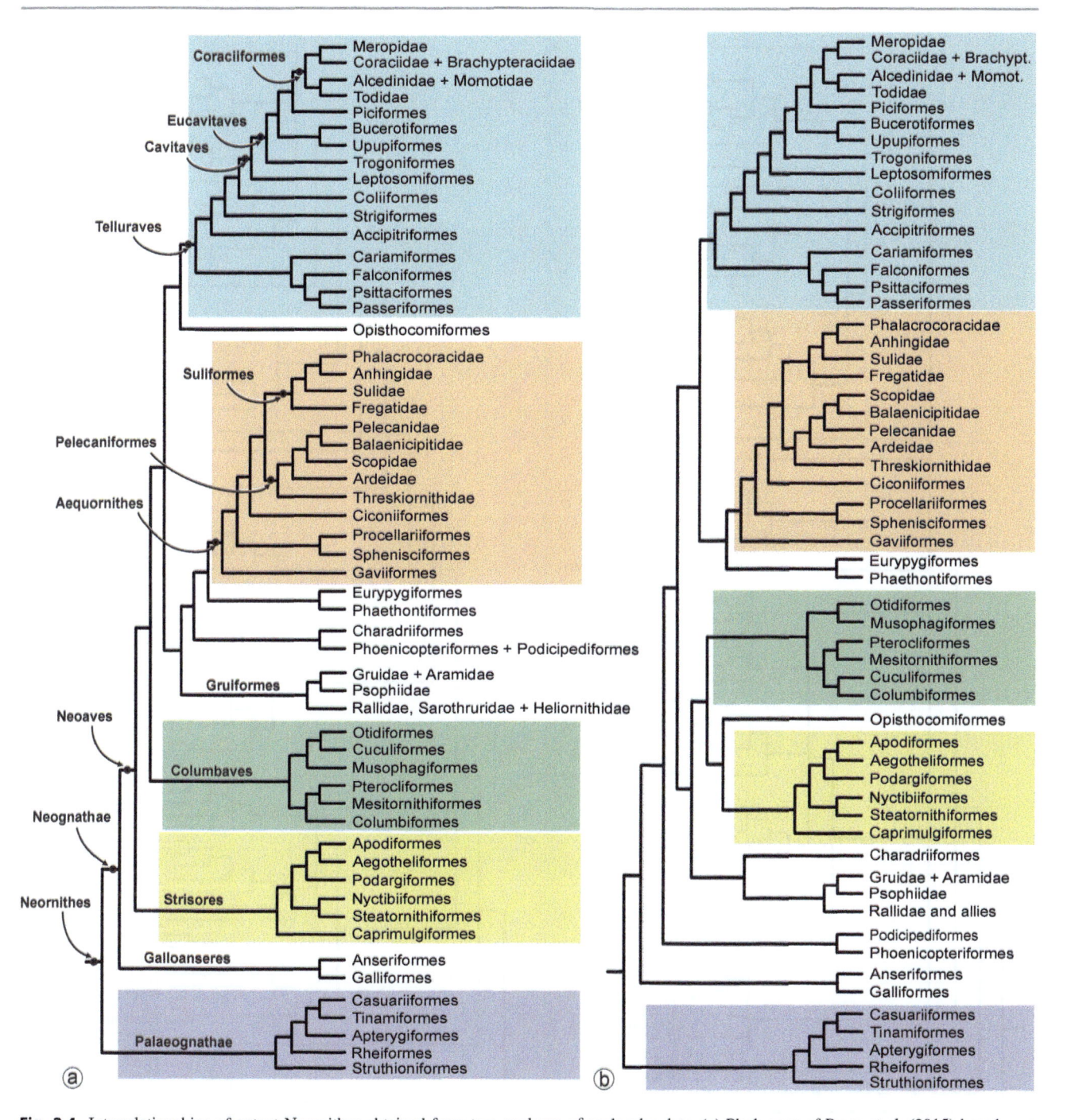

Fig. 2.4 Interrelationships of extant Neornithes obtained from two analyses of molecular data. (**a**) Phylogeny of Prum et al. (2015) based on a Bayesian analysis of nuclear sequences. (**b**) Phylogeny of Kuhl et al. (2021) based on an analysis of transcriptome data

constitute appropriate approaches or whether they conceal phylogenetic signals from the morphological data.

There is consensus among current systematists that neornithine birds can be divided into the sister taxa Palaeognathae and Neognathae. Presumably apomorphic features of palaeognathous birds are, among others, the lack of fusion between the maxillary process of the nasal bone and the maxillary bone, as well as a pair of furrows on the ventral surface of the mandibular symphysis, the dorsal surface of which is furthermore flat in palaeognathous birds. Neognathous birds exhibit an intrapterygoid joint, which develops in early ontogeny and enables a greater mobility of the palate; in addition, there are conjoint auditory tubes (tuba auditiva communis) in neognathous birds and the ilioischiadic fenestrae of the pelvis are caudally closed.

Within Neognathae, the Galloanseres—Galliformes (landfowl), Anseriformes (waterfowl), and their extinct

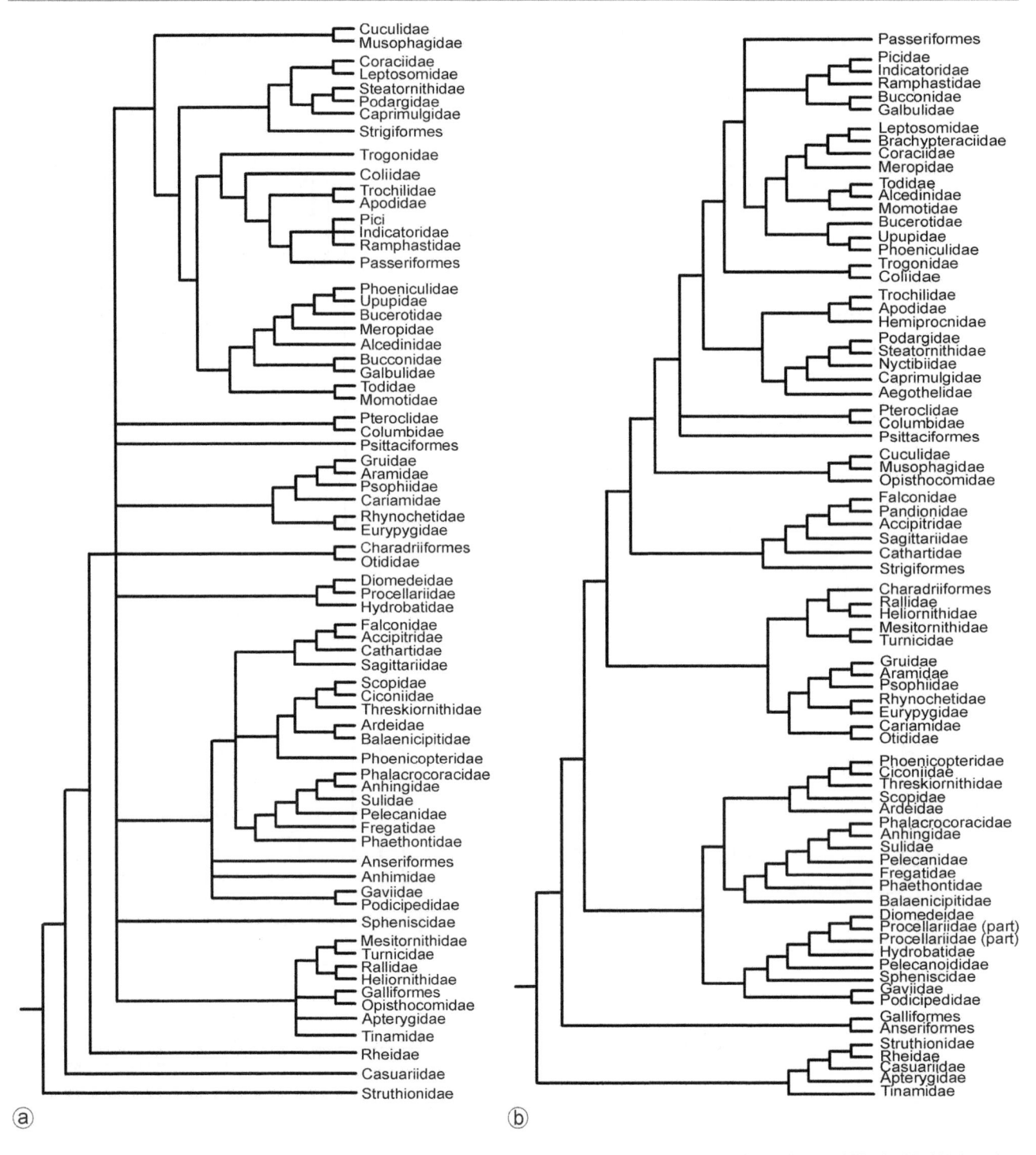

Fig. 2.5 Morphology-based hypotheses on the interrelationships of Neornithes. (**a**) Summary phylogeny of Fürbringer (1888: pls. 27–30), based on multiple trees shown by this author. (**b**) Phylogeny resulting from an analysis of 2954 morphological characters by Livezey and Zusi (2007)

relatives—are the sister group of the remaining taxa, the Neoaves. A sister group relationship between the Galliformes and Anseriformes was first proposed on the basis of anatomical data (Dzerzhinsky 1995) and is retained in all current molecular analyses. The taxa of the Neoaves share a reduction of the phallus, which convergently occurred in some Tinamiformes and Galliformes (Montgomerie and Briskie 2007; Mayr 2008; Brennan et al. 2008).

The early divergences within Neoaves are poorly resolved and molecular analyses yielded controversial results. Whereas Prum et al. (2015) found the Strisores (nightjars, swifts, and allies) to be the sister taxon of all other extant

Neoaves, the analysis by Kuhl et al. (2021) resulted in a sister group relationship between a clade formed by Phoenicopteriformes (flamingos) and Podicipediformes (grebes) and the remaining Neoaves, with Gruiformes (cranes and allies) and Charadriiformes (shorebirds and allies) branching next (Fig. 2.4). An early divergence of phoenicopteriform and gruiform birds is in better concordance with the morphology of the earliest neornithine fossils, but because these resemblances are likely to be due to plesiomorphic characteristics, the fossil record does not provide strong support for one of the conflicting molecular phylogenies.

Sequence-based analyses do not support the monophyly of many of the traditional (e.g., sensu Wetmore 1960) neoavian higher-level taxa. Examples, therefore, are the "Pelecaniformes" (pelicans and allies), "Ciconiiformes" (storks and allies), "Gruiformes" (cranes and allies), "Caprimulgiformes" (nightjars and allies), "Falconiformes" (diurnal birds of prey), and "Coraciiformes" (rollers and allies).

The "Pelecaniformes" were long considered to be a well-established taxon, the members of which share several derived features not found in other birds, such as totipalmate feet (all four toes connected by a web) and a gular pouch. However, these birds are now recognized as a polyphyletic assemblage and are split into the three taxa Pelecaniformes, Suliformes, and Phaethontiformes. The Pelecaniformes sensu stricto comprise the Pelecanidae (pelicans) as well as the Threskiornithidae (ibises and spoonbills), Ardeidae (herons), Balaenicipitidae (shoebill), and Scopidae (hamerkop). The latter four taxa were traditionally assigned to the "Ciconiiformes", which encompassed storks and other long-legged, semi-aquatic birds. The Suliformes include the Fregatidae (frigatebirds) and the Suloidea (Sulidae [gannets and boobies], Phalacrocoracidae [cormorants], and Anhingidae [anhingas]). The Phaethontiformes (tropicbirds) are recovered as the sister taxon of the Eurypygiformes (sunbittern and kagu) in current molecular analyses (Prum et al. 2015; Kuhl et al. 2021).

The traditional "Gruiformes" comprised various groups of rail- or crane-like birds. The monophyly of a taxon including these birds has never been well-established, and current molecular analyses merely support a clade of "core-Gruiformes," which includes the Rallidae (rails), Sarothruridae (flufftails), Heliornithidae (finfoots), Psophiidae (trumpeters), Aramidae (limpkin), and Gruidae (cranes). The affinities of most other taxa once allied with the Gruiformes are still subject to some debate, although there is consensus that the Eurypygidae (sunbittern) are the sister taxon of the Rhynochetidae (kagu) (Ericson et al. 2006; Fain et al. 2007; Hackett et al. 2008; Prum et al. 2015; Kuhl et al. 2021). As shown by multiple molecular analyses of both mitochondrial and nuclear gene sequences, the Turnicidae (buttonquails)–quail-like birds of open habitats of the Old World, which were traditionally also classified into the "Gruiformes" - are aberrant representatives of the Charadriiformes (shorebirds and allies; Paton and Baker 2006; Fain and Houde 2007).

The traditional "Caprimulgiformes" united five extant taxa of crepuscular or nocturnal birds: the Steatornithidae (oilbird), Podargidae (frogmouths), Caprimulgidae (nightjars), Nyctibiidae (potoos), and Aegothelidae (owlet-nightjars). The paraphyly of these birds was first shown by Mayr (2002b), who, on the basis of morphological data, proposed a sister group relationship between the Aegothelidae and apodiform birds (swifts and hummingbirds) but could not resolve the affinities of the Podargidae and Steatornithidae. Meanwhile, there is congruent molecular support for a clade including all of the "caprimulgiform" birds and the Apodiformes, which is termed Strisores (Mayr et al. 2003; Ericson et al. 2006; Hackett et al. 2008; Mayr 2010; Prum et al. 2015; Kuhl et al. 2021). The family-level taxa of the traditional "Caprimulgiformes" have been elevated to ordinal rank, so that the Strisores now comprise the Steatornithiformes, Podargiformes, Caprimulgiformes, Nyctibiiformes, Aegotheliformes, and Apodiformes.

All current molecular analyses furthermore recover a clade including the Podicipediformes (grebes) and Phoenicopteriformes (flamingos), for which the name Mirandornithes was proposed (van Tuinen et al. 2001; Sangster 2005; Ericson et al. 2006; Hackett et al. 2008; Prum et al. 2015; Kuhl et al. 2021). Even though grebes and flamingos are very different in their external appearance, they share a number of derived anatomical characteristics, including eleven primary wing feathers (except for the Ciconiidae all other avian taxa have ten or less primaries), nail-like ungual pedal phalanges, and a chalky layer of amorphous calcium phosphate on the eggshell (Mayr 2004; Manegold 2006). Phoenicopteriformes and Podicipediformes are furthermore parasitized by an exclusively shared taxon of cestodes, and their phthirapteran feather lice are closely related (Mayr 2004; Johnson et al. 2006). The closest extant relatives of the flamingo/grebe clade, however, remain elusive. Some molecular analyses indicated a sister group relationship to the Charadriiformes (Morgan-Richards et al. 2008; Prum et al. 2015; Braun and Kimball 2021), whereas others obtained the clade (Phoenicopteriformes + Podicipediformes) as the sister taxon of all remaining Neoaves (Kuhl et al. 2021).

Molecular analyses congruently support a clade including most aquatic or semi-aquatic neoavian taxa, namely the Gaviiformes (loons), Procellariiformes (tubenoses), Sphenisciformes (penguins), Ciconiiformes (storks), Suliformes, and Pelecaniformes. This clade is recovered in multiple analyses of different kinds of data and was termed Aequornithes (Ericson et al. 2006; Hackett et al. 2008; Morgan-Richards et al. 2008; Mayr 2011; Prum et al. 2015;

Kuhl et al. 2021). A "waterbird clade", albeit with a somewhat different composition, has already been assumed by earlier authors (e.g., Olson 1985). A similar clade was obtained in an analysis of morphological data by Livezey and Zusi (2007), even though it also included the Phoenicopteriformes and Podicipediformes in this analysis.

Earlier authors assumed close affinities of various arboreal birds, which lack the ambiens muscle (a small muscle of the leg), and a group including these birds was informally termed "higher land birds" by Olson (1985). A similar clade is congruently obtained in analyses of molecular sequence data and was named Telluraves (Yuri et al. 2013). However, in addition to the Strigiformes (owls), Coliiformes (mousebirds), Leptosomiformes (kurol), Trogoniformes (trogons), Bucerotes (hoopoes, woodhoopoes, and hornbills), Coraciiformes (rolllers, kingfishers and allies), Piciformes (woodpeckers and allies), and Passeriformes (passerines), Telluraves also comprises the Psittaciformes (parrots), Accipitriformes (New World vultures, secretary bird, and hawk-like diurnal birds of prey), Falconiformes (falcons), and Cariamiformes (seriemas), which possess the ambiens muscle.

In contrast to analyses of morphological data, most sequence-based studies do not support monophyly of diurnal birds of prey, and the Falconiformes are recovered in a clade together with the Cariamiformes, Psittaciformes, and Passeriformes (Ericson et al. 2006; Hackett et al. 2008; Prum et al. 2015; Kuhl et al. 2021). Close affinities of these four taxa have not been assumed by earlier systematists, and there exists no supporting morphological character evidence. It has to be noted, however, that the detection of a clade including the Cariamiformes, Falconiformes, Psittaciformes, and Passeriformes is sensitive to the kind of molecular data and how they are analyzed, with the positions of the Falconiformes and Cariamiformes being variable under different settings (Fig. 2.6; Kuhl et al. 2021; Braun and Kimball 2021). A sister group relationship between the Passeriformes and Psittaciformes is, however, congruently obtained in all current analyses of molecular data, and as detailed in Sect. 9.5 it conforms to the morphology of some fossil taxa with an unusual character mosaic and has important consequences for the classification of various fossil birds with zygodactyl feet. All current molecular analyses likewise support a clade including the Cathartidae (New World vultures), Sagittariidae (secretary bird), Pandionidae (osprey), and Accipitridae (hawks and allies), with these three taxa of diurnal birds of prey sharing a derived morphology of the syrinx (Griffiths 1994).

The traditional assignment of the Leptosomiformes to the rollers (Coracii) was not well founded, and in sequence-based analyses, the taxon is recovered as the sister group of a clade including the Trogoniformes, Bucerotes, Coraciiformes, and Piciformes (Mayr et al. 2003; Ericson et al. 2006; Hackett et al. 2008; Prum et al. 2015; Kuhl et al. 2021). The clade including the Leptosomiformes and the latter taxa was termed Cavitaves (Yuri et al. 2013). Molecular analyses furthermore congruently suggest a paraphyly of the remaining "coraciiform" birds, with the Trogoniformes, Bucerotes, and Coraciiformes being successive sister taxa of the Piciformes.

2.4 Mesozoic Neornithes

There were many lineages of archaic birds in the Early Cretaceous, but most Late Cretaceous taxa belong to either the arboreal Enantiornithes or the more terrestrial Ornithuromorpha, with the latter clade including the Neornithes (Chiappe 2007; O'Connor et al. 2011; Mayr 2017a). The oldest fossils of the Ornithuromorpha come from the Early Cretaceous of China, but these birds still differed in numerous plesiomorphic features from the crown group taxa. Even the closest relatives of neornithine birds, such as the aquatic Hesperornithiformes, had toothed jaws, and the complete loss of teeth is one of the apomorphies of the avian crown group.

At the time the first edition of this book was published, there was an apparent conflict between divergence dates for higher-ranked clades of neornithine birds derived from calibrated molecular data and the fossil record. Some earlier analyses not only supported the existence of extant family-level taxa in the Late Cretaceous but even suggested a divergence of some neornithine lineages in the Early Cretaceous (Mayr 2009 and references therein). Based on the avian fossil record, by contrast, it was argued that virtually all taxa of Neornithes underwent a rapid radiation immediately after the mass extinction event at the Cretaceous/Paleogene (K/Pg) boundary in what was termed the "big bang" of avian evolution (Feduccia 1995, 2003, 2014). Meanwhile, a more balanced picture emerged from refined calibrations of molecular data, which indicate that the initial diversification of neornithine birds already commenced in the Late Cretaceous, whereas most neoavian lineages originated in the early Cenozoic (Ericson et al. 2006; Prum et al. 2015; Kuhl et al. 2021).

Still, the evolutionary origins of neornithine birds remain elusive from a paleornithological point of view, and for a long time the Mesozoic fossil record of neornithine, or neornithine-like, birds was very scant and consisted of fragmentary bones of uncertain affinities. In the past years, however, this record has been substantially improved through new fossil finds and the restudy of already known specimens.

One of the earliest neornithine-like ornithuromorph birds is *Tingmiatornis arctica* from the Turonian (90 Ma) of Axel Heiberg Island in the Canadian Arctic, the description of which was based on a few isolated wing bones (Bono et al. 2016). The exact affinities of *Tingmiatornis* are unknown, but the humerus shows a resemblance to that of the early

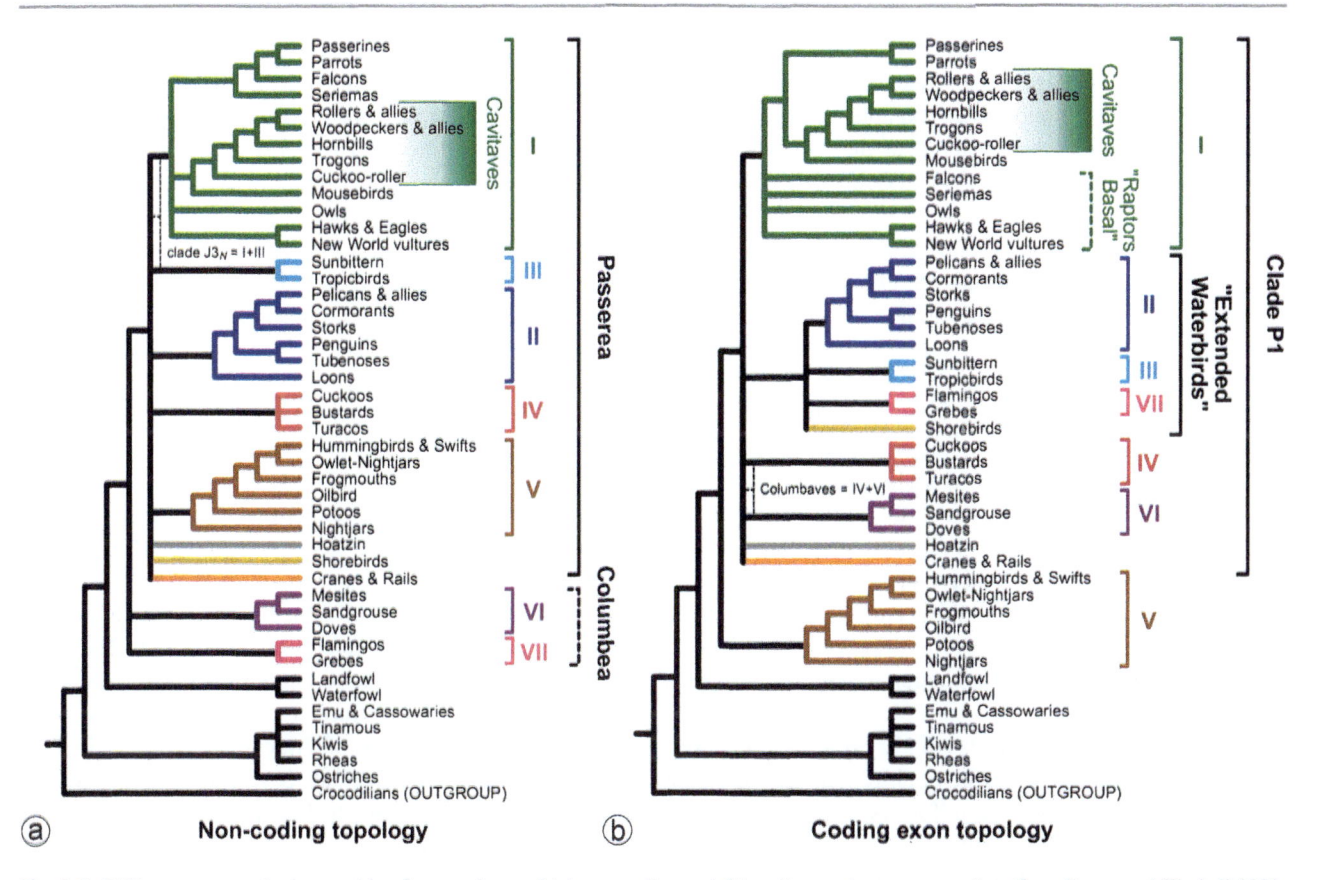

Fig. 2.6 Different tree topologies resulting from analyses of (**a**) non-coding and (**b**) coding nuclear sequence data (from Braun and Kimball 2021: fig. 1; published under a Creative Commons CC BY 4.0 license)

Paleogene Lithornithidae (Sect. 3.1), whereas the ulna appears to be less stout than that of lithornithids.

Almost 140 years before the description of Tingmiatornis, a much better represented but somewhat younger partial postcranial skeleton was reported from the Late Cretaceous (late Santonian/early Campanian; ~80–85 mya) Niobrara Formation of Kansas. This fossil is currently known as *Iaceornis marshi* and was first described by Marsh (1880), who assigned it to *Apatornis celer*, another species from the marine deposits of the Niobrara Formation. The holotype and only known specimen of *A. celer* is a synsacrum, and because this compound bone is not preserved in the aforementioned partial skeleton, it was assigned to the new species *Iaceornis marshi* by Clarke (2004). Whether this taxonomic action is justified can only be assessed once overlapping skeletal elements of both species, *I. marshi* and *A. celer*, have been found. Clarke's (2004) analysis supported a position of *I. marshi* outside crown group Neornithes, even though the character evidence for this placement is rather weak (Clarke 2004: 149). A distinctive osteological feature of *Iaceornis* is a very long and narrow acromion of the scapula. With regard to this trait and the overall morphology of most skeletal elements, *Iaceornis* also shows a resemblance to the Lithornithidae, which are considered to be either palaeognathous birds or stem group Neornithes. Unlike in lithornithids, however, the distal end of the tibiotarsus of *Iaceornis* exhibits an ossified supratendinal bridge.

Apart from fragmentary fossils from the Maastrichtian of North America (see below), the few specimens from the Niobrara Formation have long been the main source of information for modern-type birds from the Late Cretaceous. In the past 20 years, however, several well-preserved partial skeletons of Late Cretaceous neornithine birds were reported from fossil sites in the Southern Hemisphere. The first of these is a partial skeleton from the Late Cretaceous (Maastrichtian, about 66–69 Ma) López de Bertodano Formation of Vega Island (Antarctica), which was assigned to the anseriform Presbyornithidae by Noriega and Tambussi (1995). This fossil was subsequently described as *Vegavis iaai* by Clarke et al. (2005), whose analysis supported a phylogenetic position within a clade including the Presbyornithidae and Anatidae. Another partial skeleton of *Vegavis* from Vega Island was reported by Clarke et al. (2016) and includes skeletal elements unknown from the holotype. The new specimen demonstrates close affinities between *Vegavis* and another Late Cretaceous species from Antarctica, which was initially considered to be a gaviiform bird. This latter species, *Polarornis gregorii* from the Maastrichtian López de

Bertodano Formation of Seymour Island, was described on the basis of a fragmentary partial skeleton, which includes the caudal part of the bill and some incomplete hindlimb bones (Chatterjee 2002). Further material of the Vegaviidae was described by Acosta Hospitaleche and Gelfo (2015) and West et al. (2019).

The well-developed, proximally projecting cnemial crests of the tibiotarsus of *Vegavis* and *Polarornis*, as well as the shape of the short and markedly bowed femur, indicate that these taxa were foot-propelled seabirds, which foraged by swimming or diving (Acosta Hospitaleche and Worthy 2021). This is also suggested by the morphology of the distal end of the tarsometatarsus, which is mediolaterally compressed and has a very short trochlea for the second toe (Acosta Hospitaleche and Worthy 2021). Agnolin (2017) classified both *Vegavis* and *Polarornis* in the new taxon Vegaviidae (as detailed by Mayr et al. 2018a, other taxa were erroneously referred to this taxon), and the derived similarities shared by *V. iaai* and *P. gregorii* are actually so strong that even the generic distinction of both species needs to be scrutinized.

The phylogenetic affinities of *Vegavis* are not well resolved. An analysis by Clarke et al. (2016) recovered the taxon within crown group Anseriformes, in a clade together with the Presbyornithidae and Anatidae. This result was, however, considered to be poorly supported by Mayr et al. (2018a), who noted that the pterygoid of *Vegavis* exhibits stalked basipterygoid articular facets (in galloanserine birds, these facets are sessile) and that the mandible lacks long retroarticular processes, with the latter representing a key apomorphy of the Galloanseres. A more recent analysis resulted in a sister group relationship between *Vegavis* and either the Neornithes or the Neognathae (Field et al. 2020).

Another species from the early Maastrichtian (71 Ma) of Antarctica, *Antarcticavis capelambensis*, is known from a partial skeleton and was obtained as the sister taxon of a clade formed by the Vegaviidae and crown group birds in a phylogenetic analysis (Cordes-Person et al. 2020). With regard to the absence of a supratendinal bridge on the distal tibiotarsus, *Antarcticavis* agrees with *Iaceornis*, but the carpometacarpus has a lower extensor process and the coracoid lacks a foramen for the supracoracoideus nerve. The holotype of *A. capelambensis* includes the proximal end of a tarsometatarsus which, judging from the published illustrations, resembles the proximal tarsometatarsus of *Neogaeornis wetzeli*, with which it has not been compared. The holotype of *N. wetzeli* is a tarsometatarsus from the Campanian or Maastrichtian Quiriquina Formation of Chile. The hypotarsus of *N. wetzeli* corresponds to that of *A. capelambensis* in that it forms two widely separated crests, and here it is considered well possible that *Antarcticavis* and the roughly coeval *Neogaeornis* are closely related. *N. wetzeli* was assigned to the Gaviiformes by Olson (1992), but its tarsometatarsus is clearly distinguished from that of early Paleogene Gaviiformes. The hypotarsus of *Vegavis* has four crests, but the two central ones are very poorly developed (Acosta Hospitaleche and Worthy 2021), and there is a possibility that *Antarcticavis* and *Neogaeornis* are further representatives of the Vegaviidae.

Fragmentary Neornithes-like wing and pectoral girdle bones are known from several Late Cretaceous localities in South America, but most of these so far defied an unambiguous phylogenetic placement. Agnolin et al. (2006) described an incomplete, odd-looking coracoid from the Late Cretaceous of Patagonia, which they considered to be from a neornithine bird. This specimen comes from a quail-sized species and its position within Neornithes has yet to be convincingly established. Another coracoid was reported by Agnolin (2010) from the Late Cretaceous (Campanian-Maastrichtian) of Argentina and is similar to a coracoid from the Late Cretaceous of Canada, which was figured by Longrich (2009: fig. 4); with regard to the strongly developed procoracoid crest, this bone resembles the coracoid of some extant gruiform birds. A further neornithine-like coracoid from the Late Cretaceous of Argentina was described as *Kookne yeutensis* (Novas et al. 2019). The wing bones of *Limenavis patagonica* from the Late Cretaceous of Patagonia (Argentina) are likewise of essentially modern appearance, even though the species was assumed to be outside crown group Neornithes by Clarke and Chiappe (2001); owing to the very fragmentary nature of the material, the affinities of this bird are best regarded uncertain.

The majority of the above fossils are from the Southern Hemisphere, but a substantial specimen of an early neornithine bird has recently also been reported from a Northern Hemispheric fossil site: In addition to a few fragmentary limb bones, the holotype of *Asteriornis maastrichtensis* from the latest Cretaceous (66 Ma) of Belgium includes a well-preserved partial skull, which exhibits galloanserine apomorphies, such as a mandible with long retroarticular processes (Field et al. 2020). The small-sized species is characterized by a relatively short beak with an unusually straight dorsal ridge (culmen) and very large nostrils, which shows a resemblance to the beak of the anseriform Anhimidae and to an anhimid-like fossil from the London Clay in particular (Sect. 4.5.1). The original phylogenetic analysis recovered *Asteriornis* as the sister taxon of either crown group Galloanseres or Galliformes. However, the shape of the long, rostrally projecting postorbital process is actually a derived characteristic of the Anseriformes, and the similarities to galliform birds (premaxillary and nasal bones not co-ossified, absence of nasofrontal hinge) may well be plesiomorphic.

Hope (2002) reviewed the Mesozoic fossil record of putative Neornithes from the latest Cretaceous (Maastrichtian; 66 Ma) of the North American Lance Formation. These specimens were—in some cases tentatively—assigned to stem group Galliformes (*Palintropus*), the anseriform Presbyornithidae, Charadriiformes (*Graculavus*, *Cimolopteryx*), Procellariiformes (*Lonchodytes*), Pelecaniformes (*Torotix* and unnamed species, which were identified as representatives of the Phalacrocoracidae), Psittaciformes, and Neornithes incertae sedis (*Ceramornis*). All of these fossils consist of very fragmentary remains and none can be unambiguously identified. The taxon *Palintropus*, for example, is solely known from a scapula and omal extremities of the coracoid and was tentatively assigned to the galliform Quercymegapodiidae (see Sect. 4.3.5) by Hope (2002). However, *Palintropus* is at best the sister taxon of all other Galliformes, because the coracoid exhibits a foramen for the supracoracoideus nerve (Hope 2002: fig. 15.3), a plesiomorphic feature that is invariably absent in Cenozoic and extant galliform birds including the Quercymegapodiidae (see also Hope 2002: 353). Longrich (2009) described further material of *Palintropus* from the Late Cretaceous (Campanian) of Canada and even questioned its neornithine affinities. A quadrate from the Lance Formation, which was initially assigned to *Cimolopteryx*, was identified as that of a galloanserine bird by Elzanowski and Stidham (2011), who hypothesized that the bone most likely is from a stem group anseriform.

Clarke (2004) hypothesized that the incomplete distal tarsometatarsus that constitutes the holotype of "*Ichthyornis*" *lentus* is from a galliform bird and assigned the species to the new taxon *Austinornis*. The holotype of *A. lentus*, as the species is now known, lacks exact stratigraphic data. If the fossil is indeed from Late Cretaceous strata, possible affinities to the galliform-like *Palintropus* (see above) need to be taken into consideration.

Dyke and Mayr (1999) detailed that the assignment of a fragmentary mandible from the Lance Formation to the Psittaciformes (Stidham 1998) is not convincing. Hope (2002) retained this fossil in the Psittaciformes, noting that the presence of a mandibular symphysis supports its assignment to Neornithes. However, this feature also occurs in the edentulous Oviraptorosauria (Elzanowski 1999), and it is well possible that the Lance Formation mandible comes from a representative of the Caenagnathidae, a North American taxon of oviraptorosaurs.

Of similar age to the specimens from the Lance Formation are avian remains from the Late Cretaceous or early Paleocene Hornerstown Formation of New Jersey (Olson and Parris 1987; Parris and Hope 2002). Many of these fossils were assigned to the "form-family" Graculavidae by Olson and Parris (1987), which was regarded as a "convenient catch-all, intended as such" by Olson (1999: 125). The species from the New Jersey deposits included in the "Graculavidae" by Olson and Parris (1987) are *Graculavus velox*, *Palaeotringa littoralis*, *P. vagans*, *Telmatornis priscus* (which also occurs in the unambiguously Late Cretaceous Navesink Formation), *Anatalavis* ("*Telmatornis*") *rex*, and *Laornis edvardsianus*.

Although Olson and Parris (1987) already considered it likely that the "Graculavidae" constitute a polyphyletic assemblage, "graculavids" found their way into the literature as "transitional shorebirds" (e.g., Feduccia 1999). However, *A. rex* was subsequently referred to the Anseriformes by Olson (1999; Sect. 4.5.3), and Olson (1985) and Olson and Parris (1987) detailed that the other New Jersey taxa of the "Graculavidae" are very similar to the Presbyornithidae, which are now also regarded as anseriform birds (Sect. 4.5.2). The resemblances to presbyornithids are especially strong in the case of *Graculavus velox*, which is known from a proximal humerus (Olson and Parris 1987: fig. 4). Another species of *Graculavus*, *G. augustus*, was described by Hope (1999) from the Late Cretaceous Lance Formation and is also solely represented by the proximal end of a humerus. This species is larger than *G. velox*, but otherwise, the humerus is very similar. The holotype and only known specimen of the large *Laornis edvardsianus*, a distal tibiotarsus, has also been compared with the Presbyornithidae by Olson and Parris (1987).

The description of *Novacaesareala hungerfordi* from the Hornerstown Formation was based on the distal end of a humerus and associated wing fragments. The phylogenetic affinities of this species were considered uncertain by Parris and Hope (2002), who noted, however, that it resembles extant "Pelecaniformes" in the very short flexor process of the humerus. Compared to other Cretaceous taxa, and as detailed by Parris and Hope (2002), *N. hungerfordi* most closely resembles *Torotix clemensi* from the Lance Formation, which is represented by a distal humerus and was classified into the "Pelecaniformes" by Hope (2002). Mayr and Scofield (2016) noted similarities between the humerus of *N. hungerfordi* and that of stem group Phaethontiformes (tropicbirds). Other taxa from the Hornerstown Formation were tentatively assigned to the palaeognathous Lithornithidae (Parris and Hope 2002; unnamed species; see Sect. 3.1) and to the Procellariiformes (Olson and Parris 1987; *Tytthostonyx glauconiticus*; see Sect. 7.3).

Possibly also a representative of the Procellariiformes is *Maaqwi casacadensis* from the Late Cretaceous of British Columbia (Canada), the holotype of which consists of a coracoid and associated fragmentary wing bones (McLachlan et al. 2017). A phylogenetic analysis resulted in a sister group relationship between this species and the Vegaviidae. However, the coracoid of *M. casacadensis* actually has a very different morphology to that of *Vegavis* and more closely resembles the coracoid of procellariiform birds.

An enigmatic, large-sized species from the Maastrichtian of Europe, *Gargantuavis philoinos*, was described as "a bird with an advanced synsacrum, possibly as large as an extant cassowary or ostrich" (Buffetaut et al. 1995: 110). Meanwhile, multiple specimens of *Gargantuavis* have been found in France, the Iberian Peninsula, and Romania (Buffetaut et al. 1995; Angst and Buffetaut 2017; Angst et al. 2017; Mayr et al. 2020a; Pérez-Pueyo et al. 2021). However, the material assigned to this species mainly consists of pelvis fragments and it is uncertain whether referred femora and vertebrae belong to the same taxon. *Gargantuavis* was identified as a flightless ornithuromorph bird (Buffetaut and Le Loueff 2010; Pérez-Pueyo et al. 2021), but the morphology of its wide pelvis is very unlike that of more advanced ornithuromorphs and if avian, the taxon is likely to represent an archaic lineage (Mayr et al. 2020a).

In summary, the Late Cretaceous fossil record of neornithine birds provides evidence for a Mesozoic diversification of neornithine bird, even though the exact extent of this diversification is still unclear. The identification of putative neognathous and galloanserine taxa (*Asteriornis* and, possibly, *Vegavis* and *Polarornis*) implies the existence of stem group representatives of the Palaeognathae and Neoaves by the Late Cretaceous (Clarke et al. 2005). When, however, the divergence between the Palaeognathae and Neognathae occurred remains elusive on the basis of the fossil record, and it is possible that the lithornithid-like *Iaceornis* actually is a stem group representative of the Palaeognathae. The earliest unambiguous records of virtually all neoavian taxa are from Paleogene deposits, so much the more as the age of the Hornerstown Formation, where the putative procellariiform *Tytthostonyx* and the putative phaethontiform *Novacaesareala* come from, is uncertain. Given the fact that the Paleocene fossil record includes such morphologically disparate taxa as the Sphenisciformes (penguins) and Strigiformes (owls), there can, however, be little doubt that stem group representatives of some neoavian taxa were already in existence in the Late Cretaceous.

Avian diversity was severely impacted by the mass extinction event at the K/Pg boundary, which led to the extinction of major Cretaceous lineages of non-neornithine birds (Longrich et al. 2011). Although it is generally assumed that all of the latter became extinct at the Cretaceous/Paleogene (K/Pg) boundary (e.g., Feduccia 1995, 2003), the factual evidence for this hypothesis is rather weak, given the poor early Paleocene fossil record of birds. At least *Qinornis paleocenica* from the early Paleocene of China (Xue 1995) may represent a non-neornithine avian lineage, which survived the K/Pg extinction event (Mayr 2007). This species is known from a partial foot, and the incomplete fusion of the metatarsals is more reminiscent of the condition in Mesozoic taxa outside Neornithes than of that of juvenile neornithine birds (contra Xue 1995). If lithornithids are the sister taxon of neornithine birds, as suggested by some analyses (Sect. 3.1), they would be another non-neornithine taxon that survived the K/Pg extinction.

As noted above, it was suggested that neornithine birds as a whole underwent an explosive radiation after the K/Pg mass extinction event (Feduccia 1995, 2003, 2014). However, the paleontological data and calibrated molecular phylogenies published in the past decades suggest that several neornithine lineages diversified before the K/Pg boundary. A major clade that does appear to have undergone a significant post-Cretaceous radiation, however, is Telluraves, which includes most arboreal land birds.

Mayr (2014) detailed that there is a conspicuous absence of arboreal neornithine birds in the Cretaceous fossil record. Mayr (2017a: 92) considered it "very possible, if not likely, that arboreal Neornithes indeed did not diversify before the end Cretaceous extinction of the Enantiornithes (...). Predominantly terrestrial or aquatic habitat preferences of Cretaceous Neornithes would (...) be consistent with the fact that these are also assumed for non-neornithine birds close to the origin of the crown group. (...) A selective extinction of avian lineages at the K/Pg boundary could have been triggered by profound vegetation changes, such as large-scale deforestations, which have been assumed at least for North American biotas (...). This would not only have led to the extinction of arboreal enantiornithines, but, through changes of atmospheric carbon dioxide levels and accompanied oceanic acidification (...), could have also affected the food chains of marine ecosystems. Predominantly terrestrial birds that did not live in forested environments, by contrast, may have been less affected". Mayr (2017a: 204f.) furthermore concluded that "no small arboreal neornithine birds are known from Cretaceous or even early Paleocene deposits. In the early Eocene, by contrast, stem group representatives of most extant arboreal lineages were already present. It therefore seems well possible that a causal correlation existed between the extinction of the arboreal Enantiornithes at the end of the Mesozoic and the radiation of arboreal Neornithines thereafter."

Virtually the same conclusions were subsequently reached by Field et al. (2018), who hypothesized that global deforestation after the K/Pg impact caused the extinction of arboreal Mesozoic birds and structured the early evolution of modern birds. Unfortunately, these authors did not cite the above reference, which, an imprinted publication date of 2017 notwithstanding, has already been released in September 2016. Even though the study of Field et al. (2018) provides additional support for the hypothesis that large-scale deforestation around the K/Pg boundary had a significant impact on avian evolution, the omission of pertinent references conveys a wrong picture regarding the novelty of the results, and the analysis does not address the end-Cretaceous extinction of the aquatic Hesperornithiformes and Ichthyornithiformes.

2.5 Stratigraphy and Major Paleogene Fossil Localities

The Paleogene period covers three epochs, the Paleocene, Eocene, and Oligocene. In this chapter, some of the major sites that yielded Paleogene fossil birds are introduced to avoid redundancies in the taxonomic sections. Dyke et al. (2007: 341) stated that "aquatic environments of preservation dominate the early Paleogene avian fossil record, because these were the habitats in which more modern birds lived at the time of the transition." Actually, however, most Paleocene avian fossils are from land birds, and a predominance of fossil lagerstätten of aquatic origin can only be ascertained for the early and middle Eocene record of the Northern Hemisphere. Large areas of this part of the globe were covered with paratropical forests in the early Paleogene, which offered unfavorable conditions for the preservation of animal carcasses on land. The predominance of fossil birds in early/middle Eocene sediments of aquatic origin therefore clearly reflects a preservational bias.

2.5.1 Europe

Undoubtedly, Europe has the most comprehensive and best-studied Paleogene record of birds, and in the following only major localities are listed, which yielded significant numbers of avian fossils. Data on other European sites, where Cenozoic bird remains have been discovered, can be found in Mlíkovský (1996a, 2002).

The European Paleogene is subdivided into biostratigraphic zones of the Mammalian Paleogene (MP), the definitions of which are based on local mammalian faunas (Fig. 2.7; Legendre and Lévêque 1997; Speijer et al. 2020). According to this stratigraphy, the Paleocene covers the units MP 1–6 (66.0–56.0 Ma). The Eocene comprises the units MP 7–20, with the early Eocene (Ypresian) encompassing MP 7–10 (56.0–48.1 Ma), the middle Eocene being split into the Lutetian (MP 11–13; 48.1–41.0 Ma) and Bartonian (MP 14–16; 41.0–37.7 Ma), and the late Eocene (Priabonian) covering MP 17–20 (37.7–33.8 Ma). The Oligocene contains the units MP 21–30, of which the early Oligocene (Rupelian) encompasses MP 21–24 (33.8–27.3 Ma) and the late Oligocene (Chattian) MP 25–30 (27.3–23.0 Ma).

Little is still known about Paleocene avifaunas of Europe. Although hundreds of Paleocene bird bones were found in the Walbeck fissure filling in northern Germany, these remained unstudied for almost 70 years (Weigelt 1939; Mayr 2007). The Walbeck avifauna is comparatively species-poor and its exact age uncertain but probably late middle Paleocene (?MP 5, ~61 Ma; Mayr and Smith 2019a).

More recently, a coeval avifauna was reported from the Belgian locality Maret (Mayr and Smith 2019a).

Already in the mid-nineteenth century, late Paleocene (MP 6) birds were furthermore described from the localities Cernay-lès-Reims and Mont Berru in the Reims area in France (e.g., Lemoine 1878, 1881). The fossils from these sites also consist of isolated bones, and the avifauna is dominated by very large species.

Another long-known Paleocene site is Menat (Puy-de-Dôme, France). Even though the first bird fossils from this locality were reported a century ago, they remained unstudied until recently (Mayr et al. 2020b). The exact age of the site is controversial, but recent authors considered it to be from the Selandian stage (ca. 60–61 Ma).

By contrast, the early and middle Eocene European fossil record of birds is comprehensive, and numerous new taxa were described in the last two decades. Of particular significance for early Eocene birds are two localities of the London Clay in southern England, the Isle of Sheppey, and Walton-on-the-Naze. The fossiliferous sediments from Walton-on-the-Naze belong to the biostratigraphic unit MP 8 (54.6–55 Ma), and most specimens from this locality are in the private collection of Michael Daniels (1987, 1988, 1989, 1990, 1993, 1994; Feduccia 1999: table 4.1). The exposures from the Isle of Sheppey covers several horizons that span the units MP 8–9 (51.5–53.5 Ma). Fossil birds from the London Clay have already been collected in the nineteenth century, and the first two fossil avian species to have been scientifically named, *Halcyornis toliapicus* Koenig, 1825 and *Lithornis vulturinus* Owen, 1840, are actually from the Isle of Sheppey. Although the London Clay Formation is of marine origin, it yielded large numbers of terrestrial and arboreal birds, which lived in the forests near the shoreline. Most fossil specimens consist of three-dimensionally preserved bones, and often partial skeletons are found in association. The taxonomy of these fossils is, however, much complicated by the fact that numerous very fragmentary and non-comparable bones have been named by earlier authors.

Exquisitely preserved fossils were also found in the early Eocene (MP 7; 55–54 Ma) Fur Formation ("Mo-Clay") of northwestern Jutland in Denmark (Fig. 2.8c, d). Although these deposits consist of marine diatomites, most avian taxa represent land birds. Several of the fossils are represented by three-dimensional and hardly crushed skeletons (Lindow and Dyke 2006; Bertelli et al. 2010, 2013; Bourdon and Lindow 2015). A fairly diverse avifauna was furthermore reported from early Eocene (MP 8–9; 50.5–52 Ma) marine sediments of Egem in Belgium (Mayr and Smith 2019b).

Arguably the best studied and most productive locality for Eocene bird fossils, however, is Messel near Darmstadt in Hessen, Germany. The Messel "oilshale" was deposited

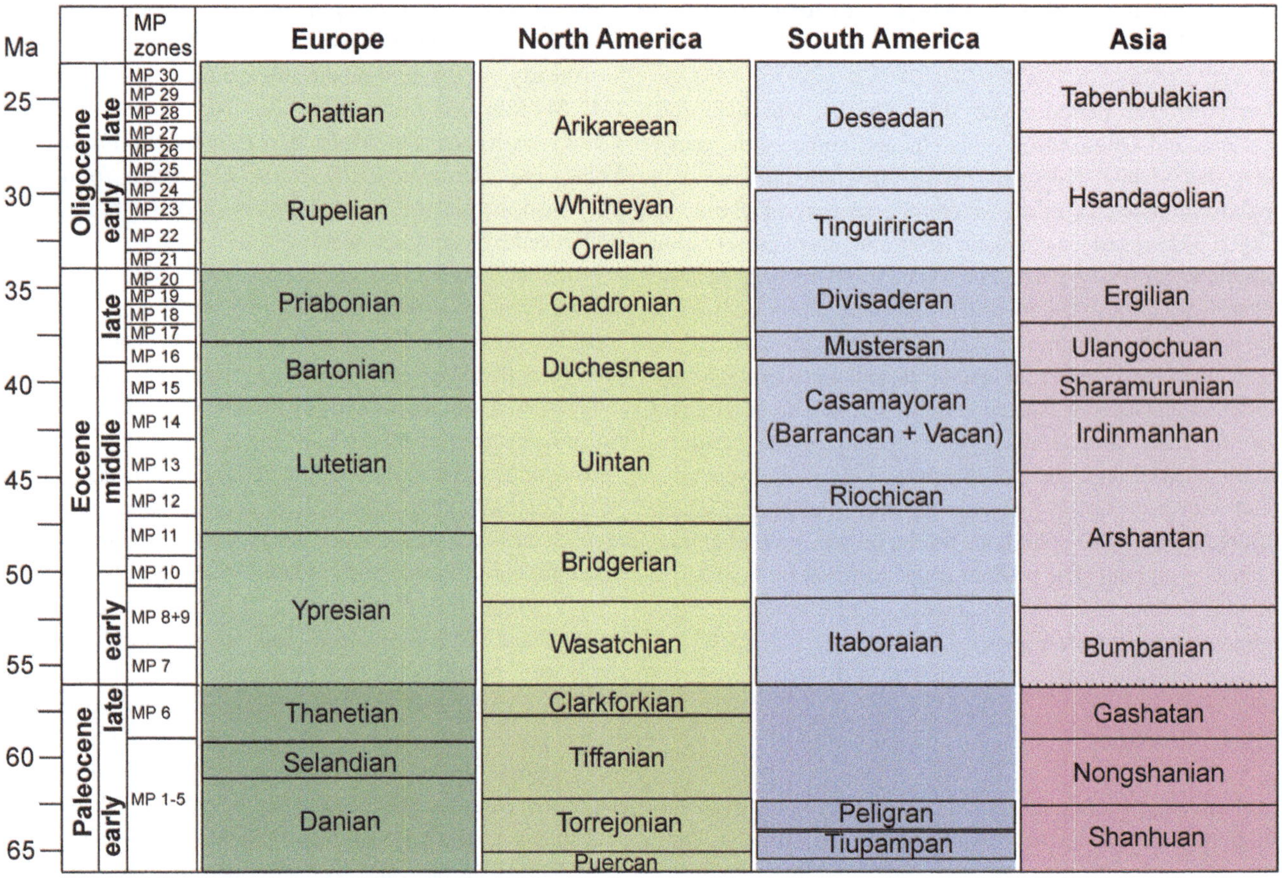

Ma	MP zones	Europe	North America	South America	Asia
25 —	Oligocene late: MP 30, MP 29, MP 28, MP 27, MP 26; Oligocene early: MP 25, MP 24, MP 23, MP 22, MP 21	Chattian / Rupelian	Arikareean / Whitneyan / Orellan	Deseadan / Tinguirirican	Tabenbulakian / Hsandagolian
30 —					
35 —	Eocene late: MP 20, MP 19, MP 18, MP 17; Eocene middle: MP 16, MP 15, MP 14, MP 13, MP 12; Eocene early: MP 11, MP 10, MP 8+9, MP 7	Priabonian / Bartonian / Lutetian / Ypresian	Chadronian / Duchesnean / Uintan / Bridgerian / Wasatchian	Divisaderan / Mustersan / Casamayoran (Barrancan + Vacan) / Riochican / Itaboraian	Ergilian / Ulangochuan / Sharamurunian / Irdinmanhan / Arshantan / Bumbanian
40 —					
45 —					
50 —					
55 —					
60 —	Paleocene late: MP 6; Paleocene early: MP 1-5	Thanetian / Selandian / Danian	Clarkforkian / Tiffanian / Torrejonian / Puercan	Peligran / Tiupampan	Gashatan / Nongshanian / Shanhuan
65 —					

Fig. 2.7 Stratigraphic chart relating the European standard stages to the North American, South American, and Asian land mammal ages (after Speijer et al. 2020). The approximate ages in million years (Ma) and the biostratigraphic zones of the Mammalian Paleogene (MP) are indicated

during the latest early or earliest middle Eocene (MP 11; 48 Ma; Lenz et al. 2015) in a small crater lake surrounded by paratropical forests (Schaal et al. 2018). Birds are among the most abundant land vertebrates in Messel and several hundred skeletons have so far been found. As yet, more than 70 species were distinguished, many of which are known from multiple specimens (Mayr 2017b, 2018). The fossil avifauna of Messel is biased toward small to medium-sized birds and only a few specimens of large species were reported. Some fossils exhibit well-preserved soft tissue remains, including feathers (Mayr 2018), foot scales (Peters 1988), and even uropygial gland lipids (Fig. 2.8a, b; Mayr 2006; O'Reilly et al. 2017). Specimens from Messel are prepared with the resin transfer method, and usually, these fossils are therefore embedded in slabs of synthetic resin.

Already in the early twentieth century, middle Eocene fossil birds were found in the now-abandoned opencast brown coal mines of the Geisel Valley (Geiseltal) near Halle in eastern Germany (Mayr 2002a, 2020). The fossiliferous deposits of these localities pertain to different stratigraphic horizons (MP 11–13; 44–46 Ma) and originated in swamps, sinkholes, or small ponds (Krumbiegel et al. 1983; Mayr 2020). Some fossils from the Geisel Valley also show exceptional soft tissue preservation, including even epithelial cells with nuclei as well as red blood corpuscles (Voigt 1988). With regard to the taxonomic composition of its avifauna, the Geisel Valley shows strong parallels to Messel.

A recently discovered middle Eocene locality, which yielded significant numbers of avian fossils, is the Ikovo site in eastern Ukraine (Mayr and Zvonok 2011, 2012; Zvonok and Gorobets 2016). These specimens are from marine sands, which originated in the Tethys Sea, and most belong to seabirds.

The "Phosphorites du Quercy" in France yielded a large number of middle Eocene to late Oligocene (MP 10/11–28) bird bones, which accumulated in karstic fissure fillings (Gaillard 1908; Mourer-Chauviré 2006). Many fossils were found during phosphate mining activities in the nineteenth century and lack accurate information on the locality and stratigraphic horizon they come from, but numerous bones from later excavations in the second half of the twentieth century are stratigraphically well constrained. The deposits of the Quercy fissure fillings originated in different palaeoenvironments. Most avian taxa were either terrestrial or arboreal, whereas aquatic birds are very rare. The more than

Fig. 2.8 Examples of exceptional preservation of avian fossils in some Paleogene sites. (**a**), (**b**) Uropygial gland waxes in a bird fossil (cf. *Messelirrisor*, Upupiformes) from the latest early or earliest middle Eocene of Messel (Senckenberg Research Institute Frankfurt, SMF-ME 11593A); (**a**) shows the prepared specimen, in (**b**) the uropygial gland can be seen before the fossil was transferred to the resin slab; the arrow indicates an enlarged detail of the preserved lipid residues. (**c**), (**d**) Feet from the early Eocene Fur Formation in Denmark with soft tissue preservation of the skin and toe pads (Geological Museum, Copenhagen, Denmark, **c**: MGUH DK 291, **d**: MGUH VP 34366b). (**e**) Skeleton of a small bird from the McAbee site of the Okanagan Highlands in British Columbia, Canada (Royal British Columbia Museum, Victoria, British Columbia, Canada, RBCM. EH2008.009.0001); note the yellow coloration of the bones, which may have been caused by the formation of sulfur during the decay process. Photos in (**a**), (**c**) and (**d**) by Sven Tränkner, in (**b**) by Michael Ackermann, and in (**e**) by the author

70 named species were listed by Mourer-Chauviré (1995, 2006). The Quercy localities are of particular importance not just because the avian fossils are very well studied but also because of the high number of three-dimensionally preserved bones and the wide stratigraphic range they cover. Virtually all specimens consist, however, of isolated bones, and an unambiguous assignment of different postcranial elements to the same taxon is sometimes difficult.

Late Eocene (MP 19; Mlíkovský 2002) birds were described from the Paris Gypsum in France ("Gypses de Montmartre"). Most of these are represented by articulated skeletons, but all are so poorly preserved that just a few osteological comparisons are possible (e.g., Brunet 1970; Harrison 1979). A late Eocene (MP 17, Mlíkovský 1996b) avifauna is also known from the Hampshire Basin in southern England, and a number of early Oligocene (MP 21–23, Mlíkovský 1996b) birds were discovered in the Hamstead Beds of the Isle of Wight (Harrison and Walker 1976, 1977).

Europe also has an extensive fossil record from the early Oligocene (Rupelian). At this time, large parts of the continent were covered by shallow marine seaways, which offered favorable conditions for the preservation of avian fossils in near-shore environments. Avian specimens from the Rupelian stratotype of the marine Boom Formation in Belgium have already been published in the nineteenth century. At this time, however, just a few bones were studied, and the majority of the bird fossils from the comprehensive old collections has only recently been studied (e.g., Mayr and Smith 2012, 2013).

Earliest Oligocene (MP 21) avifaunas were also reported from the fluvio-lacustrine Boutersem Member of the Borgloon Formation in Belgium (Mayr and Smith 2001, 2002). Well-preserved skeletons of early Oligocene birds were furthermore found in the locality Wiesloch-Frauenweiler in southern Germany (in the literature, the site is also referred to as Rauenberg). The former clay pit of this locality is well-known for its Rupelian (32 Ma) marine fish fauna, but the sediments are from a near-shore environment and have yielded a diverse array of marine and arboreal birds (Maxwell et al. 2016). Equally diversified avifaunas are known from Rupelian sediments of the Carpathian Basin in Poland (Bocheński et al. 2013, 2021). Other important localities for early Oligocene birds are situated in the Luberon in southern France. Most fossils from this area are from fine-grained limestone (Calcaires de Vachères) in the region around the villages Vachères and Céreste, which originated in brackish coastal lagoons (e.g., Mayr 1999; Roux 2002; Louchart et al. 2008, 2011; Duhamel et al. 2020).

2.5.2 Asia

There is an outstanding Cretaceous fossil record of non-neornithine birds from China and Mongolia (e.g., Chiappe 2007; Mayr 2017a), but little is still known about Paleogene Neornithes from Asia. Earlier revisions were published by Kurochkin (1976), Rich et al. (1986; China, Japan, and southeast Asia), Nessov (1992; area of the former Soviet Union), and Hou (2003; China). These are now much outdated and more recent reviews were provided by Zelenkov and Kurochkin (2015; Russia and adjacent countries) and Hood et al. (2019).

Significant numbers of Paleogene bird fossils were found in localities in Mongolia and Inner Mongolia (China). Paleocene and Eocene strata are exposed in various sites of the Gobi desert, with the Naranbulak Formation of the locality Tsagaan Khushuu in the Nemegt Basin having produced a particularly rich early Eocene (Bumbanian) avifauna, which includes species of the Galliformes, Anseriformes, Mirandornithes, Gruiformes, and Strigiformes (Hood et al. 2019; Zelenkov 2021a, b). Mongolia and Inner Mongolia also feature important late Eocene and early Oligocene sites, which mainly yielded remains of the cursorial Eogruidae and Ergilornithidae (Hood et al. 2019; Mayr and Zelenkov 2021).

Well-preserved early Eocene fossils were also found in Chinese localities outside Inner Mongolia (Wang et al. 2012; Zhao et al. 2015). A large number of avian specimens were furthermore found in the Vastan Lignite Mine, an early Eocene (about 52 Ma) locality in India (Mayr et al. 2010, 2013).

2.5.3 North America

The North American Paleogene is divided into eleven land mammal ages (Fig. 2.7; Speijer et al. 2020): The Paleocene comprises the Puercan (66–65 Ma), Torrejonian (65–62.5 Ma), Tiffanian (62.5–57 Ma), and Clarkforkian (57–56 Ma). The Eocene is subdivided into the early Eocene Wasatchian (56–52 Ma), the early/middle Eocene Bridgerian (52–47.5 Ma), the middle Eocene Uintan (47.5–41 Ma) and Duchesnean (41–37.5 Ma), and the late Eocene Chadronian (37.5–33.9 Ma). The Oligocene includes the early Oligocene Orellan (33.9–32 Ma) and Whitneyan (32–30 Ma), and the late Oligocene Arikareean land mammal age, which begins at about 30 Ma and extends into the Neogene.

North America has a comprehensive Paleogene fossil record of birds. The Cretaceous-Paleogene transition is covered by the marine sediments of the Navesink and Hornerstown formations of New Jersey. The Navesink Formation is of Late Cretaceous (Maastrichtian) age, whereas the upper parts of the Hornerstown Formation were deposited in the Paleocene (Parris and Hope 2002). Unfortunately, most avian specimens come from the basal part of the Hornerstown Formation, the age of which is more controversial and either latest Cretaceous or earliest Paleocene (Olson 1994; Parris and Hope 2002). Bird bones were also reported from the late Paleocene (Tiffanian) marine Aquia Formation of Maryland and Virginia (Olson 1994), and from early Tiffanian fluvio-lacustrine beds of the Wannagan Creek Quarry of North Dakota (Benson 1999).

After refinement of the collection techniques (Houde 1988), the early Eocene (latest Wasatchian to early Bridgerian) sediments of the Willwood Formation in the Bighorn Basin (Wyoming) yielded large numbers of fossil birds, many of which are still undescribed (Feduccia 1999: table 4.2). Isolated avian bones were also described from the early Eocene (ca. 53 Ma) Nanjemoy Formation of Virginia in eastern North America (Olson 1999; Mayr 2016; Mayr et al. 2022). Even though the fossiliferous deposits of this formation are of marine origin, most avian remains belong to land birds. Early Eocene (Ypresian) bird fossils from various sites of the Okanagan Highlands in British Columbia (Canada) are from higher paleoaltitudes and provide insights into the diversity of early Eocene birds, which lived in habitats with microthermal climates and mild winters (Fig. 2.8e; Mayr et al. 2019).

The most renowned North American locality for early Eocene bird finds, however, is the Green River Formation of Wyoming, Colorado, and Utah (Grande 2013). Its deposits represent the sediments of three lakes, Lake Gosiute, Lake Uinta, and Fossil Lake, which lasted, at least in parts, from the late Paleocene to the middle Eocene. Most specimens are from Fossil Lake, which was smaller and deeper than the other two lakes and only existed in the early Eocene (late Wasatchian).

The sediments of the middle Eocene (Bridgerian) Bridger Formation of the Green River Basin in southwestern Wyoming originated in different depositional environments, including river channels, floodplains, and lakes. Most birds found in these localities are represented by isolated bones.

Already a while ago, the stratigraphy of the Eocene/Oligocene boundary in North America was revised based on refined temporal correlations and calibrations (e.g., Prothero 1994). As a result, taxa originally described as "early Oligocene" (Chadronian land mammal age) are now considered to be from the late Eocene, and those from "middle" or "late" Oligocene localities (Orellan and Whitneyan land mammal ages) are of early Oligocene age. Important North American localities for middle Eocene to late Oligocene bird fossils are situated in the extensive badlands of South Dakota, Colorado, Wyoming, Nebraska, and Utah. According to the revised stratigraphy, the White River Group exposed at these sites is formed by the middle Eocene (Uintan) Uinta Formation, the late Eocene (Chadronian) Chadron Formation, and the Oligocene Brule Formation, which is divided into the Orellan Scenic Member and the Whitneyan Poleslide Member (Prothero and Emry 1996; Terry et al. 1998; Stoffer 2003). The palaeoenvironment of these localities was characterized by an increasing aridity toward the Oligocene. In the late Eocene, forested environments predominated, whereas wooded grassland occurred in the early Oligocene, and dry open grasslands prevailed in the late Oligocene (Prothero 1994: 153). A few avian fossils were also found in the latest Eocene (34 Ma) fluvio-lacustrine deposits of the Florissant Fossil Beds National Monument in Colorado and are of particular interest, because they consist of articulated skeletons (e.g., Chandler 1999; Ksepka and Clarke 2009).

Marine avifaunas were found in various sites of Oregon and Washington State (USA). Most fossils are from the latest Eocene or early Oligocene Lincoln Creek and Makah formations and the early Oligocene/early late Oligocene Pysht Formation of the Olympic Peninsula and southwest Washington State. Most fossils belong to the penguin-like Plotopteridae, but remains of other seabirds and a few land birds have also been collected. Most of these fossils were found by amateur collectors, with James Goedert being pivotal in their discovery and study (Goedert and Cornish 2002; Mayr et al. 2015; Mayr and Goedert 2016, 2018).

2.5.4 Central and South America

Apart from a humerus fragment of a pelagornithid from the middle Eocene of Mexico (González-Barba et al. 2002), there are no published bird fossils from the Paleogene of Central America. The Cenozoic avian fossil record of South America was summarized by Tambussi and Degrange (2013). The earliest specimens are from early Eocene (Itaboraian) marl fillings of the Bacia Calcária of Itaboraí in Brazil. These belong to a palaeognathous bird (*Diogenornis*), a representative of the Phorusrhacidae (*Paleopsilopterus*), a possible relative of the European taxon *Elaphrocnemus* (*Itaboravis*), and a small, long-legged taxon (*Eutreptodactylus*). The fossiliferous sediments of the Itaboraí site were long considered to be from the late Paleocene (e.g., Mayr 2009, Mayr et al. 2011), but a revised stratigraphy (Fig. 2.7) suggests that they are of early Eocene age (Woodburne et al. 2014).

Some bird fossils were described from the Casamayor Formation of the Chubut Province in the Argentinean part of Patagonia. This formation was considered to be of early Eocene age by earlier authors but is now assigned to the middle Eocene.

Avian fossils are more abundant in the late Oligocene (Deseadan land mammal age) of the Santa Cruz Province of Patagonia (the Deseadan was formerly assigned to the early Oligocene; see Berggren and Prothero 1992; Rose 2006). These localities are situated in temperate latitudes, and during the Oligocene their palaeoenvironment was dominated by an open woodland savanna with river banks and floodplains (Webb 1978).

Several avian taxa were also reported from the late Oligocene or early Miocene of the Tremembé Formation of the Taubaté Basin of São Paulo State in Brazil. These lacustrine deposits consist of bituminous shales, in which articulated skeletons were found, and layers of montmorillonitic clay that yielded isolated bones; the sediments were deposited in a shallow, alkaline lake (Alvarenga 1999). The first avian fossil, an isolated feather, was mentioned by Shufeldt (1916).

2.5.5 Africa

Paleogene bird fossils from Africa have long been very scarce, but in the past years, the fossil record was much expanded. A substantial but rather taxon-poor avifauna was reported from the late Paleocene and early Eocene phosphate beds of the Ouled Abdoun Basin in Morocco, the fossiliferous deposits of which originated in an epicontinental sea (Bourdon et al. 2005, 2010). All avian species described so far represent marine taxa, and remains of the Pelagornithidae and Prophaethontidae predominate.

Noteworthy numbers of avian fossils were also reported from the early or middle Eocene of Algeria (Mourer-Chauviré et al. 2011a), the late early or early middle Eocene of Tunisia (Mourer-Chauviré et al. 2013, 2016), and the middle Eocene of Namibia (Mourer-Chauviré et al. 2011b, 2015, 2017). A few bird fossils were furthermore found in the middle Eocene of Nigeria (Andrews 1916) and Togo (Bourdon and Cappetta 2012).

Otherwise, the only other Paleogene African avifaunas of significance come from the late Eocene/early Oligocene deposits of the Jebel Qatrani Formation of the Fayum Province in Egypt (Rasmussen et al. 1987, 2001; Stidham and Smith 2015). This formation is of fluvial origin and stretches over the Eocene/Oligocene boundary. Although the age of the lower sequences has been debated, it is now considered to be late Eocene (Rasmussen et al. 2001). The avian specimens from the Jebel Qatrani Formation consist of isolated and often fragmentary bones.

2.5.6 Australia, New Zealand, and Antarctica

Likewise very little is known about the Paleogene avifaunas of Australia, and most birds from this period were described after the reviews by Vickers-Rich (1991) and Boles (1991). The continent has no Paleocene fossil record of birds, and the only Eocene avian specimens are from deposits of the early Eocene Tingamarra Local Fauna near Murgon (Queensland), the sediments of which are of fluvio-lacustrine origin and have a minimum age of 54.6 million years (Boles 1999). The Oligocene fossil record is more comprehensive, and a fair number of late Oligocene bird fossils come from the Riversleigh site in Queensland. The more than 200 named sites of this locality were deposited in large lakes, shallow pools, and caves, and cover a wide stratigraphic range, from the late Oligocene into the Neogene (Boles 1997, 2001a, 2005). The described birds include both aquatic and terrestrial taxa. Late Oligocene (24–26 Ma) birds were also found in the Namba and Etadunna formations of the Lake Eyre Basin in South Australia, which were deposited in a primarily lacustrine environment and yielded specimens of the Casuariiformes, Anatidae, Anhingidae, Phalacrocoracidae, Accipitridae, Rallidae, Palaelodidae, Phoenicopteridae, and Columbidae (Boles 2001b; Worthy 2009, 2011, 2012).

Despite an abundant Quaternary fossil record and except for penguins, Paleogene birds are also rare in New Zealand (Worthy and Holdaway 2002). A notable exception, however, are the glauconitic sediments of the Waipara Greensand, which were deposited in deep water in a mid-shelf marine environment. Bird fossils from these strata include well-preserved specimens of mid to late Paleocene (about 58–61 Ma) stem group representatives of the Sphenisciformes as well as a few remains of other seabirds (Slack et al. 2006; Mayr and Scofield 2014, 2016; Mayr et al. 2018b, 2020c, d, 2021). Virtually all these fossils were collected by amateur collectors, with Leigh Love having found the majority of the recent specimens. Eocene and Oligocene localities in New Zealand, which yielded penguin fossils, were summarized by Simpson (1971).

The Paleogene record of neornithine birds from Antarctica was reviewed by Tambussi and Acosta Hospitaleche (2007) and Acosta Hospitaleche et al. (2019). Most specimens are from stem group Sphenisciformes and come from Seymour Island. Except for one Paleocene record from the Cross Valley Formation (Tambussi et al. 2005), these were found in the early to late Eocene La Meseta and Submeseta formations, which were deposited in a nearshore deltaic environment.

References

Acosta Hospitaleche C, Gelfo JN (2015) New Antarctic findings of Upper Cretaceous and lower Eocene loons (Aves: Gaviiformes). Ann Paléontol 101:315–324

Acosta Hospitaleche C, Worthy TH (2021) New data on the *Vegavis iaai* holotype from the Maastrichtian of Antarctica. Cret Res 124:104818

Acosta Hospitaleche C, Jadwiszczak P, Clarke JA, Cenizo M (2019) The fossil record of birds from the James Ross basin, West Antarctica. Adv Polar Sci 30:251–273

Agnolin FL (2010) An avian coracoid from the Upper Cretaceous of Patagonia, Argentina. Stud Geol Salmant 46:99–119

Agnolin FL (2017) Unexpected diversity of ratites (Aves, Palaeognathae) in the early Cenozoic of South America: palaeobiogeographical implications. Alcheringa 41:101–111

Agnolin FL, Novas FE, Lio G (2006) Neornithine bird coracoid from the Upper Cretaceous of Patagonia. Ameghiniana 43:245–248

Alvarenga HMF (1999) A fossil screamer (Anseriformes: Anhimidae) from the middle Tertiary of southeastern Brazil. In: Olson SL (ed) Avian paleontology at the close of the 20th century. Proceedings of the 4th international meeting of the Society of Avian Paleontology and Evolution, Washington, DC, 4–7 June 1996. Smithson Contrib Paleobiol 89:223–230

References

Andrews CW (1916) Note on the sternum of a large carinate bird from the (?) Eocene of southern Nigeria. Proc Zool Soc Lond 1916:519–524

Angst D, Buffetaut E (2017) Paleobiology of giant flightless birds. ISTE Press, London

Angst D, Buffetaut E, Corral JC, Pereda-Suberbiola X (2017) First record of the Late Cretaceous giant bird *Gargantuavis philoinos* from the Iberian Peninsula. Ann Paléontol 103:135–139

Baumel JJ, Witmer LM (1993) Osteologia. In: Baumel JJ, King AS, Breazile JE, Evans HE, Vanden Berge JC (eds) Handbook of avian anatomy: Nomina anatomica avium. Pub Nuttall Ornithol Club 23:45–132

Benson RD (1999) *Presbyornis isoni* and other late Paleocene birds from North Dakota. In: Olson SL (ed) Avian paleontology at the close of the 20th century: Proceedings of the 4th international meeting of the Society of Avian Paleontology and Evolution, Washington, DC, 4–7 June 1996. Smithson Contrib Paleobiol 89:253–259

Berggren WA, Prothero DR (1992) Eocene-Oligocene climatic and biotic evolution: an overview. In: Prothero DR, Berggren WA (eds) Eocene-Oligocene climatic and biotic evolution. Princeton University Press, Princeton, NJ, pp 1–28

Bertelli S, Lindow BEK, Dyke GJ, Chiappe LM (2010) A well-preserved 'charadriiform-like' fossil bird from the early Eocene Fur Formation of Denmark. Palaeontology 53:507–531

Bertelli S, Lindow BEK, Dyke GJ, Mayr G (2013) Another charadriiform-like bird from the lower Eocene of Denmark. Paleontol J 47:1282–1301

Bochenski ZM, Tomek T, Swidnicka E (2013) A review of avian remains from the Oligocene of the Outer Carpathians and Central Paleogene Basin. In: Göhlich UB, Kroh A (eds) Paleornithological research 2013—Proceedings of the 8th international meeting of the Society of Avian Paleontology and Evolution. Natural History Museum Vienna, Vienna, pp 37–42

Bochenski ZM, Tomek T, Bujoczek M, Salwa G (2021) A new passeriform (Aves: Passeriformes) from the early Oligocene of Poland sheds light on the beginnings of Suboscines. J Ornithol 162:593–604

Boles WE (1991) The origin and radiation of Australasian birds: perspectives from the fossil record. In: Bell BD, Cossee RO, Flux JEC, Heather BD, Hitchmough RA, Robertson CJR, Williams MJ (eds) Acta XX Congressus Internationalis Ornithologici. New Zealand Ornithological Congress Trust Board, Christchurch, pp 383–391

Boles WE (1997) Riversleigh birds as palaeoenvironmental indicators. Mem Qld Mus 41:241–246

Boles WE (1999) Early Eocene shorebirds (Aves: Charadriiformes) from the Tingamarra Local Fauna, Murgon, Queensland, Australia. Rec West Austral Mus, Suppl 57:229–238

Boles WE (2001a) A swiftlet (Apodidae: Collocaliini) from the Oligo-Miocene of Riversleigh, northwestern Queensland. Mem Assoc Austral Palaeontol 25:45–52

Boles WE (2001b) A new emu (Dromaiinae) from the Late Oligocene Etadunna Formation. Emu 101:317–321

Boles WE (2005) A new flightless gallinule (Aves: Rallidae: *Gallinula*) from the Oligo-Miocene of Riversleigh, northwestern Queensland, Australia. Rec Austral Mus 57:179–190

Bono RK, Clarke J, Tarduno JA, Brinkman D (2016) A large ornithurine bird (*Tingmiatornis arctica*) from the Turonian High Arctic: climatic and evolutionary implications. Sci Rep 6:38876

Bourdon E, Cappetta H (2012) Pseudo-toothed birds (Aves, Odontopterygiformes) from the Eocene phosphate deposits of Togo, Africa. J Vertebr Paleontol 32:965–970

Bourdon E, Lindow B (2015) A redescription of *Lithornis vulturinus* (Aves, Palaeognathae) from the Early Eocene Fur Formation of Denmark. Zootaxa 4032:493–514

Bourdon E, Bouya B, Iarochène M (2005) Earliest African neornithine bird: a new species of Prophaethontidae (Aves) from the Paleocene of Morocco. J Vertebr Paleontol 25:157–170

Bourdon E, Amaghzaz M, Bouya B (2010) Pseudotoothed birds (Aves, Odontopterygiformes) from the early Tertiary of Morocco. Am Mus Novit 3704:1–71

Braun EL, Kimball RT (2021) Data types and the phylogeny of Neoaves. Birds 2:1–22

Brennan PLR, Birkhead TR, Zyskowski K, van der Waag J, Prum RO (2008) Independent evolutionary reductions of the phallus in basal birds. J Avian Biol 39:487–492

Brunet J (1970) Oiseaux de l'Éocène supérieur du bassin de Paris. Ann Paléontol 56:3–57

Buffetaut E, Le Loueff L (2010) *Gargantuavis philoinos*: Giant bird or giant pterosaur? Ann Paléontol 96:135–141

Buffetaut E, Le Loeuff J, Mechin P, Mechin-Salessy A (1995) A large French Cretaceous bird. Nature 377:110

Chandler RM (1999) Fossil birds of Florissant, Colorado: with a description of a new genus and species of cuckoo. Geol Resour Div Techn Rep NPS/NRGRD/GRDTR-99:49–53

Chatterjee S (2002) The morphology and systematics of *Polarornis*, a Cretaceous loon (Aves: Gaviidae) from Antarctica. In: Zhou Z, Zhang F (eds) Proceedings of the 5th symposium of the Society of Avian Paleontology and Evolution, Beijing, 1–4 June 2000. Science Press, Beijing, pp 125–155

Chiappe LM (2007) Glorified dinosaurs: the origin and early evolution of birds. Wiley and Sons, Hoboken

Clarke JA (2004) Morphology, phylogenetic taxonomy, and systematics of *Ichthyornis* and *Apatornis* (Avialae: Ornithurae). Bull Am Mus Nat Hist 286:1–179

Clarke JA, Chiappe LM (2001) A new carinate bird from the Late Cretaceous of Patagonia (Argentina). Am Mus Novit 3323:1–22

Clarke JA, Tambussi CP, Noriega JI, Erickson GM, Ketcham RA (2005) Definitive fossil evidence for the extant avian radiation in the Cretaceous. Nature 433:305–308

Clarke JA, Chatterjee S, Li Z, Riede T, Agnolin F, Goller F, Isasi MP, Martinioni DR, Mussel FJ, Novas FE (2016) Fossil evidence of the avian vocal organ from the Mesozoic. Nature 538:502–505

Cordes-Person A, Hospitaleche CA, Case J, Martin J (2020) An enigmatic bird from the lower Maastrichtian of Vega Island, Antarctica. Cret Res 108:104314

Daniels M (1987) The Waltonian avifauna of Britain. Society of Avian Paleontology and Evolution, information letter 1. http://www.sapesociety.org/wp-content/uploads/2014/10/newsletter1987.pdf

Daniels M (1988) [Untitled report]. Society of Avian Paleontology and Evolution, information letter 2. http://www.sapesociety.org/wp-content/uploads/2014/10/newsletter1988.pdf

Daniels M (1989) [Untitled report]. Society of Avian Paleontology and Evolution, information letter 3. http://www.sapesociety.org/wp-content/uploads/2014/10/newsletter1989.pdf

Daniels M (1990) [Untitled report]. Society of Avian Paleontology and Evolution, information letter 4. http://www.sapesociety.org/wp-content/uploads/2014/10/newsletter1990.pdf

Daniels M (1993) [Untitled report]. Society of Avian Paleontology and Evolution, information letter 7. http://www.sapesociety.org/wp-content/uploads/2014/10/newsletter1993.pdf

Daniels M (1994) [Untitled report]. Society of Avian Paleontology and Evolution, information letter 8. http://www.sapesociety.org/wp-content/uploads/2014/10/newsletter1994.pdf

Duhamel A, Balme C, Legal S, Riamon S, Louchart A (2020) An early Oligocene stem Galbulae (jacamars and puffbirds) from southern France, and the position of the Paleogene family Sylphornithidae. Auk 137:ukaa023

Dyke GJ, Mayr G (1999) Did parrots exist in the Cretaceous period? Nature 399:317–318

Dyke GJ, Nudds RL, Benton MJ (2007) Modern avian radiation across the Cretaceous-Paleogene boundary. Auk 124:339–341

Dzerzhinsky FY (1995) Evidence for common ancestry of the Galliformes and Anseriformes. Cour Forsch-Instit Senckenberg 181:325–336

Elzanowski A (1999) A comparison of the jaw skeleton in theropods and birds, with a description of the palate in the Oviraptoridae. In: Olson SL (ed) Avian paleontology at the close of the 20th century: Proceedings of the 4th international meeting of the Society of Avian Paleontology and Evolution, Washington, DC, 4–7 June 1996. Smithson Contrib Paleobiol 89:311–323

Elzanowski A, Stidham TA (2011) A galloanserine quadrate from the Late Cretaceous Lance Formation of Wyoming. Auk 128:138–145

Ericson PGP, Anderson CL, Britton T, Elzanowski A, Johansson US, Källersjö M, Ohlson JI, Parsons TJ, Zuccon D, Mayr G (2006) Diversification of Neoaves: integration of molecular sequence data and fossils. Biol Lett 2:543–547

Fain MG, Houde P (2007) Multilocus perspectives on the monophyly and phylogeny of the order Charadriiformes (Aves). BMC Evol Biol 7:35

Fain MG, Krajewski C, Houde P (2007) Phylogeny of "core Gruiformes" (Aves: Grues) and resolution of the Limpkin-Sungrebe problem. Mol Phylogenet Evol 43:515–529

Feduccia A (1995) Explosive evolution in Tertiary birds and mammals. Science 267:637–638

Feduccia A (1999) The origin and evolution of birds, 2nd edn. Yale University Press, New Haven

Feduccia A (2003) 'Big bang' for Tertiary birds? Trends Ecol Evol 18:172–176

Feduccia A (2014) Avian extinction at the end of the Cretaceous: assessing the magnitude and subsequent explosive radiation. Cret Res 50:1–15

Field DJ, Bercovici A, Berv JS, Dunn R, Fastovsky DE, Lyson TR, Vajda V, Gauthier JA (2018) Early evolution of modern birds structured by global forest collapse at the end-Cretaceous mass extinction. Curr Biol 28:1825–1831

Field DJ, Benito J, Chen A, Jagt JW, Ksepka DT (2020) Late Cretaceous neornithine from Europe illuminates the origins of crown birds. Nature 579:397–401

Fürbringer M (1888) Untersuchungen zur Morphologie und Systematik der Vögel, zugleich ein Beitrag zur Anatomie der Stütz- und Bewegungsorgane, vol 2. Van Holkema, Amsterdam

Gaillard C (1908) Les oiseaux des Phosphorites du Quercy. Ann Univ Lyon (Nouv Sér) 23:1–178

Goedert JL, Cornish J (2002) A preliminary report on the diversity and stratigraphic distribution of the Plotopteridae (Pelecaniformes) in Paleogene rocks of Washington State, USA. In: Zhou Z, Zhang F (eds) Proceedings of the 5th symposium of the Society of Avian Paleontology and Evolution, Beijing, 1–4 June 2000. Science Press, Beijing, pp 63–76

González-Barba G, Schwennicke T, Goedert JL, Barne LG (2002) Earliest Pacific Basin record of the Pelagornithidae (Aves: Pelecaniformes). J Vertebr Paleontol 22:722–725

Grande L (2013) The lost world of Fossil Lake. Snapshots from deep time. University of Chicago Press, Chicago

Griffiths CS (1994) Monophyly of the Falconiformes based on syringeal morphology. Auk 111:787–805

Hackett SJ, Kimball RT, Reddy S, Bowie RCK, Braun EL, Braun MJ, Chojnowski JL, Cox WA, Han K-L, Harshman J, Huddleston CJ, Marks BD, Miglia KJ, Moore WS, Sheldon FH, Steadman DW, Witt CC, Yuri T (2008) A phylogenomic study of birds reveals their evolutionary history. Science 320:1763–1767

Harrison CJO (1979) The Upper Eocene birds of the Paris basin: a brief re-appraisal. Tert Res 2:105–109

Harrison CJO, Walker CA (1976) Birds of the British Upper Eocene. Zool J Linnean Soc 59:323–351

Harrison CJO, Walker CA (1977) Birds of the British Lower Eocene. Tert Res Spec Pap 3:1–52

Hood SC, Torres CR, Norell MA, Clarke JA (2019) New fossil birds from the earliest Eocene of Mongolia. Am Mus Novit 3934:1–24

Hope S (1999) A new species of *Graculavus* from the Cretaceous of Wyoming (Aves: Neornithes). In: Olson SL (ed) Avian Paleontology at the close of the 20th century. Proceedings of the 4th international meeting of the Society of Avian Paleontology and Evolution, Washington, DC, 4–7 June 1996. Smithson Contrib Paleobiol 89:251–259

Hope S (2002) The Mesozoic radiation of Neornithes. In: Chiappe LM, Witmer LM (eds) Mesozoic birds: above the heads of dinosaurs. University of California Press, Berkeley, pp 339–388

Hou L-H (2003) Fossil birds of China. Yunnan Science and Technology Press, Kunming

Houde P (1988) Palaeognathous birds from the early Tertiary of the Northern Hemisphere. Publ Nuttall Ornithol Club 22:1–148

Johnson KP, Kennedy M, McCracken KG (2006) Reinterpreting the origins of flamingo lice: cospeciation or host-switching? Biol Lett 2:275–278

Krumbiegel G, Rüffle L, Haubold H (1983) Das eozäne Geiseltal. Neue Brehm-Bücherei, vol 237. Ziemsen, Wittenberg

Ksepka DT, Clarke JA (2009) Affinities of *Palaeospiza bella* and the phylogeny and biogeography of mousebirds (Coliiformes). Auk 126:245–259

Kuhl H, Frankl-Vilches C, Bakker A, Mayr G, Nikolaus G, Boerno ST, Klages S, Timmermann B, Gahr M (2021) An unbiased molecular approach using 3′UTRs resolves the avian family-level tree of life. Mol Biol Evol 38:108–121

Kurochkin EN (1976) A survey of the Paleogene birds of Asia. Smithson Contrib Paleobiol 27:5–86

Legendre S, Lévêque F (1997) Etalonnage de l'échelle biochronologique mammalienne du Paléogène d'Europe occidentale: vers une intégration à l'échelle globale. In: Aguilar J-P, Legendre S, Michaux J (eds) Actes du Congrès BiochroM'97. Mém Trav Ec Prat Hautes Etudes, Inst Montpellier 21:461–473

Lemoine V (1878) Recherches sur les oiseaux fossiles des terrains tertiaires inférieurs des environs de Reims. Imprimerie Keller, Reims

Lemoine V (1881) Recherches sur les oiseaux fossiles des terrains tertiaires inférieurs des environs de Reims, deuxième partie. Imprimerie Matot-Braine, Reims, pp 75–170

Lenz OK, Wilde V, Mertz DF, Riegel W (2015) New palynology-based astronomical and revised 40 Ar/39 Ar ages for the Eocene maar lake of Messel (Germany). Int J Earth Sci 104:873–889

Lindow BEK, Dyke GJ (2006) Bird evolution in the Eocene: climate change in Europe and a Danish fossil fauna. Biol Rev 81:483–499

Livezey BC, Zusi RL (2007) Higher-order phylogeny of modern birds (Theropoda, Aves: Neornithes) based on comparative anatomy: II.—analysis and discussion. Zool J Linnean Soc 149:1–94

Longrich N (2009) An ornithurine-dominated avifauna from the Belly River Group (Campanian, Upper Cretaceous) of Alberta, Canada. Cret Res 30:161–177

Longrich NR, Tokaryk T, Field DJ (2011) Mass extinction of birds at the Cretaceous-Paleogene (K-Pg) boundary. Proc Natl Acad Sci U S A 108:15253–15257

Louchart A, Tourment N, Carrier J, Roux T, Mourer-Chauviré C (2008) Hummingbird with modern feathering: an exceptionally well-preserved Oligocene fossil from southern France. Naturwissenschaften 95:171–175

Louchart A, Tourment N, Carrier J (2011) The earliest known pelican reveals 30 million years of evolutionary stasis in beak morphology. J Ornithol 152:15–20

Manegold A (2006) Two additional synapomorphies of grebes Podicipedidae and flamingos Phoenicopteridae. Acta Ornithol 41:79–82

References

Marsh OC (1880) Odontornithes: a monograph on the extinct toothed birds of North America. U.S. Government Printing Office, Washington

Maxwell EE, Alexander S, Bechly G, Eck K, Frey E, Grimm K, Kovar-Eder J, Mayr G, Micklich N, Rasser M, Roth-Nebelsick A, Salvador RB, Schoch R, Schweigert G, Stinnesbeck W, Wolf-Schwenninger-K, Ziegler R (2016) The Rauenberg fossil Lagerstätte (Baden-Württemberg, Germany): a window into early Oligocene marine and coastal ecosystems of Central Europe. Palaeogeogr Palaeoclimatol Palaeoecol 463:238–260

Mayr G (1999) A new trogon from the middle Oligocene of Céreste, France. Auk 116:427–434

Mayr G (2002a) Avian remains from the middle Eocene of the Geiseltal (Sachsen-Anhalt, Germany). In: Zhou Z, Zhang F (eds) Proceedings of the 5th symposium of the Society of Avian Paleontology and Evolution, Beijing, 1–4 June 2000. Science Press, Beijing, p 77–96

Mayr G (2002b) Osteological evidence for paraphyly of the avian order Caprimulgiformes (nightjars and allies). J Ornithol 143:82–97

Mayr G (2004) Morphological evidence for sister group relationship between flamingos (Aves: Phoenicopteridae) and grebes (Podicipedidae). Zool J Linnean Soc 140:157–169

Mayr G (2006) New specimens of the Eocene Messelirrisoridae (Aves: Bucerotes), with comments on the preservation of uropygial gland waxes in fossil birds from Messel and the phylogenetic affinities of Bucerotes. Paläontol Z 80:405–420

Mayr G (2007) The birds from the Paleocene fissure filling of Walbeck (Germany). J Vertebr Paleontol 27:394–408

Mayr G (2008) Avian higher-level phylogeny: well-supported clades and what we can learn from a phylogenetic analysis of 2954 morphological characters. J Zool Syst Evol Res 46:63–72

Mayr G (2009) Paleogene fossil birds, 1st edn. Springer, Heidelberg

Mayr G (2010) Phylogenetic relationships of the paraphyletic "caprimulgiform" birds (nightjars and allies). J Zool Syst Evol Res 48:126–137

Mayr G (2011) Metaves, Mirandornithes, Strisores and other novelties—a critical review of the higher-level phylogeny of neornithine birds. J Zool Syst Evol Res 49:58–76

Mayr G (2014) The origins of crown group birds: molecules and fossils. Palaeontology 57:231–242

Mayr G (2016) The world's smallest owl, the earliest unambiguous charadriiform bird, and other avian remains from the early Eocene Nanjemoy Formation of Virginia (USA). Paläontol Z 90:747–763

Mayr G (2017a) Avian evolution: the fossil record of birds and its paleobiological significance. Wiley-Blackwell, Chichester

Mayr G (2017b) The early Eocene birds of the Messel fossil site: a 48 million-year-old bird community adds a temporal perspective to the evolution of tropical avifaunas. Biol Rev 92:1174–1188

Mayr G (2018) Birds—the most species-rich vertebrate group in Messel. In: Schaal SKF, Smith K, Habersetzer J (eds) Messel—An ancient greenhouse ecosystem. Schweitzerbart, Stuttgart, pp 169–214

Mayr G (2020) An updated review of the middle Eocene avifauna from the Geiseltal (Germany), with comments on the unusual taphonomy of some bird remains. Geobios 62:45–59

Mayr G, Clarke J (2003) The deep divergences of neornithine birds: a phylogenetic analysis of morphological characters. Cladistics 19:527–553

Mayr G, Goedert JL (2016) New late Eocene and Oligocene remains of the flightless, penguin-like plotopterids (Aves, Plotopteridae) from western Washington State. USA J Vertebr Paleontol 36:e1163573

Mayr G, Goedert JL (2018) First record of a tarsometatarsus of *Tonsala hildegardae* (Plotopteridae) and other avian remains from the late Eocene/early Oligocene of Washington State (USA). Geobios 51:51–59

Mayr G, Scofield RP (2014) First diagnosable non-sphenisciform bird from the early Paleocene of New Zealand. J Roy Soc New Zealand 44:48–56

Mayr G, Scofield RP (2016) New avian remains from the Paleocene of New Zealand: the first early Cenozoic Phaethontiformes (tropicbirds) from the Southern Hemisphere. J Vertebr Paleontol 36:e1031343

Mayr G, Smith R (2001) Ducks, rails, and limicoline waders (Aves: Anseriformes, Gruiformes, Charadriiformes) from the lowermost Oligocene of Belgium. Geobios 34:547–561

Mayr G, Smith R (2002) Avian remains from the lowermost Oligocene of Hoogbutsel (Belgium). Bull Inst Roy Sci Nat Belg 72:139–150

Mayr G, Smith T (2012) Phylogenetic affinities and taxonomy of the Oligocene Diomedeoididae, and the basal divergences amongst extant procellariiform birds. Zool J Linnean Soc 166:854–875

Mayr G, Smith T (2013) Galliformes, Upupiformes, Trogoniformes, and other avian remains (?Phaethontiformes and ?Threskiornithidae) from the Rupelian stratotype in Belgium, with comments on the identity of "*Anas*" *benedeni* Sharpe, 1899. In: Göhlich UB, Kroh A (eds) Paleornithological research 2013—Proceedings of the 8th international meeting of the Society of Avian Paleontology and Evolution. Natural History Museum Vienna, Vienna, pp 23–35

Mayr G, Smith T (2019a) New Paleocene bird fossils from the North Sea Basin in Belgium and France. Geol Belg 22:35–46

Mayr G, Smith T (2019b) A diverse bird assemblage from the Ypresian of Belgium furthers knowledge of early Eocene avifaunas of the North Sea Basin. N Jb Geol Paläontol, Abh 291:253–281

Mayr G, Zelenkov N (2021) Extinct crane-like birds (Eogruidae and Ergilornithidae) from the Cenozoic of Central Asia are ostrich precursors. Ornithology 138:1–15

Mayr G, Zvonok E (2011) Middle Eocene Pelagornithidae and Gaviiformes (Aves) from the Ukrainian Paratethys. Palaeontology 54:1347–1359

Mayr G, Zvonok E (2012) A new genus and species of Pelagornithidae with well-preserved pseudodentition and further avian remains from the middle Eocene of the Ukraine. J Vertebr Paleontol 32:914–925

Mayr G, Manegold A, Johansson U (2003) Monophyletic groups within "higher land birds"—comparison of morphological and molecular data. J Zool Syst Evol Res 41:233–248

Mayr G, Rana RS, Rose KD, Sahni A, Kumar K, Singh L, Smith T (2010) *Quercypsitta*-like birds from the early Eocene of India (Aves, ?Psittaciformes). J Vertebr Paleontol 30:467–478

Mayr G, Alvarenga H, Clarke J (2011) An *Elaphrocnemus*-like landbird and other avian remains from the late Paleocene of Brazil. Acta Palaeontol Polon 56:679–684

Mayr G, Rana RS, Rose KD, Sahni A, Kumar K, Smith T (2013) New specimens of the early Eocene bird *Vastanavis* and the interrelationships of stem group Psittaciformes. Paleontol J 4:1308–1314

Mayr G, Goedert JL, Vogel O (2015) Oligocene plotopterid skulls from western North America and their bearing on the phylogenetic affinities of these penguin-like seabirds. J Vertebr Paleontol 35:e943764

Mayr G, Scofield P, De Pietri V, Worthy T (2018a) On the taxonomic composition and phylogenetic affinities of the recently proposed clade Vegaviidae Agnolín et al., 2017—neornithine birds from the Upper Cretaceous of the Southern Hemisphere. Cretac Res 86:178–185

Mayr G, De Pietri VL, Love L, Mannering AA, Scofield RP (2018b) A well-preserved new mid-Paleocene penguin (Aves, Sphenisciformes) from the Waipara Greensand in New Zealand. J Vertebr Paleontol 37:e1398169I5

Mayr G, Archibald SB, Kaiser GW, Mathewes RW (2019) Early Eocene (Ypresian) birds from the Okanagan Highlands, British Columbia (Canada) and Washington State (USA). Canad J Earth Sci 56:803–813

Mayr G, Codrea V, Solomon A, Bordeianu M, Smith T (2020a) A well-preserved pelvis from the Maastrichtian of Romania suggests that the enigmatic *Gargantuavis* is neither an ornithurine bird nor an insular endemic. Cret Res 106:104271

Mayr G, Hervet S, Buffetaut E (2020b) On the diverse and widely ignored Paleocene avifauna of Menat (Puy-de-Dôme, France): new taxonomic records and unusual soft tissue preservation. Geol Mag 156:572–584

Mayr G, De Pietri VL, Love L, Mannering AA, Bevitt JJ, Scofield RP (2020c) First complete wing of a stem group sphenisciform from the Paleocene of New Zealand sheds light on the evolution of the penguin flipper. Diversity 12(2):46

Mayr G, De Pietri VL, Love L, Mannering AA, Scofield RP (2020d) Leg bones of a new penguin species from the Waipara Greensand add to the diversity of very large-sized Sphenisciformes in the Paleocene of New Zealand. Alcheringa 44:194–201

Mayr G, De Pietri VL, Love L, Mannering A, Scofield RP (2021) Oldest, smallest and phylogenetically most basal pelagornithid, from the early Paleocene of New Zealand, sheds light on the evolutionary history of the largest flying birds. Pap Palaeontol 7:217–233

Mayr G, De Pietri V, Scofield RP (2022) New bird remains from the early Eocene Nanjemoy Formation of Virginia (USA), including the first records of the Messelasturidae, Psittacopedidae, and Zygodactylidae from the Fisher/Sullivan site. Hist Biol 34:322–334

McLachlan SM, Kaiser GW, Longrich NR (2017) *Maaqwi cascadensis*: a large, marine diving bird (Avialae: Ornithurae) from the Upper Cretaceous of British Columbia, Canada. PLoS One 12:e0189473

Mlíkovský J (ed) (1996a) Tertiary avian localities of Europe. Acta Univ Carolinae, Geol 39:1–852

Mlíkovský J (1996b) Tertiary avian localities of the United Kingdom. In: Mlíkovský J (ed) Tertiary Avian Localities of Europe. Acta Univ Carolinae, Geol 39:759–771

Mlíkovský J (2002) Cenozoic birds of the world. Part 1: Europe. Ninox Press, Praha

Montgomerie R, Briskie JV (2007) Anatomy and evolution of copulatory structures. In: Jamieson BGM (ed) Reproductive biology and phylogeny of birds. Science Publishers Inc, Enfield, NH, pp 115–148

Morgan-Richards M, Trewick SA, Bartosch-Härlid A, Kardailsky O, Phillips MJ, McLenachan PA, Penny D (2008) Bird evolution: testing the Metaves clade with six new mitochondrial genomes. BMC Evol Biol 8:20

Mourer-Chauviré C (1995) Dynamics of the avifauna during the Paleogene and the early Neogene of France. Settling of the recent fauna. Acta zool crac 38:325–342

Mourer-Chauviré C (2006) The avifauna of the Eocene and Oligocene Phosphorites du Quercy (France): an updated list. Strata, sér 1 13: 135–149

Mourer-Chauviré C, Tabuce R, Mahboubi MH, Adaci M, Bensalah M (2011a) A Phororhacoid bird from the Eocene of Africa. Naturwiss 98:815–823

Mourer-Chauviré C, Pickford M, Senut B (2011b) The first Palaeogene galliform from Africa. J Ornithol 152:617–622

Mourer-Chauviré C, Tabuce R, El Mabrouk E, Marivaux L, Khayati H, Vianey-Liaud M, Ben Haj Ali M (2013) A new taxon of stem group Galliformes and the earliest record for stem group Cuculidae from the Eocene of Djebel Chambi, Tunisia. In: Göhlich UB, Kroh A (eds) Paleornithological research 2013—Proceedings of the 8th international meeting of the Society of Avian Paleontology and Evolution. Natural History Museum Vienna, Vienna, pp 1–15

Mourer-Chauviré C, Pickford M, Senut B (2015) Stem group galliform and stem group psittaciform birds (Aves, Galliformes, Paraortygidae, and Psittaciformes, family incertae sedis) from the middle Eocene of Namibia. J Ornithol 156:275–286

Mourer-Chauviré C, Ammar HK, Marivaux L, Marzougui W, Temani R, Vianey-Liaud M, Tabuce R (2016) New remains of the very small cuckoo, *Chambicuculus pusillus* (Aves, Cuculiformes, Cuculidae) from the late Early or early Middle Eocene of Djebel Chambi, Tunisia. Palaeovertebr 40(1):e2:1–4

Mourer-Chauviré C, Pickford M, Senut B (2017) New data on stem group Galliformes, Charadriiformes, and Psittaciformes from the middle Eocene of Namibia. Contrib MACN 7:99–131

Nessov LA (1992) Mesozoic and Paleogene birds of the USSR and their paleoenvironments. In: Campbell KE (ed) Papers in avian paleontology honoring Pierce Brodkorb. Nat Hist Mus Los Angeles Cty, Sci Ser 36:465–478

Noriega JI, Tambussi CP (1995) A Late Cretaceous Presbyornithidae (Aves: Anseriformes) from Vega Island, Antarctic Peninsula: Palaeobiogeographic implications. Ameghiniana 32:57–61

Novas FE, Agnolin F, Rozadilla S, Aranciaga-Rolando AM, Brisson-Egli F, Motta MJ, Cerroni M, Ezcurra MD, Martinelli AG, D'Angelo JD, Avarez-Herrera G, Gentil AR, Bogan S, Chimento NR, García-Marsa JA, Lo Coco G, Miqwuel SE, Brito FF, Vera EI, Perez Loinaze VS, Fernández MS, Salgado L (2019) Paleontological discoveries in the Chorrillo Formation (Upper Campanian-Lower Maastrichtian, Upper Cretaceous), Santa Cruz Province, Patagonia, Argentina. Rev Mus Argent Cienc Nat 21:217–293

O'Connor JK, Chiappe LM, Bell A (2011) Pre-modern birds: avian divergences in the Mesozoic. In: Dyke G, Kaiser G (eds) Living dinosaurs. The evolutionary history of modern birds. John Wiley & Sons, Chichester, pp 39–114

O'Reilly S, Summons R, Mayr G, Vinther J (2017) Preservation of uropygial gland lipids in a 48-million-year-old bird. Proc Roy Soc London, Ser B 284:20171050

Olson SL (1985) The fossil record of birds. In: Farner DS, King JR, Parkes KC (eds) Avian biology, vol 8. Academic Press, New York, pp 79–238

Olson SL (1992) *Neogaeornis wetzeli* Lambrecht, a Cretaceous loon from Chile (Aves: Gaviidae). J Vertebr Paleontol 12:122–124

Olson SL (1994) A giant *Presbyornis* (Aves: Anseriformes) and other birds from the Paleocene Aquia Formation of Maryland and Virginia. Proc Biol Soc Wash 107:429–435

Olson SL (1999) Early Eocene birds from eastern North America: a faunule from the Nanjemoy Formation of Virginia. In: Weems RE, Grimsley GJ (eds) Early Eocene vertebrates and plants from the Fisher/Sullivan site (Nanjemoy Formation) Stafford County, Virginia. Virginia Div Mineral Resour Publ 152:123–132

Olson SL, Parris DC (1987) The Cretaceous birds of New Jersey. Smithson Contrib Paleobiol 63:1–22

Parris DC, Hope S (2002) New interpretations of the birds from the Navesink and Hornerstown Formations, New Jersey, USA (Aves: Neornithes). In: Zhou Z-H, Zhang F-Z (eds) Proceedings of the 5th symposium of the Society of Avian Paleontology and Evolution, 1–4 June 2000. Science Press, Beijing, pp 113–124

Paton TA, Baker AJ (2006) Sequences from 14 mitochondrial genes provide a well-supported phylogeny of the charadriiform birds congruent with the nuclear RAG-1 tree. Mol Phylogenet Evol 39:657–667

Pérez-Pueyo M, Puértolas-Pascual E, Moreno-Azanza M, Cruzado-Caballero P, Gasca JM, Núñez-Lahuerta C, Canudo JI (2021, in press) First record of a giant bird (Ornithuromorpha) from the uppermost Maastrichtian of the Southern Pyrenees, northeast Spain. J Vertebr Paleontol:e1900210

Peters DS (1988) Die Messel-Vögel—eine Landvogelfauna. In: Schaal S, Ziegler W (eds) Messel—Ein Schaufenster in die Geschichte der Erde und des Lebens. Kramer, Frankfurt/M, pp 135–151

Prothero DR (1994) The late Eocene-Oligocene extinctions. Annu Rev Earth Planet Sci 22:145–165

Prothero DR, Emry RJ (1996) The terrestrial Eocene-Oligocene transition in North America. Cambridge University Press, Cambridge

Prum RO, Berv JS, Dornburg A, Field DJ, Townsend JP, Lemmon EM, Lemmon AR (2015) A comprehensive phylogeny of birds (Aves) using targeted next-generation DNA sequencing. Nature 526:569–573

Rasmussen DT, Olson SL, Simons EL (1987) Fossil birds from the Oligocene Jebel Qatrani Formation, Fayum Province, Egypt. Smithson Contrib Paleobiol 62:1–20

Rasmussen DT, Simons EL, Hertel F, Judd A (2001) Hindlimb of a giant terrestrial bird from the Upper Eocene, Fayum, Egypt. Palaeontology 44:325–337

Rich PV, Hou L-H, Ono K, Baird RF (1986) A review of the fossil birds of China, Japan and Southeast Asia. Geobios 19:755–772

Rose KD (2006) The beginning of the age of mammals. Johns Hopkins University Press, Baltimore

Roux T (2002) Deux fossiles d'oiseaux de l'Oligocène inférieur du Luberon. Courr sci Parc nat rég Luberon 6:38–57

Sangster G (2005) A name for the flamingo-grebe clade. Ibis 147:612–615

Schaal SKF, Smith K, Habersetzer J (eds) (2018) Messel—An ancient greenhouse ecosystem. Schweitzerbart, Stuttgart

Shufeldt RW (1916) A fossil feather from Taubaté. Auk 33:206–207

Simpson GG (1971) A review of the pre-Pliocene penguins of New Zealand. Bull Am Mus Nat Hist 144:319–378

Slack KE, Jones CM, Ando T, Harrison GL, Fordyce RE, Arnason U, Penny D (2006) Early penguin fossils, plus mitochondrial genomes, calibrate avian evolution. Mol Biol Evol 23:1144–1155

Speijer RP, Pälike H, Hollis CJ, Hooker JJ, Ogg JG (2020) The Paleogene period. In: Gradstein FM, Ogg JG, Schmitz MD, Ogg GM (eds) Geologic Time Scale 2020. Elsevier, Amsterdam, pp 1087–1140

Stidham TA (1998) A lower jaw from a Cretaceous parrot. Nature 396:29–30

Stidham TA, Smith NA (2015) An ameghinornithid-like bird (Aves, Cariamae, ?Ameghinornithidae) from the early Oligocene of Egypt. Palaeontol Electron 18.1(5A):1–8

Stoffer P (2003) Geology of badlands national park: a preliminary report. US Geological Survey, OFR 03-35:1–63

Tambussi CP, Acosta Hospitaleche C (2007) Antarctic birds (Neornithes) during the Cretaceous-Eocene times. Rev Asoc Geol Argent 62:604–617

Tambussi CP, Degrange F (2013) South American and Antarctic continental Cenozoic birds: Paleobiogeographic affinities and disparities. Springer, Dordrecht

Tambussi CP, Reguero MA, Marenssi SA, Santillana SN (2005) *Crossvallia unienwillia*, a new Spheniscidae (Sphenisciformes, Aves) from the late Paleocene of Antarctica. Geobios 38:667–675

Terry DO, LaGarry HE, Hunt RM (eds) (1998) Depositional environments, lithostratigraphy, and biostratigraphy of the White River and Arikaree groups (late Eocene to early Miocene, North America). Geol Soc Am, Spec Pap 325

Torres CR, Norell MA, Clarke JA (2021) Bird neurocranial and body mass evolution across the end-Cretaceous mass extinction: the avian brain shape left other dinosaurs behind. Sci Adv 7(31):eabg7099

van Tuinen M, Butvill DB, Kirsch JAW, Hedges SB (2001) Convergence and divergence in the evolution of aquatic birds. Proc Roy Soc London, Ser B 268:1345–1350

Vickers-Rich PV (1991) The Mesozoic and Tertiary history of birds on the Australian plate. In: Vickers-Rich P, Monaghan TM, Baird RF, Rich TH (eds) Vertebrate palaeontology of Australasia. Pioneer Design Studio and Monash University Publications Committee, Melbourne, pp 721–808

Voigt E (1988) Preservation of soft tissues in the Eocene lignite of the Geiseltal near Halle/S. Cour Forsch-Inst Senckenberg 107:325–343

Wang M, Mayr G, Zhang J, Zhou Z (2012) Two new skeletons of the enigmatic, rail-like avian taxon *Songzia* Hou, 1990 (Songziidae) from the early Eocene of China. Alcheringa 36:487–499

Webb SD (1978) A history of savanna vertebrates in the New World. Part II: South America and the great interchange. Ann Rev Ecol Syst 9:393–426

Weigelt J (1939) Die Aufdeckung der bisher ältesten tertiären Säugetierfauna Deutschlands. Nova Acta Leopold, N F 7:515–528

West AR, Torres CR, Case JA, Clarke JA, O'Connor PM, Lamanna MC (2019) An avian femur from the Late Cretaceous of Vega Island, Antarctic Peninsula: removing the record of cursorial landbirds from the Mesozoic of Antarctica. PeerJ 7:e7231

Wetmore A (1960) A classification for the birds of the world. Smithson Misc Collects 139:1–37

Woodburne MO, Goin FJ, Raigemborn MS, Heizler M, Gelfo JN, Oliveira EV (2014) Revised timing of the south American early Paleogene land mammal ages. J S Am Earth Sci 54:109–119

Worthy TH (2009) Descriptions and phylogenetic relationships of two new genera and four new species of Oligo-Miocene waterfowl (Aves: Anatidae) from Australia. Zool J Linnean Soc 156:411–454

Worthy TH (2011) Descriptions and phylogenetic relationships of a new genus and two new species of Oligo-Miocene cormorants (Aves: Phalacrocoracidae) from Australia. Zool J Linnean Soc 163:277–314

Worthy TH (2012) A new species of Oligo-Miocene darter (Aves: Anhingidae) from Australia. Auk 129:96–104

Worthy TH, Holdaway RN (2002) The lost world of the moa. Prehistoric life of New Zealand. Indiana University Press, Bloomington

Xue X (1995) *Qinornis paleocenica*—a Paleocene bird discovered in China. Cour Forsch-Inst Senckenberg 181:89–93

Yuri T, Kimball RT, Harshman J, Bowie RC, Braun MJ, Chojnowski JL, Han K-L, Hackett SJ, Huddleston CJ, Moore WS, Reddy S, Sheldon FH, Steadman DW, Witt CC, Braun EL (2013) Parsimony and model-based analyses of indels in avian nuclear genes reveal congruent and incongruent phylogenetic signals. Biol 2:419–444

Zelenkov N (2021a) New bird taxa (Aves: Galliformes, Gruiformes) from the early Eocene of Mongolia. Paleontol J 55:438–446

Zelenkov N (2021b) A revision of the Palaeocene-Eocene Mongolian Presbyornithidae (Aves: Anseriformes). Paleontol J 55:323–330

Zelenkov NV, Kurochkin EN (2015) Class aves. In Kurochkin EN, Lopatin AV, Zelenkov NV (eds) Fossil vertebrates of Russia and neighbouring countries. Fossil Reptiles and Birds. Part 2. GEOS, Moscow, p 86–290 [in Russian]

Zhao T, Mayr G, Wang M, Wang W (2015) A trogon-like arboreal bird from the early Eocene of China. Alcheringa 39:287–294

Zvonok E, Gorobets L (2016) A record of a landbird (Telluraves) from the Eocene Ikovo locality (East Ukraine). Acta zool crac 59:37–45

Palaeognathous Birds

Extant Palaeognathae include the volant tinamous (Tinamiformes; South and Central America) and the flightless Apterygiformes (kiwis; New Zealand), Rheiformes (rheas; South America), Struthioniformes (ostrich; Africa), and Casuariiformes (cassowaries and emus; Australian region). Recently extinct flightless taxa are the Aepyornithiformes (elephant birds) of Madagascar and the Dinornithiformes (moas) of New Zealand.

The Tinamiformes were long considered to be the sister group of the flightless palaeognathous birds, which were classified as "ratites" (e.g., Sibley and Ahlquist 1990; Livezey and Zusi 2007). However, current molecular analyses congruently supported a clade including the Tinamiformes, Casuariiormes, and Apterygiformes (Hackett et al. 2008; Harshman et al. 2008; Prum et al. 2015; Kuhl et al. 2021). Accordingly, flightlessness must have evolved independently within several palaeognathous lineages, as has already been assumed by some earlier authors (e.g., Olson 1985; Houde 1988; Feduccia 1999).

The distribution of extant palaeognathous birds is mainly restricted to the Southern Hemisphere, but several fossil taxa were reported from the Paleogene of Europe and North America. Most interesting from an evolutionary point of view are various long-legged, crane-like birds from Eurasia and North America, which are likely to be stem group representatives of the Struthioniformes.

Rheiformes and Casuariiformes have a Paleogene fossil record in their current ranges, whereas Paleogene fossils of the Tinamiformes, Dinornithiformes, Apterygiformes, and Aepyornithiformes have not yet been found. Disregarding the superficially tinamou-like Lithornithiformes, the affinities of which are uncertain (see below), the earliest unambiguous Tinamiformes are from the Miocene of Argentina (Bertelli and Chiappe 2005). Eggshell fragments of putative Dinornithiformes were reported from the early or middle Miocene of New Zealand (Tennyson et al. 2010), and the same deposits also yielded a putative stem group representative of the Apterygiformes (Worthy et al. 2013). Fossil Aepyornithiformes are known from Quaternary sites only.

3.1 Lithornithidae

These superficially tinamou-like birds were reported from late Paleocene to middle Eocene strata of North America and Europe (Fig. 3.1). Even though some specimens from the London Clay are among the earliest bird fossils to have been scientifically described, lithornithids were first recognized as a distinctive palaeognathous taxon by Houde and Olson (1981) and Houde (1986, 1988).

Houde (1988) included the following six species in the Lithornithidae: *Lithornis vulturinus* (early Eocene of the London Clay and the Fur Formation), *L. plebius* (early Eocene of the Willwood Formation and possibly the London Clay), *L. promiscuus* (early Eocene of the Willwood Formation), *L. celetius* (late Paleocene [Tiffanian] of Montana and Wyoming), ?*L. hookeri* (early Eocene of the London Clay), *Pseudocrypturus cercanaxius* (early Eocene of the Green River Formation and the London Clay), and *Paracathartes howardae* (early Eocene of the Willwood Formation). A further species considered valid by Houde (1988), *Lithornis nasi* from the early Eocene of the London Clay of Walton-in-the-Naze, was synonymized with *L. vulturinus* by Bourdon and Lindow (2015). The latter authors also described an exceptionally well-preserved partial skeleton of *L. vulturinus* from the Danish Fur Formation, which was initially studied by Leonard et al. (2005); further lithornithid remains from the Fur Formation were reported by Kristoffersen (1999).

Since the publication of Houde's (1988) monograph, various other new lithornithid fossils were described, which furthered our knowledge of the taxonomic diversity and stratigraphic occurrence of these birds. Parris and Hope (2002) tentatively assigned a scapula from the latest

Cretaceous/earliest Paleocene of New Jersey to the Lithornithidae but noted that the similarities shared with lithornithids may be plesiomorphic. A fragmentary humerus of a lithornithid was also reported from the late Paleocene of California (Stidham et al. 2014). Nesbitt and Clarke (2016) described a second lithornithid species from the Green River Formation as *Calciavis grandei*; this species is represented by multiple skeletons, some of which exhibit well-preserved feather remains (Torres et al. 2020).

An omal extremity of a coracoid from the late middle Paleocene of Walbeck in Germany, which was described as *Fissuravis weigelti* (Fig. 3.1g), was tentatively assigned to the Lithornithidae (Mayr 2007) but differs from typical lithornithids in a proportionally smaller acrocoracoid process. The earliest definitive European fossils of the Lithornithidae are from the latest Paleocene (latest Thanetian) of Rivecourt-Petit Pâtis in France (Mayr and Smith 2019). The latest fossil record of these birds is from the latest early or earliest middle Eocene of Messel, where a postcranial skeleton and a skull were found (Mayr 2008, 2009a). This unnamed species is one of the smallest lithornithids and of similar size to the poorly known ?*L. hookeri*.

Lithornithid species exhibited very different sizes, with the turkey-sized *Paracathartes howardae* having been about twice as large as *Pseudocrypturus cercanaxius*. In the morphology of the bones of the palate, they most closely resemble the Apterygiformes (according to Houde 1988, *Pseudocrypturus* is, however, distinguished from other lithornithids in the absence of a pterygoid fossa, which the other species share with the Apterygiformes). As in other palaeognathous birds, the upper beak bears a pair of furrows rostral of the nostrils and the ilioschiadic foramina of the pelvis are caudally open, which represents the plesiomorphic condition for Neornithes. Lithornithids have tinamou-like overall limb proportions, a long rhynchokinetic beak, and a well-developed sternal keel. As in extant Tinamiformes the fronto-parietal suture of the skull remained open in the adult birds (in all other extant birds the frontal and parietal bones are fully co-ossified). Other diagnostic features of the lithornithid skeleton include a very long and narrow acromion of the scapula, and the absence of incisions in the caudal margin of the sternum. Unlike in extant palaeognathous birds, the lateral surfaces of the bodies of the thoracic vertebrae exhibit large pneumatic openings. The distal end of the tibiotarsus lacks a supratendinal bridge (which is present in the Tinamiformes and Dinornithiformes), and contrary to all extant palaeognathous birds the hallux of at least the smaller lithornithid species is well developed. Based on the morphology of the caudal vertebrae, Houde (1988: 113) assumed that the tail of lithornithids was short as in extant Tinamiformes.

Several lithornithid fossils were found in association with eggshells, the structure of which resembles that of the Tinamiformes both microscopically and macroscopically (Houde 1988; Grellet-Tinner and Dyke 2005). According to Houde (1988); the large accumulations of eggshells in some sites indicate large clutches.

Lithornithids may have used their long and narrow beak for probing along shorelines or other bodies of water (Houde 1988). In contrast to extant Tinamiformes, they probably were strong flyers capable of sustained flight (Torres et al. 2020), and the well-developed hallux and curved ungual phalanges indicate perching capabilities (Houde 1988).

The interrelationships between the lithornithid taxa and the phylogenetic affinities between lithornithids and other palaeognathous birds are poorly understood. Houde (1988) considered it possible that the taxon Lithornithidae is actually not monophyletic, but at this time the interrelationships of extant Palaeognathae were hardly resolved. In discussing the phylogenetic affinities of lithornithids, Houde (1988) not only assumed a sister group relationship between the Tinamiformes and neognathous birds but also considered the flightless palaeognathous taxa to be monophyletic. The cladogram he regarded as best supported by the morphological data (Houde 1988: fig. 39) shows *Pseudocrypturus*, *Lithornis*, and *Paracathartes* as successive sister taxa of extant flightless Palaeognathae, but more recent analyses by Nesbitt and Clarke (2016) supported lithornithid monophyly.

Lithornithids were considered to be the sister taxon of the Tinamiformes by Leonard et al. (2005) who did, however, not detail the character evidence for this hypothesis. Although lithornithids share with extant Tinamiformes an open frontoparietal suture, this feature also occurs in Mesozoic stem group representatives of the Neornithes (Elzanowski and Galton 1991) and may be a plesiomorphic trait. An analysis by Livezey and Zusi (2007) supported a sister group relationship between lithornithids and all other neornithine birds; again, however, the character evidence for this placement was not listed. Analyses by Nesbitt and Clarke (2016) and Worthy et al. (2017) did not conclusively resolve the affinities of lithornithids, which under different settings resulted either as the sister taxon of the Tinamiformes or as that of all other palaeognaths (one analysis by Worthy et al. 2017 even placed lithornithids within the Neognathae).

3.2 Palaeotididae, Geranoididae, and Eogruidae—Northern Hemispheric Stem Group Representatives of the Struthioniformes

This section includes three taxa, which have not been united before, and their combination is one of the most radical systematic rearrangements compared to the first edition of this book. I believe, however, that the reasons for this action are coherent and will stand future scrutiny.

3.2 Palaeotididae, Geranoididae, and Eogruidae—Northern Hemispheric Stem Group...

Fig. 3.1 Fossils of the Lithornithidae. (**a**) Skull of a lithornithid from the latest early or earliest middle Eocene of Messel (Royal Belgian Institute of Natural Sciences, Brussels, Belgium, IRSNB Av 82). (**b**) Skeleton of *Pseudocrypturus cercanaxius* from the early Eocene Green River Formation (private collection in Switzerland, cast in Senckenberg Research Institute). (**c**), (**d**) Partial skeleton of *Lithornis vulturinus* from the early Eocene Fur Formation in Denmark (Geological Museum, Copenhagen, Denmark, MGUH 26770); (**d**) shows a detail of the skull in ventral view. (**e**), (**f**) Two partial skeletons from the early Eocene London Clay of Walton-on-the-Naze (collection of Michael Daniels, Holland-on-Sea, UK, **e**: WN 84474; **f**: WN 80280); the arrow indicates an enlarged detail of the coracoid. (**g**) Coracoid of *Fissuravis weigelti* from the late middle Paleocene of Walbeck in Germany (holotype, Institut für Geologische Wissenschaften of Martin-Luther-Universität Halle-Wittenberg, Halle/Saale, Germany IGWuG WAL346.2007). (Photo in (**b**) courtesy of Peter Houde; all others by Sven Tränkner)

Ostriches are the sole extant didactyl (two-toed) birds and the survivors of an ancient lineage, which diverged early from other palaeognathous birds. The oldest modern-type stem group representative of the Struthioniformes is *Struthio coppensi* from the early Miocene of Namibia. This species was somewhat smaller than the extant species but otherwise very similar in the known skeletal elements. Mourer-Chauviré et al. (1996) hypothesized that the fossils of this modern-type species suggest that ostriches originated in Africa before the Miocene, from where they dispersed into Eurasia in the Neogene. However, ostriches have a comprehensive fossil record in the Neogene of Eurasia (Mayr 2017), and as detailed in the following, stem group representatives of the Struthioniformes appear to have been widespread in the Northern Hemisphere.

3.2.1 The European Palaeotididae

This distinctive group of early Paleogene palaeognathous birds was originally established for *Palaeotis weigelti* from the early/middle Eocene of the Geisel Valley and Messel in Germany, which is represented by several skeletons and had a standing height of slightly less than 1 m (Fig. 3.2; Houde and Haubold 1987; Peters 1988; Mayr 2015, 2019). *P. weigelti* was a long-legged, flightless bird, which certainly had cursorial habits although it lived in a forested palaeoenvironment. Kohring and Hirsch (1996) described eggshell with a "palaeognathous morphotype" from the Geisel Valley as *Medioolithus geiselensis*; with a diameter of 9 cm, these eggs probably are from *P. weigelti*.

Another representative of the Palaeotididae is *Galligeranoides boriensis* from the early Eocene of France (Fig. 3.3c; Bourdon et al. 2016). This species was assigned to the Geranoididae (see next section) in the original description, and palaeotidid affinities were established by Mayr (2015, 2019). The holotype of *G. boriensis* is a tarsometatarsus, but an ulna from the type locality, which was compared to the suliform taxon *Limnofregata* by Bourdon et al. (2016), probably also belongs to the species (Mayr 2015, 2019). As noted by Mayr (2019), palaeotidid affinities are furthermore likely for *Palaeogrus princeps* from the middle Eocene (MP 11–13; Mlíkovský 2002) of Italy. Initially described as a species of the gruiform Gruidae by Portis (1885), *P. princeps* is just known from a distal tibiotarsus, which is indistinguishable from that of *Palaeotis*.

A distinctive trait of the Palaeotididae is the presence of a supratendinal bridge on the distal end of the tibiotarsus (Mayr 2019). In extant or recently extinct palaeognathous birds, this bridge is only present in the Tinamiformes and Dinornithiformes. The distal tibiotarsus of palaeotidids furthermore exhibits a well-developed tubercle next to the supratendinal bridge and a notch in the rim of the medial condyle, which are not found in palaeognathous birds but are derived characteristics of various long-legged neognathous birds, including some taxa of the Gruiformes and the Phoenicopteriformes. The "gruiform" appearance of the distal end of the tibiotarsus may have contributed to the initial misidentifications of *Galligeranoides boriensis* and *Palaeogrus princeps*. An isolated leg of *Palaeotis weigelti* was also assigned to the taxon *Palaeogrus* by Lambrecht (1935), who considered the fossil to be from a crane-like gruiform bird, and there even exists a possibility that *Palaeotis* is a junior synonym of *Palaeogrus* (Mayr 2019).

The holotype of *Palaeotis weigelti* was originally described as a bustard (Otidiformes) by Lambrecht (1928) on the basis of fossils from the Geisel Valley. Palaeognathous affinities of *P. weigelti* were first proposed by Houde (1986) and were further substantiated by Houde and Haubold (1987). The main characteristics shared with extant Palaeognathae include the eponymous morphology of the palate, a pair of furrows along the ventral surface of the mandibular symphysis, as well as a coracoid, which is co-ossified with the scapula to form a scapulocoracoid and has a distinctive shape only found in palaeognathous birds (Houde and Haubold 1987; Mayr 2015). *P. weigelti* has a narrower beak than all extant Palaeognathae except the Tinamiformes, Apterygiformes, and the casuariiform Casuariidae (cassowaries). As in extant Struthioniformes and Rheiformes, the humerus and the ulna of *P. weigelti* are fairly long, whereas the wing skeleton is much more reduced in the Casuariiformes, Apterygiformes, and Dinornithiformes (Peters 1988). In contrast to all extant Palaeognathae, the tarsometatarsus of *Palaeotis* furthermore exhibits deep furrows along the midlines of its dorsal and plantar surfaces, which give the bone an almost H-shaped cross-section (Mayr 2002). Unlike in extant Palaeognathae, there are ossified tendons along the plantar surface of the tarsometatarsus (Mayr 2019). The first phalanx of the fourth pedal digit bears a proximally directed process, which led Lambrecht (1928) to assume a ground-scratching habit for *Palaeotis*.

Palaeotis shares a number of derived characteristics with the Struthioniformes, Rheiformes, and Casuariiformes, including a skull with well-developed supraorbital processes, a scapulocoracoid, a mediolaterally compressed pelvis, as well as the reduction of the sternal keel and the loss of the hallux. As noted above, these extant taxa do not form a clade, so that the above traits must have evolved several times independently, which diminishes their phylogenetic significance. Houde (1986) and Houde and Haubold (1987) considered a sister group relationship between the Palaeotididae and the Struthioniformes, whereas Peters (1988) assumed a sister group relationship to the Rheiformes. Phylogenetic analyses by Mayr (2015) and Nesbitt and Clarke (2016) resulted in a sister group relationship between *Palaeotis* and a clade formed by the Struthioniformes, Rheiformes, and

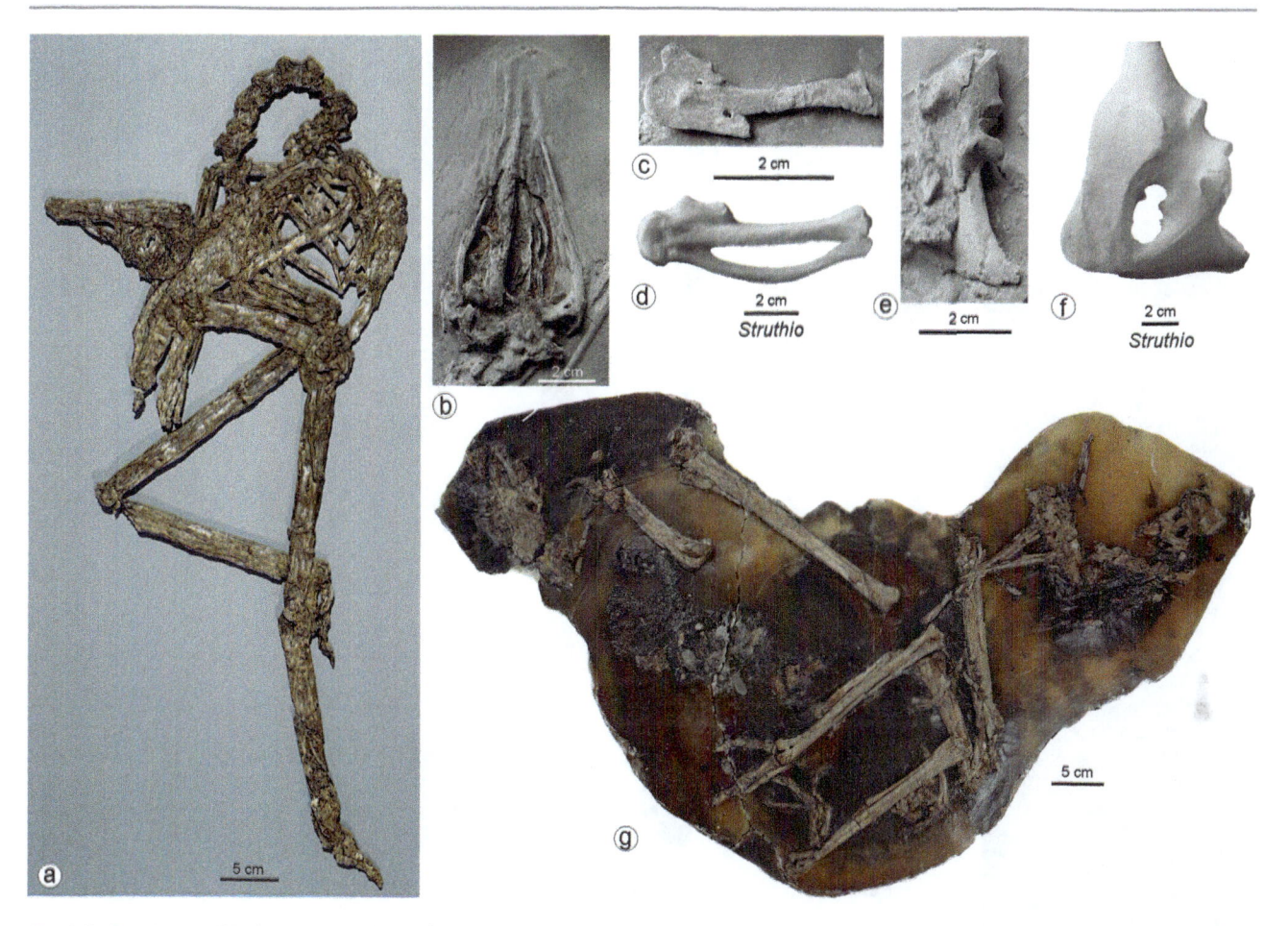

Fig. 3.2 Specimens of *Palaeotis weigelti*. (**a**) Skeleton from the latest early or earliest middle Eocene of Messel (Senckenberg Research Institute Frankfurt, SMF-ME 1578). (**b**) Skull from Messel in ventral view (Hessisches Landesmuseum Darmstadt, HLMD Me 771). (**c**) Carpometacarpus (HLMD Me 771) in comparison to (**d**) that of the extant *Struthio camelus* (Struthionidae). (**e**) Coracoid (HLMD Me 771) in comparison to (**f**) that of the extant *Struthio camelus* (Struthionidae). (**g**) Partial skeleton from the middle Eocene of the Geisel Valley (Geiseltalsammlung, Martin-Luther Universität of Halle-Wittenberg, Germany, GMH 4362). Specimens in (**b**), (**c**), and (**e**) were coated with ammonium chloride. (Photo in (**a**) by Erwin Haupt, all others by Sven Tränkner)

Casuariiformes. However, this clade is not obtained in analyses of molecular data, which congruently supported a sister group relationship between the Struthioniformes and all other extant Palaeognathae. From a biogeographic point of view, struthioniform affinities of the Palaeotididae appear most plausible. As detailed in the following, they are also supported by the close resemblances between palaeotidids and three extinct groups of birds, the Geranoididae, Eogruidae, and Ergilornithidae, which were considered ostrich ancestors by earlier authors.

3.2.2 The North American Geranoididae

This North American taxon was originally introduced for *Geranoides jepseni*, a species represented by a distal tibiotarsus and a partial tarsometatarsus from the early Eocene Willwood Formation of Wyoming (Wetmore 1933). Cracraft (1969, 1973) included six further species in the Geranoididae, which are also known from leg bones only, namely *Paragrus prentici* (known from distal tibiotarsus, distal femur, pedal phalanges), *P. shufeldti* (distal ends of tibiotarsus and tarsometatarsus), *Eogeranoides campivagus* (distal tibiotarsus and proximal tarsometatarsus), *Palaeophasianus meleagroides* (distal tibiotarsus, incomplete femur, and tarsometatarsus), *P. incompletus* (fragmentary distal tarsometatarsus), and *Geranodornis aenigma* (distal tibiotarsus). In a more recent revision, Mayr (2016) concluded that *Geranoides jepseni* is a junior synonym of *Palaeophasianus meleagroides* and that *Eogeranoides campivagus* is poorly differentiated from *Paragrus prentici*, which reduces the number of geranoidid species to four or five.

The species included in the Geranoididae are medium-sized to large birds and as shown by a nearly complete leg of *Eogeranoides campivagus/Paragrus prentici*, they had a very long and slender tarsometatarsus (Fig. 3.3d; Mayr 2016). Except for *G. aenigma*, which is from the middle

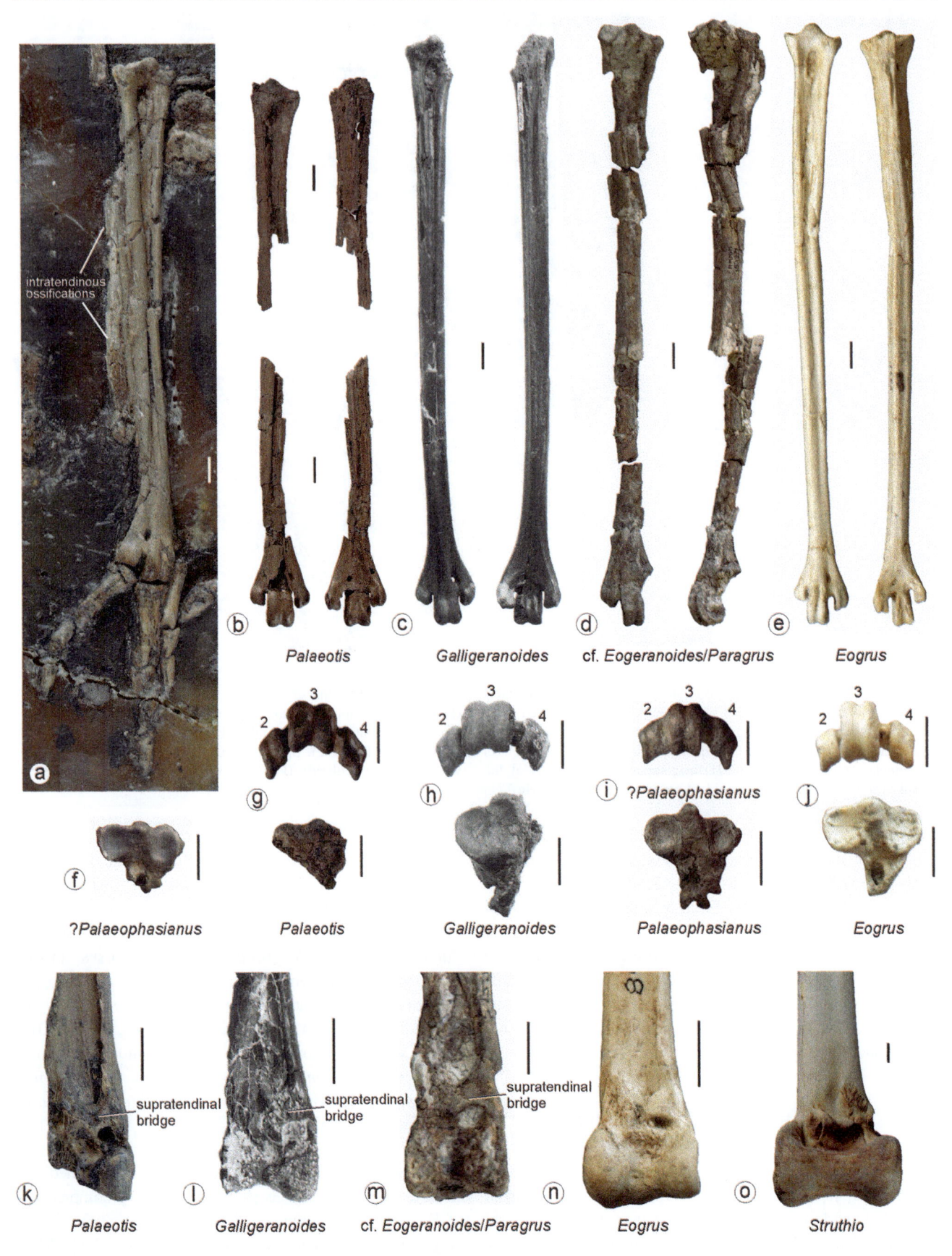

Fig. 3.3 Tarsometatarsi (**a–j**) and tibiotarsi (**k–o**) of the Palaeotididae, Geranoididae, Eogruidae, and Struthionidae (modified after Mayr 2019; published under a Creative Commons CC BY 4.0 license). (**a**) *Palaeotis weigelti* (Palaeotididae), right tarsometatarsus (Geiseltalsammlung, Martin-Luther Universität of Halle-Wittenberg, Germany, GMH 4362). (**b**) *P. weigelti*, proximal end of right tarsometatarsus (GMH

Eocene Bridger Formation, all geranoidid species come from the Willwood Formation. Fossils of *P. meleagroides* were found in both the Willwood and Bridger formations.

Cracraft (1969) considered geranoidids to be similar to the late Eocene and Oligocene Eogruidae and Ergilornithidae of Central Asia (see next section). As in the latter and extant Gruoidea (the gruiform clade including the Psophiidae, Aramidae, and Gruidae), the distal end of the tibiotarsus of the Geranoididae exhibits a prominent tubercle lateral of the supratendinal bridge. Mayr (2016) hypothesized that geranoidids are stem group representatives of the Gruoidea, but this hypothesis had to be revised after the putative geranoidid *Galligeranoides* was identified as a member of the Palaeotididae (see the previous section). As detailed by Mayr (2019), a Gruoidea-like morphology also characterizes the tibiotarsus of the Palaeotididae, to which the Geranoididae show a great resemblance in the known bones. Close affinities between the Geranoididae and Palaeotididae are also suggested by an undescribed proximal end of a geranoidid humerus mentioned by Olson (1985: 157), who noted that this bone "is unquestionably from a flightless bird."

Following the proposal of Mayr (2019), it is here assumed that geranoidids are palaeognathous birds, which are closely related to the Palaeotididae. The latter appear to have been a common faunal element in the early Eocene of Europe, whereas flightless representatives of the Palaeognathae have long been unknown from the early Eocene of North America. As evidenced by the large Gastornithidae, land corridors must have existed between both continents, and the recognition of close affinities between palaeotidids and geranoidids, therefore, closes an important gap in the North American fossil record.

3.2.3 The Asian Eogruidae and Ergilornithidae

Hindlimb bones of large, long-legged birds are comparatively abundant in some middle and late Eocene localities in Central Asia and were assigned to the taxa Eogruidae and Ergilornithidae (e.g., Kurochkin 1976, 1981, 1982; Olson 1985; Zelenkov and Kurochkin 2015). The taxon Eogruidae was established by Wetmore (1934) for *Eogrus aeola*, a species named on the basis of tibiotarsi and tarsometatarsi from the middle Eocene Irdin Manha Formation of Inner Mongolia (China). New material of *E. aeola* from the late Eocene of Mongolia was identified by Clarke et al. (2005) and also consists of tibiotarsi and tarsometatarsi. Two further species of *Eogrus*, *E. crudus*, and *E.* ("*Progrus*") *turanicus*, are known from the late Eocene of Mongolia and Kazakhstan, respectively. Another species assigned to the Eogruidae is *Sonogrus gregalis* from the late Eocene (Priabonian) Ergilin Dzo Formation in Mongolia, which may have been sexually dimorphic in size (Kurochkin 1976, 1981). Of *E. crudus*, *E. turanicus*, and *S. gregalis* likewise only hindlimb bones are known.

Paleogene species assigned to the Ergilornithidae are *Proergilornis minor* and *Ergilornis rapidus* from late Eocene and early Oligocene strata of the Ergilin Dzo Formation in Mongolia (Kurochkin 1976, 1981). Other ergilornithid taxa were described from the Miocene and Pliocene of Eurasia (Zelenkov and Kurochkin 2015; Musser et al. 2020).

Eogruids and ergilornithids were cursorial birds with a very long and slender tarsometatarsus, which bears a distinct crest along the lateral side of its plantar surface. The species assigned to these taxa show a progressive reduction of the tarsometatarsal trochlea for the second toe. This trochlea is somewhat reduced in the late Eocene *Eogrus*, even smaller in *Sonogrus gregalis*, vestigial in *Proergilornis minor*, and completely lost in *Ergilornis rapidus* and the Neogene taxa *Amphipelargus*, *Sinoergilornis*, and *Urmiornis*.

Wetmore (1934) assumed that eogruids are related to the Gruidae, and they were considered to be the sister taxon of the Gruoidea by Cracraft (1973: fig. 46). Because of the progressive reduction and ultimate loss of the trochlea for the second toe, however, a few earlier authors already hypothesized that eogruids and ergilornithids are stem group representatives of the Struthioniformes, which include the only other didactyl birds. Possible affinities between ergilornithids and the Struthioniformes were first considered

Fig. 3.3 (continued) IX-566-1953) and distal end of left tarsometatarsus (holotype, GMH 4416) in dorsal and plantar view. (**c**) *Galligeranoides boriensis* (Palaeotididae), right tarsometatarsus (mirrored to ease comparisons) in dorsal and plantar view (holotype, from Bourdon et al. 2016; published under a Creative Commons CC BY 4.0 license). (**d**) cf. *Eogeranoides campivagus* (Geranoididae, American Museum of Natural History, New York, AMNH 5127), left tarsometatarsus in dorsal and plantar view (the fossil consists of several fragments, which were assembled for the photo). (**e**) *Eogrus aeola* (Eogruidae, AMNH 2937), right tarsometatarsus in dorsal and plantar view (image mirrored to ease comparisons). (**f**)–(**j**) Proximal (lower row) and distal ends (upper row) of the tarsometatarsi of (**f**) ?*Palaeophasianus* sp. (AMNH 5156), (**g**) *P. weigelti* (distal end: GMH 4416, proximal end: GMH IX-566-1953), (**h**) *G. boriensis* (holotype, from Bourdon et al. 2016; published under a Creative Commons CC BY 4.0 license), (**i**) ?*Palaeophasianus* sp. (distal end: AMNH 5156, proximal end: AMNH 5128), and (**j**) *E. aeola* (Eogruidae, AMNH 2937). (**k**)–(**o**) Distal ends of right tibiotarsi (cranial view) of (**k**) *P. weigelti* (GMH XXXVIII-6-1964; surrounding matrix was digitally removed), (**l**) *G. boriensis* (from Bourdon et al. 2016), (**m**) cf. *Eogeranoides campivagus* (AMNH 5127; left tibiotarsus, mirrored to ease comparisons), (**n**) *Eogrus aeola* (AMNH 2946), and (**o**) extant *Struthio camelus* (Struthionidae). (Photos in (**a**), (**b**), (**g**), (**k**), and (**o**) by Sven Tränkner, others (except **c**, **h**, and **l**) by the author)

by Burchak-Abramovich (1951), based on an examination of Neogene fossils. The hypothesis of struthioniform affinities of eogruids and ergilornithids was elaborated by Feduccia (1980) and Olson (1985), but their evolutionary scenarios were linked to a polyphyly of palaeognathous birds and to an origin of the Struthioniformes from a "gruiform" stem species. Because ostriches are united with other palaeognathous birds in all current sequence-based analyses, and because the tibiotarsus of eogruids and ergilornithids exhibits presumptive derived features of gruiform birds, the idea of close affinities between the Eogruidae and the Struthioniformes did not gain much acceptance and was also considered weakly supported in the first edition of this book.

Most current authors assigned eogruids and ergilornithids to the Gruiformes (e.g, Feduccia 1999; Mayr 2009b; Zelenkov and Kurochkin 2015; Musser et al. 2020), and *Eogrus* resulted as the sister taxon of a clade including the Aramidae (limpkins) and Gruidae (cranes) in an analysis by Clarke et al. (2005). In this latter study, which did not include representatives of the Ergilornithidae, a single character (distal rim of medial condyle of tibiotarsus notched) was optimized as an apomorphy of a clade including the Eogruidae, Psophiidae, Aramidae, and Gruidae, and another one (tarsometatarsal trochlea for second toe subequal in distal projection to that for fourth toe) supported a sister group relationship between eogruids and the clade (Aramidae + Gruidae). However, both characters are now also known to be present in the Palaeotididae and the North American Geranoididae (Mayr 2019).

Many distal hindlimb elements of eogruids and ergilornithids have been discovered, but for a long time little was known about other aspects of their skeleton. Kurochkin (1976) reported a proximal humerus of an unidentified ergilornithid from the late Eocene of Mongolia, the morphology of which suggests flightlessness of the species it belonged to. Concerning *Eogrus*, however, Clarke et al. (2005) noted that a referred proximal phalanx of the major wing digit, which was assigned to *E. aeola* by Wetmore (1934), does not indicate a loss of flight capabilities.

Olson (1985) hypothesized that a coracoid from the type locality of *Eogrus aeola*, which Wetmore (1934) assigned to the Accipitridae, most likely also belongs to *Eogrus*. This bone is clearly distinguished from the coracoid of palaeognathous birds, and its morphology would conflict with struthioniform affinities of the Eogruidae. Actually, however, the putative *Eogrus* coracoid is virtually identical to the coracoid of the presumed presbyornithid *Wilaru prideauxi* from the early Miocene of Australia (De Pietri et al. 2016; Sect. 4.5.2). Most likely, this bone, therefore, does not belong to *Eogrus* but is from an anseriform bird (Mayr and Zelenkov 2021). Removal of the putative *Eogrus* coracoid from the Eogruidae eliminates a major character conflict regarding the hypothesis of struthioniform affinities of these birds and may furthermore suggest that the above-mentioned wing phalanx likewise is not from *Eogrus*.

In the late Eocene Ergilin Dzo Formation in Mongolia, fossils of both eogruids (*Sonogrus*) and ergilornithids (*Proergilornis* and *Ergilornis*) have been found. Olson (1985: 158) briefly commented on a skull of an unidentified eogruid or ergilornithid species from this formation and noted that the specimen exhibits fossae for nasal glands on the dorsal surface of the skull roof similar to those of extant Struthionidae. A re-examination of the fossil confirmed this observation (Mayr and Zelenkov 2021), which clearly conflicts with close affinities between eogruids/ergilornithids and the Gruiformes, in which nasal glands on the skull roof are absent. In the fossils skull, the temporal fossae for the jaw muscles are furthermore much deeper than those of gruiform birds. As in extant palaeognathous birds, the otic region seems to exhibit only a single well-defined articular facet for the quadrate, even though unambiguous inferences on the morphology of the otic head of the quadrate are not possible (Mayr and Zelenkov 2021). In most neognathous birds, including all Gruiformes, the otic head of the quadrate is bipartite and articulates with two well-defined facets, and an undivided head is one of the hallmark features of the Palaeognathae. The fossil skull exhibits a very narrow interorbital section and a spherical shape of the caudal portion of the cranium, and with regard to its proportions it shows a close resemblance to the skull of *Palaeotis*. Olson (1985: 158) stated that "the fossil cranium shows a distinct nasofrontal hinge, indicating that the skull was not rhynchokinetic and thus not paleognathous." However, these observations were based on a misinterpretation of the specimen, in which the area of the nasofrontal hinge is not preserved (Mayr and Zelenkov 2021).

A femur from the late Eocene of the Ergilin Dzo Formation also shows a closer resemblance to the femur of the Struthionidae than to that of gruiform birds, especially with regard to the stoutness of the bone and its asymmetric distal end (Mayr and Zelenkov 2021). In ergilornithids, the distal phalanges of the fourth toe are strongly abbreviated (Zelenkov and Kurochkin 2015: fig. 70P; Musser et al. 2020). These phalanges are likewise shortened in *Palaeotis* and extant Struthionidae, albeit not to the same degree as in ergilornithids.

Whereas eogruids and ergilornithids were considered closely related to the Geranoididae by Olson (1985), they were not compared with the Palaeotididae by earlier authors. In fact, however, palaeotidids, geranoidids, and eogruids show a strong resemblance in their hindlimb bones, and the above-mentioned skull from Ergilin Dzo is very similar to that of *Palaeotis* in its proportions. As detailed in Sect. 3.2.1., palaeotidids were considered to be stem group representatives of the Struthioniformes, and the recognition of a clade including the Palaeotididae, Geranoididae, Eogruideae, Ergilornithidae, and Struthionidae provides a coherent picture of the evolution of struthioniform birds.

A clade including the Eogruidae, Ergilornithidae, and Struthionidae is supported by the skull features outlined above (nasal gland and, possibly, single articular facet for the quadrate), the progressive reduction of the tarsometatarsal trochlea for the second toe, and by the strongly shortened distal phalanges of the fourth toe (these phalanges are also very short in *Palaeotis*). The "gruiform" traits of the tibiotarsus, such as the tubercle next to the supratendinal bridge and the notch in the distal rim of the medial condyle, are likewise present in the Geranoididae and Palaeotididae and are best interpreted as plesiomorphic traits that were lost in the Struthionidae.

Clarke et al. (2005) synonymized the Ergilornithidae with the Eogruidae. However, if these taxa are stem group representatives of the Struthioniformes, eogruids are the sister taxon of a clade including ergilornithids and struthionids, in which the tarsometatarsal trochlea for the second toe is more strongly reduced or completely lost. Under the assumption of struthioniform affinities of these birds, ergilornithids are furthermore paraphyletic, with the didactyl *Ergilornis* being more closely related to crown group Struthionidae than is *Proergilornis*, in which this trochlea is reduced but still present (contra Kurochkin 1981, who synonymized *Proergilornis* with *Ergilornis*).

Recognition of the European Palaeotididae, the North American Geranoididae, and the Asian Eogruidae as stem group representatives of the Struthioniformes (Fig. 3.4) supports a Northern Hemispheric origin of ostriches and a dispersal into Africa toward the late Paleogene or earliest Neogene. With the closest relatives of ostriches being the didactyl species of the Eogruidae, Asia is the most likely center of origin of ostriches, as has already been assumed by Feduccia (1980). Struthioniform affinities of ergilornithids furthermore suggest that "aepyornithid-type" eggshell from the early Oligocene of Mongolia (Bibi et al. 2006) may actually be from a species of the Eogruidae or Ergilornithidae (Mayr and Zelenkov 2021).

3.3 Remiornithidae

The Remiornithidae are a further group of Paleogene Northern Hemispheric palaeognathous birds. The single species included in the taxon, *Remiornis heberti* from the late Paleocene of France, is mainly represented by leg elements (tibiotarsus and tarsometatarsus; Martin 1992, Buffetaut and Angst 2014). A large cervical vertebra from the latest Paleocene of Rivecourt-Petit Pâtis in France was assigned to *Remiornis* by Buffetaut and de Ploëg (2020). If correctly identified, this fossil would constitute the youngest fossil record of the taxon.

Unlike in the Palaeotididae and as in most extant palaeognathous birds, the tibiotarsus of *R. heberti* lacks an ossified supratendinal bridge. The tarsometatarsus has about the length of that of *Palaeotis weigelti* but is much stouter, indicating a significantly larger overall size of *R. heberti* as compared to *P. weigelti*. As in the latter species, however, the tarsometatarsus of *Remiornis* exhibits a marked sulcus along its dorsal surface. Except for the more deflected trochlea for the second toe, the bone also resembles the tarsometatarsus of *Palaeotis* in the configuration of the distal trochleae (compare Martin 1992: fig. 8 and Mayr 2002: fig. 1).

In the first edition of this book (Mayr 2009b), the possibility of close affinities between the Palaeotididae and Remiornithidae was taken into consideration. New data on the leg morphology of *Palaeotis*, however, disproved this hypothesis and a better knowledge of the skeletal anatomy of *Remiornis* is required for a well-founded classification of the taxon.

3.4 Eremopezidae

The Eremopezidae occur in the late Eocene of the Jebel Qatrani Formation (Fayum) in Egypt. The original description of the rhea-sized *Eremopezus eocaenus* was based on the distal end of a tibiotarsus (Andrews 1904). Lambrecht (1929) reported a fragmentary tarsometatarsus from the same locality (Fig. 3.5b), which formed the holotype of *Stromeria fajumensis*. Lambrecht (1929) erroneously (Rasmussen et al. 2001) considered the latter specimen to be of early Oligocene age and assumed a close relationship to the Madagascan Aepyornithidae. Rasmussen et al. (1987) already supposed that these two fossils belong to a single species, but a synonymy of *E. eocaenus* and *S. fajumensis* could not be proven until Rasmussen et al. (2001) identified new and better-preserved specimens from the Fayum. Even with this new material, however, *E. eocaenus* (as the species is now known) is represented by leg bones only.

The distal end of the tibiotarsus of *E. eocaenus* lacks an ossified supratendinal bridge, which seems to have been the main reason why the species was associated with palaeognathous birds. The tarsometatarsus is distinguished from that of all palaeognathous birds in that the distal end is unusually flattened and the trochleae are splayed. The trochlea for the second toe is considerably shorter than that for the fourth toe (Rasmussen et al. 2001: figs. 1 and 4). Judging from the well-developed articular facet for the first metatarsal on the tarsometatarsus, *Eremopezus* had a hallux, which is absent in all flightless Palaeognathae apart from the Apterygiformes and Dinornithiformes.

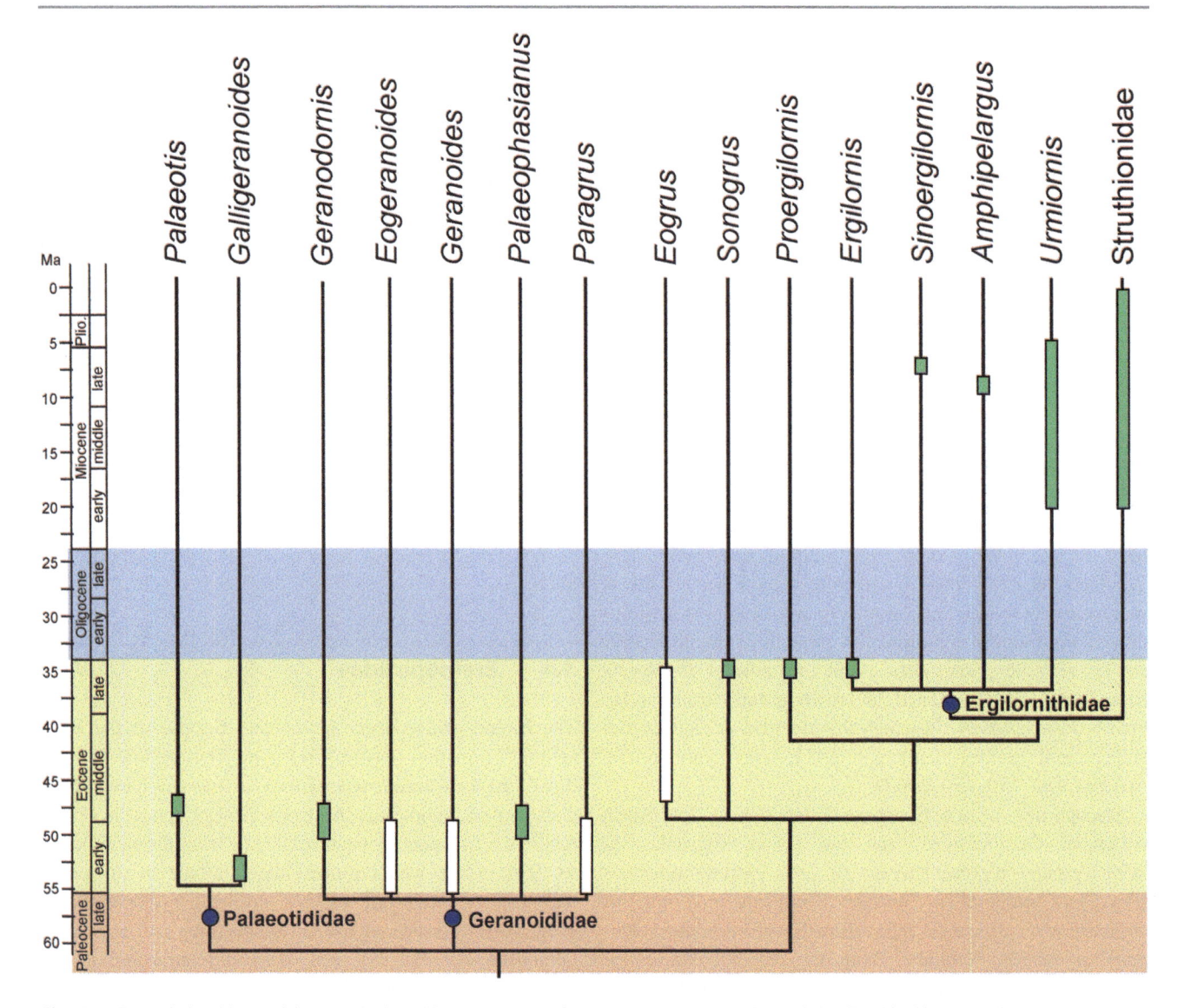

Fig. 3.4 Interrelationships and known stratigraphic occurrences of stem group representatives of the Struthioniformes. Divergence dates are hypothetical, the affinities of taxa shown in a polytomy are unresolved (see text for further details)

As noted by Rasmussen et al. (2001: 334), the hypothesis that *Eremopezus* is a flightless ancestor of the Aepyornithidae conflicts with the early (Jurassic) split of Madagascar from continental Africa and the absence of possible dispersal corridors for a large flightless bird thereafter. Rasmussen et al. (2001) even questioned palaeognathous affinities of *Eremopezus*, and the phylogenetic affinities of this remarkable bird can probably not be conclusively resolved without further material. At least the distal end of the tibiotarsus is, however, very similar to that of flightless palaeognathous birds.

Eggshells assigned to the ootaxon "*Psammornis*" were found in North Africa and Arabia and are considered to possibly date from the Eocene (Sauer 1969). If these age estimates are correct, they may well belong to *Eremopezus* or a related bird.

3.5 Rheiformes (Rheas)

Putative stem group representatives of the Rheiformes were reported from the Paleocene of South America, with the earliest fossils being pedal phalanges from the middle Paleocene of the Río Chico Formation in Argentina (Tambussi 1995). The fossil record of *Diogenornis fragilis* from the early Eocene of Itaboraí in Brazil is more comprehensive, and this flightless species reached about 2/3 the size of the extant Greater Rhea, *Rhea americana*. Various limb bones, vertebrae, and the tip of the praemaxilla of *D. fragilis* were discovered and represent a minimum of 4–5 individuals (Fig. 3.5a; Alvarenga 1983). More recently, Agnolin (2017) assigned a distal tibiotarsus from the mid-Paleocene of Argentina to *Diogenornis*, which, if correctly identified,

3.5 Rheiformes (Rheas)

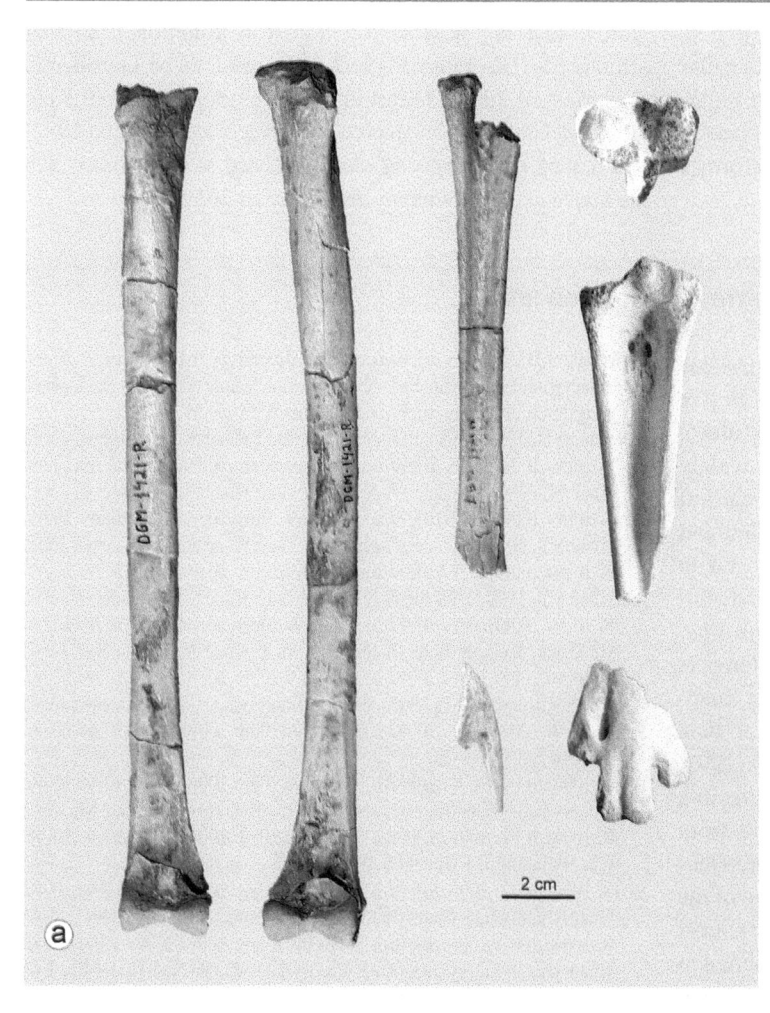

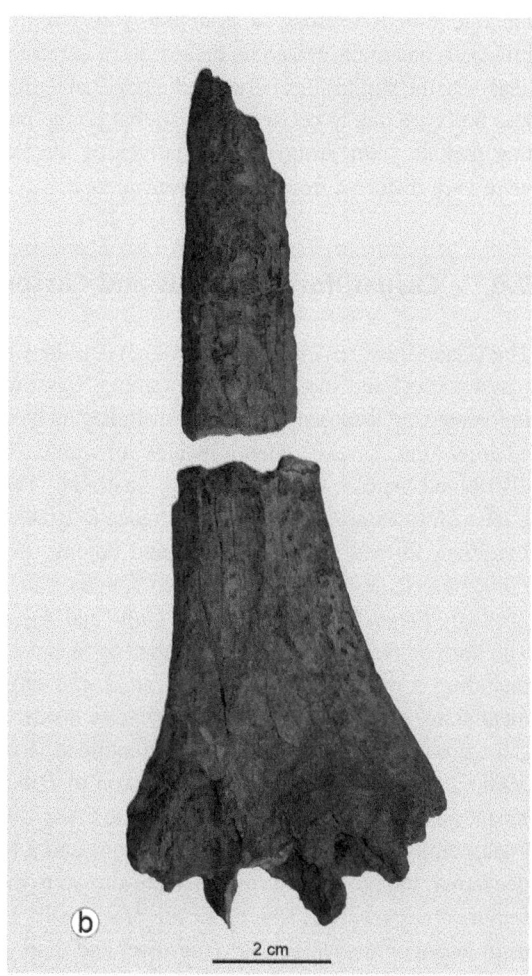

Fig. 3.5 (a) Selected bones of *Diogenornis fragilis* (Rheiformes) from the early Eocene of Itaboraí in Brazil (tibiotarsi, incomplete tarsometatarsus: holotype, Coleção da Seção de Paleontologia do Departamento Nacional da Produção Mineral, Rio de Janeiro, Brazil, DGM-1421-R; upper beak: DGM-1428-R; proximal right tarsometatarsus: DGM-1422-R; distal right tarsometatarsus: DGM-1422-R). (b) Partial tarsometatarsus of *Eremopezus eocaenus* from the late Eocene of the Jebel Qatrani Formation (Fayum) in Egypt (holotype of "*Stromeria fajumensis*", Bayerische Staatssammlung für Paläontologie und Geologie, Munich, Germany, BSPG 1914 I 53). (Photos in (**a**) courtesy of Herculano Alvarenga, (**b**) by the author)

would expand the known geographic and stratigraphic ranges of the taxon.

In the original description (Alvarenga 1983), *Diogenornis* was classified into the Opisthodactylidae, a taxon first introduced for a rheiform species from the early Miocene of Argentina. The similarities between these early forms are, however, likely to be plesiomorphic. *D. fragilis* has a narrower beak and a somewhat less reduced humerus than extant Rheiformes, but otherwise it is quite similar in the morphology of the known bones. The distal end of the tibiotarsus lacks a supratendinal bridge. Compared to other Paleogene palaeognathous birds, the tarsometatarsus morphology of *D. fragilis* is most similar to that of the Remiornithidae. As in the latter, there is a marked extensor sulcus along the dorsal surface of the tarsometatarsus.

In a conference note, Alvarenga (2010) hypothesized that *Diogenornis* is actually a stem group representative of the Australo-Papuan Casuariidae (cassowaries). In general, close affinities between South American and Australian taxa are possible from a biogeographic perspective, because land connections between South America and Australia existed across Antarctica in the earliest Paleogene (Sect. 11.2.1). However, the Casuariidae are the sister taxon of the Australian Dromaiidae (emus), with both taxa being very similar in their postcranial skeleton morphology. *Diogenornis* differs from both, Casuariidae and Dromaiidae, in plesiomorphic skeletal traits, and comparisons with other Paleogene palaeognathous birds are necessary for an informed assessment of its affinities.

Agnolin (2017) described a small and comparatively short tibiotarsus from the middle Paleocene (Riochican) of Argentina, which he considered to be distinct from that of the Rheiformes. Whether this fossil indeed indicates a higher diversity of non-rheiform palaeognathous birds in the early

Paleogene of Argentina, as assumed by Agnolin (2017), is, however, uncertain. Based on the results of current molecular analyses, the similarities between rheas and other large flightless birds are due to convergence, so that it is to be expected that archaic stem group representatives of the Rheiformes were very different from more advanced taxa.

3.6 Casuariiformes (Emus and Cassowaries)

The Casuariiformes include two extant taxa, the Casuariidae (cassowaries) and the Dromaiidae (emus). Cassowaries live in forested environments of North Australia and New Guinea, whereas the single extant species of emus is widely distributed across open areas of Australia. The earliest Casuariiformes are from the late Oligocene Riversleigh and Etadunna formations of Australia and belong to the taxon *Emuarius* (Boles 1992, 2001). Two species can be distinguished, *Emuarius* ("*Dromaius*") *gidju* and *E. guljaruba*. The former is known from a number of skeletal elements including a partial beak, but of *E. guljaruba* only a tarsometatarsus was described. *Emuarius* was considered to be a stem group representative of the Dromaiidae (Boles 1992, 2001) and also resulted as the sister taxon of *Dromaius* in a recent analysis (Worthy et al. 2014). The taxon differs from extant emus in proportionally smaller eyes and a more slender femur, the proportions of which are similar to those of the femur of cassowaries. The tip of the praemaxilla is rounded as in extant emus, whereas it is narrower and more pointed in cassowaries. The weight of *Emuarius gidju* was estimated at 19–21 kg (Boles 1997), and the species was probably less cursorial than extant emus but more so than the graviportal cassowaries (Boles 1997). If correctly assigned to the Dromaiidae, *Emuarius* indicates that the lineages leading to cassowaries and emus already separated in the Paleogene.

3.7 Putative Palaeognathous Bird from the Eocene of Antarctica

Tambussi et al. (1994) reported a fragmentary distal tarsometatarsus of a reputedly palaeognathous bird from the late Eocene of the La Meseta Formation of Seymour Island in Antarctica (see also Tambussi and Acosta Hospitaleche 2007). Just because of its size, this specimen is likely to be from a flightless species, the weight of which was estimated at 60 kg (Tambussi and Acosta Hospitaleche 2007). The fossil differs, however, from all extant Palaeognathae in the unusually large trochlea for the second toe.

Cenizo (2012) detailed that a partial upper beak from the late Eocene of Seymour Island, which was tentatively assigned to the Phorusrhacidae by previous authors (Case et al. 1987), belongs to a palaeognathous bird. The fossil is mediolaterally narrow like the beak of *Casuarius* (Casuariidae), so that affinities to *Diogenornis* (Sect. 3.5) need to be considered. Footprints of a large terrestrial bird were reported from King George Island in West Antarctica (Covacevich and Rich 1982) and maybe from a species closely related to that from which the above skeletal remains stem (Cenizo 2012).

References

Agnolin FL (2017) Unexpected diversity of ratites (Aves, Palaeognathae) in the early Cenozoic of South America: palaeobiogeographical implications. Alcheringa 41:101–111

Alvarenga HMF (1983) Uma ave ratitae do Paleoceno Brasileiro: bacia calcária de Itaboraí, Estado do Rio de Janeiro, Brasil. Bol mus nac, Geol 41:1–8

Alvarenga HMF (2010) *Diogenornis fragilis* Alvarenga, 1985, restudied: a South American ratite closely related to Casuariidae. 25th International Ornithological Congress, abstracts:143

Andrews CW (1904) On the pelvis and hind-limb of *Mullerornis betsilei* M.-Edw. & Grand.; with a note on the occurrence of a ratite bird in the Upper Eocene beds of the Fayum, Egypt. Proc Zool Soc Lond 1904:163–171

Bertelli S, Chiappe LM (2005) Earliest tinamous (Aves: Palaeognathae) from the Miocene of Argentina and their phylogenetic position. Contrib Sci 502:1–20

Bibi F, Shabel AB, Kraatz BP, Stidham TA (2006) New fossil ratite (Aves: Palaeognathae) eggshell discoveries from the late Miocene Baynunah Formation of the United Arab Emirates, Arabian Peninsula. Palaeontol Electron 9.1(2A):1–13

Boles WE (1992) Revision of *Dromaius gidju* Patterson and Rich 1987 from Riversleigh, Northwestern Queensland, Australia, with a reassessment of its generic position. In: Campbell KE (ed) Papers in avian paleontology honoring Pierce Brodkorb. Nat Hist Mus Los Angeles Cty, Sci Ser 36:195–208

Boles WE (1997) Hindlimb proportions and locomotion of *Emuarius gidju* (Patterson & Rich, 1987) (Aves: Casuariidae). Mem Qld Mus 41:235–240

Boles WE (2001) A new emu (Dromaiinae) from the Late Oligocene Etadunna Formation. Emu 101:317–321

Bourdon E, Lindow B (2015) A redescription of *Lithornis vulturinus* (Aves, Palaeognathae) from the early Eocene Fur Formation of Denmark. Zootaxa 4032:493–514

Bourdon E, Mourer-Chauviré C, Laurent Y (2016) Early Eocene birds from La Borie, southern France. Acta Palaeontol Polon 61:175–190

Buffetaut E, Angst D (2014) Stratigraphic distribution of large flightless birds in the Palaeogene of Europe and its palaeobiological and palaeogeographical implications. Earth-Sci Rev 138:394–408

Buffetaut E, de Ploëg G (2020) Giant birds from the uppermost Paleocene of Rivecourt (Oise, northern France). Bol Centro Português Geo-Hist Pré-Hist 2:29–33

Burchak-Abramovich NI (1951) [*Urmiornis* (*Urmiornis maraghanus* Mecq.) ostrich-like bird of the *Hipparion* fauna of Transcaucasia and southern Ukraine.] Izv Akad. Nauk Az SSR 6:83–94. [In Russian]

Case JA, Woodburne MO, Chaney DS (1987) A gigantic phororhacoid (?) bird from Antarctica. J Paleontol 61:1280–1284

Cenizo MM (2012) Review of the putative Phorusrhacidae from the Cretaceous and Paleogene of Antarctica: new records of ratites and pelagornithid birds. Pol Polar Res 33:239–258

Clarke JA, Norell MA, Dashzeveg D (2005) New avian remains from the Eocene of Mongolia and the phylogenetic position of the Eogruidae (Aves, Gruoidea). Am Mus Novit 3494:1–17

References

Covacevich V, Rich PV (1982) New bird ichnites from Fildes Peninsula, King George Island, West Antarctica. In: Craddock C (ed) Antarctic Geoscience. University of Wisconsin Press, Madison, pp 245–254

Cracraft J (1969) Systematics and evolution of the Gruiformes (class Aves). 1. The Eocene family Geranoididae and the early history of the Gruiformes. Am Mus Novit 2388:1–41

Cracraft J (1973) Systematics and evolution of the Gruiformes (class Aves). 3. Phylogeny of the suborder Grues. Bull Am Mus Nat Hist 151:1–127

De Pietri VL, Scofield RP, Zelenkov N, Boles WE, Worthy TH (2016) The unexpected survival of an ancient lineage of anseriform birds into the Neogene of Australia: the youngest record of Presbyornithidae. Roy Soc Open Sci 3:150635

Elzanowski A, Galton PM (1991) Braincase of *Enaliornis*, an early Cretaceous bird from England. J Vertebr Paleontol 11:90–107

Feduccia A (1980) The age of birds. Harvard University Press, Cambridge, Massachusetts

Feduccia A (1999) The origin and evolution of birds, 2nd edn. Yale University Press, New Haven

Grellet-Tinner G, Dyke GJ (2005) The eggshell of the Eocene bird *Lithornis*. Acta Palaeontol Pol 50:831–835

Hackett SJ, Kimball RT, Reddy S, Bowie RCK, Braun EL, Braun MJ, Chojnowski JL, Cox WA, Han K-L, Harshman J, Huddleston CJ, Marks BD, Miglia KJ, Moore WS, Sheldon FH, Steadman DW, Witt CC, Yuri T (2008) A phylogenomic study of birds reveals their evolutionary history. Science 320:1763–1767

Harshman J, Braun EL, Braun MJ, Huddleston CJ, Bowie RCK, Chojnowski JL, Hackett SJ, Han K-L, Kimball RT, Marks BD, Miglia KJ, Moore WS, Reddy S, Sheldon FH, Steadman DW, Steppan SJ, Witt CC, Yuri T (2008) Phylogenomic evidence for multiple losses of flight in ratite birds. Proc Natl Acad Sci U S A 36:13462–13467

Houde P (1986) Ostrich ancestors found in the Northern Hemisphere suggest new hypothesis of ratite origin. Nature 324:563–565

Houde P (1988) Palaeognathous birds from the early Tertiary of the Northern Hemisphere. Publ Nuttall Ornithol Club 22:1–148

Houde P, Haubold H (1987) *Palaeotis weigelti* restudied: a small middle Eocene ostrich (Aves: Struthioniformes). Palaeovertebr 17:27–42

Houde P, Olson SL (1981) Palaeognathous carinate birds from the early Tertiary of North America. Science 214:1236–1237

Kohring R, Hirsch KF (1996) Crocodilian and avian eggshells from the middle Eocene of the Geiseltal, eastern Germany. J Vertebr Paleontol 16:67–80

Kristoffersen AV (1999) Lithornithid birds (Aves, Palaeognathae) from the lower Palaeogene of Denmark. Geol Mijnb 78:375–381

Kuhl H, Frankl-Vilches C, Bakker A, Mayr G, Nikolaus G, Boerno ST, Klages S, Timmermann B, Gahr M (2021) An unbiased molecular approach using 3'UTRs resolves the avian family-level tree of life. Mol Biol Evol 38:108–121

Kurochkin EN (1976) A survey of the Paleogene birds of Asia. Smithson Contrib Paleobiol 27:5–86

Kurochkin EN (1981) [New representatives and evolution of two archaic gruiform families in Eurasia]. Trudy Sovmest Sovetsko-Mongolskaja Paleontol Ekspedit 15:59–85 [in Russian]

Kurochkin EN (1982) On the evolutionary pathways of didactylous Tertiary gruids under increasing aridization. In: Novak VJA, Mlíkovský J (eds) Evolution and environment. CSAV, Prague, pp 731–736

Lambrecht K (1928) Palaeotis Weigelti n. g. n. sp., eine fossile Trappe aus der mitteleozänen Braunkohle des Geiseltales. Jahrb Halleschen Verb Erforsch mitteldtsch Bodenschätze. N F 7:1–11

Lambrecht K (1929) Ergebnisse der Forschungsreisen Prof. E. Stromers in den Wüsten Ägyptens. V. Tertiäre Wirbeltiere. 4. *Stromeria fajumensis* n. g., n. sp., die kontinentale Stammform der Aepyornithidae, mit einer Übersicht über die fossilen Vögel Madagaskars und Afrikas. Abh Bayer Akad Wiss Math-naturwiss Abt, N F, 4:1–18

Lambrecht K (1935) Drei neue Vogelformen aus dem Lutétian des Geiseltales. Nova Acta Leopold, N F 3:361–367

Leonard L, Dyke GJ, van Tuinen M (2005) A new specimen of the fossil palaeognath *Lithornis* from the lower Eocene of Denmark. Am Mus Novit 3491:1–11

Livezey BC, Zusi RL (2007) Higher-order phylogeny of modern birds (Theropoda, Aves: Neornithes) based on comparative anatomy: II.— analysis and discussion. Zool J Linnean Soc 149:1–94

Martin LD (1992) The status of the Late Paleocene birds *Gastornis* and *Remiornis*. In: Campbell KE (ed) Papers in avian paleontology honoring Pierce Brodkorb. Nat Hist Mus Los Angeles Cty, Sci Ser 36:97–108

Mayr G (2002) Avian remains from the Middle Eocene of the Geiseltal (Sachsen-Anhalt, Germany). In: Zhou Z, Zhang F (eds) Proceedings of the 5th symposium of the Society of Avian Paleontology and Evolution, Beijing, 1–4 June 2000. Science Press, Beijing, pp 77–96

Mayr G (2007) The birds from the Paleocene fissure filling of Walbeck (Germany). J Vertebr Paleontol 27:394–408

Mayr G (2008) First substantial Middle Eocene record of the Lithornithidae (Aves): a postcranial skeleton from Messel (Germany). Ann Paléontol 94:29–37

Mayr G (2009a) Towards the complete bird—the skull of the middle Eocene Messel lithornithid (Aves, Lithornithidae). Bull Inst Roy Sci Nat Belg 79:169–173

Mayr G (2009b) Paleogene fossil birds, 1st edn. Springer, Heidelberg

Mayr G (2015) The middle Eocene European "ratite" *Palaeotis* (Aves, Palaeognathae) restudied once more. Paläontol Z 89:503–514

Mayr G (2016) On the taxonomy and osteology of the early Eocene North American Geranoididae (Aves, Gruoidea). Swiss J Palaeontol 135:315–325

Mayr G (2017) Avian evolution: the fossil record of birds and its paleobiological significance. Wiley-Blackwell, Chichester

Mayr G (2019) Hindlimb morphology of *Palaeotis* suggests palaeognathous affinities of the Geranoididae and other "crane-like" birds from the Eocene of the Northern Hemisphere. Acta Palaeontol Polon 64:669–678

Mayr G, Smith T (2019) New Paleocene bird fossils from the North Sea Basin in Belgium and France. Geol Belg 22:35–46

Mayr G, Zelenkov N (2021) Extinct crane-like birds (Eogruidae and Ergilornithidae) from the Cenozoic of Central Asia are ostrich precursors. Ornithology 138:1–15

Mlíkovský J (2002) Cenozoic birds of the world. Part 1: Europe. Ninox Press, Praha

Mourer-Chauviré C, Senut B, Pickford M, Mein P (1996) Le plus ancien représentant du genre *Struthio* (Aves, Struthionidae), *Struthio coppensi* n. sp., du Miocène inférieur de Namibie. C R Acad Sci Paris 322:325–332

Musser G, Li Z, Clarke JA (2020) A new species of Eogruidae (Aves: Gruiformes) from the Miocene of the Linxia Basin, Gansu, China: Evolutionary and climatic implications. Auk 137:ukz067

Nesbitt SJ, Clarke JA (2016) The anatomy and taxonomy of the exquisitely preserved Green River Formation (early Eocene) lithornithids (Aves) and the relationships of Lithornithidae. Bull Am Mus Nat Hist 406:1–91

Olson SL (1985) The fossil record of birds. In: Farner DS, King JR, Parkes KC (eds) Avian biology, vol 8. Academic Press, New York, pp 79–238

Parris DC, Hope S (2002) New interpretations of the birds from the Navesink and Hornerstown Formations, New Jersey, USA (Aves: Neornithes). In: Zhou Z, Zhang F (eds) Proceedings of the 5th symposium of the Society of Avian Paleontology and Evolution, 1–4 June 2000. Science Press, Beijing, pp 113–124

Peters DS (1988) Ein vollständiges Exemplar von *Palaeotis weigelti* (Aves, Palaeognathae). Cour Forsch-Inst Senckenberg 107:223–233

Portis A (1885) Contribuzioni alla ornitolitologia italiana. Mem Reale Accad SciTorino 36:361–384

Prum RO, Berv JS, Dornburg A, Field DJ, Townsend JP, Lemmon EM, Lemmon AR (2015) A comprehensive phylogeny of birds (Aves) using targeted next-generation DNA sequencing. Nature 526:569–573

Rasmussen DT, Olson SL, Simons EL (1987) Fossil birds from the Oligocene Jebel Qatrani Formation, Fayum Province, Egypt. Smithson Contrib Paleobiol 62:1–20

Rasmussen DT, Simons EL, Hertel F, Judd A (2001) Hindlimb of a giant terrestrial bird from the Upper Eocene, Fayum, Egypt. Palaeontology 44:325–337

Sauer EGF (1969) Evidence and evolutionary interpretation of *Psammornis*. Bonner zool Beitr 20:290–310

Sibley CG, Ahlquist JE (1990) Phylogeny and classification of birds: a study in molecular evolution. Yale University Press, New Haven

Stidham TA, Lofgren D, Farke AA, Paik M, Choi R (2014) A lithornithid (Aves: Palaeognathae) from the Paleocene (Tiffanian) of southern California. PaleoBios 31:1–7

Tambussi CP (1995) The fossil Rheiformes from Argentina. Cour Forsch-Inst Senckenberg 181:121–129

Tambussi CP, Acosta Hospitaleche C (2007) Antarctic birds (Neornithes) during the Cretaceous-Eocene times. Rev Asoc Geol Argent 62:604–617

Tambussi CP, Noriega JI, Gaździcki A, Tatur A, Reguero MA, Vizcaino SF (1994) Ratite bird from the Paleogene La Meseta Formation, Seymour Island, Antarctica. Pol Polar Res 15:15–20

Tennyson AJ, Worthy TH, Jones CM, Scofield RP, Hand SJ (2010) Moa's ark: Miocene fossils reveal the great antiquity of moa (Aves: Dinornithiformes) in Zealandia. Rec Austral Mus 62:105–114

Torres CR, Norell MA, Clarke JA (2020) Estimating flight style of early Eocene stem palaeognath bird *Calciavis grandei* (Lithornithidae). Anat Rec 303:1035–1042

Wetmore A (1933) An Oligocene eagle from Wyoming. Smithson Misc Collect 87:1–9

Wetmore A (1934) Fossil birds from Mongolia and China. Am Mus Novit 711:1–16

Worthy T, Worthy JP, Tennyson AJD, Salisbury SW, Hand SJ, Scofield RP (2013) Miocene fossils show that kiwi (*Apteryx*, Apterygidae) are probably not phyletic dwarves. In: Göhlich UB, Kroh A (eds) Paleornithological research 2013—Proceedings of the 8th international meeting of the Society of Avian Paleontology and Evolution. Natural History Museum Vienna, Vienna, pp 63–80

Worthy TH, Hand SJ, Archer M (2014) Phylogenetic relationships of the Australian Oligo-Miocene ratite *Emuarius gidju* Casuariidae. Integrat Zool 9:148–166

Worthy TH, Degrange FJ, Handley WD, Lee MS (2017) The evolution of giant flightless birds and novel phylogenetic relationships for extinct fowl (Aves, Galloanseres). Roy Soc Open Sci 4(10): 170975

Zelenkov NV, Kurochkin EN (2015) Class Aves. In Kurochkin EN, Lopatin AV, Zelenkov NV (eds) Fossil vertebrates of Russia and neighbouring countries. Fossil reptiles and birds. Part 2. GEOS, Moscow, p 86–290 [in Russian]

Pelagornithidae, Gastornithidae, and Crown Group Galloanseres

4

The clade including galliform and anseriform birds is often termed "Galloanserae," but here the grammatically correct term Galloanseres is used. This clade is supported by virtually all analyses of different kinds of molecular data, including DNA-DNA hybridization as well as mitochondrial and nuclear gene sequences, and it also resulted from analyses of morphological data (e.g., Sibley and Ahlquist 1990; Mayr and Clarke 2003; Ericson et al. 2006; Livezey and Zusi 2007; Prum et al. 2015; Kuhl et al. 2021). Prum et al. (2015) obtained an early Eocene date for the split of galliform and anseriform birds, some 55 million years. However, this divergence estimate is in clear conflict with the fossil record, which includes morphologically disparate stem group representatives of both Galliformes and Anseriformes from deposits of that age (anseriform birds have an even earlier fossil record).

Extant Galloanseres are mainly characterized by morphological apomorphies that concern skull features, such as sessile basipterygoid processes (in palaeognathous birds and most Neoaves these processes, if present, are stalked), as well as long, blade-like retroarticular processes formed by the caudal end of the mandible and a characteristic morphology of the quadrate, which has just two condyles for the articulation with the mandible (e.g., Mayr and Clarke 2003; Livezey and Zusi 2007). The postcranial skeleton of extant Galliformes and Anseriformes is quite different, but the morphology of Paleogene stem group Galliformes bridges the morphological gap between the extant taxa.

Various ecomorphologically disparate Paleogene taxa were assigned to the Galloanseres. If the classification of all of them is corroborated by future studies, the morphological and ecological diversity within Paleogene Galloanseres was extraordinarily high, including giant flightless ground birds with greatly reduced wings, long-legged filter-feeders, and pelagic taxa with wingspans of 4–5 m.

4.1 Pelagornithidae (Bony-Toothed Birds)

These large to very large soaring seabirds are characterized by the possession of superficially tooth-like, spiny projections along the cutting edges of the long beak ("bony teeth" or "pseudoteeth"), greatly elongated wings, and extremely thin-walled limb bones. Pelagornithids existed throughout most of the Cenozoic, from the Paleocene to the Pliocene, and already appear to have achieved a global distribution in the early Paleogene.

The earliest representative of the taxon is *Protodontopteryx ruthae* from the early Paleocene (61.5–62 Ma) of New Zealand (Fig. 4.1; Mayr et al. 2021). This species is also the smallest bony-toothed bird known so far and the most archaic one, which lacks derived traits characterizing geologically younger species. *P. ruthae* was the size of an average gull, and even though its long beak resembles that of other pelagornithids in proportions, the bony teeth are less pronounced. Most postcranial bones also show a less derived morphology, which is particularly true for the wing and pectoral girdle elements. The proportionally shorter humerus indicates that *P. ruthae* was less specialized for sustained soaring than geologically younger pelagornithids.

The holotype of *P. ruthae* is a partial skeleton in a sedimentary concretion. The specimen is remarkable in that it appears to represent an association of two individuals of different species, with the left humerus being markedly shorter than the right one and both radii being distinctly longer than the single ulna preserved in the specimen (in birds, ulna and radius are always subequal in length, and if two individuals are involved, one is not a pelagornithid). The preservation of the *P. ruthae* holotype is furthermore unusual in that the disarticulated bones form a dense tangle, which may indicate that the specimen represents feeding remains of a marine scavenger, which were dragged into a burrow (Mayr et al. 2021).

As the oldest and least specialized representative of the Pelagornithidae, *Protodontopteryx* suggests that pelagornithids originated in the Southern Hemisphere. Toward the late Paleocene, however, these birds had a wide distribution and also occurred in the Northern Hemisphere. Harrison (1985) described a partial mandible from the late Paleocene of England as *Pseudodontornis tenuirostris*, and another jaw fragment from the latest Paleocene of Kazakhstan was described as *Pseudodontornis tshulensis* by Averianov et al. (1991). Although these specimens clearly are from pelagornithids, their taxonomic distinctness is questionable, because little is known about the intraspecific variability of the "bony teeth" in pelagornithids. Fossils of indeterminable medium-sized bony-toothed birds were also reported from the late Paleocene of France (Mayr and Smith 2019a).

The taxon *Pseudodontornis* was initially established for a very large skull described as *Pseudodontornis longirostris* (Fig. 4.2a). The age and provenance of the fossil, which seems to have been destroyed in the Second Word War (Olson 1985), are unknown. It was said to have been brought to Europe by a Brazilian sailor, but Lambrecht (1930) rightly considered a Brazilian origin unlikely. The time of its initial description in the early twentieth century as well as the preservation in what appears to be a fairly solid matrix may indicate that the fossil comes from the London Clay of the Isle of Sheppey (see below). If so, it possibly belongs to *Dasornis emuinus*, which is similar in size and morphology.

Numerous isolated bones of bony-toothed birds were found in late Paleocene and early Eocene sediments of the Ouled Abdun Basin in Morocco. Bourdon et al. (2010) distinguished three species of *Dasornis* ("*Odontopteryx*") from this locality, which were referred to *D. toliapica* (wingspan 2–3 m), *D. emuinus* (wingspan 3.5–4.5 m), and *D. abdoun* (wingspan 1.5–1.7 m). The former two species were originally described from the early Eocene London Clay of the Isle of Sheppey, where the first specimens of pelagornithids have already been found in the mid-nineteenth century. Unfortunately, a revision of the London Clay Pelagornithidae by Harrison and Walker (1976a) resulted in formidable taxonomic confusion, with these authors recognizing no less than six species in five genera and three families. Many of these taxa were based on non-comparable skeletal elements, and in fact, all London Clay pelagornithids can probably be assigned to two species (Fig. 4.2b, c): the small *Dasornis toliapica*, the skull of which was first described by Owen (1873), and the large *D. emuinus* (Mayr 2008a; Bourdon et al. 2010). The description of *D. emuinus* was originally based on a cranium and its affinities with bony-toothed birds went long unrecognized (the pelagornithid species *Argillornis longipennis* and the alleged procellariiform *Neptuniavis miranda* are junior synonyms of *D. emuinus*, which is likely to be also true for the pelagornithid *Pseudodontornis longidentata*; Mayr 2008a). A species of a size in between that of *D. toliapica* and *D. emuinus* was described as *Macrodontopteryx oweni* by Harrison and Walker (1976a), but the fossil may actually also belong to *D. toliapica* (Bourdon et al. 2010).

A few fragmentary remains of indeterminable early Eocene pelagornithids were reported from Egem in Belgium (Mayr and Smith 2019b) and from the Nanjemoy Formation in Virginia, USA (Olson 1999a; Mayr 2016). Pelagornithids likewise occurred in early to late Eocene deposits of the La Meseta and Submeseta formations of Seymour Island, Antarctica (Tonni and Tambussi 1985; Tambussi and Acosta Hospitaleche 2007; Rubilar-Rogers et al. 2011; Cenizo et al. 2015; Acosta Hospitaleche and Reguero 2020; Kloess et al. 2020), where medium-sized and very large species coexisted. Stilwell et al. (1998) furthermore assigned a fragment of a humerus shaft from the middle Eocene of East Antarctica to the Pelagornithidae (see also Jones 2000).

The fossil record of *Lutetodontopteryx tethyensis* from the middle Eocene of Ukraine is quite substantial and includes partial jaws and limb bones (Fig. 4.2d–h; Mayr and Zvonok 2011, 2012). This species was initially tentatively assigned to *Dasornis* ("*Odontopteryx*") *toliapica*, but further fossil material revealed distinct differences to the latter species and showed the Ukrainian species to be more closely related to late Paleogene and Neogene pelagornithids. *L. tethyensis* coexisted with a much larger species, of which various skeletal elements including a sternum have been found. The latter bone closely resembles a very large fragmentary sternum from the middle Eocene of Nigeria, which was described as *Gigantornis eaglesomei* by Andrews (1916). The pelagornithid affinities of this specimen were recognized by Harrison and Walker (1976a), but the exact interrelationships of *Gigantornis* and *Dasornis*, the sternum of which is unknown, still need to be determined. The occurrence in the middle Eocene of Ukraine of a very large pelagornithid with *Dasornis*-like limb bones and a *Gigantornis*-like sternum (Mayr and Zvonok 2012) may indicate that the latter is a junior synonym of the former. Bones of a large pelagornithid species from the middle Eocene of Togo were tentatively referred to *Gigantornis* by Bourdon and Cappetta (2012).

A partial upper beak of another large-sized middle Eocene pelagornithid, from the Aridal Formation in southwestern Morocco, was assigned to the taxon *Pelagornis* (Zouhri et al. 2021). This latter taxon encompasses most Neogene pelagornithids and exhibits a more derived morphology than *Dasornis*, *Lutetodontopteryx*, and other Eocene pelagornithids. The assignment of the Moroccan fossil to *Pelagornis* was based on the presence of a transverse furrow across the dorsal surface of the tip of the beak. If this classification is confirmed by future finds, the unnamed Moroccan species would predate other *Pelagornis* fossils by more than 10 million years.

A partial skeleton of a medium-sized pelagornithid from the middle Eocene of Belgium, which was first tentatively

4.1 Pelagornithidae (Bony-Toothed Birds)

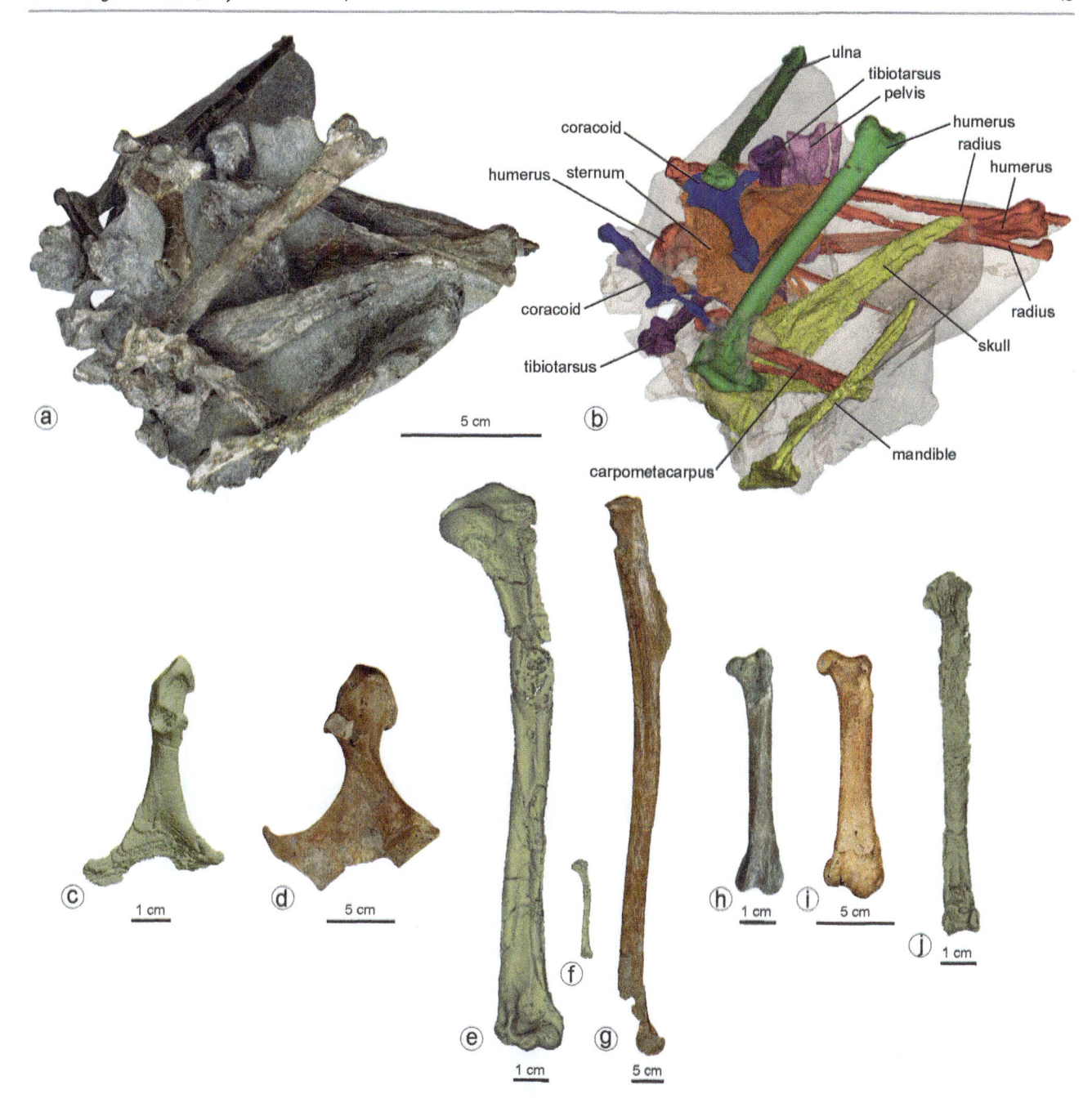

Fig. 4.1 The holotype of the bony-toothed bird *Protodontopteryx ruthae* (Pelagornithidae) from the early Paleocene Waipara Greensand in New Zealand. (**a**) Main block of matrix containing the fossil, with (**b**) a color-coded X-ray computed tomography image showing the bones (Canterbury Museum, Christchurch, New Zealand, CM 2018.124.8); the bones in green probably belong to a different, non-pelagornithid species (Mayr et al. 2021). (**c**) Left coracoid of *P. ruthae* (holotype, X-ray computed tomography image). (**d**) Right coracoid (mirrored) of the late Miocene *Pelagornis chilensis* (holotype, Museo Nacional de Historia Natural, Santiago, Chile, MNHN SGO.PV 1061). (**e**)–(**g**) Humerus of (**e**) *P. ruthae* (holotype, X-ray computed tomography image) and (**g**) *P. chilensis* (holotype, mirrored); (**f**) shows the actual size of the *P. ruthae* humerus compared to that of *P. chilensis*. (**h**), (**i**) Femora of (**h**) *P. ruthae* (holotype, X-ray computed tomography image) and (**i**) *P. chilensis* (holotype). (**j**) Tibiotarsus of *P. ruthae* (holotype, X-ray computed tomography image). (X-ray computed tomography images courtesy of Paul Scofield, photos in (**d**), (**g**), and (**i**) by Sven Tränkner; (**a**) and (**h**) by the author)

assigned to "*Macrodontopteryx oweni*" (Mayr and Smith 2010), is more likely to belong to *Lutetodontopteryx* (Fig. 4.2j; Mayr and Zvonok 2012). A very large fragmentary humerus from another middle Eocene site in Belgium was tentatively assigned to *Dasornis emuinus* by Mayr and Smith (2010). The putatively gruiform *Zheroia kurochkini*, the holotype of which is a worn distal tibiotarsus from the late Eocene of Kazakhstan, may also be a representative of the Pelagornithidae, of which partial mandibles were found in the type locality of *Z. kurochkini* (Nessov 1992).

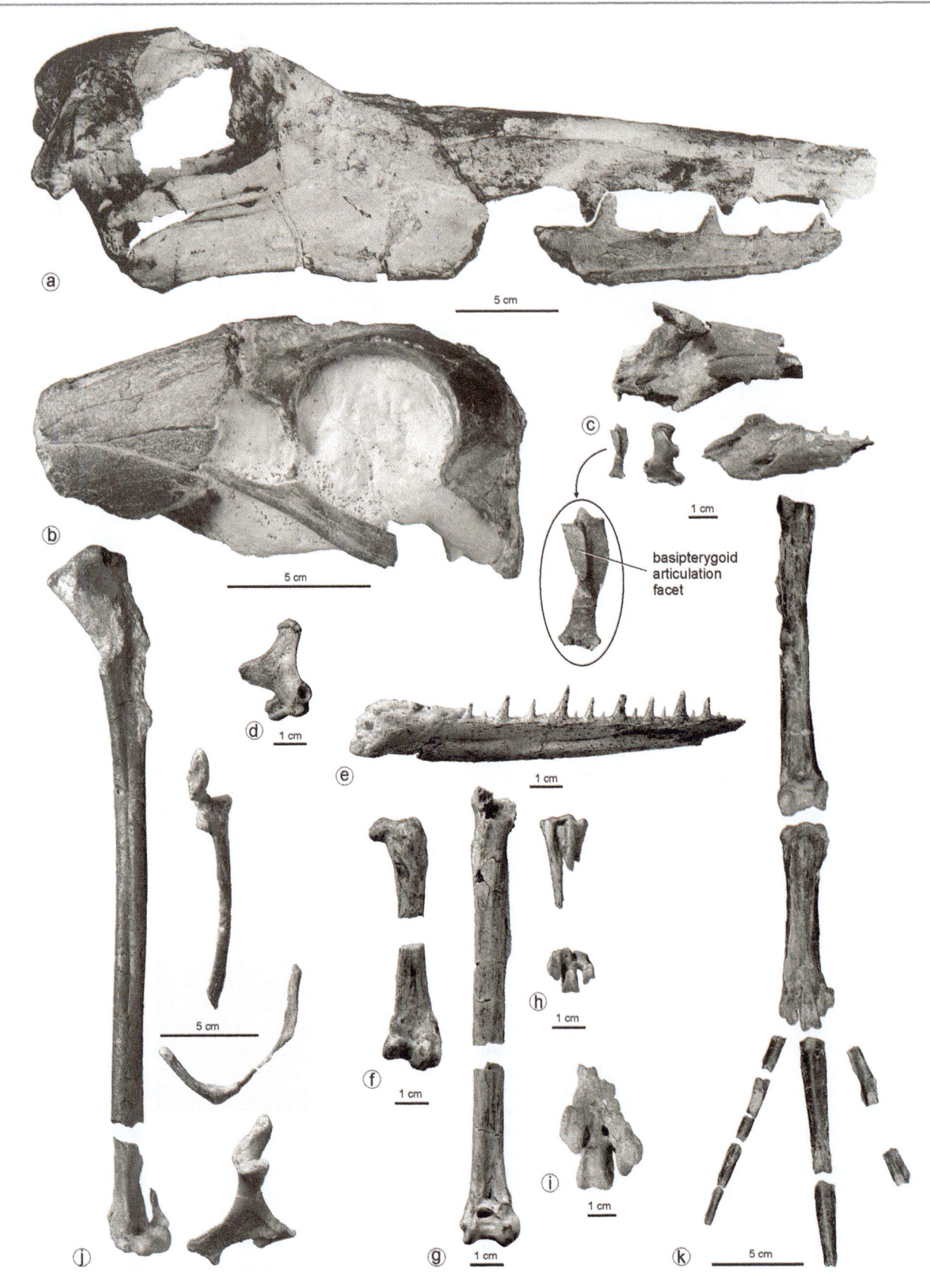

Fig. 4.2 Bones of bony-toothed birds (Pelagornithidae). (**a**) Skull of *Pseudodontornis longirostris* (from Lambrecht 1930); the provenance and whereabouts of this fossil are unknown. (**b**) Partial skull of *Dasornis emuinus* from the early Eocene of the London Clay of the Isle of Sheppey (Staatliches Museum für Naturkunde Karlsruhe, Germany, SMNK-PAL 4017). (**c**) Partial skull, pterygoid, quadrate, and mandible

The earliest fossil record of a pelagornithid from the Pacific Basin is a fragmentary distal humerus from the middle Eocene of Mexico, which was tentatively referred to *Dasornis* ("*Odontopteryx*"; González-Barba et al. 2002). Possibly the earliest South American record of the Pelagornithidae is a partial femur from the middle to late Eocene of Chile described by Yury-Yáñez et al. (2012), who compared the fossil with the Procellariiformes and Sphenisciformes. Goedert (1989) reported specimens of giant pelagornithids from the late Eocene to early Oligocene of Oregon (Pacific coast), which may belong to *Dasornis* ("*Argillornis*").

Compared with this fairly comprehensive Eocene fossil record, relatively few pelagornithid remains were found in Oligocene deposits. *Caspiodontornis kobystanicus* from the "middle" Oligocene of Azerbaijan is known from a largely complete skull with a length of some 30 cm (Aslanova and Burchak-Abramovich 1999). Aslanova and Burchak-Abramovich (1999) emphasized the fact that the specimen is strongly flattened dorsoventrally, which is, however, likely to be an artifact of preservation. *C. kobystanicus* may be a junior synonym of *Guguschia nailiae* from the same locality, which was misidentified as a swan (Sect. 4.5.4).

Fossils of unnamed species of bony-toothed birds are also known from the Oligocene of Japan (Okazaki 1989, 2006). *Palaeochenoides mioceanus* and *Tympanonesiotes wetmorei* from the late Oligocene of South Carolina are further late Paleogene representatives of the Pelagornithidae (Olson 1985), but the exact taxonomic affinities of these species still need to be clarified. The tarsometatarsus of a partial skeleton from the late Oligocene/early Miocene of Oregon (Fig. 4.2k) shows a resemblance to a tarsometatarsus, which was referred to *Palaeochenoides mioceanus* (Mayr et al. 2013).

Pelagornis sandersi from the late Oligocene (25–28 Ma) of South Carolina, USA, is known from a partial skeleton including the skull and major limb bones (Ksepka 2014). With an estimated wingspan of 6.4 m, this species is the largest pelagornithid species known to date.

Paleogene representatives of the Pelagornithidae cover a size range from small (*Protodontopteryx ruthae*) to giant (*Pelagornis sandersi*) species, and in the early and middle Eocene, medium-sized (*Dasornis toliapica*) and very large species (*D. emuinus*) with wingspans of more than four meters occurred in the same localities. By contrast, all Neogene Pelagornithidae had a very large size, reaching wingspans of five to six meters (Mayr and Rubilar-Rogers 2010).

Pelagornithids have very long and slender wing bones and short legs. Early Paleogene pelagornithids were very different from the highly specialized giant species that lived in the Neogene period. The tarsometatarsus of the early Eocene *Dasornis*, for example, exhibits a proportionally shorter and more plantarly directed trochlea for the second toe than the tarsometatarsus of the middle Eocene *Lutetodontopteryx* and other geologically younger pelagornithids (Mayr and Rubilar-Rogers 2010; Mayr and Zvonok 2012). This plesiomorphic morphology suggests a sister group relationship between *Dasornis* and all other pelagornithids except *Protodontopteryx*, which is the phylogenetically earliest diverging pelagornithid taxon (Fig. 4.3). Comparisons with similar tarsometatarsi of extant seabirds suggest that *Dasornis* used its feet for aquatic locomotion to a greater extent than geologically younger pelagornithids and may have been foraging near the sea surface while swimming. The tarsometatarsus of *Protodontopteryx* is unknown, but the morphology of the wing bones suggests that the taxon was less adapted to sustained soaring and retained some capabilities of flapping flight.

The proximal end of the humerus of more advanced and geologically younger pelagornithid species exhibits a distinctive and highly derived morphology, which impeded rotation of the bone. This suggests a specialization for long-distance soaring, and at least the very large Neogene species were not capable of sustained flapping flight (Olson 1985). In *Gigantornis*, the tip of the sternal keel exhibits a large articular facet for the furcula. Among extant birds, a similar syndesmotic or synostotic joint between the furcula and the sternum mainly occurs in larger soaring birds (Mayr et al. 2008). The pelagornithid sternum is furthermore characterized by a large, steep-walled opening in the cranial portion of the visceral (dorsal) surface, and a strongly vaulted body (Mayr et al. 2008). The pedal phalanges of pelagornithids are unusual in that they are dorsoplantarly flattened and in that their shaft has an unusual "bloated" appearance in some species (Fig. 4.2k; Mayr and Zvonok 2011; Mayr et al. 2013).

As evidenced by the Paleocene *Protodontopteryx*, pseudoteeth evolved before pelagornithids became highly

Fig. 4.2 (Continued) fragment of *D. toliapica* (holotype, Natural History Museum, London, NHMUK A 44096); the arrow indicates an enlarged detail of the pterygoid. (**d**)–(**h**) Bones of *Lutetodontopteryx tethyensis* from the middle Eocene of Ukraine (Senckenberg Research Institute, Frankfurt, **d**: quadrate, SMF Av 559; **e**: mandible, SMF Av 555; **f**: femur, SMF Av 576; **g**: tibiotarsus, SMF Av 578a; **h**: holotype tarsometatarsus, SMF Av 553). (**i**) Distal tarsometatarsus of *D. emuinus* (NHMUK A 894). (**j**) Humerus, coracoid, furcula, and scapula of a *Lutetodontopteryx*-like species from the middle Eocene of Belgium (Royal Belgian Institute of Natural Sciences, Brussels, Belgium, IRSNB Av 86). (**k**) Leg of *Pelagornis* sp. from the late Oligocene or early Miocene Nye Mudstone of Oregon, USA (Natural History Museum of Los Angeles County, LACM 128424). (Photo in (**k**) by Samuel McLeod; all others except (**a**) by Sven Tränkner)

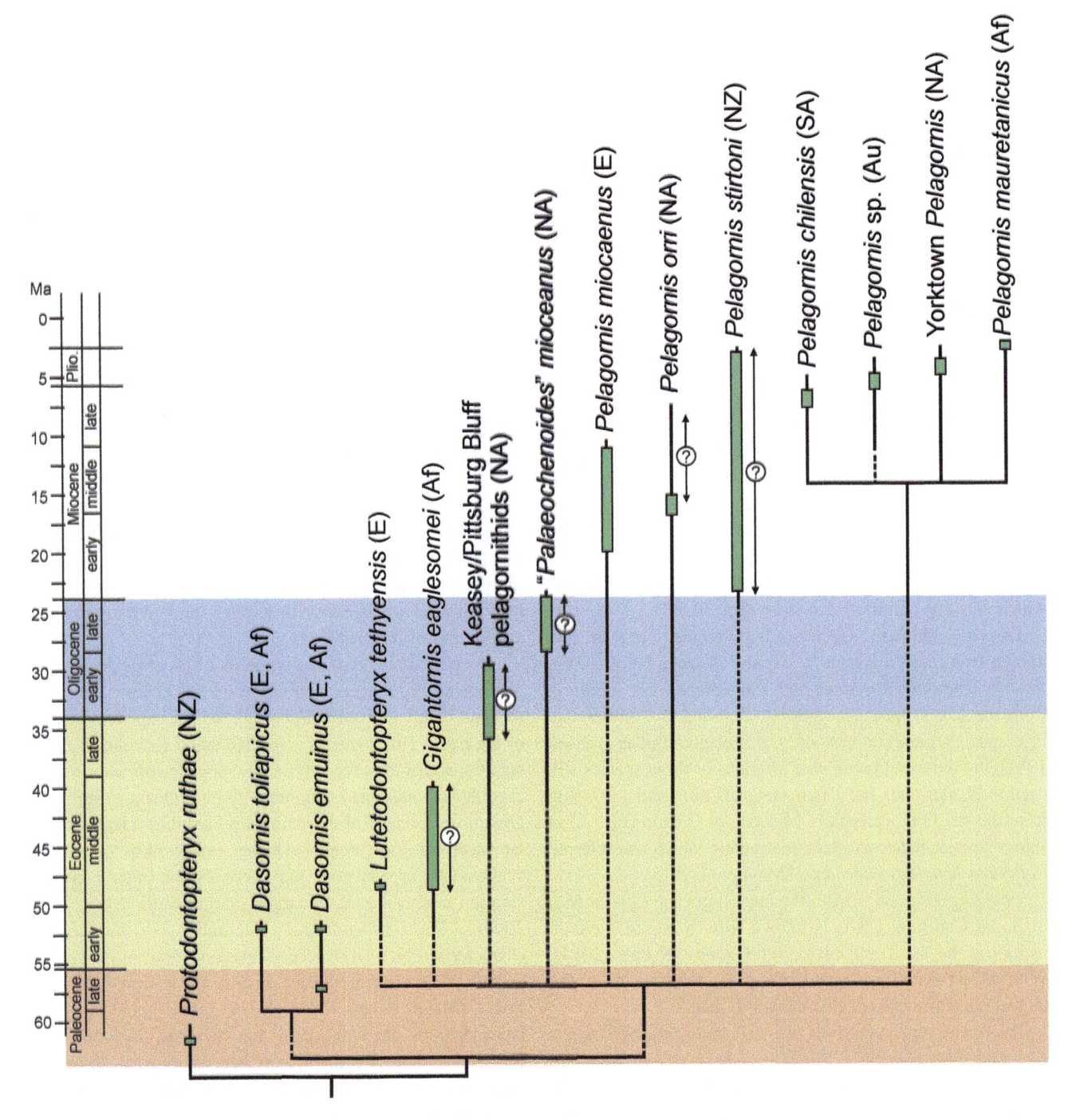

Fig. 4.3 Interrelationships and stratigraphic occurrences of Paleogene and Neogene Pelagornithidae (redrawn and modified after Mayr et al. 2013)

specialized soaring birds. Mayr et al. (2021) hypothesized that *Protodontopteryx* was piscivorous and targeted selected prey items, whereas the giant Neogene species probably were skimmers and mainly fed on squid. Pelagornithids captured their prey, larger fish or squid, near the water surface, and the morphology of the cranial cervical vertebrae indicates that the beak was held in a vertical position during foraging (Mayr and Rubilar-Rogers 2010; Mayr and Zvonok 2011). Virtual endocasts of the skull of *Dasornis* ("*Odontopteryx*") were studied by Milner and Walsh (2009), who found the brain shape of this taxon to be similar to that of extant seabirds.

The pseudoteeth of pelagornithids are very different from true avian teeth in that they are hollow outgrowths of the jaws (Fig. 4.4), whereas teeth are situated in alveoles and are covered with a layer of enamel. However, and as first hypothesized by Mayr and Rubilar-Rogers (2010),

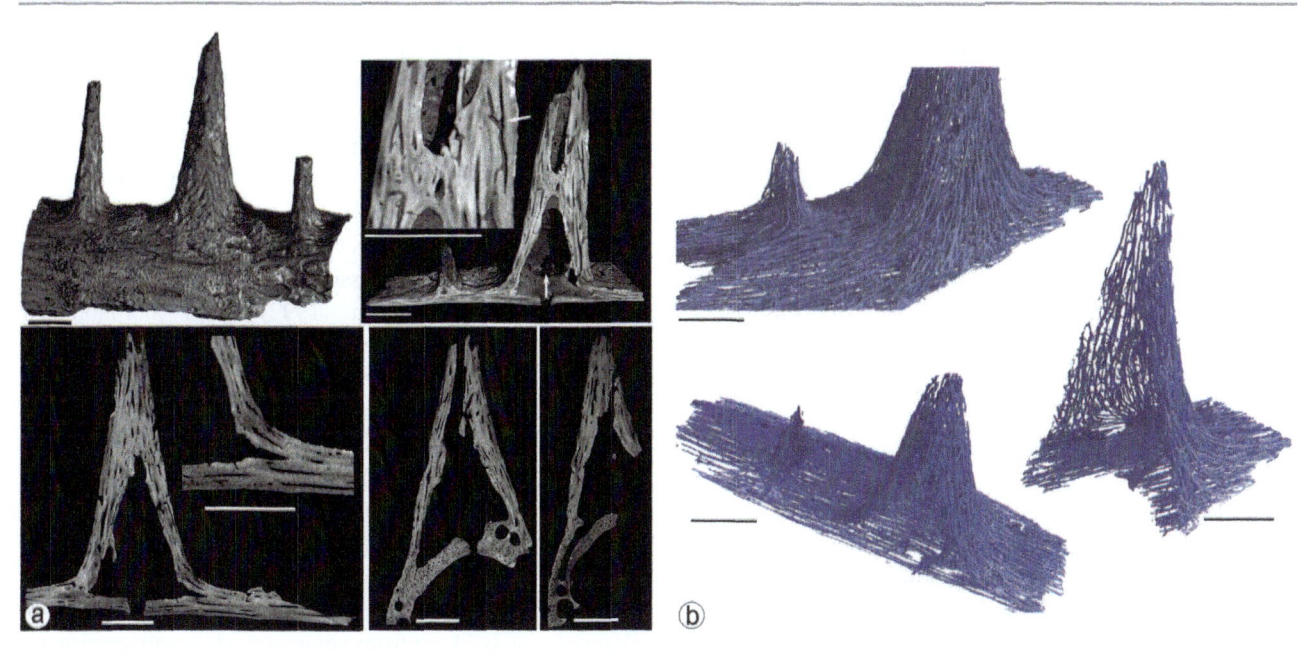

Fig. 4.4 Microstructure of the pseudoteeth of the late Pliocene pelagornithid *Pelagornis mauretanicus* (from Louchart et al. 2013, lettering digitally removed; published under a Creative Commons CC BY 4.0 license). (**a**) Computed tomographic images showing the hollow interior of the pseudoteeth. (**b**) Reconstruction of the vascular network within the pseudoteeth. Scale bars equal 2 mm

pseudoteeth may be homologous to true teeth on an early developmental level, and Mayr (2017: 124) detailed that "if pseudoteeth go back on incompletely expressed tooth-specific developmental programs, they may nevertheless be homologous to true archosaur teeth on a molecular level, because early ontogenetic stages of alligator teeth are also mere outgrowths of the epithelial surface layers, and the pathways for tooth development can still be induced in chicken embryos" (see also Mayr 2011). Similar conclusions were subsequently drawn by Louchart et al. (2018), who proposed a developmental model for the evolution of pseudoteeth but did not cite any of the above references in the main text of their paper.

Some distinctive skull features of pelagornithids are related to these unique bony tooth-like projections. For example, there is a pair of distinct furrows along the ventral surface of the upper beak, and these bear marked pits for the reception of the pseudoteeth of the mandible. The deep jugal bars and the strongly ventrally projecting palatine bone also seem to be functionally correlated with the specialized feeding technique of bony-toothed birds. The mandible has an intraramal joint and lacks a mandibular symphysis. Both features are unknown from extant birds but are also found in Mesozoic birds with true teeth (Zusi and Warheit 1992; Mayr and Rubilar-Rogers 2010).

If pseudotheeth are homologous to avian teeth at an early developmental level, their occurrence in pelagornithids may suggest an early divergence of these birds from other Neornithes. In addition to the lack of a mandibular symphysis, the skull of pelagornithids also exhibits an incompletely ossified frontoparietal suture, another plesiomorphic condition, which is present in Mesozoic non-neornithine taxa but in extant birds only occurs in the palaeognathous Tinamidae (Bourdon et al. 2010; Mayr 2011). However, neognathous affinities of pelagornithids are suggested by the lack of co-ossification between the pterygoid and the palatine bone (unlike palaeognathous birds), a supratendinal bridge on the tibiotarsus, and a hypotarsus, which exhibits distinct sulci and crests (a supratendinal bridge is, however, also present in the palaeognathous Palaeotididae as well as in extant Tinamidae).

Most earlier authors either assigned pelagornithids to the traditional "Pelecaniformes" (Olson 1985; Olson and Rasmussen 2001) or classified them into a taxon Odontopterygiformes, which was considered closely related to the "Pelecaniformes" and Procellariiformes (e.g., Harrison and Walker 1976a). However, a position of bony-toothed birds outside Neoaves is supported by the plesiomorphic absence of a ventral crest on the palatine bone (Mayr 2008a) and the presence of only two instead of three condyles on the mandibular process of the quadrate. Harrison and Walker (1976a) noted that pelagornithids possess the derived basipterygoid articulation of the Galloanseres, and an analysis by Bourdon (2005) resulted in a sister group relationship between pelagornithids and the Anseriformes. In this regard, it is notable that the limb bones of pelagornithids resemble those of putative galloanserines from the Late Cretaceous and Paleocene of Antarctica, with the coracoid of *Protodontopteryx* being particularly similar to that of *Vegavis* (Sect. 2.4) and *Conflicto* (Sect. 4.5). Still, anseriform affinities are not strongly supported and pelagornithids lack a characteristic galloanserine apomorphy, that is, a mandible with long retroarticular processes (Mayr and Rubilar-Rogers 2010). An analysis by Mayr (2011) resulted in a sister group relationship between pelagornithids and the Galloanseres (Mayr 2011). By contrast, pelagornithids were obtained as the sister taxon of the Anseriformes in an analysis by Field et al. (2020), whereas an analysis by Mayr et al. (2021) recovered them in a polytomy together with the Galloanseres and Neoaves.

4.2 Gastornithidae

The Gastornithidae (Fig. 4.5) are large, flightless, and graviportal birds, which occur in the Paleocene to middle Eocene of Europe and in the early Eocene of North America and Asia. The first fossils were reported from the late Paleocene of France by Hébert (1855), who assigned them to the taxon *Gastornis*. Owing to the fragmentary nature of the then known specimens, early reconstructions of these European gastornithids were grossly erroneous (Martin 1992). This prevented recognition of their close similarity to the much better represented North American species, which were described as *Diatryma*, within the taxon Diatrymidae (Cope 1876; Matthew and Granger 1917). The Eocene European gastornithids were also assigned to *Diatryma* by earlier authors. Although Martin (1992) noted morphological differences between the latter taxon and *Gastornis*, synonymization of *Gastornis* and *Diatryma* was suggested by Buffetaut (1997) and formalized by Mlíkovský (2002).

The earliest records of the Gastornithidae are a femur from the late Paleocene of the Belgian locality Maret (Mayr and Smith 2019a) and a coracoid from the Paleocene of Walbeck in Germany (Mayr 2007a); Weigelt (1939) mentioned other gastornithid remains from this latter locality, which, however, seem to have been lost. The Maret femur and the Walbeck coracoid are from comparatively small species and

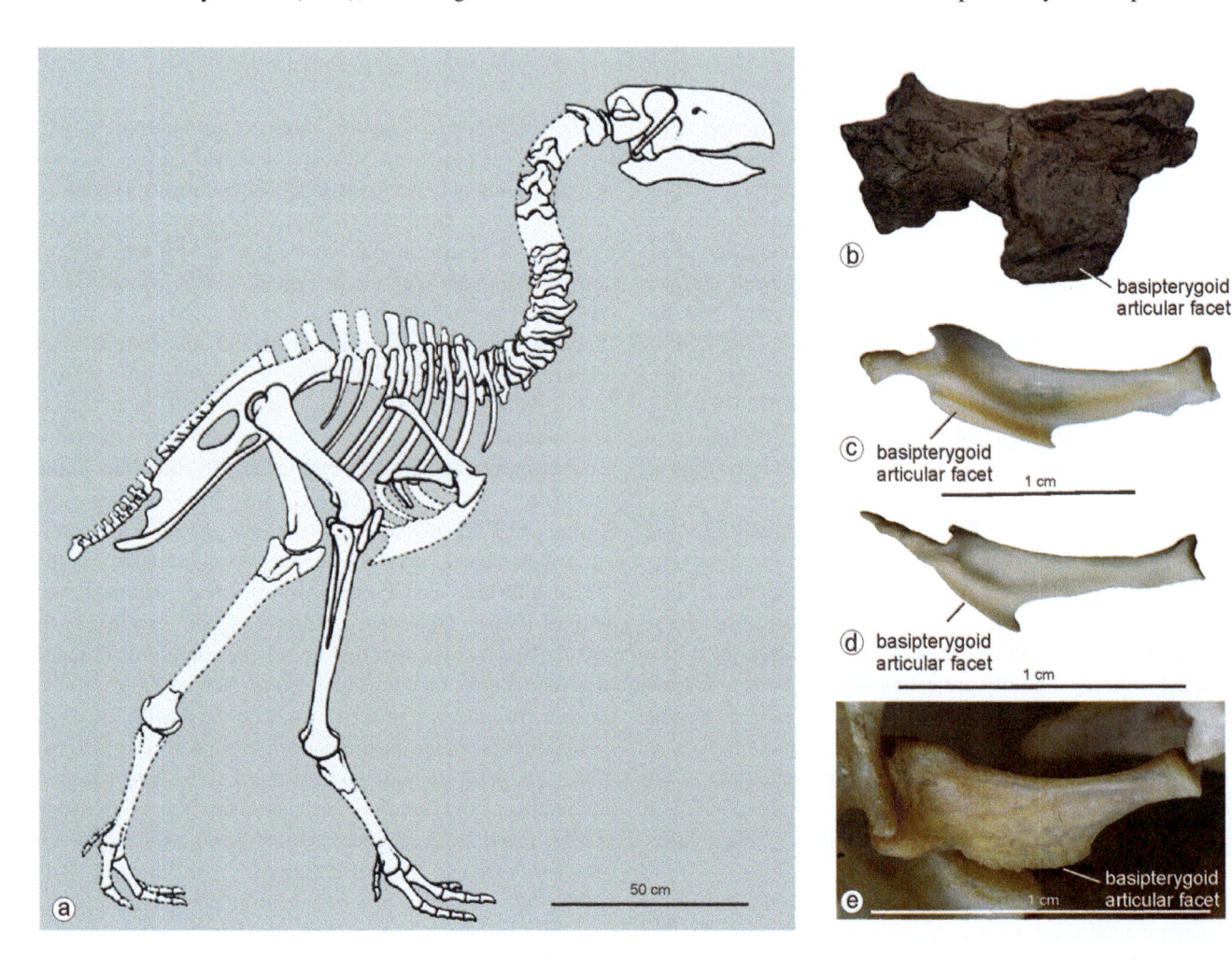

Fig. 4.5 (**a**) Skeletal reconstruction of the early Eocene *Gastornis gigantea* (Gastornithidae, redrawn after Matthew and Granger 1917). (**b**) Pterygoid of *G. gigantea* (American Museum of Natural History, New York, USA, AMNH 6169). (**c**)–(**e**) Pterygoids of extant Anseriformes (**c**: *Anseranas semipalmata*, Anseranatidae; **d**: *Bucephala clangula*, Anatidae; **e**: *Chauna torquata*, Anhimidae). (Photo in (**b**) by the author and not to scale; photos in (**c**)–(**e**) by Sven Tränkner)

may belong to *Gastornis russelli*, which is the smallest named gastornithid species. The description of *G. russelli* was based on a tarsometatarsus and beak fragment from the late Paleocene of the Reims area in France (Martin 1992) and the species has a more slender tarsometatarsus than other gastornithids; it is probably conspecific with *Gastornis minor* from the same locality, which was considered a nomen dubium by Martin (1992). Bones of an unnamed larger species of *Gastornis* were described from the late Paleocene site Louvois in France (Mourer-Chauviré and Bourdon 2016).

G. russelli measured less than half of the ostrich-sized and much better known late Paleocene to early Eocene *Gastornis parisiensis*. According to Martin (1992), the latter species is synonymous with "*G. edwardsi*" from the Paleocene of France and "*G. klaasseni*" from the early Eocene of England. Mlíkovský (2002) regarded *Gastornis* ("*Diatryma*") *geiselensis* from the latest early or earliest middle Eocene of Messel and the middle Eocene of the Geisel Valley (Fischer 1962, 1978; Hellmund 2013; Mayr 2018) as a junior synonym of *Gastornis sarasini* from the early Eocene (MP 10; Mlíkovský 2002) of France (Schaub 1929). An in-depth taxonomic revision of these species is, however, overdue and requires the direct comparison of the type specimens. The most recently described European gastornithid species is *Gastornis laurenti* from the early Eocene of southern France (Mourer-Chauviré and Bourdon 2020).

The North American Gastornithidae were revised by Andors (1992), who recognized two species, *Gastornis* ("*Diatryma*") *gigantea* and *G.* ("*D.*") *regens*, which differ in toe proportions. Both occur in the Rocky Mountain region of western North America and are restricted to the early Eocene (mainly Wasatch and Willwood formations). Remains of gastornithids were also reported from the early Eocene (52–53 Ma) of Ellesmere Island in the Canadian Arctic (Stidham and Eberle 2016).

Of *Zhongyuanus xichuanensis* from the early Eocene of China only the distal end of a tibiotarsus was found (Hou 1980, 2003). According to Andors (1992), this species is morphologically distinct from the European and North American Gastornithidae, but Buffetaut (2013) considered the differences to be of minor nature and synonymized *Zhongyuanus* with *Gastornis*. Therefore, the Chinese species is now classified as *Gastornis xichuanensis*.

Mustoe et al. (2012) assigned large, tridactyl footprints from the early Eocene of Washington State (USA) to *Gastornis*. These authors contested correct identification of putative *Gastornis* footprints from the middle Eocene (45 Ma) of Washington State (USA), which were described by Patterson and Lockley (2004) and are about 5 million years younger than the known North American skeletal remains of the Gastornithidae; if indeed from a species of *Gastornis*, these footprints would establish the presence of hoof-like ungual phalanges in this taxon. Buffetaut (2004) commented on footprints of giant, tridactyl birds from the late Eocene of France, which were also assigned to *Gastornis*; again this identification conflicts with the fact that gastornithids have no late Eocene skeletal record (Buffetaut and Angst 2021 hypothesized that these tracks could have been made by *Macrornis tanaupus*; see Sect. 8.5.2).

The discovery of gastornithid bones in the early Eocene of southern France supports earlier hypotheses (Dughi and Sirugue 1959; Fabre-Taxy and Touraine 1960) that large eggs from coeval deposits of Provence, which were described as *Ornitholithus arcuatus*, belong to the Gastornithidae (Buffetaut 2008; Angst et al. 2015). According to Fabre-Taxy and Touraine (1960) some of these eggs measure 24 × 15 cm and would therefore be of a size to be expected for gastornithids; Dughi and Sirugue (1959) even estimated the size of the largest eggs at 40 × 20 cm (see Buffetaut 2008). The eggshell assigned to *Gastornis* is distinctive in that its surface exhibits a tubercular ornamentation (Angst and Buffetaut 2017). Putative feathers of *Gastornis* ("*Diatryma*") from the early Eocene of Colorado were described by Cockerell (1923), but identification of these hair-like filaments was disputed by Wetmore (1930).

Gastornithids are characterized by a huge, bilaterally compressed beak, which has a convex culmen (dorsal ridge) and lacks a hooked tip. The cranium exhibits upper temporal fenestrae. The wings are greatly reduced and the sternum has no keel. At least in the larger species, scapula and coracoid are co-ossified and form a scapulocoracoid. Such fusion is absent in the Walbeck coracoid, which, unlike in other gastornithids, also lacks a foramen for the supracoracoideus nerve. The palate is of neognathous structure, and in further contrast to palaeognathous birds, the ilioischiadic foramina of the pelvis are closed. The ribs do not bear ossified uncinate processes. The legs are very robust, with a short tarsometatarsus and a greatly reduced hallux. The largest species were nearly two meters tall and had an estimated weight of about 175 kg (Andors 1992).

Andors (1992) revised the phylogenetic affinities of gastornithids and concluded that they are the sister group of the Anseriformes. Osteological characteristics shared with galloanserine birds include retroarticular processes on the caudal ends of the mandible and features of the quadrate. Unlike in extant Galloanseres, however, the pterygoid exhibits stalked basipterygoid articular facets (Fig. 4.5b–e).

Because of its smaller size and more slender tarsometatarsus, *G. russelli* may be the sister taxon of the other gastornithids. If true, this would argue for a European origin of the group, which is also in concordance with the temporal occurrence of the known species (Andors 1992; Buffetaut 1997). Current evidence suggests that gastornithids dispersed into North America in the early Eocene (Andors 1992; Mayr 2009; Buffetaut and Angst 2014).

Gastornithids lived in a forested environment, and most earlier authors assumed that they were carnivorous. However, Andors (1992: 117) elaborated the hypothesis that these birds actually were folivores and noted that the absence of a terminal hook on the beak and the weak ungual phalanges impeded a raptorial way of living. Witmer and Rose (1991: 95) countered that "excessively high safety factors in the construction of the skull" contradicted the hypothesis of herbivory. A folivorous diet may, however, also include very hard food items, such as seeds and twigs, and the evidence presented by Andors is the more compelling one. A herbivorous diet of gastornithids was also deduced from analyses of carbon isotopes (Angst et al. 2014, 2015) and is furthermore supported by possible gastroliths found in association with *Gastornis* remains (Angst and Buffetaut 2017).

The evolutionary origins of gastornithids are unknown, but given their high degree of specialization, their stem species almost certainly diverged from other neornithine birds in the Cretaceous. Of the known Late Cretaceous birds, only the skull of the putative anseriform *Asteriornis* (Field et al. 2020) shows a remote similarity to gastornithids in its proportions, and it is conceivable, that a similar taxon gave rise to the Gastornithidae. Why gastornithids reached such a large body size in the earliest Cenozoic is another aspect of their evolutionary history, which remains unaddressed. There was a low predation pressure in early Cenozoic ecosystems of Europe, where gastornithids appear to have originated (see Sect. 11.3.1), and possibly the large size of these birds represents an adaptation to their herbivorous diet.

4.3 Galliformes (Landfowl)

Extant Galliformes include the Australasian Megapodiidae (megapodes), which are the sister taxon of a clade formed by the New World Cracidae (guans, chachalacas, and curassows) and the globally distributed Phasianoidea (grouse, quails, pheasants, and allies). Even though stem group Galliformes may have already existed in the Late Cretaceous (see Sect. 2.4), the identification of these fragmentary remains needs to be corroborated with further material. The Paleogene record of galliform birds, however, is quite extensive and has been substantially improved in the past years, with these fossils providing interesting insights into the evolutionary history of the galliform birds. For a long time, early Paleogene Galliformes were mainly known from Europe and North America, but meanwhile, relevant fossils were also reported from Asia and Africa.

4.3.1 The Eocene Gallinuloididae—Stem Group Galliformes from North America and Europe

The taxon Gallinuloididae comprises some of the earliest galliform birds and was initially established for *Gallinuloides wyomingensis* from the North American Green River Formation (Fig. 4.6a), which is the first avian species described from this early Eocene locality (Eastman 1900; Lucas 1900; Mayr and Weidig 2004). Representatives of the Gallinuloididae were also reported from Europe, namely *Paraortygoides messelensis* from Messel (Fig. 4.6e, f) and *P. radagasti* from the London Clay (Fig. 4.7a; Mayr 2000, 2006; Dyke and Gulas 2002). Other species assigned to the Gallinuloididae by earlier authors have meanwhile been removed from the taxon (Mayr 2009).

Gallinuloidids are the only early Paleogene stem group representatives of the Galliformes, which are known from well-preserved skeletons. Although the skull of *Gallinuloides wyomingensis* resembles that of extant Galliformes in overall morphology, the postorbital processes are less developed and there are no ossified zygomatic aponeuroses, that is, bony connections between the zygomatic and postorbital processes, which are a derived characteristic of many extant Galliformes (Mayr and Weidig 2004). As in the Paraortygidae (Sect. 4.3.3) and some extant galliform taxa, the humerus exhibits a well-developed second pneumotricipital fossa. Otherwise, however, the pectoral girdle and wing skeleton of gallinuloidids display a plesiomorphic morphology. The humerus is more elongate and less robust than that of crown group Galliformes, and its proximal end is more similar to that of ducks (Anatidae) in its proportions. The shorter and more protruding deltopectoral crest is another plesiomorphic feature that shows gallinuloidids to be the sister taxon of all other galliform birds (Fig. 4.8; Mayr 2006). As in other stem group Galliformes (see below) and extant Anseriformes, the coracoid of gallinuloidids exhibits a plesiomorphic, cup-like articular facet for the scapula. The very long scapula and the long and slender carpometacarpus are also more similar to the corresponding bones of anseriform birds than to those of crown group Galliformes (Mayr 2000; Mayr and Weidig 2004). The shafts of the furcula are much wider than in crown group Galliformes, and the tip of the sternal keel reaches farther cranially (Mayr 2006). The bodies of the thoracic vertebrae bear marked lateral depressions (Fig. 4.7a; Dyke and Gulas 2002), which also occur in Mesozoic stem group representatives of the Neornithes (e.g., *Ichthyornis*; Clarke 2004) and in the anseriform Presbyornithidae (Sect. 4.5.2) but are absent in extant Galliformes.

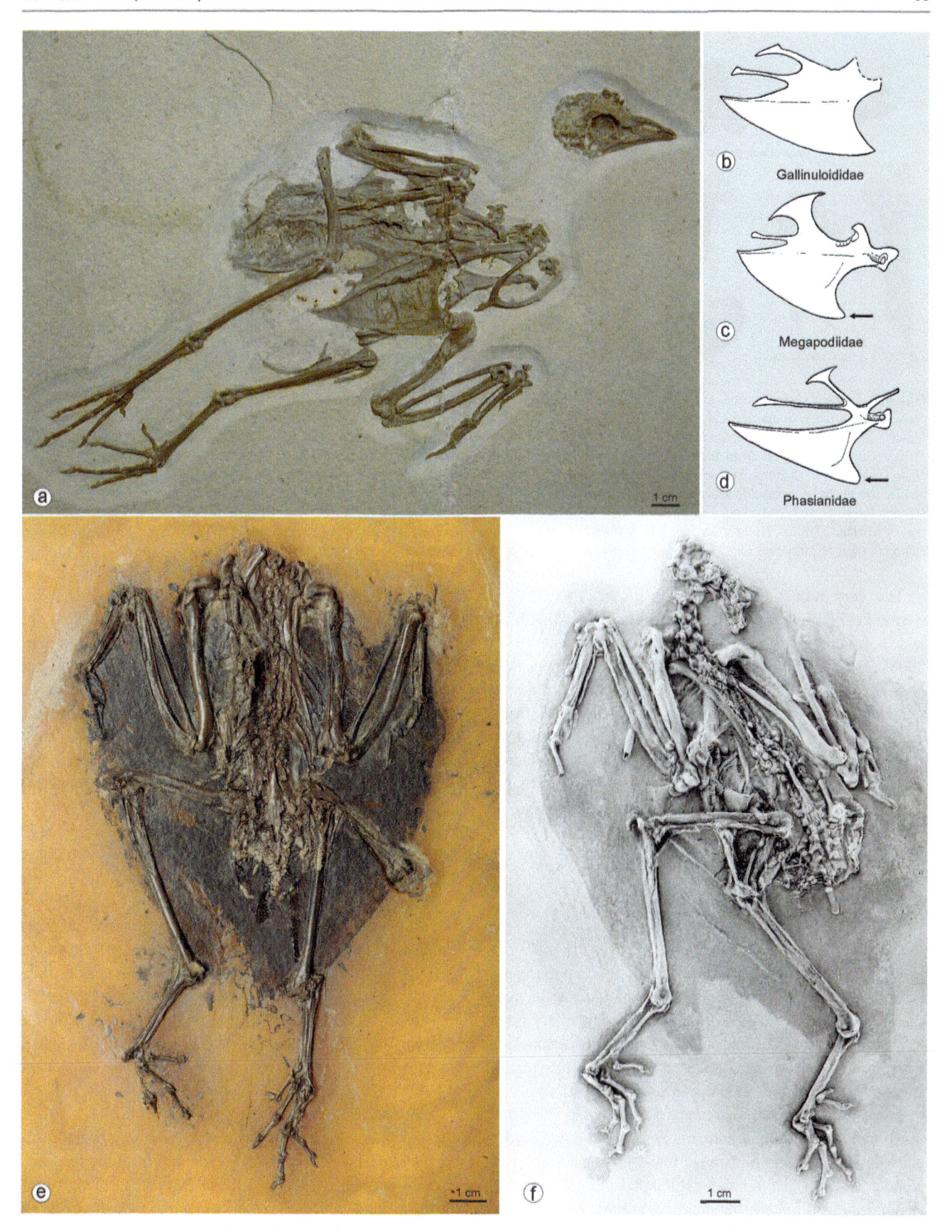

Fig. 4.6 Skeletons of the early Eocene Gallinuloididae (Galliformes). (**a**) Skeleton of *Gallinuloides wyomingensis* (Gallinuloididae) from the early Eocene Green River Formation (Wyoming Dinosaur Center, Thermopolis, USA, WDC CGR-012). (**b**)–(**d**) Sterna of (**b**) *G. wyomingensis*, (**c**) *Alectura lathami* (Megapodiidae), and (**d**) *Rollulus rouloul* (Phasianidae); the arrows point to the caudally displaced tip of the sternal keel of extant Galliformes (after Mayr 2006, not to scale). (**e**), (**f**) Skeletons of *Paraortygoides messelensis* (Gallinuloididae) from the latest early or earliest middle Eocene of Messel (Senckenberg Research Institute, Frankfurt, **e**: holotype, SMF-ME 1303a; **f**: SMF-ME 11112b, coated with ammonium chloride). (Photo in (**a**) courtesy of Ilka Weidig, (**e**), (**f**) by Sven Tränkner)

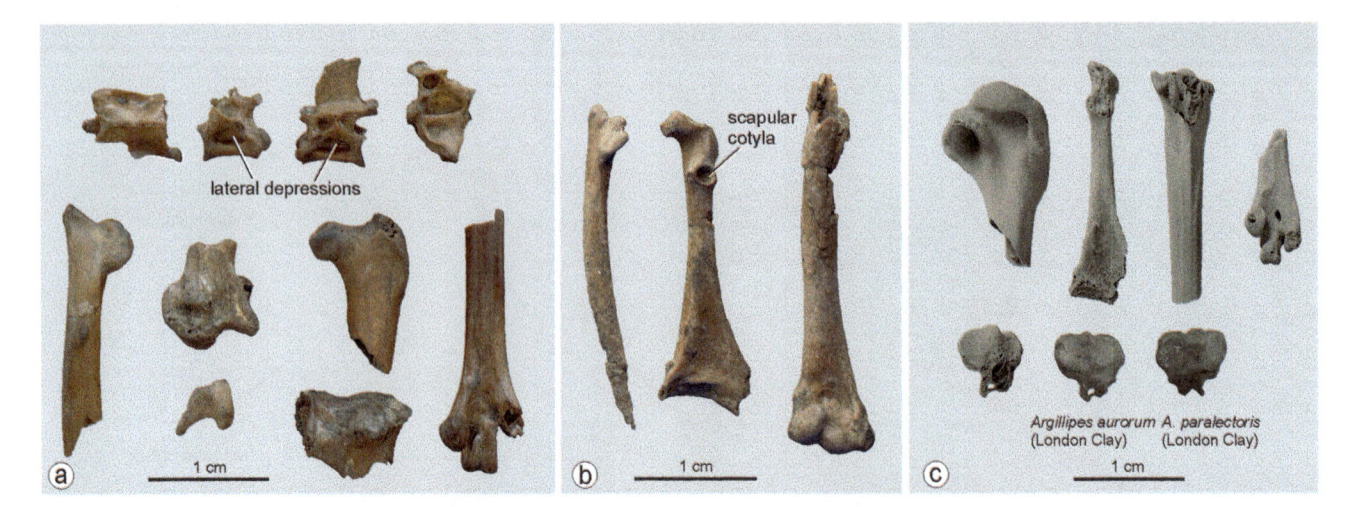

Fig. 4.7 Further early Eocene galliform birds. (**a**) Selected bones (vertebrae, partial scapula and carpometacarpus, ulnar carpal, partial femur, tibiotarsus, and tarsometatarsus) of the holotype of *Paraortygoides radagasti* from the London Clay of Walton-on-the-Naze, Essex, UK (Natural History Museum, London, NHMUK A 6217). (**b**) Scapula, coracoid, and partial humerus of an undescribed smaller *Paraortygoides*-like species from Walton-on-the-Naze (collection of Michael Daniels, Holland-on-Sea, UK, WN 92715). (**c**) Skeletal elements (partial humerus, coracoid, partial tarsometatarsi) of a small species (aff. *Argillipes aurorum*) from the early Eocene of Egem in Belgium (Royal Belgian Institute of Natural Sciences, Brussels, Belgium; proximal humerus: IRSNB Av 163, coracoid: IRSNB Av 167, proximal tarsometatarsus: IRSNB Av 166, distal tarsometatarsus: IRSNB Av 165); the proximal end of the tarsometatarsus is shown in comparison to proximal views of the tarsometatarsi of *Argillipes aurorum* (holotype, Natural History Museum, London, NHMUK A 3130) and *A. paralectoris* (holotype, NHMUK 3604) from the London Clay. (All photos by Sven Tränkner)

The very robust furcula and the projecting tip of the sternal keel (Fig. 4.6b–d) indicate that gallinuloidids had a less voluminous crop than their extant relatives (Mayr 2006). Their vegetable food component therefore probably consisted of fruits and other easily digestible plant matter, rather than coarse material such as seeds. This assumption is in concordance with the fact that in none of the known specimens of the Gallinuloididae gastroliths are preserved, whereas extant Galliformes regularly ingest grit and small pebbles to mechanically break down plant matter in the gizzard.

4.3.2 Other Early and Middle Eocene Stem Group Galliforms

In the past years, galliform fossils were also reported from early and middle Eocene sites of Asia and Africa. In many cases, however, these are represented by fragmentary bones only, and their exact affinities to the better-known taxa from Europe and North America are uncertain. Such is also true for some poorly represented European galliform fossils.

In the early Eocene (Bumbanian) Naranbulag Formation in Mongolia, a proximal humerus and a partial coracoid of unidentified galliform birds were found (Hwang et al. 2010; Hood et al. 2019). Both bones exhibit plesiomorphic traits of stem group Galliformes, but the exact affinities of the fossils are uncertain. The coracoid shows a resemblance to the corresponding bone of the Quercymegapodiidae (Sect. 4.3.5), whereas the humerus is clearly distinguished from that of quercymegapodiids. Whether these bones are from the same or closely related species cannot be said, but they document the earliest occurrence of Galliformes on the Asian continent. Most recently, another small stem group galliform from early Eocene strata of the Tsagaan-Khushuu locality of Mongolia was described as *Bumbanortyx transitoria* by Zelenkov (2021a). This species is represented by a partial coracoid (the holotype), as well as a referred proximal humerus and a proximal tarsometatarsus. The coracoid of *B. transitoria* resembles the corresponding bone of the Gallinuloididae and Quercymegapodiidae. *Bumbanipodius magnus*, another and slightly larger stem group galliform from Tsagaan-Khushuu, is known from a proximal tarsometatarsus and a referred partial coracoid (Zelenkov 2021a). This species was likened to *Argillipes aurorum* from the London Clay (see below), but the fragmentary material defies a robust phylogenetic placement.

The phylogenetic relationships of some small galliform species from the Eocene of Africa are likewise elusive. *Namaortyx sperrgebietensis* from of the middle Eocene (47–49 Ma) of Namibia (Mourer-Chauviré et al. 2011) is only known from a tarsometatarsus, which does not permit a well-founded classification of this stem group galliform. This is also true for *Chambiortyx cristata* from the late early or early middle Eocene of Tunisia (Mourer-Chauviré et al. 2013), a small bird the size of a Common Quail (*Coturnix coturnix*), the holotype of which is the distal end of a tarsometatarsus. Remains of unnamed small galliforms were also

4.3 Galliformes (Landfowl)

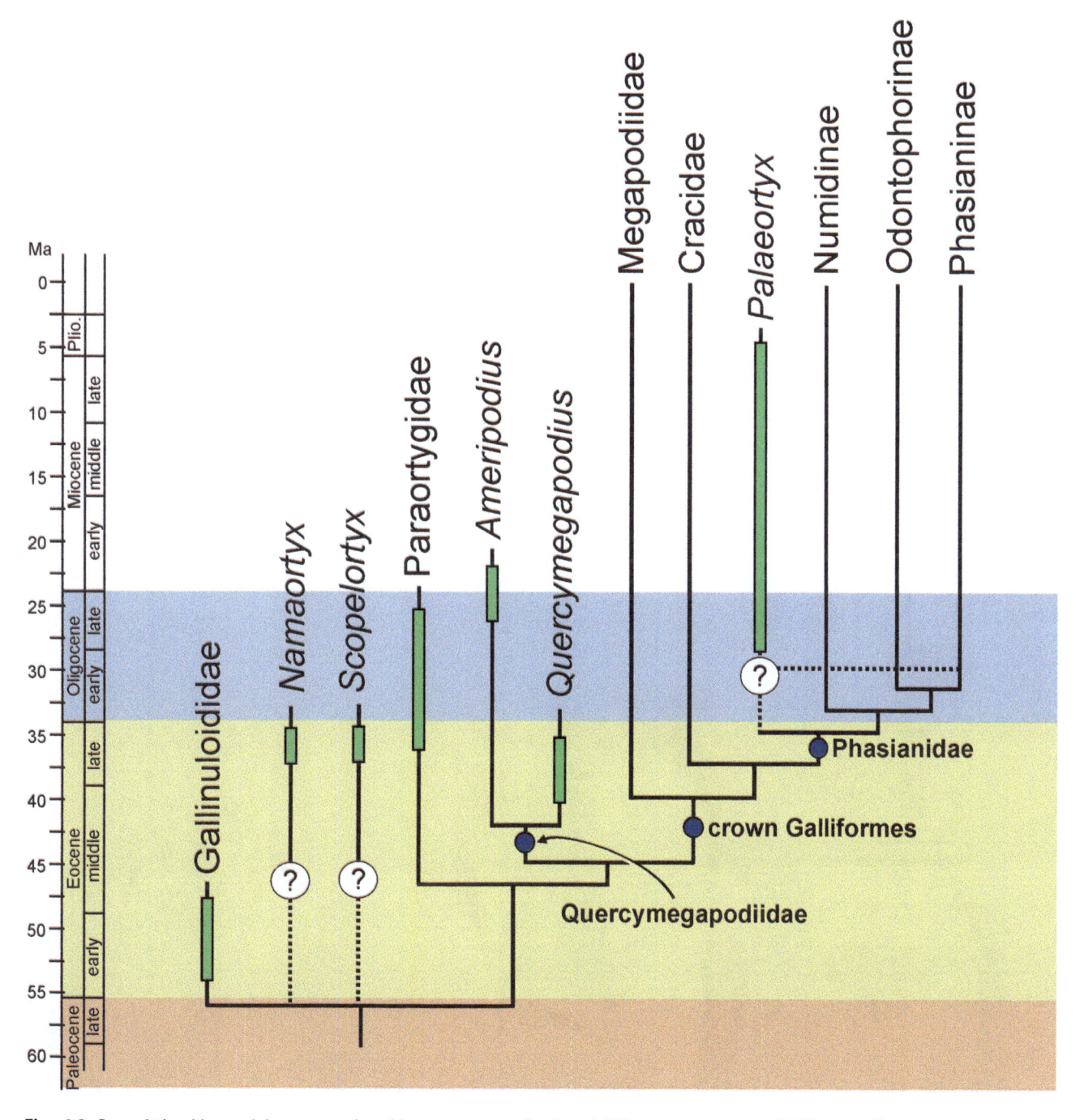

Fig. 4.8 Interrelationships and known stratigraphic occurrences of selected Paleogene and extant Galliformes. The divergence dates are hypothetical

found in the late early or early middle Eocene of Algeria (Garcia et al. 2020).

Small stem group galliforms of unresolved affinities are furthermore known from the early Eocene of Europe. The most substantial fossils come from the early Eocene of Egem in Belgium and include a number of diagnostic elements (Fig. 4.7c; Mayr and Smith 2019b). These fossils are clearly distinguished from the coeval Gallinuloididae, and the tarsometatarsus shows a resemblance to that of *Argillipes aurorum* from the London Clay, a species described by Harrison and Walker (1977) on the basis of very fragmentary material. Legs of a putative galliform bird were also described by Lindow and Dyke (2007) from the Danish Fur Formation.

4.3.3 The Paraortygidae—A Well-Represented Group of Stem Group Galliformes

A further taxon of Paleogene stem group Galliformes are the Paraortygidae. These birds are well-known from numerous well-preserved postcranial bones from middle Eocene to late Oligocene deposits of the Quercy fissure fillings, where three species have been recognized: the late Eocene and early Oligocene *Paraortyx brancoi* and *P. lorteti*, as well as *Pirortyx major*, which was found in late Oligocene deposits (Mourer-Chauviré 1992). *Paraortyx brancoi* and *Pirortyx major* were also reported from the early Oligocene of Germany (Fischer 1990, 2003). Specimens of *Paraortyx lorteti* and *P. brancoi* have furthermore been found in the early Oligocene of Belgium, from where also tentative records of *Pirortyx major* exist (Fig. 4.9; Mayr and Smith 2013).

Geographical and temporal distribution as well as morphology (carpometacarpus with bowed minor metacarpal but without intermetacarpal process) suggest that *Taoperdix pessieti* from the late Oligocene (MP 25) of France is a further representative of the Paraortygidae (Mayr and Weidig 2004). This species was originally assigned to the Gallinuloididae (Brodkorb 1964), but in size and morphology, the two known skeletons of *T. pessieti* (Milne-Edwards 1867–1871; Eastman 1905) correspond well to *Paraortyx brancoi*. The affinities of a distal tarsometatarsus from the late Eocene of the Quercy fissure fillings, which was identified as *Taoperdix* sp. by Mourer-Chauviré (1988, 1992), need to be restudied.

As in gallinuloidids but in contrast to the Quercymegapodiidae (Sect. 4.3.5), the humerus of *Paraortyx* exhibits a marked second pneumotricipital fossa. In the larger taxon *Pirortyx*, this fossa is less developed. Paraortygids differ from gallinuloidids and quercymegapodiids in the less elongated and less slender carpometacarpus, which more closely resembles the corresponding bone of extant Galliformes in its proportions. Unlike in the Quercymegapodiidae and extant Galliformes, there is no transverse ridge in the capital incision (incisura capitis) of the humerus. As in *Gallinuloides*, the sternum of *Paraortyx* lacks an internal spine (spina

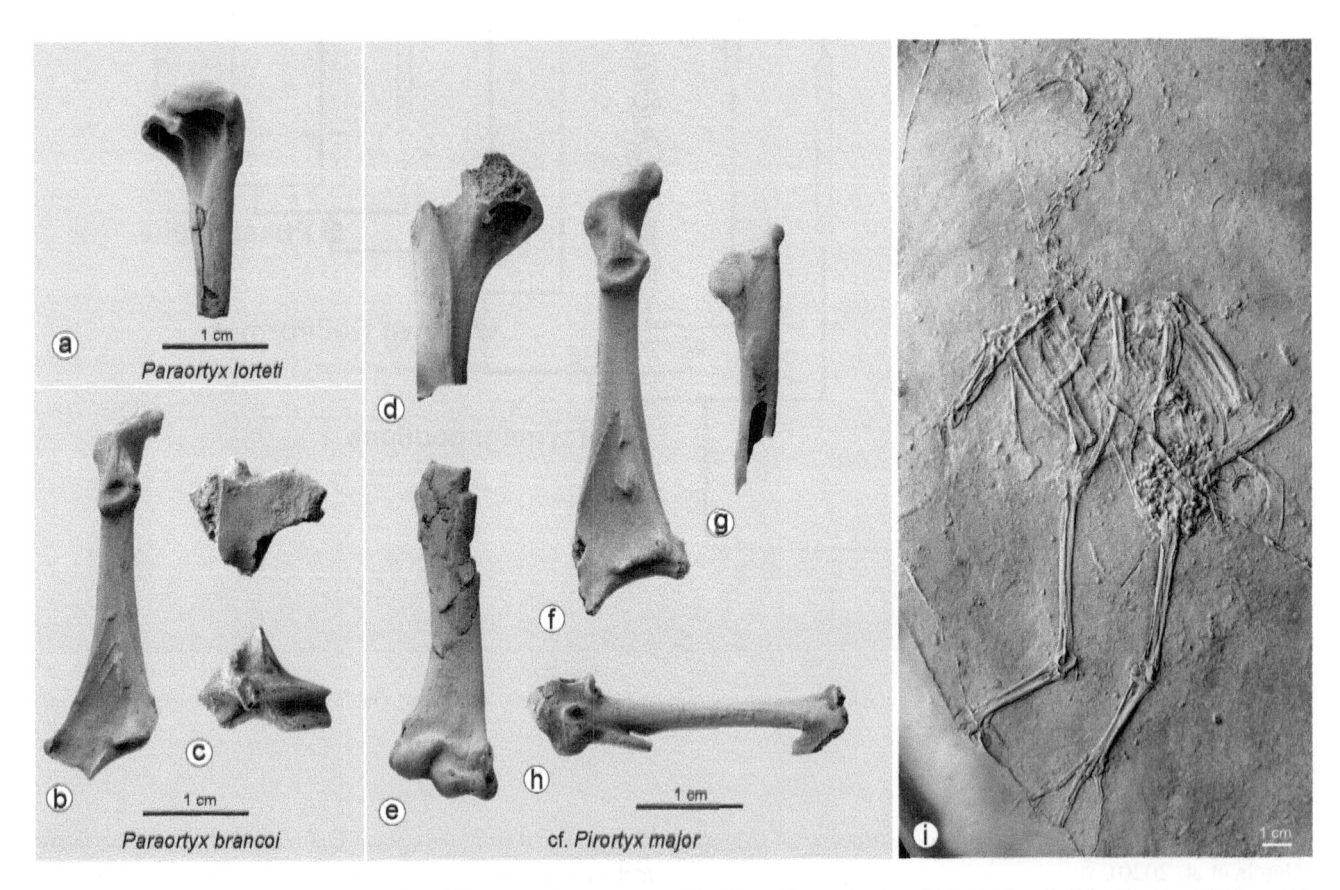

Fig. 4.9 Oligocene fossils of phasianid Galliformes. (**a**) *Paraortyx lorteti* from the early Oligocene Boom Formation in Belgium (Royal Belgian Institute of Natural Sciences, Brussels, Belgium, IRSNB Av 115). (**b**) Coracoid and (**c**) sternum (dorsal and cranial view) of *Paraortyx brancoi* from the early Oligocene Boom Formation in Belgium (**b**: IRSNB Av 116a; **c**: IRSNB Av 116b). (**d**)–(**h**) cf. *Pirortyx major* from the early Oligocene Boom Formation in Belgium: (**d**) proximal and (**e**) distal end of the humerus, (**f**) coracoid, (**g**) cranial end of the scapula, and (**h**) carpometacarpus (**d**: IRSNB Av 119, **e**: IRSNB Av 120, **f**: IRSNB Av 118a; **g**: IRSNB Av 118a, **h**: IRSNB Av 122). (**i**) Skeleton of *Palaeortyx* cf. *gallica* from the late Oligocene of Enspel in Germany with preserved gastroliths (Landesamt für Denkmalpflege Rheinland-Pfalz, Mainz, Germany, PW 2005/5023a-LS). (All photos by Sven Tränkner)

interna), which constitutes further evidence for a stem group position of the taxon and possibly suggests that it still lacked an enlarged crop (Mayr and Smith 2013).

In the past years, putative representatives of the Paraortygidae were also reported from fossil sites outside Europe. One of these is *Scopelortyx klinghardtensis* from the middle Eocene (Bartonian; 47–49 Ma) of Namibia (Mourer-Chauviré et al. 2015, 2017). This small species is well represented in the fossil material from the type locality, and most major limb bones are known. The humerus exhibits a well-developed second pneumotricipital fossa and the coracoid has a concave articular facet for the scapula. A coracoid of a similar stem group galliform, *Xorazmortyx turkestanensis*, was reported from the middle Eocene (Bartonian) of Uzbekistan by Zelenkov and Panteleyev (2019), who commented on potentially high dispersal capabilities of stem group galliforms. Another coracoid of a putative representative of the Paraortygidae was discovered in the middle Eocene of Utah (Stidham et al. 2020). The assignment of all of the aforementioned non-European fossils to the Paraortygidae is based on overall similarity, which may well be due to the retention of plesiomorphic features, and more data on the skeletal morphology of these birds are needed for a robust phylogenetic placement.

4.3.4 The North American Taxa *Procrax*, *Archaealectrornis*, and *Palaeonossax*

As first noted by Mayr (2009), the taxa *Procrax*, *Archaealectrornis*, and *Palaeonossax* from the late Eocene and early Oligocene of North America likewise show a resemblance to the species of the Paraortygidae. The holotype of *Procrax brevipes* is a dissociated postcranial skeleton from the late Eocene (Chadron Formation) of South Dakota (Tordoff and Macdonald 1957). The species was initially assigned to the Gallinuloididae, but the robust humerus of *Procrax* lacks a well-developed second pneumotricipital fossa, and there are also numerous other differences in the postcranial skeleton. As far as comparisons are possible, the skeletal morphology of *Procrax* resembles that of the Paraortygidae. *P. brevipes* was somewhat larger than the males of the extant Chaco Chachalaca, *Ortalis canicollis*, and differs from all extant Cracidae except *Pipile* in the proportionally shorter legs and ungual phalanges (Tordoff and Macdonald 1957). In contrast to the Gallinuloididae and Quercymegapodiidae, the minor metacarpal of the carpometacarpus is markedly bowed, and unlike in Paleogene Phasianoidea this bone lacks an intermetacarpal process.

P. brevipes is likely to be closely related to, if not conspecific with, the equally-sized *Archaealectrornis sibleyi*. The latter species is known from a humerus from the early Oligocene Brule Formation of Nebraska and was not differentiated from *P. brevipes* in the original description (Crowe and Short 1992). Based on morphometric comparisons, *A. sibleyi* was considered to be most closely related to the Phasianidae by Crowe and Short (1992). However, the absence of a transverse ridge in the capital incision clearly shows the species to be outside crown group Galliformes. Apart from being somewhat stouter, the humeri of *Procrax* and *Archaealectrornis* match well with that of the paraortygid species *Pirortyx major*.

Palaeonossax senectus from the early Oligocene Brule Formation of South Dakota was described as a member of the Cracidae by Wetmore (1956). This species is represented by a distal humerus and also needs to be compared with *Procrax* and *Archaealectrornis*, to which, judging from the published drawings, it appears to be quite similar. The Cracidae, therefore, do not have an unambiguously identified Paleogene record.

4.3.5 Quercymegapodiidae

The taxon Quercymegapodiidae was first introduced for two species from the late Eocene (MP 16–19) of the Quercy fissure fillings, that is, *Quercymegapodius depereti* and *Q. brodkorbi*. Another late Eocene species, *Ludiortyx hoffmanni* from the Paris Gypsum, is probably also a quercymegapodiid (Mayr 2005); possibly because of its slender carpometacarpus, a plesiomorphic feature shared with gallinuloidids, the species was classified into the Rallidae (rails) by Brunet (1970) and Cracraft (1973). The same is true for *Taubacrex granivora* from the late Oligocene/early Miocene of the Taubaté Basin in Brazil, which was likewise assigned to the Rallidae in the original description (Alvarenga 1988) but was identified as a quercymegapodiid galliform by Mourer-Chauviré (2000). Also from the Oligo-Miocene of the Taubaté Basin, Alvarenga (1995) described *Ameripodius silvasantosi* as a species of the Quercymegapodiidae. Mourer-Chauviré (2000) noted that *T. granivora* and *A. silvasantosi* differ in the shape of the coracoid, but this bone is slightly crushed in the holotype of *T. granivora*, and direct comparisons of the actual specimens are needed to exclude the possibility that the two species are conspecific. *Taubacrex* provides the earliest fossil record of gastroliths in galliform birds.

Their name and initial comparisons with megapodes (Mourer-Chauviré 1982) notwithstanding, quercymegapodiids are stem group representatives of the Galliformes, and the similarities to extant Megapodiidae are plesiomorphic (Mourer-Chauviré 1992). The coracoid of quercymegapodiids still exhibits a cup-like articular facet for the scapula. Besides other features (Mourer-Chauviré 1992; Mayr 2000), quercymegapodiids differ from gallinuloidids in the more robust humerus, which lacks a second pneumotricipital fossa. The presence of a transverse ridge in the capital groove (incisura capitis) of the humerus suggests that the Quercymegapodiidae are more closely related to crown group Galliformes than are the Gallinuloididae and Paraortygidae, which lack this ridge (Fig. 4.8).

Another species that may be a representative of the Quercymegapodiidae is *Sobniogallus albinojamrozi* from the early Oligocene of the Carpathian Basin in Poland, of which a partial skeleton has been found (Tomek et al. 2014). The cranial portion of the sternum of this species shows the derived morphology found in crown group Galliformes and exhibits well-developed internal and external spines.

4.3.6 The Australian Megapodiidae (Megapodes)

The sole Paleogene fossil record of the Megapodiidae is *Ngawupodius minya* from the late Oligocene of Central Australia (Namba Formation; Boles and Ivison 1999). The holotype of this species is a tarsometatarsus, which measures just two-thirds of that of the smallest extant megapodes, to which it is otherwise similar in its morphology. Apart from its larger size, this bone also resembles the megapode-like tarsometatarsus of the Quercymegapodiidae, but from a biogeographic point of view an assignment to the stem group of the Megapodiidae is certainly justified. Boles and Ivison (1999) assumed that *N. minya* lived in wet riparian forests and that its extinction was due to environmental changes.

4.3.7 Phasianoidea (Guinea Fowl, Quails, Pheasants, and Allies)

The Phasianoidea are the most species-rich and most widely distributed group of extant Galliformes and occur in a great variety of habitats, from semi-deserts to tropical forests and subantarctic tundra. Molecular studies indicate that the Numididae (guinea fowl) and Odontophoridae (New World quail) are successive sister taxa of the Phasianidae (grouse, quails, pheasants, and allies; Kriegs et al. 2007; Prum et al. 2015; Kuhl et al. 2021).

Amitabha urbsinterdictensis from the middle Eocene Bridger Formation of Wyoming is one of the oldest species that was considered to be a representative of crown group Phasianoidea (Gulas-Wroblewski and Wroblewski 2003). The description of this species was based on several fragmentary bones (including an incomplete humerus, scapula, sternum, and pelvis), which show little resemblance to galliform birds, let alone taxa of the Phasianoidea. Ksepka (2009) proposed rallid affinities of *A. urbsinterdictensis*, but these are likewise not well founded, and the morphology of the humerus rather suggests a relationship to the Anseriformes.

Comparisons with the purported anseriform *Eonessa anaticula* from the middle Eocene of Utah (Sect. 4.5.4) are therefore required to assess the phylogenetic position and taxonomic status of *A. urbsinterdictensis*.

Telecrex grangeri is represented by an incomplete femur from the late Eocene of Mongolia and was originally classified into the Rallidae, within the monotypic taxon Telecrecinae (Wetmore 1934). The species was assigned to the Numididae by Olson (1974), but because guinea fowl are the sister taxon of all other extant Phasianoidea (see above), the similarities between *Telecrex* and extant guineafowl may well be plesiomorphic for the Phasianoidea, and even an assignment to the Galliformes is anything but certain. An erroneous record of *Telecrex* from the late Eocene of France (Mlíkovský 1989) was based on a femur of *Elaphrocnemus phasianus*, which is a stem group representative of the Cariamiformes (Mourer-Chauviré 1992).

Putative Odontophoridae from late Eocene (Chadronian) deposits of Canada were described as *Nanortyx inexpectatus* by Weigel (1963). The assignment of these fossils, a fragmentary coracoid and a distal tarsometatarsus, to the Odontophoridae seems to have been based on the fact that the specimens are from very small galliform birds and were found in the New World. However, Mourer-Chauviré et al. (2015) noted that the coracoid of *N. inexpectatus* has a cup-like cotyla scapularis, which indicates that the species is outside crown group Galliformes. A distal tarsometatarsus from the early Oligocene (Orellan) of Colorado was assigned to the Odontophoridae by Tordoff (1951), but this identification also needs to be substantiated with additional specimens.

Unambiguous representatives of the Phasianoidea were reported from Oligocene deposits of the Quercy fissure fillings and other localities in France and belong to the taxon *Palaeortyx* (Mourer-Chauviré 1992; Mourer-Chauviré et al. 2004). The Quercy fossils can be assigned to four species: *P. brevipes*, *P. gallica*, *P. prisca*, and *P. phasianoides* (Mourer-Chauviré 2006). A nearly complete skeleton of *Palaeortyx* cf. *gallica* was found in the late Oligocene (MP 28) maar lake deposits of Enspel in Germany (Mayr et al. 2006). In this specimen (Fig. 4.9i), numerous gastroliths are preserved as stomach contents.

An assignment of *Palaeortyx* to the Phasianoidea is supported by the well-developed intermetacarpal process of the carpometacarpus. The humerus exhibits a double pneumotricipital fossa, which also occurs in the Gallinuloididae and Paraortygidae, extant Odontophoridae (New World quails), as well as in *Arborophila*, and *Ammoperdix* (both Phasianidae). If not plesiomorphic for galliform birds in general, this feature is likely to be plesiomorphic for the clade including the Odontophoridae and Phasianidae. The tarsometatarsus of *Palaeortyx* lacks a spur, a feature present in the males of several extant Phasianidae. Ballmann (1969) assumed a closer relationship between *Palaeortyx* and hill partridges of the taxon *Arborophila*, which live the tropical and subtropical regions of Asia. This hypothesis was based on overall similarity, and the plesiomorphic leg proportions (as in Paleogene stem group Galliformes, the femur is subequal to the humerus in length) rather support a position of *Palaeortyx* outside crown group Phasianoidea (Mayr et al. 2006). Zelenkov (2019) hypothesized that *Palaeortyx* is the sister taxon of the Phasianidae, but at least one of the derived characters shared by *Palaeortyx*, Odontophoridae, and Phasianidae—a large intermetacarpal process on the carpometacarpus—appears to have been secondarily reduced in the Numididae (Stegmann 1978).

Schaubortyx keltica from the late Oligocene (MP 25) of France was the size of a Northern Bobwhite (*Colinus virginianus*) and is known from a dissociated skeleton on two slabs (Eastman 1905; Schaub 1945). This species differs from *Palaeortyx* in its limb proportions, with the femur being longer than the humerus (Mourer-Chauviré 1992; contra Mlíkovský 2002 who synonymized *Schaubortyx* and *Palaeortyx*). *S. keltica* is a representative of the Phasianoidea, but its affinities within the group are unresolved.

The Meleagridinae (turkeys) and Tetraoninae (grouse), which according to molecular studies are sister taxa and nested within the Phasianidae (e.g., Kriegs et al. 2007), have no Paleogene fossil record and may not have diversified before the Neogene.

4.4 The Australian Dromornithidae

These large, flightless birds are a characteristic element of the Cenozoic avifauna of Australia. Dromornithids have a comparatively abundant Neogene fossil record (Murray and Vickers-Rich 2004), but much fewer fossils were described from Paleogene deposits.

A tentative record consisting of impressions of pedal phalanges comes from the early Eocene of Queensland, and putative dromornithid trackways were also reported from the late Oligocene of Tasmania (Vickers-Rich 1991). The first species that have been scientifically described are from late Oligocene deposits of the Riversleigh site. One of these, *Barawertornis tedfordi*, was about the size of the Southern Cassowary (*Casuarius casuarius*) and is the smallest known dromornithid, which appears to have been a forest-dwelling albeit cursorial bird (Nguyen et al. 2010). Another, much larger species from the late Oligocene of the Riversleigh site was described as *Dromornis murrayi* by Worthy et al. (2016). It is the earliest representative of the taxon *Dromornis*, which is well known from Miocene fossils sites.

With the possible exception of *Barawertornis*, dromornithids were graviportal birds, and in overall skeletal morphology, they are superficially similar to gastornithids.

Dromornithids are, however, clearly distinguished from gastornithids in numerous skeletal details, including the morphology of the bones of the palate (unlike in gastornithids, the palatine bone exhibits a marked ventral crest), the much more elongated coracoid, the absence of a hallux, and the fact that the fourth toe consists of just four phalanges. The skull of dromornithids features well-developed, Galloanseres-like articular facets for basipterygoid processes, and the mandible bears long retroarticular processes.

As detailed by Murray and Vickers-Rich (2004), there is convincing evidence for a herbivorous diet of dromornithids. Not only are some specimens preserved with gastroliths, but analyses of stable isotopes from the eggshells of a Pleistocene species also indicate plant matter as food components.

Although anseriform affinities of dromornithids are well based, their position within waterfowl and their affinities to the Gastornithidae have not yet been conclusively resolved. Murray and Vickers-Rich (2004) considered them to be the sister taxon of the Anhimidae, whereas they assumed that gastornithids are outside crown group Anseriformes. Even though a more recent analysis recovered a sister group relationship between Dromornithidae and Gastornithidae (Worthy et al. 2017), the similarities almost certainly are due to convergent evolution, because current paleogeographic reconstructions do not indicate dispersal routes for large flightless birds between Eurasian and Australia. From a morphological and biogeographic point of view, it appears more likely that the closest relatives of dromornithids are the Sylviornithidae from the Holocene of New Caledonia (and possibly Fiji, if *Megavitiornis* is also assigned to the taxon). These birds, which likewise have unusually massive beaks and a similar overall skeletal morphology to dromornithids, are currently considered stem group galliform birds (Worthy et al. 2017, but see Mayr 2011).

4.5 Anseriformes (Waterfowl)

Crown group Anseriformes include the South American Anhimidae (screamers), the Anseranatidae (magpie goose) of Australia and New Guinea, and the globally distributed Anatidae (swans, geese, and ducks). Morphological and molecular evidence supports a sister group relationship between the Anhimidae and the Anatoidea, that is, the clade (Anseranatidae + Anatidae) (e.g., Livezey 1997; Hackett et al. 2008; Prum et al. 2015).

The early evolution of the Anseriformes is poorly understood. As detailed in Sect. 2.4, remains of putative anseriforms are known from latest Cretaceous sites, even though the exact affinities of many of these fossils are controversial. A remarkably well-preserved partial skeleton from the early Paleocene (66–61 Ma) of the López de Bertodano Formation of Seymour Island in Antarctica was described as

Conflicto antarcticus by Tambussi et al. (2019). The holotype of this species includes a nearly complete skull and some of the major limb bones. The fossil exhibits galloanserine apomorphies, such as sessile basipterygoid processes and a mandible with a long retroarticular process. The long beak is only incompletely preserved but appears to have had similar proportions to that of the Late Cretaceous *Polarornis* and extant mergansers (*Mergus* spp.), which may indicate a piscivorous diet. In the original analysis (Tambussi et al. 2019), *C. antarcticus* resulted as the sister taxon of the early Eocene *Anatalavis oxfordi*, and the clade including both taxa was placed outside crown group Anseriformes; the same topology was found by Field et al. (2020: fig. 9). However, Tambussi et al. (2019) detailed that a sister group relationship between *Conflicto* and *Anatalavis* received weak statistical support and most bones of both taxa differ to such an extent that close affinities are unlikely. Not only is the beak of *Anatalavis* wide and "duck-like," but the temporal fossae for the jaw muscles are much deeper, the humerus is stouter than that of *Conflicto*, the furcula has a broader sternal extremity, and unlike in *Conflicto* the coracoid of *Anatalavis* has a foramen for the supracoracoideus nerve.

The skeletal morphology of *Conflicto* resembles that of the Vegaviidae (*Vegavis* and *Polarornis*). However, *Conflicto* does not have a short and strongly bowed femur, and whereas the species of the Vegaviidae exhibit swimming or diving adaptations, the long legs of *Conflicto* indicate a more terrestrial or wading habit. Close relationships between *Conflicto* and the Vegaviidae were not supported in the analyses of Tambussi et al. (2019) and Field et al. (2020), but the affinities of the Vegaviidae were not congruently resolved in these studies, and there remains a possibility that *Conflicto* is a vegaviid, which was less specialized for swimming or diving than *Vegavis* and *Polarornis*.

4.5.1 Anhimidae (Screamers)

The three extant species of the Anhimidae are endemic to South America. These fairly large birds exhibit a characteristic skeletal morphology and are distinguished from other Anseriformes in their galliform-like bill morphology, the incomplete webbing of the feet, and the highly pneumatized bones.

Elzanowski and Boles (2012) described a fragmentary quadrate from the early Eocene Tingamarra Local Fauna of Australia, which they regarded to be similar to the Anhimidae. They considered it possible that this bone belongs to the same species as a presbyornithid-like coracoid reported by Boles (1999) from the same locality. Initially, this coracoid and other bones were likened to the charadriiform "Graculavidae" (Boles 1999), but Mayr (2009) noted that a humerus from this material is similar to that of the Juncitarsidae, which belong to Mirandornithes (the clade including Phoenicopteriformes and Podicipediformes).

The oldest unambiguously identified fossils of the Anhimidae come from the late Oligocene/early Miocene of the Taubaté Basin in Brazil and belong to *Chaunoides antiquus* (Alvarenga 1999). This species is known from several isolated postcranial bones and was smaller than the smallest extant species of screamers, the Northern Screamer *Chauna chavaria*, and had a less pneumatized skeleton. According to Alvarenga (1999), *Loxornis clivus* from the late Oligocene (Deseadan) of Argentina, which is solely represented by a distal tibiotarsus, may also belong to the Anhimidae.

Ericson (1997), Olson (1999b), and Feduccia (1980) mentioned unpublished specimens of early Eocene anhimid-like birds from the Willwood Formation (Wyoming). Presumably closely related birds were also collected by Michael Daniels in the early Eocene London Clay of Walton-on-the-Naze and were listed as screamers by Feduccia (1980: Table 4.1). These remarkably well-preserved fossils are here for the first time figured (Fig. 4.10f, g). Features that support their assignment to the Galloanseres include a long retroarticular process on the mandible and a derived Galloanseres-like mandibular articulation for the quadrate, which has only two ventral condyles. The beak is quite short and narrow, with a rounded tip, and has similar proportions to the beak of extant Anhimidae. As in the latter and other anseriform birds, the corpus of the thoracic vertebrae exhibits large pneumatic openings. These foramina are absent in the Galliformes, from which the fossil furthermore differs in that the coracoid exhibits a foramen for the supracoracoideus nerve. As in extant screamers, the tarsometatarsus is fairly short and the hypotarsus forms a wide but shallow sulcus. The pectoral girdle and limb bones, however, show some distinct differences to the corresponding elements of extant Anhimidae, in which, for example, the extensor process of the carpometacarpus forms a marked spur. The exact phylogenetic affinities of the "anhimid-like" birds from the London Clay still need to be determined, but here it is noted that the beak also shows a resemblance to that of the recently described *Asteriornis* from the latest Cretaceous of Belgium (Field et al. 2020; Sect. 2.4).

Another species with possible anhimid affinities is *Perplexicervix microcephalon* from Messel (Fig. 4.10a–d; Mayr 2007b, 2010). This species, which was about the size of a crow (*Corvus* sp.), is characterized by a notably small skull, a short and narrow beak, as well as fairly long wings. There are distinct neurovascular furrows on the dorsal surface of the cranium, and the skull has basipterygoid processes that are similar to those of extant Anhimidae. *P. microcephalon* is known from several specimens, which exhibit a distinctive morphology in that the surface of the cervical vertebrae bears numerous small tubercles. Such tubercles also occur on the holotype skeleton of the putative cariamiform *Dynamopterus*

4.5 Anseriformes (Waterfowl)

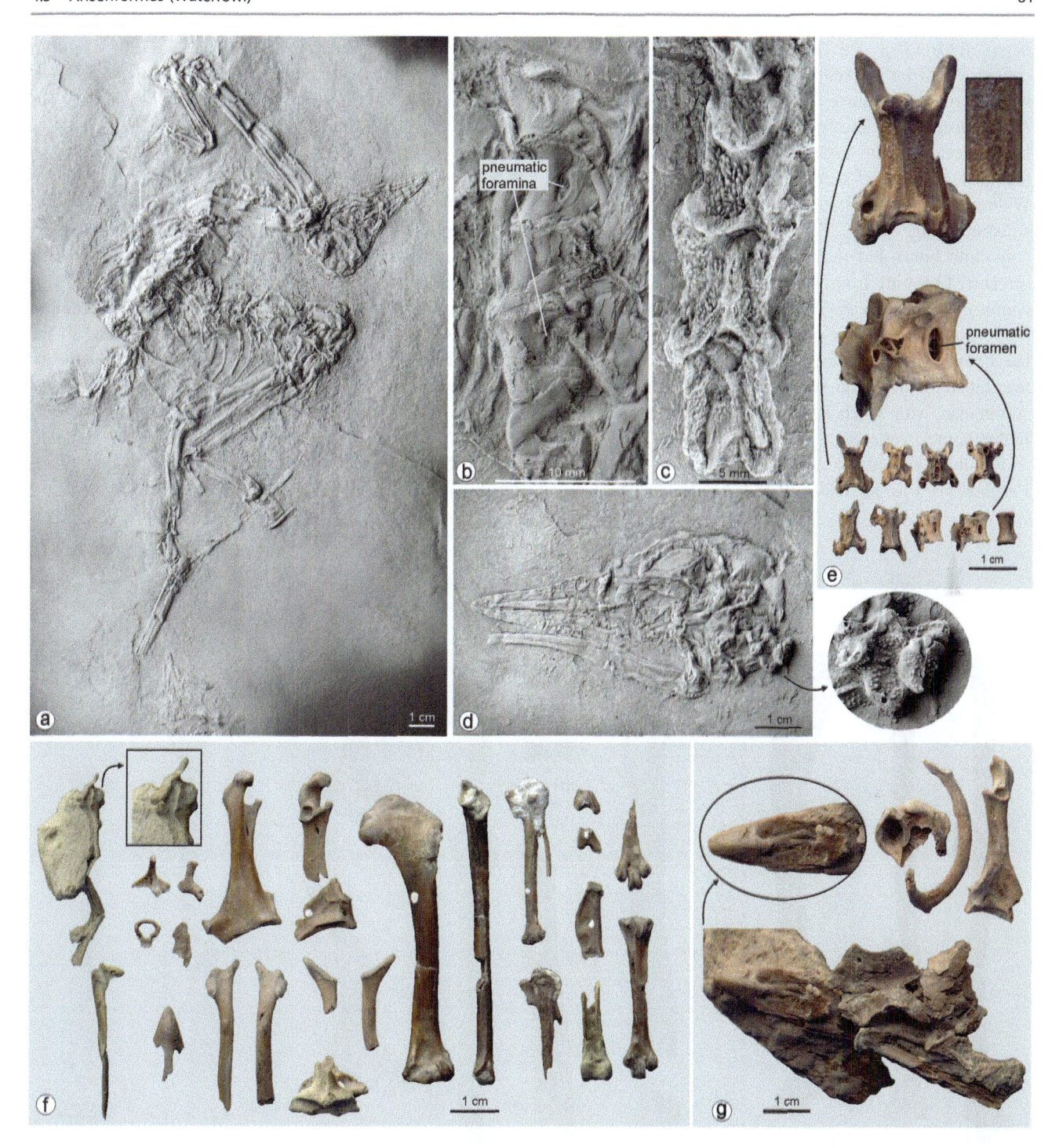

Fig. 4.10 Early Eocene anhimid-like birds with tubercles on the cervical vertebrae. (**a**) Holotype of *Perplexicervix microcephalon* from the latest early or earliest middle Eocene of Messel (Senckenberg Research Institute, Frankfurt, SMF-ME 11211a). (**b**) *P. microcephalon*, detail of thoracic vertebrae with large pneumatic openings in lateral surface of corpus (SMF-ME 2559a). (**c**) *P. microcephalon*, detail of the cervical vertebrae of the holotype. (**d**) *P. microcephalon*, skull in ventral view (SMF-ME 3548); the arrow indicates an enlarged detail of the tubercles on the cervical vertebrae. (**e**) Vertebrae from the early Eocene London Clay of Walton-on-the-Naze, which show tubercles similar to those of *Perplexicervix* (collection of Michael Daniels, Holland-on-Sea, UK, WN 82405); the arrows indicate enlarged details of the large pneumatic opening in the corpus of the thoracic vertebra and the tubercles on the surface of the cervical vertebra. (**f**), (**g**) undescribed anhimid-like birds from Walton-on-the-Naze (**f**: WN 80300, **g**: WN 85510); the arrows indicate an enlarged detail of the long retroarticular process on the caudal end of the mandible in (**f**) and the beak with the surrounding matrix digitally removed in (**g**). (All photos by Sven Tränkner)

tuberculatus (Sect. 8.4.2). These unusual structures were considered to be of pathological origin by Mayr (2007b), but their occurrence in multiple specimens of *P. microcephalon* indicates that they represent true morphological features (Mayr 2010). This is also suggested by the fact that a series of vertebrae from the London Clay of Walton-on-the-Naze in the Daniels collection shows similar structures (Fig. 4.10e). These vertebrae are not associated with limb bones but may belong to the above-mentioned screamer-like bird from the London Clay (Daniels, pers. comm.) from which *Perplexicervix* is distinguished in the proportionally longer tarsometatarsus. With regard to possible anhimid affinities of *Perplexicervix*, it is notable that a few tubercles occur on some cervical vertebrae of extant Anhimidae, with which *Perplexicervix* also shares openings in the thoracic vertebrae.

Possibly related to the London Clay fossils and *Perplexicervix* is an undescribed screamer-like bird from the Green River Formation (Grande 2013: 120). Unfortunately, no cervical vertebrae are preserved in the single skeleton known of this bird.

4.5.2 Presbyornithidae

The Presbyornithidae are fairly abundant in some early Paleogene fossil localities, but the taxonomic history and classification of these birds are quite convoluted. The first specimens, a few isolated bones from the early Eocene of Utah described by Wetmore (1926), were assigned to the charadriiform Recurvirostridae (stilts and avocets: *Presbyornis*) and Alcidae (auks: "*Nautilornis*"; Fig. 4.11b). Howard (1955) subsequently described bones from the early Eocene (Casamayoran) of Patagonia as *Telmabates antiquus* (Fig. 4.11c, d). She tentatively referred this species to the Phoenicopteriformes but also made comparisons with the

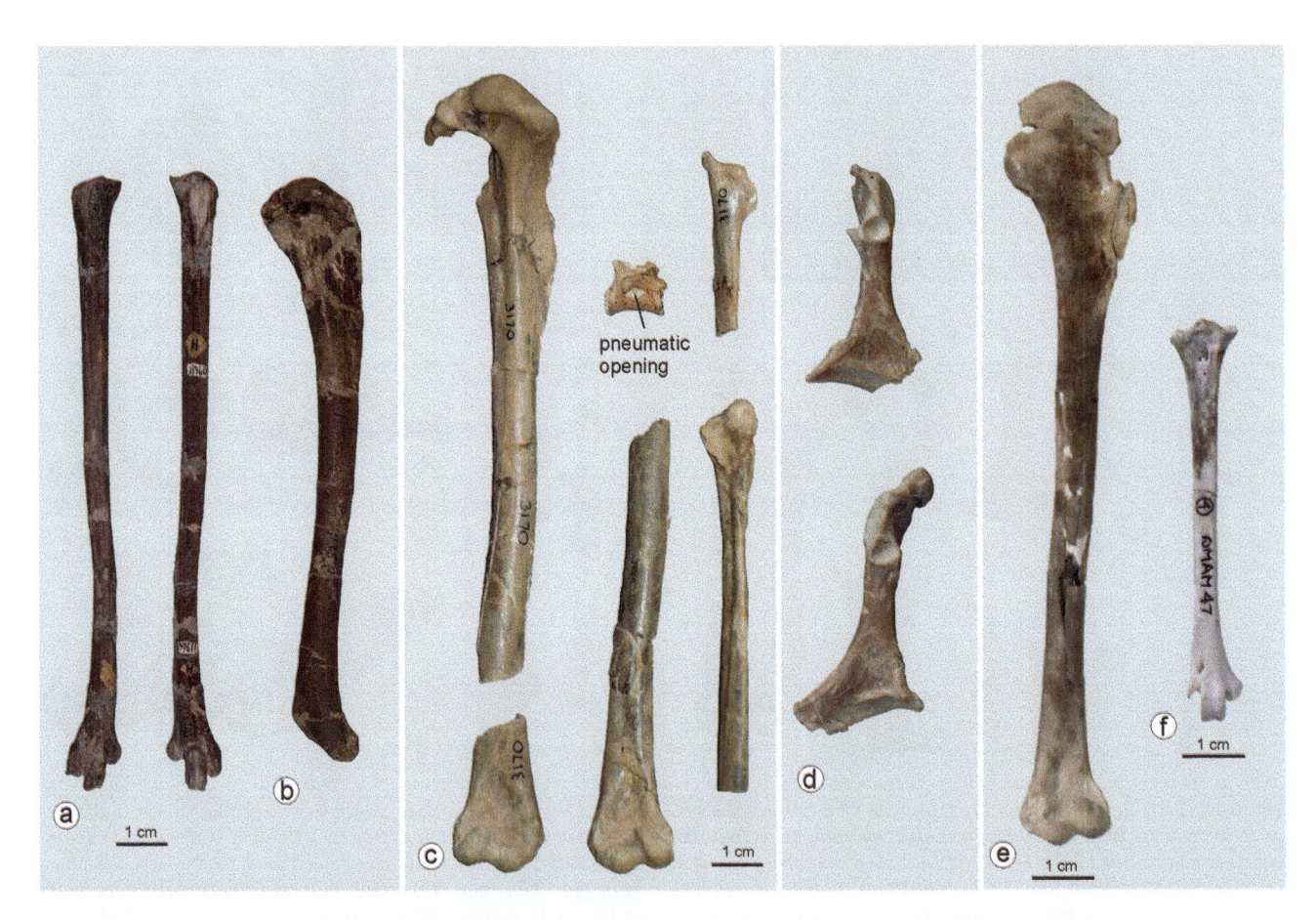

Fig. 4.11 Fossils of the anseriform Presbyornithidae. (**a**) Tarsometatarsus of *Presbyornis pervetus* from the early Eocene Green River Formation of Utah, USA, in plantar and dorsal view (holotype, Carnegie Museum, Pittsburgh, USA, CMNH 11360). (**b**) Humerus of *P. pervetus* from the early Eocene Green River Formation of Utah, USA (holotype of "*Nautilornis proavitus*"; CMNH 11358). (**c**) Limb and pectoral girdle bones as well as a thoracic vertebra of *Telmabates antiquus* from the early Eocene of Argentina (holotype, American Museum of Natural History, New York, USA, AMNH 3170). (**d**) Right and left coracoid of *T. antiquus* from the early Eocene of Argentina (AMNH 3181). (**e**) Humerus and (**f**) tarsometatarsus of *Wilaru tedfordi* from the late Oligocene of Australia (e: holotype, South Australian Museum, Adelaide, SAM P.48925, f: American Museum of Natural History, New York, USA, AMNH 11413). ((**e**) and (**f**) from De Pietri et al. (2016) with lettering digitally removed, published under a Creative Commons CC BY 4.0 license; all other images by the author)

Charadriiformes and Mesozoic stem group representatives of the Neornithes. It was not until Feduccia and McGrew (1974) reported new material of *P. pervetus* from the early Eocene of Wyoming that *Presbyornis* and *Telmabates* were classified in the same higher-level taxon.

Ericson (2000) revised the then known New World fossil record of the Presbyornithidae and confirmed the validity of four species: the large Paleocene *Presbyornis isoni* (Olson 1994; Benson 1999), as well as the early Eocene *P. pervetus*, *P. recurvirostra*, and *Telmabates antiquus*. Except for the latter, all of these species occur in North American sites. *T. howardae* from the early Eocene (Casamayoran) of Argentina was excluded from the Presbyornithidae by Ericson (2000), who regarded the affinities of this poorly known species uncertain.

Since Ericson's (2000) review, the fossil record of presbyornithids outside the Americas has been much improved. Wing bones of putative Presbyornithidae from the Late Cretaceous of Mongolia were described as *Teviornis gobiensis* by Kurochkin et al. (2002); De Pietri et al. (2016) confirmed the correct identification of this species, which was considered doubtful by Clarke and Norell (2004). A large collection of *Presbyornis* bones from the late Paleocene and early Eocene of Mongolia was described by Kurochkin and Dyke (2010), who named a new species, *Presbyornis mongoliensis*, and mentioned the occurrence of further unnamed species. The Mongolian prebyornithid material was revised by Zelenkov (2021b), who noted that it consists of different species and that the holotype of *P. mongolicus* may actually belong to a *Juncitarsus*-like representative of the Mirandornithes, the clade including the Phoenicopteriformes and Podicipediformes. Zelenkov (2021b) described a small presbyornithid species from the Mongolian material as *Bumbalavis anatoides*, but further studies are necessary to untangle the exact taxonomic composition and phylogenetic affinities of the alleged Mongolian presbyornithids. Undetermined fossils of putative presbyornithids were also reported from the late early or early middle Eocene of Algeria (Garcia et al. 2020) and from the early Eocene (52–53 Ma) of Ellesmere Island (Stidham and Eberle 2016).

Wilaru tedfordi from the late Oligocene of Australia (Fig. 4.11e, f), which was initially assigned to the charadriiform Burhinidae (Boles et al. 2013), was identified as a representative of the Presbyornithidae by De Pietri et al. (2016). Unlike in *Presbyornis*, however, the tarsometatarsus of *W. tedfordi* is not strongly elongated but rather short; hypotarsus morphology and the configuration of the tarsometatarsal trochleae are also different from presbyornithids and other anseriform birds. The carpometacarpus of *W. tedfordi* exhibits a spur-like extensor process, which is absent in *Presbyornis and Telmabates* and indicates that *Wilaru* used its wings in intraspecific combats.

The short legs suggest a more terrestrial way of living rather than a wading behavior. As long as the skull of *Wilaru* is unknown, there remains a possibility that the similarities to presbyornithids are plesiomorphic. Oddly enough, there are no unambiguously identified fossils of presbyornithids from the Paleogene of Europe, with the presumed presbyornithid affinities of *Proherodius oweni*, which is represented by a poorly preserved sternum from the early Eocene London Clay, not being well based (Olson 1985; contra Harrison and Walker 1978).

Even though the best-represented presbyornithid species, *P. pervetus* (Fig. 4.11a, b) and *P. mongoliensis*, are known from numerous bones, no articulated skeletons of presbyornithids have as yet been found. *P. pervetus* combines a duck-like skull with very long legs, and its postcranial skeleton is quite different from that of extant Anseriformes. Most likely, the very long legs of presbyornithids are an autapomorphic feature, and certain peculiarities in which presbyornithids differ from other Anseriformes may be related therewith. The caudal margin of the sternum, for example, exhibits two pairs of incisions, whereas there is only a single pair in all extant Anseriformes. Also unlike in extant Anseriformes, the pneumotricipital fossa of the humerus is very shallow and lacks pneumatic foramina, and in the pelvic girdle the ilia are not co-ossified with the synsacrum. As evidenced by *T. antiquus*, the caudalmost thoracic vertebrae of presbyornithids are furthermore opisthocoelous (Howard 1955), whereas they are heterocoelous in extant Anseriformes and most other neornithine birds. The thoracic vertebrae of *Telmabates* exhibit large pneumatic openings in the lateral surfaces of the vertebral body (Fig. 4.11c).

As yet no detailed descriptions of the skull of presbyornithids have been published. However, it was noted that *Presbyornis* differs from crown group Anseriformes in proportionally shorter postorbital processes and in the well-developed temporal fossae (Olson and Feduccia 1980), and the quadrate is distinguished from that of crown group Anseriformes in various plesiomorphic characteristics (Elzanowski and Stidham 2010; Elzanowski 2013). Even though the spatulate beak of *Presbyornis* indicates some capability for filter feeding, Zelenkov and Stidham (2018) detailed that the taxon was not able to filter very small particles. This hypothesis conflicts with earlier assumptions that presbyornithids nourished on planktonic organisms in shallow, saline lakes (Olson and Feduccia 1980; Feduccia 1980).

Judging from mass mortality layers with thousands of bones, mainly at Lake Gosiute of the Green River Formation, *Presbyornis* appears to have been highly gregarious (Olson 1985; Ericson 2000). Trackways from the Green River Formation of eastern Utah, which have been allocated to *P. pervetus* by earlier authors, were described as

Presbyorniformipes feduccii [sic] by Yang et al. (1995). If correctly identified, they provide evidence not only for fully webbed anterior toes and a long hallux in *Presbyornis*, but, through the presence of dabble marks, also inform the feeding behavior of this species. Eggshells of *Presbyornis* were reported by Leggitt and Buchheim (1997).

Before the skull was discovered, presbyornithids were regarded to be most closely related to either charadriiform birds (e.g., Wetmore 1926) or the Phoenicopteriformes (Howard 1955; Feduccia and McGrew 1974; Feduccia 1976). Feduccia (1976) hypothesized that the skeletal morphology of presbyornithids constitutes evidence for a close relationship between the Phoenicopteriformes (flamingos) and Charadriiformes (shorebirds). Anseriform affinities of presbyornithids were first assumed by Harrison and Walker (1976b, 1979) and became widely recognized through the publication of Olson and Feduccia (1980), who cited the mosaic character distribution displayed by *Presbyornis* in order to establish a charadriiform ancestry of the Anseriformes.

Olson and Feduccia (1980: 22) considered presbyornithids to represent "a charadriiform grade of morphology," and Olson (1985) listed them as "transitional shorebirds." By contrast, the Presbyornithidae resulted as the sister taxon of the Anatidae in analyses by Ericson (1997) and Livezey (1997). Ericson (1997) identified a single character as a possible apomorphy of a clade including the Presbyornithidae and Anatidae, that is, a pneumatic foramen on the medial process of the mandible. Livezey (1997) stated that fully webbed anterior toes, which can be inferred for *Presbyornis* from the above-mentioned trackways, are a synapomorphy of the Presbyornithidae and Anatidae. In current analyses, presbyornithids are recovered as either the sister taxon of the clade (Anseranatidae + Anatidae) (Worthy et al. 2017) or as the sister taxon of the Anseranatidae (Field et al. 2020).

Even though no consensus, therefore, exists on the exact affinities of presbyornithids, they are recovered as part of the Anatoidea in all analyses. Because the Anseranatidae also have comparatively long legs, the short legs of most extant Anatidae are likely to be a derived characteristic of geese, ducks, and allies. If so, anseriform birds may have evolved their filter-feeding capabilities in shallow waters, and the swimming habit of most extant short-legged Anatidae evolved comparatively late in the evolution of anseriform birds (see below).

4.5.3 Anseranatidae (Magpie Geese)

The Australo-Papuan Magpie Goose (*Anseranas semipalmata*) is the only extant member of the Anseranatidae. Putative stem group representatives of this taxon were reported from fossil sites in North America and Europe. The best represented species belong to the taxon *Anatalavis*, which was originally introduced for *A. rex* from the Late Cretaceous/early Paleocene Hornerstown Formation of New Jersey (Olson and Parris 1987). The holotype of *A. rex* is the distal portion of a humerus, but another species, *Anatalavis oxfordi* from the early Eocene London Clay of Walton-on-the-Naze, is known from a well preserved partial skeleton, which includes elements of the wing and pectoral girdle, as well as the skull and portions of the sternum and pelvis (Fig. 4.12a; Olson 1999b). Even though it is rather unlikely that the goose-sized *A. oxfordi* belongs to the same "genus" as a North American species, which lived some 10 million years earlier, the fossil material does not allow an unambiguous separation of the two species. Mlíkovský's (2002) introduction of the new taxon *Nettapterornis* for *A. oxfordi* may be correct, but currently, it is not well founded (Mourer-Chauviré 2004; Mayr 2005).

A large pneumatic foramen on the sternal end of the coracoid as well as a very wide sternal extremity of the furcula (Olson 1999b) are derived similarities of *Anatalavis* and *Anseranas*, but the two taxa otherwise have a quite different skeletal morphology. The humerus of *A. oxfordi* is proportionally shorter and more robust than that of crown group Anseriformes, and Olson (1999b) hypothesized that the species was capable of a powerful and rapid flight.

The plesiomorphic presence of a foramen for the supracoracoideus nerve (coracoid) shows *Anatalavis* to be outside crown group Anatidae. Still, the beak is very duck-like and *A. oxfordi* probably was a filter-feeder, whereas extant Anseranatidae have a goose-like beak and are macrofeeders (Olson 1999b). In contrast to all extant Anseriformes, the temporal fossae of *A. oxfordi* are well developed. The shape of the pterygoid also differs from that of crown group Anseriformes, which led Olson (1999b) to assume that the skull of *A. oxfordi* departed from that of extant waterfowl in functional aspects. *A. oxfordi* is furthermore distinguished from all crown group Anseriformes in the short and blunt postorbital process (skull) and in the short and broad preacetabular portion of the pelvis (Olson 1999b).

Olson's (1999b) classification of *A. oxfordi* in the Anseranatidae was challenged by Dyke (2001), whose analysis supported a sister group relationship between *Anatalavis* and a clade including the Presbyornithidae and Anatidae. However, this analysis did not include all potential synapomorphies of *Anseranas* and *A. oxfordi* listed by Olson (1999b), and a revision and reanalysis of the data supported a sister group relationship between *A. oxfordi* and *Anseranas* (Mayr 2008b). By contrast, *A. oxfordi* was recovered as a stem group anseriform in more recent analyses by Tambussi et al. (2019) and Field et al. (2020).

Here it is noted that there is a tarsometatarsus of a juvenile anseriform bird from the London Clay of Walton-on-the-

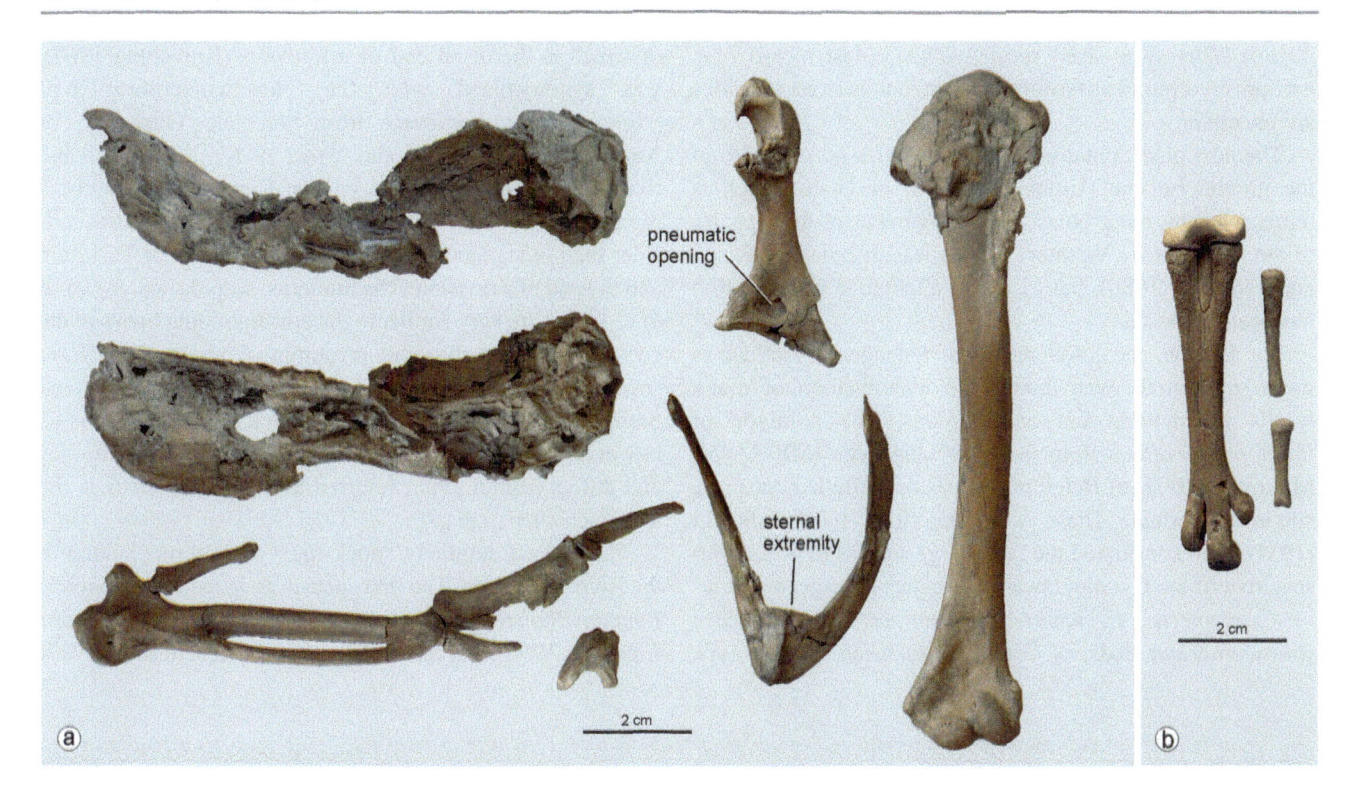

Fig. 4.12 The early Eocene anseriform bird *Anatalavis oxfordi*. (**a**) Selected elements (skull, coracoid, furcula, humerus, and distal wing elements) of the holotype from the early Eocene London Clay of Walton-on-the-Naze (Natural History Museum, London, NHMUK A 5922); the skull is shown in two different views. (**b**) Tarsometatarsus and pedal phalanges of a juvenile specimen from Walton-on-the-Naze (collection of Michael Daniels, Holland-on-Sea, UK, WN 86520). (All photos by Sven Tränkner)

Naze in the collection of Michael Daniels, which is likely to belong to *A. oxfordi*. This fossil (Fig. 4.12b) is very similar to the tarsometatarsus of the Anatidae in its proportions and the short trochlea for the second toe. Together with the duck-like beak, it may indicate a sister group relationship between *A. oxfordi* and the Anatidae, and this so much the more as all of the presumed anseranatid characteristics are rather weak characters. A pneumatic opening on the coracoid, for example, also occurs in the Anhimidae and some galliform birds and is, therefore, either plesiomorphic for the Anseriformes or evolved several times independently within the Galloanseres. The sternal extremity of the furcula is furthermore a thin sheet in *A. oxfordi* but thickened and pneumatized in *Anseranas*. However, and as noted above, *A. oxfordi* exhibits several plesiomorphic traits by which it is distinguished from all crown group Anseriformes. Irrespective of its exact position, the species, therefore, demonstrates a high degree of homoplasy in the evolution of anseriform birds.

Another putative anseranatid, *Anserpica kiliani*, was described from the late Oligocene of France by Mourer-Chauviré et al. (2004). This species is just known from a coracoid, which is very similar to that of *Geranopsis hastingsiae* from the late Eocene and early Oligocene of England (Lydekker 1891), a presumptive representative of the Gruoidea, with which *A. kiliani* was not compared (see Sect. 5.3.5; Mayr 2005).

The earliest Australian stem group representative of the Anseranatidae was reported by Worthy and Scanlon (2009) from the late Oligocene/early Miocene of the Riversleigh site. This species, *Eoanseranas handae*, is represented by a coracoid and referred scapulae. It mainly differs from the extant *A. semipalmata* by its smaller size and the presence of pneumatic openings in the acrocoracoid process.

4.5.4 Anatidae (Ducks, Geese, and Swans)

Most extant anseriform species belong to the Anatidae, which are widespread in all kinds of aquatic and semi-aquatic environments. Anatids typically are short-legged birds. The crown group representatives are characterized by the reduction of the coracoidal foramen for the supracoracoideus nerve, which in some extant representatives of early diverging lineages still forms a notch in the medial margin of the bone.

The Paleogene fossil record of duck- or goose-like birds is rather scant. *Naranbulagornis khun* from the latest Paleocene of Mongolia was described on the basis of a partial carpometacarpus and a referred partial femur (Zelenkov

2018b). This swan-sized species is the oldest record of a non-presbyornithid anseriform in Asia, but its exact affinities are uncertain.

The next oldest putative anatid is *Eonessa anaticula* from the middle Eocene (Uinta Formation) of Utah, which is represented by wing bones. The assignment of this species to the Anatidae by Wetmore (1938) was disputed by Olson and Feduccia (1980), but as yet no alternative classification has been proposed.

For Europe, the fossil record of Paleogene Anatidae is more substantial, even though the identification of many fossils is anything but unambiguous. The holotype of *Cygnopterus affinis* from the early Oligocene (MP 23–24; Mlíkovský 2002) of Belgium consists of multiple bones of a single individual. These were described by Lambrecht (1931), who considered the fairly large species to be a representative of the Cygnini (swans). Another Paleogene species of *Cygnopterus*, *C. lambrechti* from the late Oligocene (*Indricotherium* Beds) of Central Kazakhstan, the holotype of which is the distal end of a humerus (Kurochkin 1968), was synonymized with the phoenicopteriform bird "*Agnopterus*" *turgaiensis* from the late Oligocene of Kazakhstan (Mlíkovský and Švec 1986). Mlíkovský and Švec (1986: 266) hypothesized that *Cygnopterus* may be a representative of the Phoenicopteriformes, and I also consider the phylogenetic affinities of the taxon to be uncertain and in need of a revision. The humerus, scapula, and coracoid of *C. affinis* are very similar to the corresponding bones of the coeval and similarly-sized presumptive phoenicopteriform species ?*Agnopterus hantoniensis* (Sect. 5.1; Mayr and Smith 2002; Mayr 2005, 2008b). The femur of *C. affinis* is, however, more slender than that of the Phoenicopteriformes and differs from a femur referred to ?*A. hantoniensis* in the low trochanteric crest.

The earliest definitely "duck-like" anseriforms belong to the Romainvilliinae. The first species assigned to this taxon is *Romainvillia stehlini* from the late Eocene (MP 20; Mourer-Chauviré 1996) of France (Fig. 4.13a). Lebedinsky (1927)

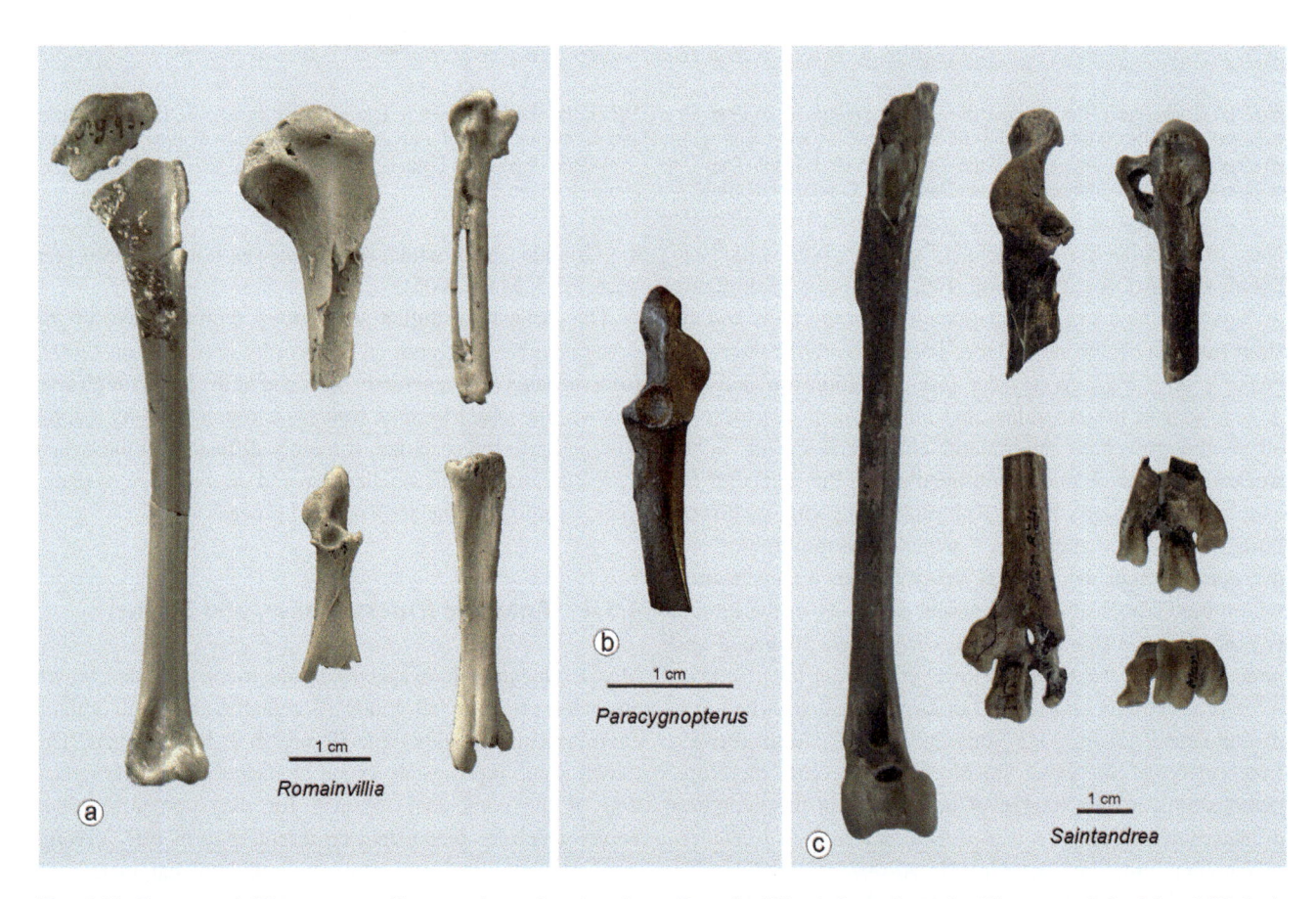

Fig. 4.13 Eocene and Oligocene anseriform species assigned to the Romainvilliinae. (**a**) Skeletal elements of *Romainvillia stehlini* from the late Eocene of Romainville in France (Naturhistorisches Museum Basel, Switzerland; left humerus: NMB P.G.77 and NMB P.G.93, proximal right humerus: NMB P.G.53.c, left carpometacarpus: NMB P.G.53, right tarsometatarsus: NMB P.G.25.l, left coracoid: NMB P.G.38.a, lectotype). (**b**) The holotype coracoid of *Paracygnopterus scotti* (Romainvilliinae) from the early Oligocene of the Isle of Wight in England (Natural History Museum, London, NHMUK A 4404). (**c**) Selected elements (tibiotarsus, partial coracoid, carpometacarpus, and tarsometatarsi) of the holotype of *Saintandrea chenoides* from the late Oligocene of southern France (NMB Mar. 874). (All photos by Sven Tränkner)

based the description of this species on a coracoid, furcula, carpometacarpus, and tarsometatarsus, but the material from the type locality also includes humeri that were not mentioned in the original description (Mayr 2008b). *R. stehlini* was a small species, the size of the extant Silver Teal (*Anas versicolor*). The coracoid is distinguished from that of crown group Anatidae in that it still exhibits a small foramen for the supracoracoideus nerve and in that the articular facet for the furcula is ovate, essentially flat, and in line with the medial margin of the shaft of the coracoid. The tip of the omal extremity of the furcula differs from crown group Anatidae in being wider and blunt. The elongate humerus is similar to that of the Presbyornithidae; as in the latter the pneumotricipital fossa is very shallow and lacks pneumatic foramina. The short tarsometatarsus of *R. stehlini* is distinguished from crown group Anatidae in the shape of the narrower trochlea for the second toe. The differences in the shape of the tarsometatarsal trochlea may suggest that *Romainvillia* was less adapted to a swimming way of living than extant Anatidae, which while paddling bring the toes together when moving their feet forward (Mayr 2008b). Mlíkovský (2002) and Olson (1999b) assumed that *R. stehlini* belongs to the Anseranatidae, but a phylogenetic analysis supported a sister group relationship between *Romainvillia* and the Anatidae (Mayr 2008b). Derived characters shared by *Romainvillia* and extant Anatidae include the lack of a tarsometatarsal fossa for the first metatarsal, which indicates that the hallux was small as in extant Anatidae.

A record of *Romainvillia* sp. was reported from the early Oligocene of Belgium (Mayr and Smith 2001). *Paracygnopterus scotti*, from the early Oligocene of England, is another representative of the Romainvilliinae (Fig. 4.13b; Mayr 2008b). The description of *Romainvillia kazakhstanensis* from the late Eocene of Kazakhstan was based on a partial coracoid, which exhibits the distinctive morphology of the taxon *Romainvillia* (Zelenkov 2018a). Another species from the type locality of *R. kazakhstanensis* was described as *Cousteauvia kustovia* by Zelenkov (2020) and considered to be a diving anseriform; this species is solely represented by the proximal portion of a tarsometatarsus, which is so different from that of other anseriform birds in the morphology of the hypotarsus that the affinities of *C. kustovia* need to be corroborated with further material. An incomplete distal tarsometatarsus of a large anseriform species from the late Eocene of western China was tentatively assigned to the Romainvilliinae by Stidham and Ni (2014). However, the fragmentary nature of this specimen impedes an unambiguous identification and this all the more so as the distal end of a femur of a similar-sized bird from the same locality was tentatively referred to *Cygnopterus* (Stidham and Ni 2014). Another putative species of the Romainvilliinae, *Saintandrea chenoides* from the late Oligocene of southern France, was described by Mayr and De Pietri (2013). This species is distinctly larger than *R. stehlini* and is known from various fragmentary bones of a single individual (Fig. 4.13c).

A putative anatid, *Palaeopapia eous*, is represented by a sternum fragment and a referred coracoid from the late Eocene and early Oligocene, respectively, of England (Harrison and Walker 1976b, 1979). The oldest modern-type anatids are from the earliest Oligocene of Belgium (Mayr and Smith 2001) and were tentatively assigned to *Paracygnopterus* in the original description. This identification is no longer upheld, and the specimens are now regarded as indeterminate Anatidae (Mayr 2008b).

Crown group representatives of the Anatidae were reported from the late Oligocene of France and belong to *Mionetta blanchardi* and *M. natator* (Hugueney et al. 2003). According to Livezey and Martin (1988: 196), *Mionetta* "diverged from the rest of the Anatidae after *Dendrocygna* but before *Stictonetta*," but an analysis by Worthy and Lee (2008) resulted in a sister group relationship between *Mionetta* and a clade including *Stictonetta*, *Malacorhynchus*, and *Oxyura*.

The holotype and only known specimen of the alleged swan *Cygnavus formosus* from the early Oligocene of Kazakhstan is the distal end of a tibiotarsus (Kurochkin 1968). Because this bone is damaged in *C. senckenbergi*, the early Miocene type species of the taxon *Cygnavus*, identification of the Kazakh specimen is arguable. Kurochkin (1968) furthermore assigned a distal ulna from the late Oligocene (*Indricotherium* Beds) of Kazakhstan to the extant taxon *Somateria*, but this assignment was likewise poorly established. There is a greater possibility that the fragmentary bone belongs to one of the other taxa described by Kurochkin (1968).

Louchart et al. (2005) detailed that *Guguschia nailiae* from the Oligocene of Azerbaijan is not a swan (Cygnini), as which it was originally described (Aslanova and Burchak-Abramovich 1968). The species is known from wing bones, including a nearly complete humerus, and needs to be restudied. It is well possible that it is conspecific with the pelagornithid *Caspiodontornis kobystanicus* from the same locality (Aslanova and Burchak-Abramovich 1999; see Sect. 4.1).

In contradiction to Olson's (1989) assumption that the Anatidae originated in the Southern Hemisphere, the Paleogene fossil record of duck-like birds is rather scarce in the southern continents and restricted to the late Oligocene. From Africa, Paleogene representatives of the Anatidae are altogether unknown. *Teleornis impressus* from the Deseadan of Patagonia (Argentina) was assumed to be a member of the Tadornini (shelducks) by Agnolin (2004). However, this species is represented by a distal humerus only, and although it indeed seems to be an anseriform, possibly even an anatid, additional specimens are required to establish its exact affinities. Agnolin (2004) also assigned to the Anseriformes another species from the Deseadan of Patagonia, *Aminornis excavatus*, which was established on the basis of an omal extremity of a coracoid and was previously considered to be a representative of the Aramidae (limpkins). The affinities of this species also have to be established with further material.

Worthy (2009) described crown group Anatidae from the late Oligocene Etadunna and Namba formations in Australia. These are represented by many well-preserved limb bones and were assigned to new taxa of the Oxyurini (stiff-tailed ducks; *Pinpanetta tedfordi*, *P. vickersrichae*, *P. fromensis*) and the Tadornini (shelducks; *Australotadorna alecwilsoni*).

As in the case of other freshwater avifaunas (Sect. 11.1.2), the evolution of crocodilians may have had an impact on the diversity of Anseriformes in lacustrine or fluvial habitats. Most early Paleogene fossils of waterfowl are known from marine palaeoenvironments, and only after crocodilians disappeared from the temperate regions, anseriforms are abundantly represented in fossils sites of lacustrine origin.

References

Acosta Hospitaleche C, Reguero M (2020) Additional Pelagornithidae remains from Seymour Island, Antarctica. J S Am Earth Sci 99: 102504
Agnolin FL (2004) Revisión systemática de algunas aves deseadenses (Oligoceno Medio) descriptas por Ameghino en 1899. Rev Mus Argent Cienc Nat, n s 6:239–244
Alvarenga HMF (1988) Ave fóssil (Gruiformes: Rallidae) dos folhelhos da Bacia de Taubaté, estado de São Paulo, Brasil. An Acad Bras Cienc 60:321–328
Alvarenga HMF (1995) Um primitivo membro da ordem Galliformes (Aves) do Terciário médio da Bacia de Taubaté, estado de São Paulo, Brasil. An Acad Bras Cienc 67:33–44
Alvarenga HMF (1999) A fossil screamer (Anseriformes: Anhimidae) from the middle Tertiary of southeastern Brazil. In: Olson SL (ed) Avian paleontology at the close of the 20th century. Proceedings of the 4th international meeting of the Society of Avian Paleontology and Evolution, Washington, DC, 4–7 June 1996. Smithson Contrib Paleobiol 89:223–230
Andors A (1992) Reappraisal of the Eocene groundbird *Diatryma* (Aves: Anserimorphae). In: Campbell KE (ed) Papers in avian paleontology honoring Pierce Brodkorb. Nat Hist Mus Los Angeles Cty, Sci Ser 36:109–125
Andrews CW (1916) Note on the sternum of a large carinate bird from the (?) Eocene of southern Nigeria. Proc Zool Soc Lond 1916:519–524
Angst D, Buffetaut E (2017) Paleobiology of giant flightless birds. ISTE Press, London
Angst D, Lécuyer C, Amiot R, Buffetaut E, Fourel F, Martineau F, Legendre S, Abourachid A, Herrel A (2014) Isotopic and anatomical evidence of an herbivorous diet in the Early Tertiary giant bird *Gastornis*. Implications for the structure of Paleocene terrestrial ecosystems. Naturwissenschaften 101:313–322
Angst D, Buffetaut E, Lécuyer C, Amiot R, Smektala F, Giner S, Méchin A, Méchin P, Amoros A, Leroy L, Guiomar M, Tomg H, Martinez A (2015) Fossil avian eggs from the Palaeogene of southern France: new size estimates and a possible taxonomic identification of the egg-layer. Geol Mag 152:70–79
Aslanova SM, Burchak-Abramovich NI (1968) A fossil swan from the Maykopian series of Azerbaydzhan. Acta zool crac 14:325–338
Aslanova SM, Burchak-Abramovich NI (1999) A detailed description of *Caspiodontornis kobystanicus* from the Oligocene of the Caspian seashore. Acta zool crac 42:423–433
Averianov AO, Panteleyev AV, Potapova OR, Nessov LA (1991) Bony-toothed birds (Aves: Pelecaniformes: Odontopterygia) of the late Paleocene and Eocene of the western margin of ancient Asia. Tr zool Inst 239:3–12. [in Russian]
Ballmann P (1969) Die Vögel aus der altburdigalen Spaltenfüllung von Wintershof (West) bei Eichstätt in Bayern. Zitteliana 1:5–60
Benson RD (1999) *Presbyornis isoni* and other late Paleocene birds from North Dakota. In: Olson SL (ed) Avian Paleontology at the Close of the 20th Century: Proceedings of the 4th international meeting of the Society of Avian Paleontology and Evolution, Washington, DC, 4–7 June 1996. Smithson Contrib Paleobiol 89:253–259
Boles WE (1999) Early Eocene shorebirds (Aves: Charadriiformes) from the Tingamarra Local Fauna, Murgon, Queensland, Australia. Rec West Austral Mus Suppl 57:229–238
Boles WE, Ivison TJ (1999) A new dwarf megapode (Galliformes: Megapodiidae) from the late Oligocene of Central Australia. In: Olson SL (ed) Avian paleontology at the close of the 20th century: Proceedings of the 4th international meeting of the Society of Avian Paleontology and Evolution, Washington, DC, 4–7 June 1996. Smithson Contrib Paleobiol 89:199–206
Boles WE, Finch MA, Hofheins RH, Vickers-Rich P, Walters M, Rich TH (2013) A fossil stone-curlew (Aves: Burhinidae) from the late Oligocene/early Miocene of South Australia. In: Göhlich UB, Kroh A (eds) Paleornithological research 2013—Proceedings of the 8th international meeting of the Society of Avian Paleontology and Evolution. Natural History Museum Vienna, Vienna, pp 43–62
Bourdon E (2005) Osteological evidence for sister group relationship between pseudo-toothed birds (Aves: Odontopterygiformes) and waterfowls (Anseriformes). Naturwiss 92:586–591
Bourdon E, Cappetta H (2012) Pseudo-toothed birds (Aves, Odontopterygiformes) from the Eocene phosphate deposits of Togo, Africa. J Vertebr Paleontol 32:965–970
Bourdon E, Amaghzaz M, Bouya B (2010) Pseudotoothed birds (Aves, Odontopterygiformes) from the early Tertiary of Morocco. Am Mus Novit 3704:1–71
Brodkorb P (1964) Catalogue of fossil birds. Part 2 (Anseriformes through Galliformes). Bull Fla State Mus, Biol Sci 8(3):195–335
Brunet J (1970) Oiseaux de l'Éocène supérieur du bassin de Paris. Ann Paléontol 56:3–57
Buffetaut E (1997) New remains of the giant bird *Gastornis* from the upper Paleocene of the eastern Paris Basin and the relationships between *Gastornis* and *Diatryma*. Neues Jahrb Geol Paläontol, Mh 1997:179–190
Buffetaut E (2004) Footprints of giant birds from the Upper Eocene of the Paris Basin: an ichnological enigma. Ichnos 11:357–362
Buffetaut E (2008) First evidence of the giant bird *Gastornis* from southern Europe: a tibiotarsus from the Lower Eocene of Saint-Papoul (Aude, southern France). Oryctos 7:75–82
Buffetaut E (2013) The giant bird *Gastornis* in Asia: a revision of *Zhongyuanus xichuanensis* Hou, 1980, from the early Eocene of China. Paleontol J 47:1302–1307
Buffetaut E, Angst D (2014) Stratigraphic distribution of large flightless birds in the Palaeogene of Europe and its palaeobiological and palaeogeographical implications. Earth-Sci Rev 138:394–408
Buffetaut E, Angst D (2021) *Macrornis tanaupus* Seeley, 1866: an enigmatic giant bird from the upper Eocene of England. Geol Mag 158:1129–1134
Cenizo M, Hospitaleche CA, Reguero M (2015) Diversity of pseudo-toothed birds (Pelagornithidae) from the Eocene of Antarctica. J Paleontol 89:870–881

Clarke JA (2004) Morphology, phylogenetic taxonomy, and systematics of *Ichthyornis* and *Apatornis* (Avialae: Ornithurae). Bull Am Mus Nat Hist 286:1–179

Clarke JA, Norell MA (2004) New avialan remains and a review of the known avifauna from the Late Cretaceous Nemegt Formation of Mongolia. Am Mus Novit 3447:1–12

Cockerell TDA (1923) The supposed plumage of the Eocene bird Diatryma. Am Mus Novit 62:1–4

Cope ED (1876) On a gigantic bird from the Eocene of New Mexico. Proc Acad Nat Sci Philadelphia 28:10–11

Cracraft J (1973) Systematics and evolution of the Gruiformes (class Aves). 3. Phylogeny of the suborder Grues. Bull Am Mus Nat Hist 151:1–127

Crowe TM, Short LL (1992) A new gallinaceous bird from the Oligocene of Nebraska, with comments on the phylogenetic position of the Gallinuloididae. In: Campbell KE (ed) Papers in avian paleontology honoring Pierce Brodkorb. Nat Hist Mus Los Angeles Cty, Sci Ser 36:179–185

De Pietri VL, Scofield RP, Zelenkov N, Boles WE, Worthy TH (2016) The unexpected survival of an ancient lineage of anseriform birds into the Neogene of Australia: the youngest record of Presbyornithidae. Roy Soc Open Sci 3:150635

Dughi R, Sirugue F (1959) Sur des fragments de coquilles d'oeufs fossils de l'Eocène de Basse-Provence. C R Acad Sci, Paris 249:959–961

Dyke GJ (2001) The fossil waterfowl (Aves: Anseriformes) from the Eocene of England. Am Mus Novit 3354:1–15

Dyke GJ, Gulas BE (2002) The fossil galliform bird *Paraortygoides* from the Lower Eocene of the United Kingdom. Am Mus Novit 3360:1–14

Eastman CR (1900) New fossil bird and fish remains from the Middle Eocene of Wyoming. Geol Mag 7:54–58

Eastman CR (1905) Fossil avian remains from Armissan. Mem Carnegie Mus 2:131–138

Elzanowski A (2013) More evidence for plesiomorphy of the quadrate in the Eocene anseriform avian genus *Presbyornis*. Acta Palaeontol Polon 59:821–825

Elzanowski A, Boles WE (2012) Australia's oldest anseriform fossil: a quadrate from the early Eocene Tingamarra Fauna. Palaeontology 55:903–911

Elzanowski A, Stidham TA (2010) Morphology of the quadrate in the Eocene anseriform *Presbyornis* and extant galloanserine birds. J Morphol 271:305–323

Ericson PGP (1997) Systematic relationships of the palaeogene family Presbyornithidae (Aves: Anseriformes). Zool J Linnean Soc 121:429–483

Ericson PGP (2000) Systematic revision, skeletal anatomy, and paleoecology of the New World early Tertiary Presbyornithidae (Aves: Anseriformes). PaleoBios 20:1–23

Ericson PGP, Anderson CL, Britton T, Elzanowski A, Johansson US, Källersjö M, Ohlson JI, Parsons TJ, Zuccon D, Mayr G (2006) Diversification of Neoaves: integration of molecular sequence data and fossils. Biol Lett 2:543–547

Fabre-Taxy S, Touraine F (1960) Gisements d'oeufs d'oiseaux de très grande taille dans l'Èocène de Provence. C R Hebd Acad Sci (Paris) 250:3870–3871

Feduccia A (1976) Osteological evidence for shorebird affinities of the flamingos. Auk 93:587–601

Feduccia A (1980) The age of birds. Harvard University Press, Cambridge, MA

Feduccia A, McGrew PO (1974) A flamingolike wader from the Eocene of Wyoming. Contrib Geol, Univ Wyoming 113:49–61

Field DJ, Benito J, Chen A, Jagt JW, Ksepka DT (2020) Late Cretaceous neornithine from Europe illuminates the origins of crown birds. Nature 579:397–401

Fischer K (1962) Der Riesenlaufvogel *Diatryma* aus der eozänen Braunkohle des Geiseltales. Hallesches Jahrb mitteldtsch Erdgesch 4:26–33

Fischer K (1978) Neue Reste des Riesenlaufvogels *Diatryma* aus dem Eozän des Geiseltals bei Halle (DDR). Mitt Zool Mus Berlin 54, Suppl: Ann Ornithol 2:133–144

Fischer K (1990) Der Hühnervogel *Pirortyx major* (Gaillard, 1939) aus dem marinen Mitteloligozän bei Leipzig (DDR). Mitt Zool Mus Berlin 66, Suppl: Ann Ornithol 14:133–136

Fischer K (2003) Weitere Vogelknochen von *Diomedeoides* (Diomedeoididae, Procellariiformes) und *Paraortyx* (Paraortygidae, Galliformes) aus dem Unteroligozän des Weißelsterbeckens bei Leipzig (Sachsen). Mauritiana 18:387–395

Garcia G, Mourer-Chauviré C, Adaci M, Bensalah M, Mebrouk F, Valentin X, Mahboubi M, Tabuce R (2020) First discovery of avian egg and bone remains (Presbyornithidae) from the Gour Lazib (Eocene, Algeria). J Afric Earth Sci 162:103666

Goedert JL (1989) Giant late Eocene marine birds (Pelecaniformes: Pelagornithidae) from northwestern Oregon. J Paleontol 63:939–944

González-Barba G, Schwennicke T, Goedert JL, Barne LG (2002) Earliest Pacific Basin record of the Pelagornithidae (Aves: Pelecaniformes). J Vertebr Paleontol 22:722–725

Grande L (2013) The lost world of Fossil Lake. Snapshots from deep time. University of Chicago Press, Chicago

Gulas-Wroblewski BE, Wroblewski AF-J (2003) A crown-group galliform bird from the Middle Eocene Bridger Formation of Wyoming. Palaeontology 46:1269–1280

Hackett SJ, Kimball RT, Reddy S, Bowie RCK, Braun EL, Braun MJ, Chojnowski JL, Cox WA, Han K-L, Harshman J, Huddleston CJ, Marks BD, Miglia KJ, Moore WS, Sheldon FH, Steadman DW, Witt CC, Yuri T (2008) A phylogenomic study of birds reveals their evolutionary history. Science 320:1763–1767

Harrison CJO (1985) A bony-toothed bird (Odontopterygiformes) from the Palaeocene of England. Tert Res 7:23–25

Harrison CJO, Walker CA (1976a) A review of the bony-toothed birds (Odontopterygiformes): with descriptions of some new species. Tert Res Spec Pap 2:1–72

Harrison CJO, Walker CA (1976b) Birds of the British Upper Eocene. Zool J Linnean Soc 59:323–351

Harrison CJO, Walker CA (1977) Birds of the British Lower Eocene. Tert Res Spec Pap 3:1–52

Harrison CJO, Walker CA (1978) *Proherodius oweni* as a species of Presbyornithidae. Tert Res 2:1–3

Harrison CJO, Walker CA (1979) Birds of the British Lower Oligocene. In: Harrison CJO, Walker CA (eds) Studies in Tertiary avian paleontology. Tert Res Spec Pap 5:29–43

Hébert E (1855) Note sur le tibia du *Gastornis pariensis* [sic]. C R Acad Sci, Paris 40:579–582

Hellmund M (2013) Reappraisal of the bone inventory of *Gastornis geiselensis* (Fischer, 1978) from the Eocene "Geiseltal Fossillagerstätte"(Saxony-Anhalt, Germany). N Jb Geol Paläontol, Abh 269:203–220

Hood SC, Torres CR, Norell MA, Clarke JA (2019) New fossil birds from the earliest Eocene of Mongolia. Am Mus Novit 3934:1–24

Hou L-H (1980) New form of the Gastornithidae from the Lower Eocene of the Xichuan, Honan. Vertebr PalAsiat 18:111–115. [in Chinese]

Hou L-H (2003) Fossil birds of China. Yunnan Science and Technology Press, Kunming

Howard H (1955) A new wading bird from the Eocene of Patagonia. Am Mus Novit 1710:1–25

Hugueney M, Berthet D, Bodergat A-M, Escuillié F, Mourer-Chauviré C, Wattinne A (2003) La limite Oligocène-Miocène en Limagne: changements fauniques chez les mammifères, oiseaux et ostracodes des différents niveaux de Billy-Créchy (Allier, France). Geobios 36:719–731

Hwang SH, Mayr G, Minjin B (2010) The earliest record of a galliform bird in Asia, from the late Paleocene/early Eocene of the Gobi Desert, Mongolia. J Vertebr Paleontol 30:1642–1644

Jones CM (2000) The first record of a fossil bird from East Antarctica. Antarct Res Ser 76:359–364

Kloess PA, Poust AW, Stidham TA (2020) Earliest fossils of giant-sized bony-toothed birds (Aves: Pelagornithidae) from the Eocene of Seymour Island, Antarctica. Sci Rep 10:18286

Kriegs JO, Matzke A, Churakov G, Kuritzin A, Mayr G, Brosius J, Schmitz J (2007) Waves of genomic hitchhikers shed light on the evolution of gamebirds (Aves: Galliformes). BMC Evol Biol 7:190

Ksepka DT (2009) Broken gears in the avian molecular clock: new phylogenetic analyses support stem galliform status for *Gallinuloides wyomingensis* and rallid affinities for *Amitabha urbsinterdictensis*. Cladistics 25:173–197

Ksepka DT (2014) Flight performance of the largest volant bird. Proc Natl Acad Sci U S A 111:10624–10629

Kuhl H, Frankl-Vilches C, Bakker A, Mayr G, Nikolaus G, Boerno ST, Klages S, Timmermann B, Gahr M (2021) An unbiased molecular approach using 3'UTRs resolves the avian family-level tree of life. Mol Biol Evol 38:108–121

Kurochkin EN (1968) New Oligocene birds from Kazakhstan. Paleontol Ž 1:92–101. [in Russian]

Kurochkin EN, Dyke GJ (2010) A large collection of *Presbyornis* (Aves, Anseriformes, Presbyornithidae) from the late Paleocene and early Eocene of Mongolia. Geol J 45:375–387

Kurochkin EN, Dyke GJ, Karhu AA (2002) A new presbyornithid bird (Aves, Anseriformes) from the Late Cretaceous of Southern Mongolia. Am Mus Novit 3386:1–11

Lambrecht K (1930) Studien über fossile Riesenvögel. Geol Hung, Ser Palaeontol 7:1–37

Lambrecht K (1931) *Cygnopterus* und *Cygnavus*, zwei fossile Schwäne aus dem Tertiär Europas. Bull Mus Roy Hist Nat Belg 7 (31):1–6

Lebedinsky NG (1927) *Romainvillia Stehlini* n.g. n.sp., canard éocène provenant des marnes blanches du Bassin de Paris. Mém Soc Paléontol Suisse 47 (2):1–8

Leggitt VL, Buchheim HP (1997) *Presbyornis* (Aves: Anseriformes) eggshell from three avian mass mortality sites: Eocene Fossil Lake, Lincoln County, Wyoming. J Vertebr Paleontol 17:60A

Lindow BEK, Dyke GJ (2007) A small galliform bird from the Lower Eocene Fur Formation. Bull Geol Soc Denmark 55:59–63

Livezey BC (1997) A phylogenetic analysis of basal Anseriformes, the fossil *Presbyornis*, and the interordinal relationships of waterfowl. Zool J Linnean Soc 121:361–428

Livezey BC, Martin LD (1988) The systematic position of the Miocene anatid *Anas*[?] *blanchardi* Milne-Edwards. J Vertebr Paleontol 8:196–211

Livezey BC, Zusi RL (2007) Higher-order phylogeny of modern birds (Theropoda, Aves: Neornithes) based on comparative anatomy: II.—analysis and discussion. Zool J Linnean Soc 149:1–94

Louchart A, Vignaud P, Likius A, MacKaye HT, Brunet M (2005) A new swan (Aves: Anatidae) in Africa, from the latest Miocene of Chad and Libya. J Vertebr Paleontol 25:384–392

Louchart A, Sire JY, Mourer-Chauviré C, Geraads D, Viriot L, De Buffrénil V (2013) Structure and growth pattern of pseudoteeth in *Pelagornis mauretanicus* (Aves, Odontopterygiformes, Pelagornithidae). PLoS One 8:e80372

Louchart A, De Buffrénil V, Bourdon E, Dumont M, Viriot L, Sire JY (2018) Bony pseudoteeth of extinct pelagic birds (Aves, Odontopterygiformes) formed through a response of bone cells to tooth-specific epithelial signals under unique conditions. Sci Rep 8:12952

Lucas FA (1900) Characters and relations of *Gallinuloides wyomingensis* Eastman, a fossil gallinaceous bird from the Green River shales of Wyoming. Bull Mus Comp Zool 36:79–84

Lydekker R (1891) Catalogue of the fossil birds in the British Museum (Natural History). British Museum (Natural History), London

Martin LD (1992b) The status of the late Paleocene birds *Gastornis* and *Remiornis*. In: Campbell KE (ed) Papers in avian paleontology honoring Pierce Brodkorb. Nat Hist Mus Los Angeles Cty, Sci Ser 36:97–108

Matthew WD, Granger W (1917) The skeleton of *Diatryma*, a gigantic bird from the Lower Eocene of Wyoming. Bull Am Mus Nat Hist 37:307–326

Mayr G (2000) A new basal galliform bird from the Middle Eocene of Messel (Hessen, Germany). Senckenb Lethaea 80:45–57

Mayr G (2005) The Paleogene fossil record of birds in Europe. Biol Rev 80:515–542

Mayr G (2006) New specimens of the early Eocene stem group galliform *Paraortygoides* (Gallinuloididae), with comments on the evolution of a crop in the stem lineage of Galliformes. J Ornithol 147:31–37

Mayr G (2007a) The birds from the Paleocene fissure filling of Walbeck (Germany). J Vertebr Paleontol 27:394–408

Mayr G (2007b) Bizarre tubercles on the vertebrae of Eocene fossil birds indicate an avian disease without modern counterpart. Naturwiss 94: 681–685

Mayr G (2008a) A skull of the giant bony-toothed bird *Dasornis* (Aves: Pelagornithidae) from the lower Eocene of the Isle of Sheppey. Palaeontology 51:1107–1116

Mayr G (2008b) Phylogenetic affinities and morphology of the late Eocene anseriform bird *Romainvillia stehlini* Lebedinsky, 1927. Neues Jahrb Geol Palaontol Abh 248:365–380

Mayr G (2009) Paleogene fossil birds, 1st edn. Springer, Heidelberg

Mayr G (2010) A new avian species with tubercle-bearing cervical vertebrae from the Middle Eocene of Messel (Germany). Rec Austral Mus 62:21–28

Mayr G (2011) Cenozoic mystery birds—on the phylogenetic affinities of bony-toothed birds (Pelagornithidae). Zool Scr 40:448–467

Mayr G (2016) The world's smallest owl, the earliest unambiguous charadriiform bird, and other avian remains from the early Eocene Nanjemoy Formation of Virginia (USA). Paläontol Z 90:747–763

Mayr G (2017) Avian evolution: the fossil record of birds and its paleobiological significance. Wiley-Blackwell, Chichester

Mayr G (2018) Birds—the most species-rich vertebrate group in Messel. In: Schaal SKF, Smith K, Habersetzer J (eds) Messel—An ancient greenhouse ecosystem. Schweitzerbart, Stuttgart, pp 169–214

Mayr G, Clarke J (2003) The deep divergences of neornithine birds: a phylogenetic analysis of morphological characters. Cladistics 19:527–553

Mayr G, De Pietri VL (2013) A goose-sized anseriform bird from the late Oligocene of France: the youngest record and largest species of Romainvilliinae. Paläontol Z 87:423–430

Mayr G, Rubilar-Rogers D (2010) Osteology of a new giant bony-toothed bird from the Miocene of Chile, with a revision of the taxonomy of Neogene Pelagornithidae. J Vertebr Paleontol 30:1313–1330

Mayr G, Smith R (2001) Ducks, rails, and limicoline waders (Aves: Anseriformes, Gruiformes, Charadriiformes) from the lowermost Oligocene of Belgium. Geobios 34:547–561

Mayr G, Smith R (2002) Avian remains from the lowermost Oligocene of Hoogbutsel (Belgium). Bull Inst Roy Sci Nat Belg 72:139–150

Mayr G, Smith T (2010) Bony-toothed birds (Aves: Pelagornithidae) from the middle Eocene of Belgium. Palaeontology 53:365–376

Mayr G, Smith T (2013) Galliformes, Upupiformes, Trogoniformes, and other avian remains (?Phaethontiformes and ?Threskiornithidae) from the Rupelian stratotype in Belgium, with comments on the identity of "*Anas*" *benedeni* Sharpe, 1899. In: Göhlich UB, Kroh A (eds) Paleornithological research 2013—Proceedings of the 8th international meeting of the Society of Avian Paleontology and Evolution. Natural History Museum Vienna, Vienna, pp 23–35

Mayr G, Smith T (2019a) New Paleocene bird fossils from the North Sea Basin in Belgium and France. Geol Belg 22:35–46

Mayr G, Smith T (2019b) A diverse bird assemblage from the Ypresian of Belgium furthers knowledge of early Eocene avifaunas of the North Sea Basin. N Jb Geol Paläontol, Abh 291:253–281

References

Mayr G, Weidig I (2004) The early Eocene bird *Gallinuloides wyomingensis*—a stem group representative of Galliformes. Acta Palaeontol Pol 49:211–217

Mayr G, Zvonok E (2011) Middle Eocene Pelagornithidae and Gaviiformes (Aves) from the Ukrainian Paratethys. Palaeontology 54:1347–1359

Mayr G, Zvonok E (2012) A new genus and species of Pelagornithidae with well-preserved pseudodentition and further avian remains from the middle Eocene of the Ukraine. J Vertebr Paleontol 32:914–925

Mayr G, Poschmann M, Wuttke M (2006) A nearly complete skeleton of the fossil galliform bird *Palaeortyx* from the late Oligocene of Germany. Acta Ornithol 41:129–135

Mayr G, Hazevoet CJ, Dantas P, Cachão M (2008) A sternum of a very large bony-toothed bird (Pelagornithidae) from the Miocene of Portugal. J Vertebr Paleontol 28:762–769

Mayr G, Goedert JL, McLeod SA (2013) Partial skeleton of a bony-toothed bird from the late Oligocene/early Miocene of Oregon (USA) and the systematics of Neogene Pelagornithidae. J Paleontol 87:922–929

Mayr G, De Pietri VL, Love L, Mannering A, Scofield RP (2021) Oldest, smallest and phylogenetically most basal pelagornithid, from the early Paleocene of New Zealand, sheds light on the evolutionary history of the largest flying birds. Pap Palaeontol 7:217–233

Milne-Edwards A (1867–1871) Recherches anatomiques et paléontologiques pour servir à l'histoire des oiseaux fossiles de la France. Victor Masson et fils, Paris

Milner AC, Walsh SA (2009) Avian brain evolution: new data from Palaeogene birds (lower Eocene) from England. Zool J Linnean Soc 155:198–219

Mlíkovský J (1989) A new guineafowl (Aves: Phasianidae) from the late Eocene of France. Ann Naturhist Mus Wien, Ser A 90:63–66

Mlíkovský J (2002) Cenozoic birds of the world. Part 1: Europe. Ninox Press, Praha

Mlíkovský J, Švec P (1986) Review of the Tertiary waterfowl (Aves: Anseridae) of Asia. Věstn Českoslov Spol Zool 50:259–272

Mourer-Chauviré C (1982) Les oiseaux fossiles des Phosphorites du Quercy (Eocène supérieur à Oligocène supérieur): implications paléobiogéographiques. Geobios, mém spéc 6:413–426

Mourer-Chauviré C (1988) Le gisement du Bretou (Phosphorites du Quercy, Tarn-et-Garonne, France) et sa faune de vertébrés de l'Eocène supérieur. II Oiseaux Palaeontographica (A) 205:29–50

Mourer-Chauviré C (1992) The Galliformes (Aves) from the Phosphorites du Quercy (France): systematics and biostratigraphy. In: Campbell KE (ed) Papers in avian paleontology honoring Pierce Brodkorb. Nat Hist Mus Los Angeles Cty, Sci Ser 36:67–95

Mourer-Chauviré C (1996) Paleogene avian localities of France. In: Mlíkovský J (ed) Tertiary Avian Localities of Europe. Acta Univ Carolinae, Geol 39:567–598

Mourer-Chauviré C (2000) A new species of *Ameripodius* (Aves: Galliformes: Quercymegapodiidae) from the lower Miocene of France. Palaeontology 43:481–193

Mourer-Chauviré C (2004) Review of: Cenozoic birds of the world, part 1: Europe. Auk 121:623–627

Mourer-Chauviré C (2006) The avifauna of the Eocene and Oligocene Phosphorites du Quercy (France): an updated list. Strata, sér 1(13):135–149

Mourer-Chauviré C, Berthet D, Hugueney M (2004) The late Oligocene birds of the Créchy quarry (Allier, France), with a description of two new genera (Aves: Pelecaniformes: Phalacrocoracidae, and Anseriformes: Anseranatidae). Senckenb Lethaea 84:303–315

Mourer-Chauviré C, Bourdon E (2016) The *Gastornis* (Aves, Gastornithidae) from the Late Paleocene of Louvois (Marne, France). Swiss J Palaeontol 135:327–341

Mourer-Chauviré C, Bourdon E (2020) Description of a new species of *Gastornis* (Aves, Gastornithiformes) from the early Eocene of La Borie, southwestern France. Geobios 63:39–46

Mourer-Chauviré C, Pickford M, Senut B (2011) The first Palaeogene galliform from Africa. J Ornithol 152:617–622

Mourer-Chauviré C, Tabuce R, El Mabrouk E, Marivaux L, Khayati H, Vianey-Liaud M, Ben Haj Ali M (2013) A new taxon of stem group Galliformes and the earliest record for stem group Cuculidae from the Eocene of Djebel Chambi, Tunisia. In: Göhlich UB, Kroh A (eds) Palaeornithological research 2013—Proceedings of the 8th international meeting of the Society of Avian Paleontology and Evolution. Natural History Museum Vienna, Vienna, pp 1–15

Mourer-Chauviré C, Pickford M, Senut B (2015) Stem group galliform and stem group psittaciform birds (Aves, Galliformes, Paraortygidae, and Psittaciformes, family incertae sedis) from the middle Eocene of Namibia. J Ornithol 156:275–286

Mourer-Chauviré C, Pickford M, Senut B (2017) New data on stem group Galliformes, Charadriiformes, and Psittaciformes from the middle Eocene of Namibia. Contrib MACN 7:99–131

Murray PF, Vickers-Rich P (2004) Magnificent Mihirungs. The colossal flightless birds of the Australian dreamtime. Indiana University Press, Bloomington

Mustoe GE, Tucker DS, Kemplin KL (2012) Giant Eocene bird footprints from Northwest Washington, USA. Palaeontology 55:1293–1305

Nessov LA (1992) Mesozoic and Paleogene birds of the USSR and their paleoenvironments. In: Campbell KE (ed) Papers in avian paleontology honoring Pierce Brodkorb. Nat Hist Mus Los Angeles Cty, Sci Ser 36:465–478

Nguyen JMT, Boles WE, Hand S (2010) New material of *Barawertornis tedfordi*, a dromornithid bird from the Oligo-Miocene of Australia, and its phylogenetic implications. Rec Austral Mus 62:45–60

Okazaki Y (1989) An occurrence of fossil bony-toothed bird (Odontopterygiformes) from the Ashiya Group (Oligocene), Japan. Bull Kitakyushu Mus Nat Hist 9:123–126

Okazaki Y (2006) An occurrence of Oligocene pseudodontorn bird from the Kishima Group, Kyushu, Japan. Bull Kitakyushu Mus Nat Hist Hum Hist, ser A 4:111–114

Olson SL (1974) *Telecrex* restudied: a small Eocene guineafowl. Wilson Bull 86:246–250

Olson SL (1985) The fossil record of birds. In: Farner DS, King JR, Parkes KC (eds) Avian biology, vol 8. Academic Press, New York, pp 79–238

Olson SL (1989 [1988]) Aspects of global avifaunal dynamics during the Cenozoic. In: Ouellet H (ed) Acta XIX Congressus Internationalis Ornithologici. University of Ottawa Press, Ottawa, pp 2023–2029

Olson SL (1994) A giant *Presbyornis* (Aves: Anseriformes) and other birds from the Paleocene Aquia Formation of Maryland and Virginia. Proc Biol Soc Wash 107:429–435

Olson SL (1999a) Early Eocene birds from eastern North America: a faunule from the Nanjemoy Formation of Virginia. In: Weems RE, Grimsley GJ (eds) Early Eocene vertebrates and plants from the Fisher/Sullivan site (Nanjemoy Formation) Stafford County, Virginia. Virginia Div Mineral Resour Publ 152:123–132

Olson SL (1999b) The anseriform relationships of *Anatalavis* Olson and Parris (Anseranatidae), with a new species from the Lower Eocene London Clay. In: Olson SL (ed) Avian paleontology at the close of the 20th century: Proceedings of the 4th international meeting of the Society of Avian Paleontology and Evolution, Washington, DC, 4–7 June 1996. Smithson Contrib Paleobiol 89:231–243

Olson SL, Feduccia A (1980) *Presbyornis* and the origin of the Anseriformes (Aves: Charadriomorphae). Smithson Contrib Zool 323:1–24

Olson SL, Parris DC (1987) The Cretaceous birds of New Jersey. Smithson Contrib Paleobiol 63:1–22

Olson SL, Rasmussen PC (2001) Miocene and Pliocene birds from the Lee Creek Mine, North Carolina. Smithson Contrib Paleobiol 90:233–365

Owen R (1873) Description of the skull of a dentigerous bird (*Odontopteryx toliapicus* [sic], ow.) from the London Clay of Sheppey. Q J Geol Soc Lond 29:511–521

Patterson J, Lockley MG (2004) A probable *Diatryma* track from the Eocene of Washington: an intriguing case of controversy and skepticism. Ichnos 11:341–347

Prum RO, Berv JS, Dornburg A, Field DJ, Townsend JP, Lemmon EM, Lemmon AR (2015) A comprehensive phylogeny of birds (Aves) using targeted next-generation DNA sequencing. Nature 526:569–573

Rubilar-Rogers D, Yury-Yáñez R, Mayr G, Gutstein C, Otero R (2011) A humerus of a giant late Eocene pseudo-toothed bird from Antarctica. J Vertebr Paleontol 2:182A

Schaub S (1929) Über eocäne Ratitenreste in der osteologischen Sammlung des Basler Museums. Verh Naturforsch Ges Basel 40:588–598

Schaub S (1945) Bemerkungen zum Typus von Taoperdix keltica Eastman. Eclogae Geol Helv 38:616–621

Sibley CG, Ahlquist JE (1990) Phylogeny and classification of birds: a study in molecular evolution. Yale University Press, New Haven

Stegmann B (1978) Relationships of the superorders Alectoromorphae and Charadriimorphae (Aves): a comparative study of the avian hand. Pub Nuttall Ornithol Club 17:1–119

Stidham TA, Eberle JJ (2016) The palaeobiology of high latitude birds from the early Eocene greenhouse of Ellesmere Island, Arctic Canada. Sci Rep 6:20912

Stidham TA, Ni X-J (2014) Large anseriform (Aves: Anatidae: Romainvilliinae?) fossils from the late Eocene of Xinjiang, China. Vertebr PalAsiat 52:98–111

Stidham TA, Townsend KE, Holroyd PA (2020) Evidence for wide dispersal in a stem galliform clade from a new small-sized middle Eocene pangalliform (Aves: Paraortygidae) from the Uinta Basin of Utah (USA). Diversity 12(3):90

Stilwell JD, Jones CM, Levy RH, Harwood DM (1998) First fossil bird from East Antarctica. Antarct J US 33:12–16

Tambussi CP, Acosta Hospitaleche C (2007) Antarctic birds (Neornithes) during the Cretaceous-Eocene times. Rev Asoc Geol Argent 62:604–617

Tambussi CP, Degrange FJ, De Mendoza RS, Sferco E, Santillana S (2019) A stem anseriform from the early Palaeocene of Antarctica provides new key evidence in the early evolution of waterfowl. Zool J Linnean Soc 186:673–700

Tomek T, Bochenski ZM, Wertz K, Swidnicka E (2014) A new genus and species of a galliform bird from the Oligocene of Poland. Palaeontol Electron 17.3(38A):1–15

Tonni EP, Tambussi CP (1985) New remains of Odontopterygia (Aves: Pelecaniformes) from the early Tertiary of Antarctica. Ameghiniana 21:121–124

Tordoff HB (1951) A quail from the Oligocene of Colorado. Condor 53:203–204

Tordoff HB, Macdonald JR (1957) A new bird (family Cracidae) from the early Oligocene of South Dakota. Auk 74:174–184

Vickers-Rich PV (1991) The Mesozoic and Tertiary history of birds on the Australian plate. In: Vickers-Rich P, Monaghan TM, Baird RF, Rich TH (eds) Vertebrate Palaeontology of Australasia. Pioneer Design Studio and Monash University Publications Committee, Melbourne, pp 721–808

Weigel RD (1963) Oligocene birds from Saskatchewan. Q J Fla Acad Sci 26:257–262

Weigelt J (1939) Die Aufdeckung der bisher ältesten tertiären Säugetierfauna Deutschlands. Nova Acta Leopold, N F 7:515–528

Wetmore A (1926) Fossil birds from the Green River deposits of eastern Utah. Ann Carnegie Mus 16:391–402

Wetmore A (1930) The supposed plumage of the Eocene *Diatryma*. Auk 47:579–580

Wetmore A (1934) Fossil birds from Mongolia and China. Am Mus Novit 711:1–16

Wetmore A (1938) A fossil duck from the Eocene of Utah. J Paleontol 12:280–283

Wetmore A (1956) A fossil guan from the Oligocene of South Dakota. Condor 58:234–235

Witmer LM, Rose KD (1991) Biomechanics of the jaw apparatus of the gigantic Eocene bird *Diatryma*: implications for diet and mode of life. Paleobiol 17:95–120

Worthy TH (2009) Descriptions and phylogenetic relationships of two new genera and four new species of Oligo-Miocene waterfowl (Aves: Anatidae) from Australia. Zool J Linnean Soc 156:411–454

Worthy TH, Lee MSY (2008) Affinities of Miocene waterfowl (Anatidae: *Manuherikia*, *Dunstanetta* and *Miotadorna*) from the St. Bathans fauna, New Zealand. Palaeontology 51:677–708

Worthy TH, Scanlon JD (2009) An Oligo-Miocene magpie goose (Aves: Anseranatidae) from Riversleigh, Northwestern Queensland, Australia. J Vertebr Paleontol 29:205–211

Worthy TH, Handley WD, Archer M, Hand SJ (2016) The extinct flightless mihirungs (Aves, Dromornithidae): cranial anatomy, a new species, and assessment of Oligo-Miocene lineage diversity. J Vertebr Paleontol 36:e1031345

Worthy TH, Degrange FJ, Handley WD, Lee MS (2017) The evolution of giant flightless birds and novel phylogenetic relationships for extinct fowl (Aves, Galloanseres). Roy Soc Open Sci 4(10):170975

Yang S-Y, Lockley MG, Greben R, Erickson BR, Lim S-K (1995) Flamingo and duck-like bird tracks from the Late Cretaceous and Early Tertiary: evidence and implications. Ichnos 4:21–34

Yury-Yáñez RE, Otero RA, Soto-Acuña S, Suárez ME, Rubilar-Rogers-D, Sallaberry M (2012) First bird remains from the Eocene of Algarrobo, Central Chile. Andean Geol 39:548–557

Zelenkov NV (2018a) The earliest Asian duck (Anseriformes: *Romainvillia*) and the origin of Anatidae. Dokl Biol Sci 483:225–227

Zelenkov NV (2018b) A swan-sized anseriform bird from the late Paleocene of Mongolia. J Vertebr Paleontol 38:e1531879

Zelenkov NV (2019) Systematic position of *Palaeortyx* (Aves, ? Phasianidae) and notes on the evolution of Phasianidae. Paleontol J 53:194–202

Zelenkov N (2020) The oldest diving anseriform bird from the late Eocene of Kazakhstan and the evolution of aquatic adaptations in the intertarsal joint of waterfowl. Acta Palaeontol Polon 65:733–742

Zelenkov N (2021a) New bird taxa (Aves: Galliformes, Gruiformes) from the early Eocene of Mongolia. Paleontol J 55:438–446

Zelenkov N (2021b) A revision of the Palaeocene-Eocene Mongolian Presbyornithidae (Aves: Anseriformes). Paleontol J 55:323–330

Zelenkov NV, Panteleyev AV (2019) A small stem-galliform bird (Aves: Paraortygidae) from the Eocene of Uzbekistan. C R Palevol 18:517–523

Zelenkov NV, Stidham TA (2018) Possible filter-feeding in the extinct *Presbyornis* and the evolution of Anseriformes (Aves). Zool J 97:943–956

Zouhri S, Gingerich PD, Khalloufi B, Bourdon E, Adnet S, Jouve S, Elbouldali N, Amane A, Rage J-C, Tabuce R, De Broin FDL (2021) Middle Eocene vertebrate fauna from the Aridal Formation, Sabkha of Gueran, southwestern Morocco. Geodiversitas 43:121–150

Zusi RL, Warheit KI (1992) On the evolution of intraramal mandibular joints in pseudodontorns (Aves: Odontopterygia). In: Campbell KE (ed) Papers in avian paleontology honoring Pierce Brodkorb. Nat Hist Mus Los Angeles Cty, Sci Ser 36:351–360

5. Mirandornithes (Grebes and Flamingos), Charadriiformes (Shorebirds and Allies), and Gruiformes (Rails, Cranes, and Allies)

A sister group relationship between the Phoenicopteriformes and the Podicipediformes is congruently supported by multiple analyses of morphological and molecular data sets, and the clade including both taxa was termed Mirandornithes (Sangster 2005). This clade has not been recognized by earlier systematists, but most of the taxa that form the Charadriiformes and Gruiformes have long been considered closely related, even though the exact contents of these clades and the interrelationships of the included taxa were disputed.

The three clades discussed in the following are obtained in varying positions in current phylogenetic analyses, and their inclusion in the present chapter is not meant to reflect close affinities. It should be mentioned, however, that some analyses supported a sister group relationship between the Mirandornithes and the Charadriiformes (Prum et al. 2015), and others recovered a clade including the Charadriiformes and the Gruiformes (Kuhl et al. 2021). Some extinct taxa of the Gruiformes are well represented, but the Mirandornithes and Charadriiformes have a rather scant early Paleogene fossil record even though various Mesozoic and early Paleogene fossils were identified as flamingos or "transitional Charadriiformes" by earlier authors.

5.1 Podicipediformes (Grebes) and Phoenicopteriformes (Flamingos)

Flamingos and grebes distinctly differ in many ecomorphological features, which prevented earlier morphologists from taking close affinities of both groups into account. The globally distributed grebes are short-legged, foot-propelled diving birds, which predominantly occur in freshwater environments (even though some species winter in marine habitats). Flamingos, by contrast, are characterized by greatly elongated hindlimbs and a derived beak morphology, which both constitute adaptations for filter-feeding in shallow saline lakes.

Mirandornithes, the clade including grebes and flamingos, appears to have had a long evolutionary history, even though the divergence date of flamingos and grebes is difficult to constrain on the basis of the fossil record. The holotype of *Scaniornis lundgreni* from the Paleocene of Sweden consists of a few poorly preserved wing and pectoral girdle bones, which were considered flamingo-like in the original description (Dames 1890). However, the long and slender humerus is equally similar to that of some Late Cretaceous and Paleocene taxa of putative galloanserine affinities, such as *Vegavis* (Sect. 2.4) and *Conflicto* (Sect. 4.5). A partial quadrate from the early Eocene (Bumbanian; ~55 Ma) Naranbulag Formation in Mongolia was assigned to the stem group of the Mirandornithes by Hood et al. (2019). If correctly identified, this fossil would be the earliest record of the group, but more skeletal elements are needed for a reliable classification of the species it belonged to.

The recognition of a sister group relationship between the Phoenicopteriformes and Podicipediformes led to a revision of the phylogenetic affinities of the early/middle Eocene Juncitarsidae, which include two named species. The description of *Juncitarsus gracillimus* from the Bridger Formation of Wyoming was based on a number of isolated bones including a complete tarsometatarsus (Olson and Feduccia 1980; Ericson 1999). *J. merkeli* is represented by a skeleton from Messel (Fig. 5.1a) and tentatively referred hindlimbs from the Green River Formation (Peters 1987). Zelenkov (2021a) noted the presence of *Juncitarsus*-like fossils in the early Eocene of Mongolia, which were before assigned to the anseriform Presbyornithidae.

Initially, *Juncitarsus* was assigned to the Phoenicopteriformes and was considered to be an evolutionary link between flamingos and charadriiform birds (Olson and Feduccia 1980; Peters 1987). As evidenced by the Messel skeleton of *J. merkeli*, the beak of *Juncitarsus* is straight and very different from that of extant Phoenicopteriformes. Peters (1987) even considered it to have been schizorhinal, but the beak of the holotype specimen of *J. merkeli* is crushed and

© The Author(s), under exclusive license to Springer Nature Switzerland AG 2022
G. Mayr, *Paleogene Fossil Birds*, Fascinating Life Sciences, https://doi.org/10.1007/978-3-030-87645-6_5

the exact caudal extent of the nostrils is difficult to ascertain (extant Podicipediformes and Phoenicopteriformes are holorhinal). In concordance with extant Phoenicopteriformes, the frontal bones of *Juncitarsus* exhibit fossae for nasal glands (Olson and Feduccia 1980), and the tarsometatarsus is extremely long and slender. However, *Juncitarsus* lacks several derived characters shared by extant Phoenicopteriformes and Podicipediformes (Mayr 2004a, 2014a). In particular, the proximal phalanx of the major wing digit is not as narrow and the hallux is proportionally longer. The notarium (co-ossified thoracic vertebrae) of *Juncitarsus* is furthermore formed by just two vertebrae, whereas there are four co-ossified thoracic vertebrae in extant flamingos and grebes (Peters 1987; Mayr 2014a). Apart from the strongly elongated legs, no derived features are exclusively shared by *Juncitarsus* and the Phoenicopteriformes, and the skeletal morphology of *Juncitarsus* suggests a sister group relationship to the Mirandornithes (Mayr 2014a).

The Podicipediformes have no published Paleogene fossil record, even though Kurochkin (1976) mentioned undescribed specimens from the late Oligocene of Kazakhstan. Vickers-Rich (1991) indicated the presence of as yet undescribed podicipediform fossils in the late Oligocene or early Miocene of the Namba Formation of Australia. The earliest formally described grebe is *Miobaptus walteri* from the early Miocene of the Czech Republic (Švec 1982, 1984). Because this species already exhibits the highly derived skeletal morphology of extant Podicipediformes, it indicates a Paleogene stem lineage, which may have originated in one of the southern continents (see also Olson 1989).

Of particular interest regarding the well-established sister group relationship between the Phoenicopteriformes and Podicipediformes are the Palaelodidae. These distinctive birds have long been recognized as stem group representatives of the Phoenicopteriformes but have proportionally shorter legs than extant flamingos. The hindlimbs furthermore exhibit morphological features, which are typically found in aquatic birds but are absent in extant flamingos. The earliest fossils assigned to the Palaelodidae come from the early Oligocene of Belgium (Mayr and Smith 2002) and belong to a fairly large bird described as *Adelalopus hoogbutseliensis*. Rasmussen et al. (1987) furthermore reported *Palaelodus*-like bones from the early Oligocene of the Jebel Qatrani Formation in Egypt. In the late Oligocene of France, three species of *Palaelodus* coexisted, which are very abundant in early Miocene deposits of Europe, that is, *P. ambiguus*, *P. crassipes*, and *P. gracilipes* (Hugueney et al. 2003; Mourer-Chauviré et al. 2004).

Whereas early Oligocene fossils of palaelodids are known from the Old World only, these birds appear to have achieved a global distribution in the late Oligocene: Two species of *Palaelodus*, *P. wilsoni*, and *P. pledgei*, were described by Baird and Vickers-Rich (1998) from the late Oligocene/early Miocene Etadunna Formation in Australia (see also Boles 2001). Bones of *Palaelodus* cf. *ambiguus* were furthermore identified in the late Oligocene/early Miocene of the Taubaté Basin in Brazil (Alvarenga 1990).

Palaelodus has a short beak with a rounded tip, which was first misidentified as that of a representative of the Gruidae (see Cheneval and Escuillié 1992). The mandible is very deep, but distinct differences in the beak morphology suggest that *Palaelodus* may not yet have been adapted to filter feeding and instead foraged on aquatic insects, such as the larvae of caddisflies (Trichoptera), the protective cases of which are common fossils in some localities that yielded *Palaelodus* remains (Mayr 2015). The tarsometatarsus is proportionally shorter than in extant Phoenicopteriformes and more strongly mediolaterally compressed. In contrast to *Juncitarsus* and extant flamingos but as in grebes, the hypotarsus furthermore encloses canals for the digital flexor tendons. A presumably autapomorphic feature of the Palaelodidae is a marked pneumatic cavity in the sternal extremity of the furcula.

In combining a very deep mandible with leg bones that "show many similarities with those of a foot-propelled diving bird such as *Podiceps*" (Cheneval and Escuillié 1992: 218), palaelodids provide a morphological link between the Phoenicopteriformes and Podicipediformes. Although palaelodids were regarded as specialized "swimming flamingos" by earlier authors (e.g., Feduccia 1999: 209), it now appears more likely that the swimming adaptations, in particular the mediolaterally compressed tarsometatarsus, are plesiomorphic for the Phoenicopteriformes and already evolved in the stem species of the Mirandornithes. Both grebes and palaelodids are, or were, aquatic birds, which use, or used, their hindlimbs for propulsion. Therefore, it is most parsimonious to assume that the stem species of crown group Mirandornithes (i.e., the last common ancestor of extant grebes and flamingos) also had a similar ecomorphology (Mayr 2004a, 2015). Species in the stem lineage of the Phoenicopteridae then entered a new ecological zone as filter-feeders in shallow water bodies. That the evolutionary history of the Mirandornithes was complex and involved much homoplasy is, however, shown by the long-legged *Juncitarsus*, which appears to have been a wading bird and suggests that aquatic adaptations evolved in the stem lineage of crown group Mirandornithes, after the latter separated from the Juncitarsidae (Mayr 2014a).

The Paleogene fossil record of true flamingos, that is, representatives of the Phoenicopteridae, is scarce and the affinities of some species are not well established. Olson and Feduccia (1980) already noted that the phylogenetic affinities of the putative phoenicopteriform *Elornis littoralis* from the early Oligocene of France cannot be established, so much the more as the whereabouts of the specimens, various

5.1 Podicipediformes (Grebes) and Phoenicopteriformes (Flamingos)

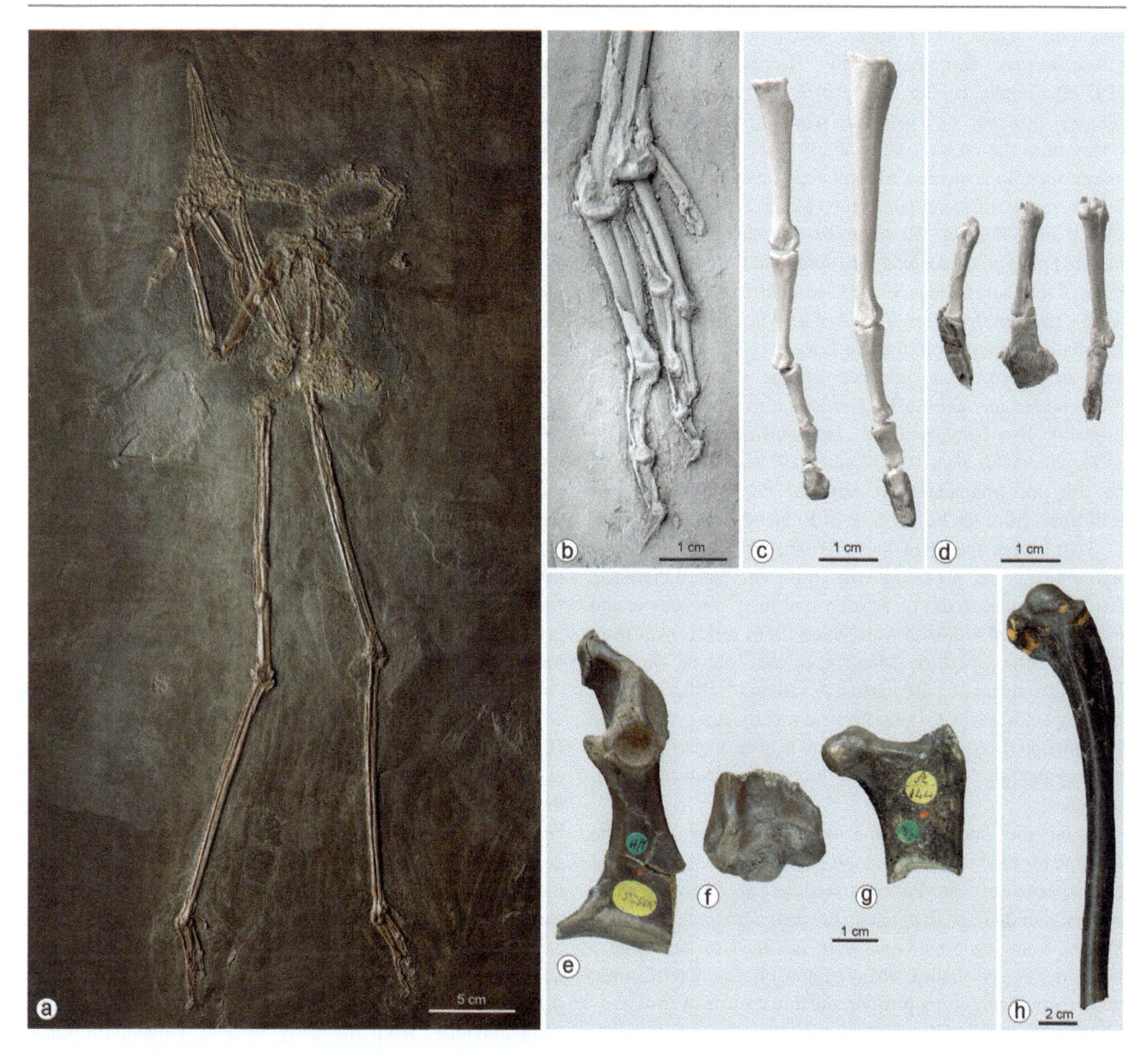

Fig. 5.1 Paleogene fossils of the Mirandornithes. (**a**) Cast of the holotype of *Juncitarsus merkeli* from the latest early or earliest middle Eocene of Messel (Juncitarsidae, Senckenberg Research Institute, Frankfurt, SMF-MEA 295). (**b**) Detail of the foot of *J. merkeli* (coated with ammonium chloride) in comparison to (**c**) the fourth (left) and third (right) toes of the extant *Phoenicopterus ruber* (Phoenicopteridae and (**d**) the penultimate and ungual phalanges of the three anterior toes of *Podiceps cristatus* (Podicipedidae). (**e**)-(**h**) Selected skeletal elements of ?*Agnopterus* ("*Headonornis*") *hantoniensis* from the late Eocene of England (**e**: holotype coracoid, Natural History Museum, London, NHMUK A 30325; **f**: distal humerus, NHMUK A 5105; **g**: cast of proximal femur, NHMUK A 144, holotype of *Gigantibis incognita*; **h**: proximal section of humerus, NHMUK A 3686). (Photo in (**a**) by Anika Vogel, all others by Sven Tränkner)

crushed postcranial bones on slabs, are unknown. The material includes a very long and slender tarsometatarsus, which suggests that at least some of the bones may indeed belong to a phoenicopteriform bird. As noted by Mlíkovský (2002: 255), the figured tibiotarsus is, however, quite different from that of the Phoenicopteriformes.

Several late Eocene and early Oligocene fossils of putative Phoenicopteriformes were assigned to the taxon *Agnopterus*, which was originally introduced for *A. laurillardi* from the late Eocene (MP 19) of the Paris Gypsum. This latter species was established on the basis of an incomplete distal tibiotarsus (see also Brunet 1970). A coracoid from the late Eocene (MP 17; Mlíkovský 2002) of England was described as ?*A. hantoniensis* by Lydekker (1891), who tentatively referred it to the Phoenicopteriformes (Fig. 5.1e–h). Harrison and Walker (1976a, 1979a) assigned humeri and a scapula from the late Eocene and early Oligocene of England to this species and classified it into the new taxon *Headonornis*, which they considered to be an Old World representative of the anseriform Presbyornithidae. Dyke (2001) even concluded that the humeri referred to *Headonornis* (?*Agnopterus*) *hantoniensis* by Harrison and Walker (1979a) actually belong to *Presbyornis isoni* (Sect. 4.5.2). There is, however, no convincing reason to disassociate these bones from the coeval and similarly-sized coracoids and to assign them to a species, which lived some 20 million years earlier on a different continent (Mayr 2008). Mlíkovský (2002) classified ?*A. hantoniensis* into the Anseranatidae, but the holotype coracoid is very similar to the corresponding bone of the Palaelodidae, and ?*A. hantoniensis* is more likely to be a stem group representative of the Phoenicopteriformes. A proximal femur from the late Eocene type locality, which was described as a member of the Threskiornithidae ("*Gigantibis incognita*") by Harrison and Walker (1976a), probably also belongs to ?*A. hantoniensis* (pro Lydekker 1891, contra Harrison and Walker 1976a). The same may be true for fragmentary remains of putative Phoenicopteriformes from the late Eocene of England (Harrison 1979).

Two further species, which were classified into the taxon *Agnopterus*, are *A. turgaiensis* from the late Oligocene (*Indricotherium* Beds) of Kazakhstan (distal tibiotarsus and distal humerus; Mlíkovský and Švec 1986) and *A. sicki* from the late Oligocene/early Miocene of the Taubaté Basin in Brazil (distal end of a tibiotarsus; Alvarenga 1990). Although these species seem to have been correctly assigned to the Phoenicopteriformes, they are known from material that is too fragmentary for a reliable phylogenetic placement within the clade.

From the late Oligocene or early Miocene of the Australian Etadunna Formation, Miller (1963) described two species of the Phoenicopteridae as *Phoeniconotius eyrensis* and *Phoenicopterus novaehollandiae*. Both are known from leg bones and were assumed to have had a better-developed hallux than extant Phoenicopteriformes, which would suggest a position outside the crown group.

Whereas the fossil record of the above species is rather sparse, numerous bones of a coeval European species, *Harrisonavis* ("*Phoenicopterus*") *croizeti*, were found in the late Oligocene and early Neogene of France. This species closely resembles crown group Phoenicopteridae, but the beak is still less downcurved than that of its extant relatives (Harrison and Walker 1976b; Torres et al. 2015). This plesiomorphic morphology shows *Harrisonavis* to be outside crown group Phoenicopteriformes, and the diversification of the latter does not appear to have commenced before the Neogene period.

5.2 Charadriiformes (Shorebirds and Allies)

Extant Charadriiformes are a diversified and species-rich group of birds and occupy many different habitats, from the open sea and coastal shores to semi-deserts. Molecular data provide a robust framework for the interrelationships of the various groups, which fall into three clades (Fig. 5.2): the Lari (gulls, auks, and allies), Charadrii (plovers and allies), and Scolopaci (sandpipers and allies; Paton and Baker 2006; Fain and Houde 2007; Hackett et al. 2008; Prum et al. 2015; Kuhl et al. 2021).

Crown group Charadriiformes share an unusually low motility of the enzyme malate dehydrogenase (Kitto and Wilson 1966), but because some taxa exhibit a highly aberrant morphology it is more difficult to characterize the group with derived osteological features. Most charadriiform birds lack pneumatic foramina in the pneumotricipital fossa of the humerus (exceptions are the Stercorariidae and Rynchopidae). In many taxa, the humerus furthermore exhibits a strongly developed dorsal supracondylar process (except Jacanidae, Burhinidae, Alcidae, and Turnicidae), and the coracoid likewise has a characteristic derived morphology with a medioventrally protruding acrocoracoid process. Charadriiform birds are furthermore characterized by distinctive proportions of the phalanges of the fourth toe, with the fourth phalanx being of subequal length to the third, whereas in most other birds the fourth phalanx of the fourth toe is longer than the third.

The Late Cretaceous/early Paleocene "Graculavidae," which were considered "transitional shorebirds" by Olson (1985) and Feduccia (1999), are discussed in Sect. 2.4 and cannot be convincingly assigned to the Charadriiformes on the basis of derived features. *Dakotornis cooperi* from the late Paleocene of North Dakota, for example, was originally described as an ibis-like bird, but the species was assigned to the "Graculavidae" by Benson (1999), who referred further fragmentary bones from the late Paleocene of North Dakota to it. None of these exhibits features, which would unambiguously support a classification into the Charadriiformes, and the same is true for fragmentary fossils of putative "Graculavidae" from the early Eocene Tingamarra Local Fauna in Australia (Boles 1999; see Sect. 5.2).

The evolutionary origin of charadriiform birds remains elusive. Two early Eocene species, however, may be archaic stem group representatives of the clade. The holotype of *Scandiavis mikkelseni* from the Danish Fur Formation is an exceptionally well-preserved skeleton, which lacks, however, the wing and pectoral girdle elements (Fig. 5.3a; Bertelli et al. 2013). The species has long nostrils, a broad pelvis with a pair of deep pits on the dorsal surface of its caudal portion and ilia that are not co-ossified with the synsacrum, as well as a rather short tarsometatarsus with a short and plantarly deflected trochlea for the second toe. *Scandiavis* resulted as the sister taxon of the Charadriiformes in an analysis by Bertelli et al. (2013), but this phylogeny is weakly based, and the fossil taxon exhibits no unambiguous charadriiform apomorphies. A very similar species from the North American Green River Formation was described as *Nahmavis grandei* (Fig. 5.3b;

5.2 Charadriiformes (Shorebirds and Allies)

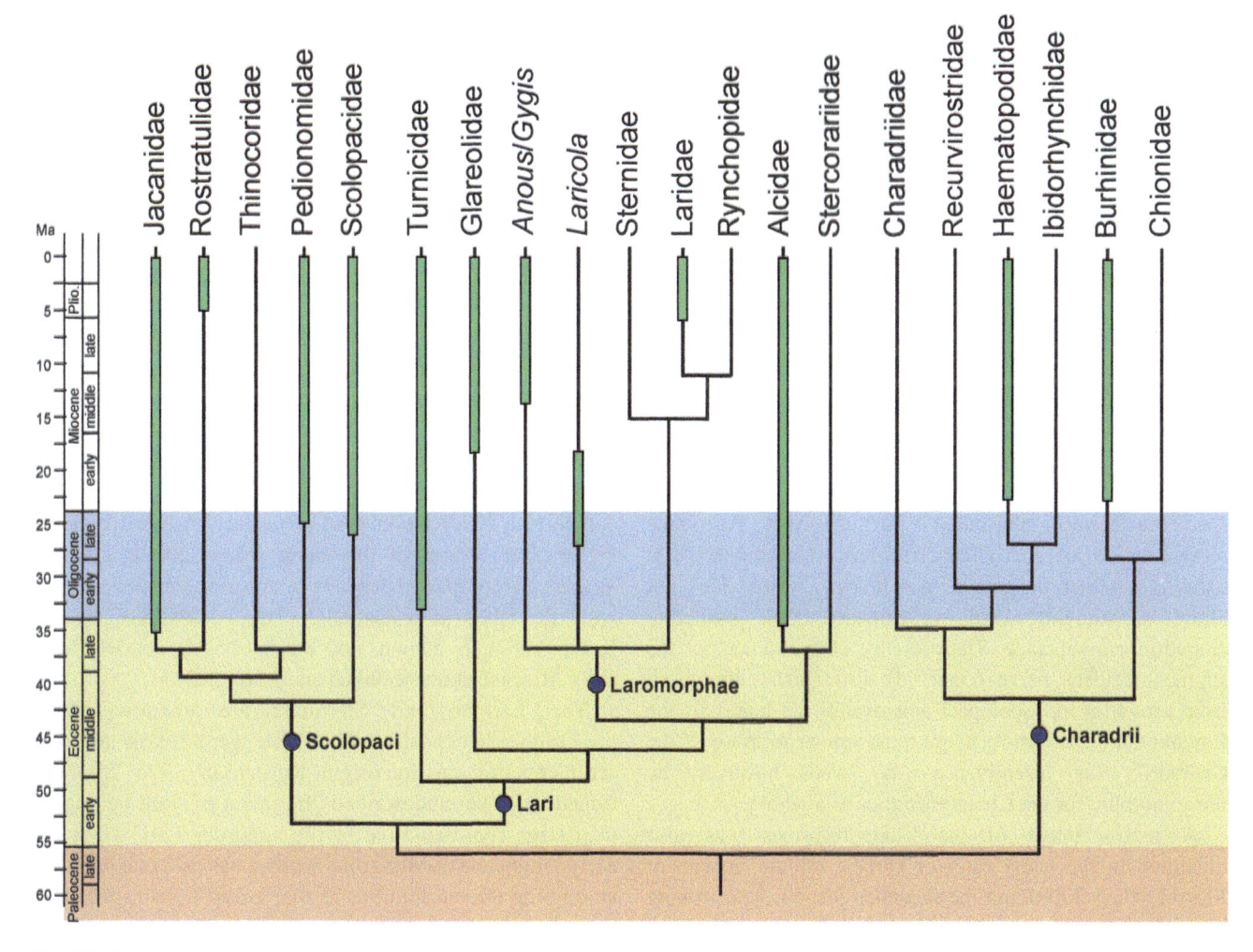

Fig. 5.2 Interrelationships and known stratigraphic occurrences of crown group Charadriiformes. The divergence dates are hypothetical

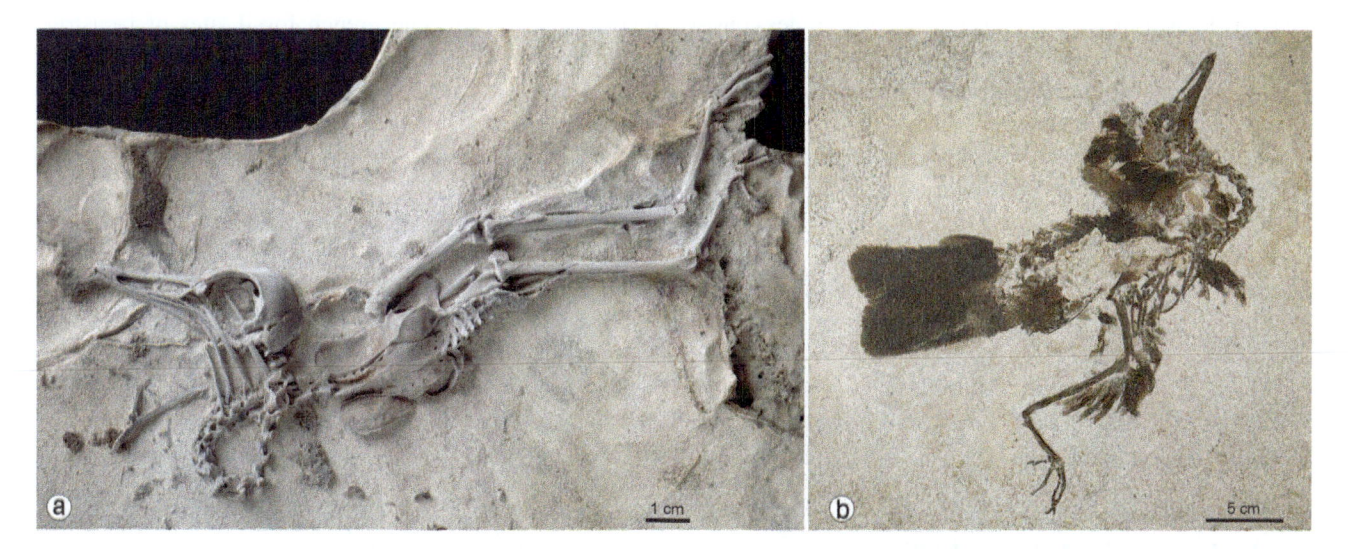

Fig. 5.3 (a) Holotype of *Scandiavis mikkelseni* from the early Eocene Fur Formation in Denmark (Moler Museet, Mors, Denmark, FU171x); photo by Sven Tränkner. (b) Holotype of *Nahmavis grandei* from the early Eocene Green River Formation, Wyoming, USA (Field Museum of Natural History, Chicago, USA, FMNH PA778; from Musser and Clarke 2020: fig. 1, published under a Creative Commons CC BY 4.0 license, lettering digitally removed)

Musser and Clarke 2020). This species is not clearly differentiated from the taxon *Scandiavis* and formed a clade with *S. mikkelseni* in some of the phylogenetic analyses of Musser and Clarke (2020). Like *S. mikkelseni*, *N. grandei* has long nostrils and fairly robust legs with a short tarsometatarsus. Unfortunately, the only known specimen of *Nahmavis* likewise lacks the wings and pectoral girdle elements.

The earliest definitive records of the Charadriiformes are likewise from the early Eocene, some 55–53 Ma: One is a distal humerus of an unnamed species from the Nanjemoy Formation in Virginia (Fig. 5.4g; Mayr 2016a), the other a distal humerus from the Naranbulag Formation of Mongolia (Hood et al. 2019). Both humeri exhibit a large supracondylar process and resemble the humerus of a charadriiform bird collected by Michael Daniels in the London Clay of Walton-on-the-Naze (Fig. 5.4i). The partial skeleton of the latter species is the most substantial record of an early Paleogene representative of the Charadriiformes. Unfortunately, it exhibits a rather unspecific morphology, which does not allow a straightforward assignment to a particular charadriiform subclade. The presence of a foramen for the supracoracoideus nerve (coracoid) distinguishes the fossil from crown group Scolopaci and overall the bones of the London Clay charadriiform are most similar to those of the Charadrii; this resemblance may well, however, be plesiomorphic for the Charadriiformes as a whole.

A representative of the Charadriiformes was also identified in the latest early or earliest middle Eocene of Messel (Mayr 2000), but the specimen just consists of wing bones and cannot be assigned to any extant charadriiform taxon with confidence. Another charadriiform-like bird from Messel, *Vanolimicola longihallucis*, exhibits a long and pointed beak and a long hallux (Mayr 2017). The holotype and only known specimen of this long-legged species consist of a poorly preserved skeleton, which agrees with extant Charadriiformes in the proportions of the phalanges of the fourth toe, the fourth phalanx of which is of subequal length to the third. Unlike in most extant Charadriiformes, the hallux is very long, but a very long hallux is also present in the Jacanidae, with which *V. longihallucis* agrees in the large distal vascular foramen of the tarsometatarsus. Otherwise, however, *Vanolimicola* distinctly differs from jacanas in its limb proportions, with the tibiotarsus being proportionally shorter and the toes not being extremely elongated. As noted by Mayr (2017), an as yet undescribed *Vanolimicola*-like species also occurred in the Green River Formation (Grande 2013: fig. 143A).

5.2.1 Lari (Gulls, Auks, and Allies)

The Lari comprise the Dromadidae (crab plover), Stercorariidae (skuas), Alcidae (auks), Laridae (gulls, terns, and skimmers), and Glareolidae (pratincoles and coursers). With the exception of some Glareolidae, all of these birds live in aquatic environments. The Turnicidae (buttonquails) are the sister taxon of the above taxa (Paton and Baker 2006; Fain and Houde 2007).

The earliest fossil Alcidae are from North American deposits. *Hydrotherikornis oregonus* from the late Eocene of Oregon was described on the basis of a distal tibiotarsus (Miller 1931). This specimen from the Pacific coast was assumed to possibly be from a procellariiform bird by Warheit (2002), but Chandler and Parmley (2003) reported a distal end of an alcid humerus from the late Eocene of Georgia (Atlantic coast), which further substantiates the occurrence of auks in the late Eocene of North America. Fossils of putative Alcidae from the late Eocene or early Oligocene of Japan (Ono and Hasegawa 1991; Mori and Miyata 2021) are in need of restudy. As already noted in the first edition of this book (Mayr 2009), *Petralca austriaca* from marine deposits of Austria, which was originally described as a late Oligocene auk (Mlíkovský and Kovar 1987), is a loon, and the species is actually from early Miocene strata (Göhlich and Mayr 2018).

The Larinae and Sterninae are unknown from pre-Oligocene deposits, and all Paleogene fossils are from sediments of lacustrine origin. Kurochkin (1976: 78) mentioned an as yet undescribed "fragment of skull belonging to a large representative of the suborder Lari" from the early Oligocene of Mongolia, which "has no exact parallels among the Recent families of that group". Two species of the taxon *Laricola*, *Laricola elegans* and *L. totanoides*, occur in late Oligocene lacustrine deposits of France (Hugueney et al. 2003; Mourer-Chauviré et al. 2004; De Pietri et al. 2011). Both species have an abundant fossil record in early Miocene French fossil sites of lacustrine origin and are very small, the size of small terns, with fairly long legs. A phylogenetic analysis by De Pietri et al. (2011) recovered *Laricola* in a basal position within Laridae, outside a clade including the Larinae, Sterninae, and Rhynchopinae.

Based on several postcranial elements, Mayr and Smith (2001) described two charadriiform species from the early Oligocene of Belgium as *Boutersemia belgica* and *B. parvula*, and tentatively assigned them to the Glareolidae. The coracoid exhibits a foramen for the supracoracoideus nerve, and the occurrence of a tarsometatarsal fossa for the first metatarsal indicates that a hallux was present. *Boutersemia* shares a large distal vascular foramen of the tarsometatarsus with extant Glareolidae, Charadriidae, and Jacanidae; this presumably derived feature is absent in other extant charadriiform taxa. Compared to extant Glareolidae, *Boutersemia* most closely resembles the taxa *Stiltia* and *Glareola* (Glareolinae), whereas in *Rhinoptilus* and *Cursorius* (Cursoriinae) the hallux is reduced and the distal

5.2 Charadriiformes (Shorebirds and Allies)

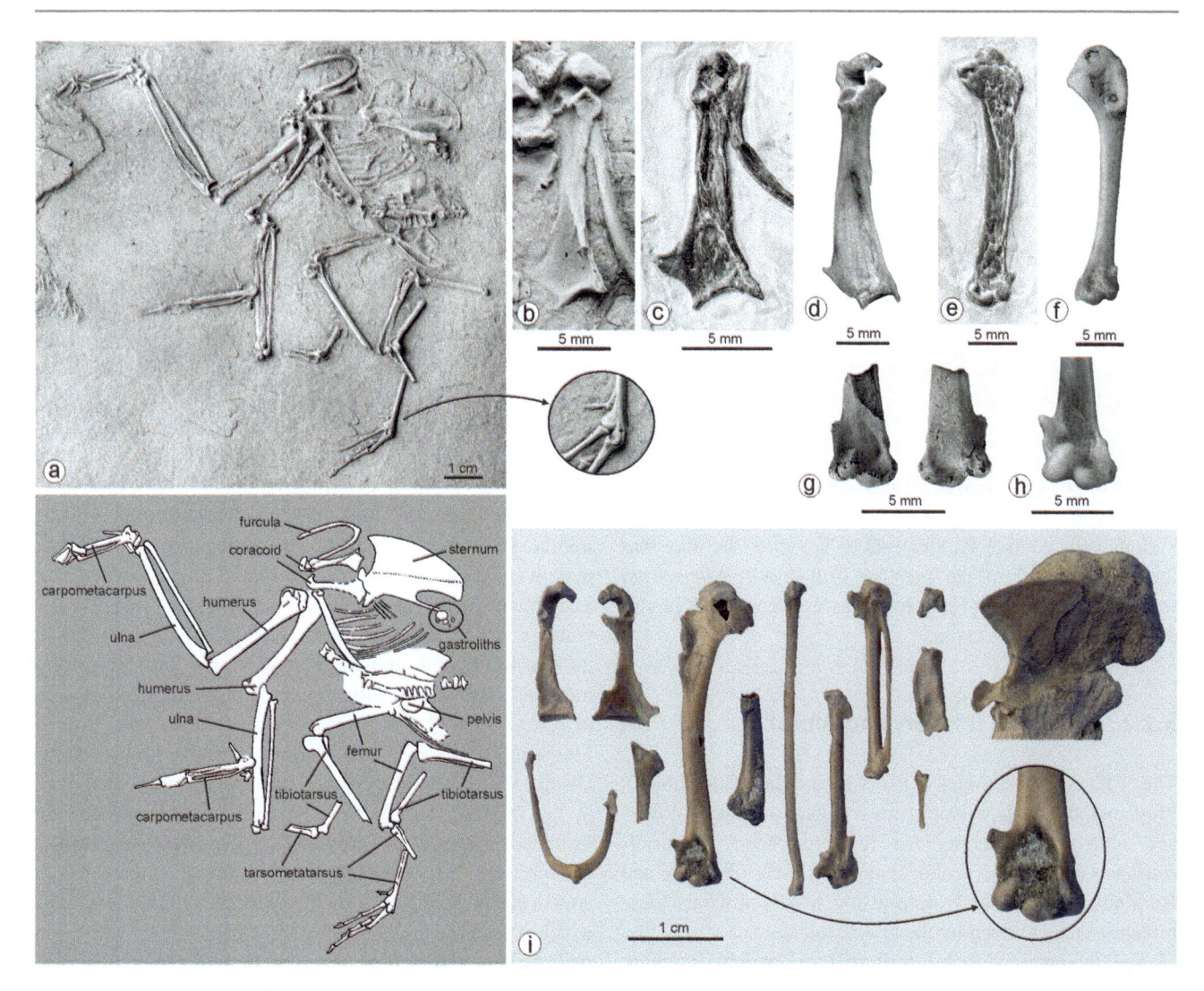

Fig. 5.4 Fossils of charadriiform birds. (**a**) Holotype of *Turnipax oechslerorum* from the early Oligocene of Wiesloch-Frauenweiler in Germany (Senckenberg Research Institute, Frankfurt, SMF Av 505a) with interpretive drawing. (**b**) Coracoid of *T. oechslerorum* (SMF Av 506a). (**c**) Coracoid of the holotype of *T. dissipata* from the early Oligocene of the Luberon area in France (SMF Av 427). (**d**) Coracoid of the extant *Turnix tanki* (Turnicidae). (**e**) Humerus of *T. dissipata* (SMF Av 427). (**f**) Humerus of the extant *Turnix hottentottus*. (**g**) Distal humerus of a charadriiform bird from the early Eocene Nanjemoy Formation, Virginia, USA (SMF Av 619) in cranial and caudal view. (**h**) Distal humerus of the extant *Charadrius tricollaris* (Charadriidae). (**i**) Partial skeleton of an undescribed charadriiform bird from the early Eocene London Clay of Walton-on-the-Naze (collection of Michael Daniels, Holland-on-Sea, UK, WN 91676); the arrow indicates an enlarged view of the distal end of the humerus. (All photos by Sven Tränkner)

vascular foramen smaller; *Cursorius* furthermore lacks a foramen for the supracoracoideus nerve.

Stem group representatives of the Turnicidae were reported from the early Oligocene of France and Germany. These fossils belong to the taxon *Turnipax* and the tentatively referred *Cerestenia pulchrapenna* (Mayr 2000). *Turnipax dissipata* was described from the Luberon area in France. A second, slightly larger species, *T. oechslerorum*, occurs in the early Oligocene of Wiesloch-Frauenweiler in Germany (Fig. 5.4a, b; Mayr and Knopf 2007). *Turnipax* can be assigned to the stem group of the Turnicidae on the basis of a characteristic morphology of the coracoid, which agrees with extant buttonquails in an unusually broad procoracoid process that almost meets the articular facet for the furcula and in a deeply excavated impression for the sternocoracoideus muscle on the sternal extremity of the bone. The skeleton of *Turnipax* combines derived characteristics of crown group Turnicidae with a plesiomorphic and more typically charadriiform overall morphology (Mayr 2000; Mayr and Knopf 2007). Crown group Turnicidae are omnivorous birds, which take seeds and other plant matter as well as various invertebrates. The presence of gastroliths in the holotype of *Turnipax oechslerorum* indicates an at least facultatively granivorous diet of this species (Mayr and Knopf 2007). *C. pulchrapenna* and extant Turnicidae lack a hallux. Its presence in *Turnipax* suggests that the habitat and way of living of this fossil taxon differed from that of crown group Turnicidae, which mainly inhabit open grasslands and arid scrub and probably did not diversify before the spread of open habitats toward the Neogene (Mayr and Knopf 2007). De Pietri et al. (in press) reported a distal humerus of a putative turnicid from the late Oligocene of France, but this fragmentary bone does not allow a determination of the exact affinities of the fossil within the Turnicidae.

A putative charadriiform with possible affinities to the Turnicidae occurs in the middle Eocene (47–49 Ma) of Namibia (Mourer-Chauviré et al. 2017). *Eocliffia primaeva* is distinguished from *Turnipax* in that the carpometacarpus and the first phalanx of the major wing digit are shorter and wider, and in that the ulna is shorter and stouter. With regard to these differences, the geologically older *Eocliffia* is more similar to crown group Turnicidae. However, more material would be desirable for corroboration of the assignment of the various isolated skeletal elements to *Eocliffia* (the ulna was initially assigned to the putative psittaciform *Namapsitta* from the same locality) and for an assessment of the phylogenetic affinities the taxon.

5.2.2 Charadrii (Plovers and Allies)

The Charadrii encompass the Chionididae (sheathbills), Pluvianellidae (Magellanic plover), Charadriidae (plovers and allies), Haematopodidae (oystercatchers), Recurvirostridae (stilts and avocets), Ibidorhynchidae (ibisbill), and Burhinidae (thick-knees). Except for the Chionididae, these birds are characterized by the loss of the hallux.

The Paleogene fossil record of the Charadrii is poor. Various fragmentary remains were assigned to the Recurvirostridae and are of doubtful identity. This is true for "*Recurvirostra*" *sanctaenebulae* from the late Eocene of France, which is known from the proximal end of an ulna (Mourer-Chauviré 1978). The taxonomic identity and affinities of this species need to be substantiated with additional bones, and even if it belonged to the Recurvirostridae, it is unlikely that it can be classified in the extant taxon *Recurvirostra*. Another putative recurvirostrid described by Harrison (1983) from the early Eocene of Portugal is just known from an incomplete femur and cannot even be reliably assigned to the Charadriiformes, let alone the Recurvirostridae (see also Mlíkovský 2002). The same is true for reputed Recurvirostridae and Burhinidae, which were reported by Harrison and Walker (1976a) from the late Eocene (MP 14–16/MP 17; Mlíkovský 2002) of England. Putative tracks of the Recurvirostridae from the Green River Formation were described by Olson (2014). Identification of these footprints was based on the fact that they appear to be from a long-legged bird with a similar foot morphology to that of extant stilts and avocets. Without osteological specimens, however, the presence of recurvirostrids in the Green River Formation remains arguable.

Chionoides australiensis from the late Oligocene Etadunna Formation of South Australia is the earliest fossil record of the Chionoidea, the clade including the Chionididae and Pluvianellidae (De Pietri et al. 2016). The holotype of this species is a partial coracoid, which displays a mosaic of features found in extant Chionididae and Pluvianellidae; for an unambiguous phylogenetic placement of this fossil, the discovery of more material is, however, needed.

The earliest record of the Burhinidae is *Genucrassum bransatensis* from the late Oligocene (23 Ma) of the Allier Basin in central France (De Pietri and Scofield 2014). The description of this species was based on fragmentary wing and pectoral girdle bones, which exhibit the distinctive morphologies found in extant Burhinidae.

Hou and Ericson (2002) described a humerus from the middle Eocene of China as *Jiliniornis huadianensis* and tentatively assigned the species to the Charadriidae. This assignment was, however, not established with derived features, and the authors themselves considered the possibility that the shared similarities may be plesiomorphic for a more inclusive charadriiform clade. The correct identification of "*Charadrius*" *sheppardianus* from the late Eocene Florissant shales of Colorado was already questioned by Olson (1985). The holotype of this species was not figured in the original description (Cope 1880) and needs to be restudied. Bessonat and Michaut (1973) briefly described a charadriiform bird from the early Oligocene of the Luberon area in southern France. The exact affinities of this specimen, which is from a species the size of a Stone-curlew (*Burhinus oedicnemus*) and is now in a private collection in France, have not been determined. Judging from its overall morphology and the absence of a hallux, it may well be a representative of the Charadrii.

5.2.3 Scolopaci (Sandpipers and Allies)

The Scolopaci include the Jacanidae (jacanas), Rostratulidae (painted snipes), Pedionomidae (plains wanderer), Thinocoridae (seed snipes), and Scolopacidae (sandpipers, snipes, and allies). Among other features, these birds are characterized by a coracoid without a foramen for the supracoracoideus nerve.

Olson (1999) described the remains of putative Scolopaci from the early Eocene of the Nanjemoy Formation of Virginia. Even though these fossils may well belong to the Charadriiformes, their fragmentary nature does not allow an unambiguous identification, and the distal humerus identified by Olson (1999) is actually from a procellariiform bird (Mayr 2016a; Mayr et al. 2022). The holotype of *Paractitis bardi* from the late Eocene (Chadronian) of Canada is the omal extremity of a coracoid (Weigel 1963), which was assigned to the Scolopacidae owing to the absence of a foramen for the supracoracoideus nerve. As noted above, however, this feature is characteristic for all representatives of the Scolopaci.

A well-preserved postcranial skeleton of an unnamed species of the Scolopacidae from the early Oligocene of the Luberon area in France was described by Roux (2002). This specimen is from a species about the size of the Eurasian Woodcock (*Scolopax rusticola*), but its tarsometatarsus is proportionally longer. The humerus bears a well-developed dorsal supracondylar process, and a short hallux is present. The assignment of the fossil to the Scolopacidae was mainly based on the shape of the acrocoracoid process of the coracoid. Although this classification would be in line with the overall morphology of the fossil, it has yet to be established with derived characters. A fragmentary coracoid of a representative of the Scolopaci was also reported from the early Oligocene of Belgium (Mayr and Smith 2001).

"*Totanus*" *edwardsi* from an unknown stratigraphic horizon of the Quercy fissure fillings (Gaillard 1908) needs to be restudied to assess its affinities within the Scolopaci. A tentative record of the Phalaropodinae (phalaropes), consisting of the omal extremity of a coracoid, comes from the late Oligocene of France (Mourer-Chauviré et al. 2004).

Oligonomus milleri from the late Oligocene (24–26 Ma) of South Australia is the earliest fossil record of the Pedionomidae (De Pietri et al. 2015). This species, of which only a partial coracoid was found, lived in a forested paleohabitat and therefore appears to have differed from the extant Plains-wanderer in its ecological preferences.

The Jacanidae today occur in the tropic regions of all continents and are the sister taxon of the Rostratulidae (e.g., Paton and Baker 2006; Fain and Houde 2007; Prum et al. 2015; Kuhl et al. 2021). These aberrant Charadriiformes have the proportionally longest toes of all extant birds, which enable them to walk on floating vegetation. The sole Paleogene fossils belong to three species from the late Eocene and early Oligocene of the Jebel Qatrani Formation in Egypt. All are just known from distal tarsometatarsi and were described by Rasmussen et al. (1987) as *Nupharanassa bulotorum*, *N. tolutaria*, and *Janipes nymphaeobates*. A derived feature shared with extant Jacanidae is the large distal vascular foramen, which opens dorsally in a very wide fossa. The early Oligocene species *N. bulotorum* and *J. nymphaeobates* are distinctly larger than extant Jacanidae, whereas the size of the late Eocene *N. tolutaria* falls within the range of the extant species (*N. tolutaria* was likewise considered to be of early Oligocene age in the original description, but see Rasmussen et al. 2001 for a revised age). Similar large-sized Jacanidae were also reported from the middle Miocene of Kenya (Mayr 2014b).

5.3 Gruiformes (Rails, Cranes, and Allies)

The traditional "Gruiformes" constitute a polyphyletic taxon (Sect. 2.3), but morphological and sequence-based analyses congruently support a sister group relationship between the Ralloidea and Gruoidea, that is, the clades (Rallidae + (Sarothruridae + Heliornithidae)) and (Psophiidae + (Aramidae + Gruidae)) (Livezey 1998; Ericson et al. 2006; Fain et al. 2007; Prum et al. 2015; Kuhl et al. 2021). Mainly because of the apomorphic morphology of the aquatic Heliornithidae, it is difficult to characterize this clade with morphological apomorphies. However, most of its representatives exhibit well-developed supraorbital processes (skull), a narrow sternum, and a characteristic pelvis morphology (Mayr 2008). The Gruiformes have a comparatively rich Paleogene fossil record, but many species were described on the basis of fragmentary remains that do not allow reliable identification.

5.3.1 Messelornithidae and *Walbeckornis*

With over 500 specimens found so far (Morlo 2004), the Messelornithidae (Messel rails) are the most abundant avian group in Messel, and these birds are also fairly common in the Green River Formation (Hesse 1990; Weidig 2010). The five species currently included in the taxon are *Messelornis cristata* from Messel (Fig. 5.5a, b; Hesse 1988a, 1990), *M. nearctica* from the Green River Formation (Fig. 5.5c; Hesse 1992; Weidig 2010), *M. russelli* from the Paleocene of France (Mourer-Chauviré 1995), *Pellornis mikkelseni* from the Fur Formation (Fig. 5.5f; Bertelli et al. 2011; Musser et al. 2019), and *Itardiornis hessae* from the late Eocene and early Oligocene (MP 17–23) of the Quercy fissure fillings (Mourer-Chauviré 1995). As yet undescribed messelornithids were also collected by Michael Daniels from the London Clay of Walton-on-the-Naze (Fig. 5.5d); these fossils may belong to the same species as a distal tarsometatarsus from the Isle of Sheppey, which was described as *Coturnipes cooperi* by Harrison and Walker (1977). Some isolated bones of unnamed early Eocene *Messelornis*-like species were furthermore reported from the Nanjemoy Formation in Virginia, USA (Mayr 2016a) and from Egem in Belgium (Fig. 5.5e; Mayr and Smith 2019a). Mayr and Zvonok (2011) described a partial *Itardiornis*-like tarsometatarsus from the middle Eocene of Ukraine. Possible messelornithid remains were also found in the mid-Paleocene (early to middle Selandian) of Maret in Belgium and in the late Paleocene of the locality Rivecourt-Petit Pâtis in France (Mayr and Smith 2019b).

Messelornis cristata, the best-known species of the Messelornithidae, was the size of a Common Moorhen (*Gallinula chloropus*), long legs with moderately sized toes,

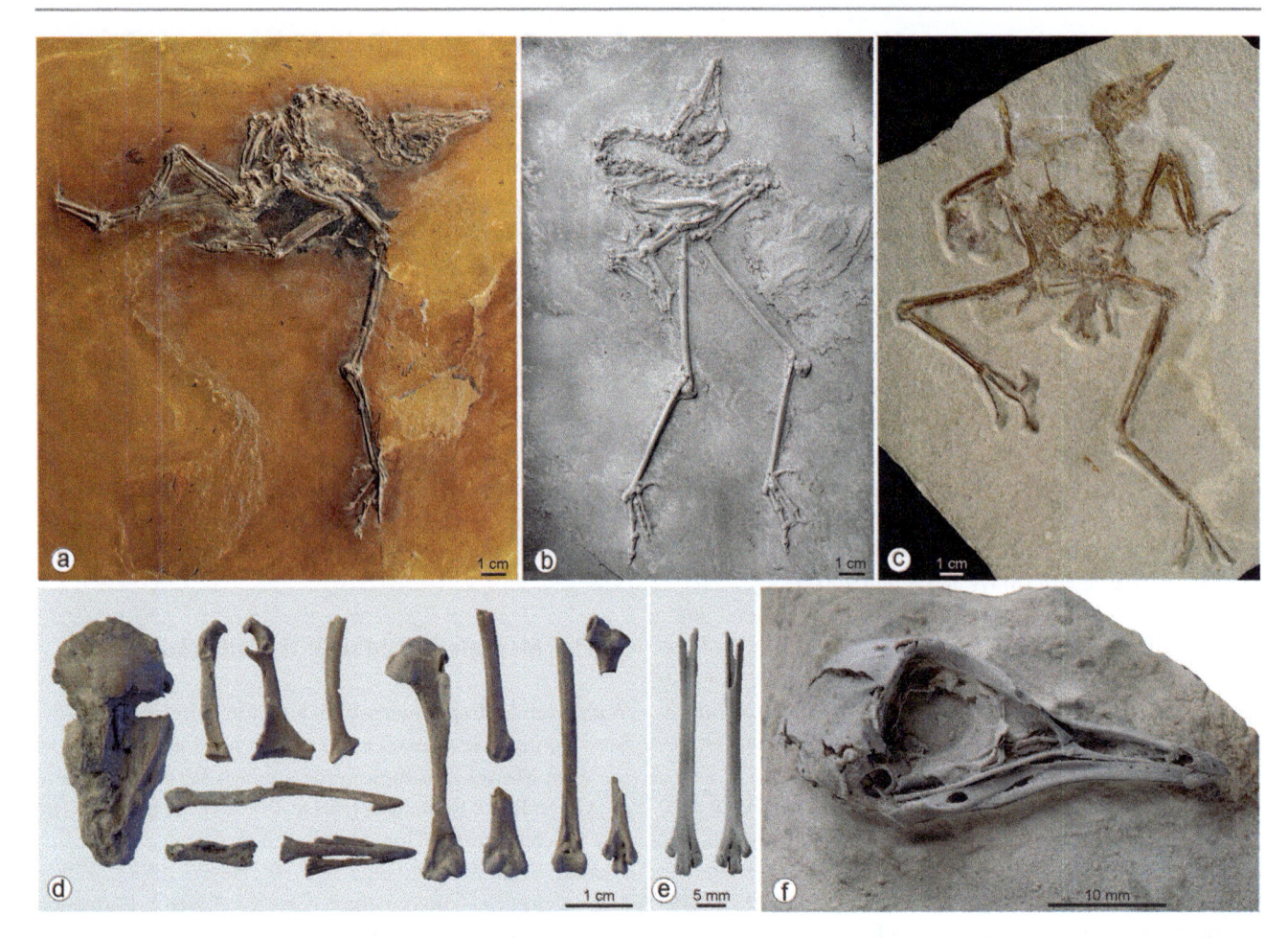

Fig. 5.5 Fossils of the gruiform Messelornithidae. (**a**), (**b**) Skeletons of *Messelornis cristata* from the latest early or earliest middle Eocene of Messel (Senckenberg Research Institute, Frankfurt, **a**: SMF-ME 1275, **b**: SMF-ME 3551); specimen in (**b**) coated with ammonium chloride. (**c**) Holotype of *M. nearctica* from the North American Green Giver Formation (SMF Av 406). (**d**) Undescribed specimen of a messelornithid from the early Eocene London Clay of Walton-on-the-Naze (collection of Michael Daniels, Holland-on-Sea, UK, WN 85505). (**e**) Partial tarsometatarsus (dorsal and plantar view) of a messelornithid from the early Eocene of Egem in Belgium (Royal Belgian Institute of Natural Sciences, Brussels, Belgium, IRSNB Av 169). (**f**) Skull of *Pellornis mikkelseni* from the early Eocene Fur Formation in Denmark (Fur Museum, Denmark, FUM-N 856). (Photo in (**a**) by Anika Vogel, all others by Sven Tränkner)

and a rather short beak (Fig. 5.5a, b). The species epithet "*cristata*" refers to the fact that in one specimen Hesse (1988a, 1990) identified the remains of what was considered to be a fleshy crest (see Feduccia 1999: 242). More likely, however, this putative crest represents accidentally associated organic matter, presumably plant material. Mayr (2004b) considered schizorhinal nostrils to be absent in *Messelornis cristata*, but for *Pellornis mikkelseni* these were established by Musser et al. (2019).

Messelornis cristata presumably was a predominantly terrestrial bird and appears to have been sexually dimorphic in size (Hesse 1990). In several specimens, the feathering is well preserved, and it can be observed that the species had a long tail (Hesse 1990: pl. 2, fig. 5). Morlo (2004) described a skeleton of *M. cristata* preserved with the remains of the percoid fish *Rhenanoperca minuta* in the area of the esophagus. The gut contents of other specimens of this species consist of seeds (Hesse 1990).

Fossils of *M. cristata* constitute about half of all avian remains discovered in Messel. Many specimens have exact stratigraphic data and can be assigned to particular horizons of the Messel deposits. Hesse and Habersetzer (1993) assumed that there is evidence for evolution toward increased perching capabilities during the sedimentation period of the Messel oilshale. This hypothesis was based on alleged differences in the position and relative size of the hallux of the numerous specimens of *Messelornis cristata*. However, the conclusions have to be viewed with caution, because it is difficult to assess such subtle differences in skeletons from the Messel site, many of which are considerably flattened and crushed.

The Quercy messelornithid *Itardiornis hessae* is represented by most major limb bones. As in the species of *Messelornis*, the proximal end of the tarsometatarsus exhibits an ossified supratendinal bridge (arcus extensorius). *I. hessae* was, however, larger than *Messelornis cristata* and *M. nearctica*, from which it is furthermore distinguished in that the articular facet on the coracoid for the scapula is shallow and not cup-like as in *Messelornis*.

Messel rails were considered to be most closely related to the South American Eurypygidae by Hesse (1988b, 1990) and Livezey (1998). However, this hypothesis conflicts with the fact that they lack several of the derived characters shared by the Eurypygidae and their sister taxon, the Rhynochetidae, such as a notarium (co-ossified thoracic vertebrae) and a deep, U-shaped incision in the caudal margin of the pelvis (Mayr 2004b). In contrast to extant Eurypygidae and Rhynochetidae, the coracoid of messelornithids furthermore exhibits a foramen for the supracoracoideus nerve (Fig. 5.6), and close affinities to the Ralloidea are much better supported by the morphological evidence. Derived characters shared by the Messelornithidae and Ralloidea include the absence of pneumatic foramina in the proximal end of the humerus and a distinctive morphology of the coracoid (Mayr 2004b). Mayr (2004b) proposed a sister group relationship between messelornithids and the Ralloidea, which was also supported in subsequent analyses (Bertelli et al. 2011; Musser et al. 2019). The hypotarsus of messelornithids shows, however, a close resemblance to that of the Sarothruridae and Heliornithidae, so that a position of messelornithids within the Ralloidea also needs to be considered (Mayr 2019).

This distinctive hypotarsus morphology, with two dorsoplantarly adjacent canals (Mayr 2004b), is known from *Itardiornis* (Mourer-Chauviré 1995) and *Pellornis* (Musser et al. 2019), and also occurs in a possible messelornithid from the Paleogene of Australia. This species, *Australlus disneyi* from late Oligocene to early Miocene deposits of the Riversleigh site in Australia, was initially considered to be a flightless rail and was assigned to the extant taxon *Gallinula* (Boles 2005). A revision of the fossil material led to its removal from *Gallinula* and classification into the new taxon *Australlus*, even though an assignment to the Rallidae was maintained (Worthy and Boles 2011). Actually, however, *A. disneyi* is more similar to the Messelornithidae in the morphology of the tarsometatarsus and that of the hypotarsus in particular (Mayr 2013, 2016b). Most major wing and leg bones of *A. disneyi* are known, and from the proportions of the limb elements, flightlessness of the species was inferred. Even though extant rails include several flightless insular species, the loss of flight capabilities in the continental *A. disneyi* would be very unusual, because the species coexisted with marsupial carnivores. However, the assumption of flightlessness was based on comparisons with extant Rallidae, and compared to the Messelornithidae, the wing elements of *Australlus* do not appear to be unusually small.

The description of *Messelornis russelli* from the Paleocene of the Reims area in France was based on two humeri, which are incomplete or poorly preserved, respectively. Owing to this sparse fossil record, the identification of the species is not uncontroversial. Although the humerus *M. russelli* resembles that of *M. cristata*, it is also very similar to the humerus of *Walbeckornis creber* from the Paleocene of Walbeck in Germany (Fig. 5.7). This latter species is the most abundant bird in the locality and is known from numerous three-dimensionally preserved bones, which represent all major limb elements (Mayr 2007). The humerus and coracoid of *W. creber* are very similar to the corresponding bones of the Messelornithidae, but the exact phylogenetic affinities of the species are unresolved (Mayr 2007). *W. creber* has a much shorter tarsometatarsus than *Messelornis*, but there seems to be another, unnamed *Walbeckornis*-like species in the Walbeck material, which has a long tarsometatarsus (Mayr 2007). As in *Messelornis*, the proximal end of the humerus of *Walbeckornis* lacks pneumatic foramina, whereas the hypotarsus does not exhibit the derived morphology found in messelornithids.

The distal end of the tibiotarsus and the proximal end of the tarsometatarsus of *W. creber* show some similarity to the corresponding bones of *Wanshuina lii* from the Paleocene of China (Hou 1994, 2003). This species was assigned to the Rallidae by Hou (1994) and its description was based on a humerus shaft, a distal tibiotarsus, and a proximal tarsometatarsus. Direct comparisons are, however, needed to evaluate the possibility of a closer relationship between *Walbeckornis* and *Wanshuina*. Zelenkov (2021b) described partial coracoids of a *Walbeckornis*-like bird from the early Eocene of the Tsagaan-Khushuu locality of Mongolia as a new species, *Bumbaniralla walbeckornithoides*.

5.3.2 Songziidae

Songzia heidangkouensis from the early Eocene Yangxi Formation of Hubei Province in China was long only known from the holotype, a partial skeleton consisting of the skull, both hindlimbs, and an incomplete wing. The species was classified in the monotypic taxon Songziidae by Hou (1990), who assumed a close relationship to the Rallidae. Further specimens of *Songzia* from the type locality were reported by Wang et al. (2012), who described a new species, *S. acutunguis* (Fig. 5.8a), and noted that *Songzia* is actually more similar to the Messelornithidae than to the Rallidae. Even though *Songzia* has rail-like proportions of the major limb bones, with a long tarsometatarsus and a short ulna, the sternum is very different from that of the Rallidae, Sarothruridae, and Heliornithidae in its shape and the presence of four caudal incisions. Unlike in the Messelornithidae, however, the hallux and other toes of *Songzia* are very long. Phylogenetic analyses performed by Wang et al. (2012) did not yield conclusive results, whereas an analysis by Musser et al. (2019) supported a sister group relationship between the Songziidae and Messelornithidae. Skeletons of unnamed

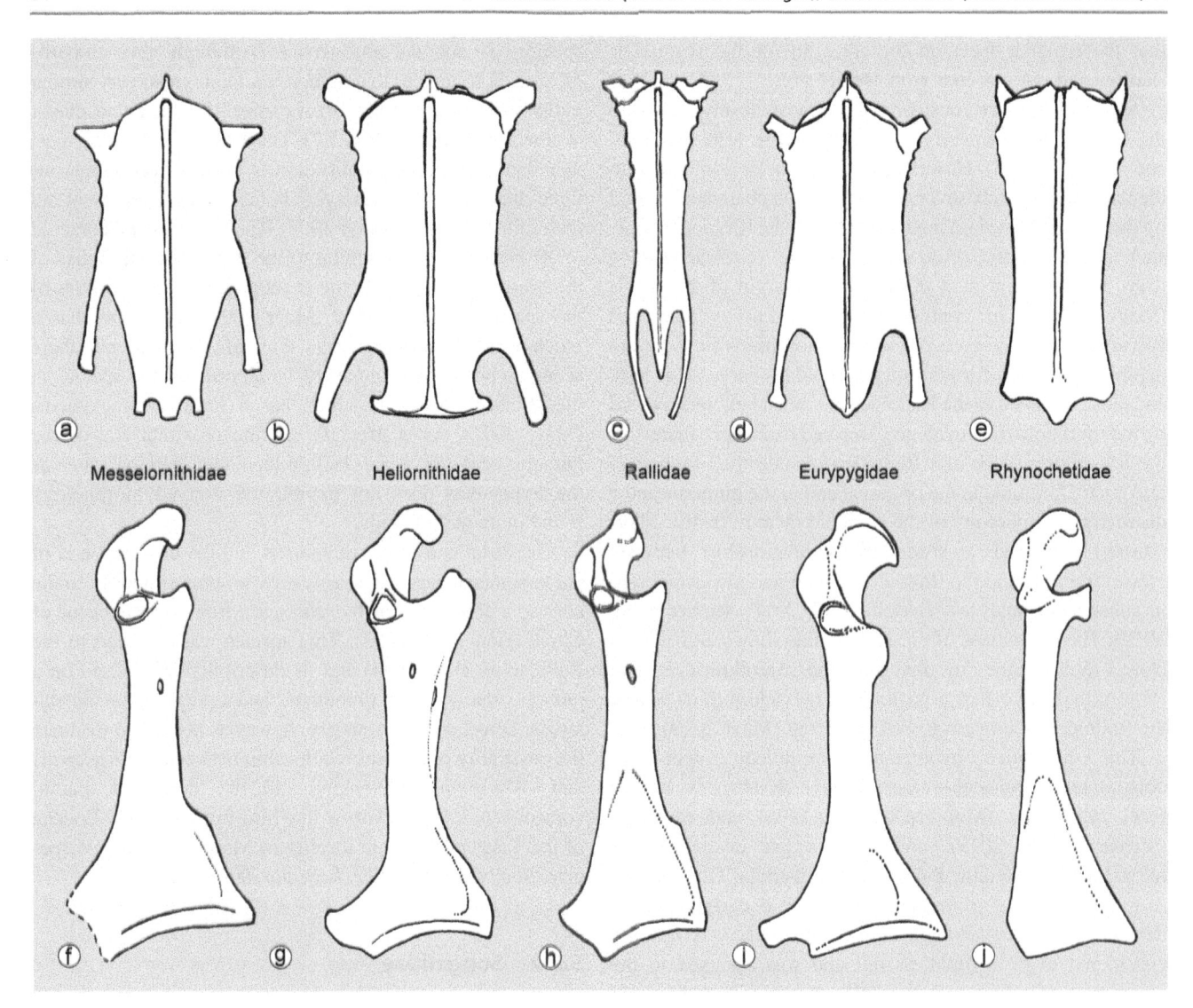

Fig. 5.6 Sterna (**a**)-(**e**) and coracoids (**f**)-(**j**) of messelornithids and extant gruiform and eurypygiform birds. (**a**), (**f**) *Messelornis cristata* (Messelornithidae). (**b**), (**g**) *Heliornis fulica* (Heliornithidae). (**c**) *Pardirallus maculatus* (Rallidae). (**d**), (**i**) *Eurypyga helias* (Eurypygidae). (**e**), (**j**) *Rhynochetos jubatus* (Rhynochetidae). (**h**) *Aramides saracura* (Rallidae). Not to scale

Songzia-like birds were also described from the late Paleocene of Menat in France (Mayr et al. 2020) and from the early Eocene of Canada (Fig. 5.8b; Mayr et al. 2019).

5.3.3 Ralloidea (Finfoots, Flufftails, and Rails)

Analyses of sequence data suggest that the traditional Rallidae (rails) are paraphyletic, with the African flufftails (Sarothruridae) being the sister taxon of the Heliornithidae (finfoots; Hackett et al. 2008, Prum et al. 2015). Extant rails are globally distributed and include ecologically disparate taxa, such a crakes (*Crex*, *Porzana*) and coots (*Fulica*). As an adaptation to their habitat, mainly dense clutter of reed, they have a mediolaterally compressed body with a very narrow sternum and pelvis. The three extant species of the Heliornithidae are foot-propelled aquatic birds, which occur in Africa south of the Sahara and in the tropical areas of South and Central America and Asia.

Finfoots have no Paleogene record, and the earliest fossil is from the middle Miocene of North America (Olson 2003). Although various early Paleogene fossils were identified as rails, many of these consist of very fragmentary or poorly preserved fossils. Without further material and before detailed comparisons with, for example, the Messelornithidae have been performed, their affinities cannot be determined. In particular, this is true for several species of putative Rallidae from early and middle Eocene deposits of North America and England (Olson 1977; Harrison and Walker 1979b; Harrison 1984). The same applies to *Ibidopsis hordwelliensis* from the late Eocene (MP 17; Mlíkovský 2002) of England, which was classified in the Rallidae by Harrison and Walker

5.3 Gruiformes (Rails, Cranes, and Allies)

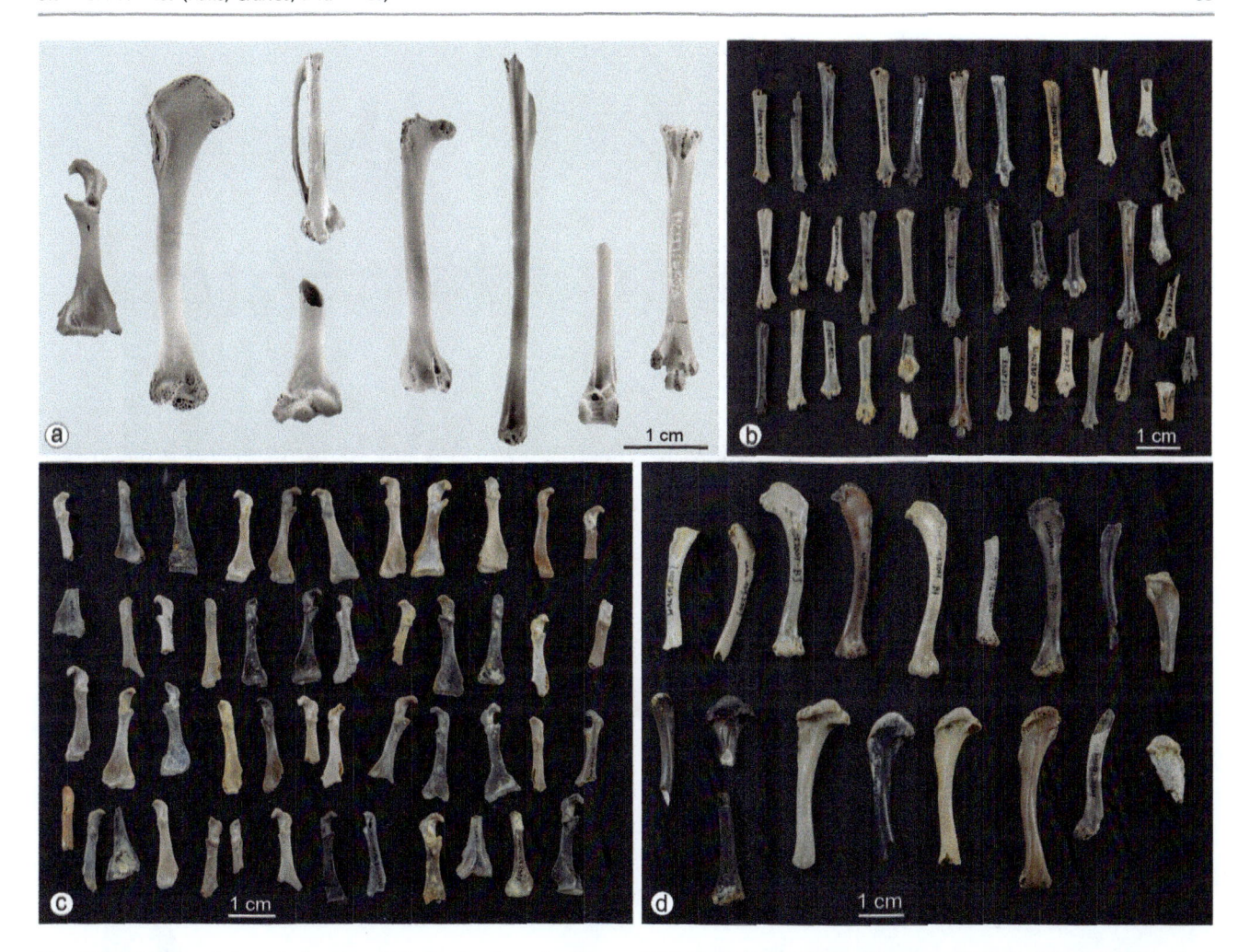

Fig. 5.7 (a) Selected skeletal elements of *Walbeckornis creber* from the late middle Paleocene of Walbeck in Germany (all Institut für Geologische Wissenschaften of Martin-Luther-Universität Halle-Wittenberg, Halle/Saale, Germany; coracoid: IGWuG WAL39.2007, humerus: IGWuG WAL102.2007, carpometacarpus: IGWuG WAL221.2007, distal humerus: IGWuG WAL162.2007, femur: IGWuG WAL372.2007, tibiotarsus: IGWuG WAL385.2007, distal tibiotarsus: IGWuG WAL390.2007, holotype tarsometatarsus: IGWuG WAL472.2007); all specimens coated with ammonium chloride. (b)–(c) Tarsometatarsi, coracoids, and humeri of *W. creber* in IGWuG to illustrate the abundance of this species in the Walbeck locality. (All photos by Sven Tränkner)

(1976a). This fairly large species is known from an incomplete coracoid and a distal tibiotarsus (Fig. 5.9d, e). Although it seems to be a representative of the Gruiformes, its assignment to the Rallidae is weakly based, and the coracoid likewise resembles the corresponding bone of the Messelornithidae (Sect. 5.3.1) and Parvigruidae (Sect. 5.3.4).

Quercyrallus arenarius and "*Quercyrallus*" *quercy* are represented by incomplete humeri from unknown stratigraphic horizons of the Quercy fissure fillings and are of uncertain affinities, just as "*Rallus*" *adelus* (= "*Quercyrallus ludianus*") from the late Eocene of the Paris Gypsum, a long-beaked species known from an incomplete and poorly preserved skeleton (see Cracraft 1973; Olson 1977). "*Quercyrallus dasypus*" was synonymized with the galliform *Palaeortyx gallica* by Mourer-Chauviré (1992). *Megagallinula harundinea* (proximal ulna) from the late Oligocene (*Indricotherium* Beds) of Kazakhstan (Kurochkin 1968a) and *Palaeorallus alienus* (distal tibiotarsus) from the "middle" Oligocene of Mongolia (Kurochkin 1968b) were excluded from the Rallidae by Cracraft (1973); *P. alienus* was assigned to the Galliformes by Zelenkov and Kurochkin (2015).

The earliest unambiguously identified fossils of the Rallidae are two species of *Belgirallus*, *B. oligocaenus* and *B. minutus*, which were described on the basis of diagnostic postcranial bones from the earliest Oligocene of Belgium (MP 21; Mayr and Smith 2001). A partial skeleton of a rail from the early Oligocene of Germany corresponds well with *Belgirallus oligocaenus* in size and morphology (Mayr 2006). This specimen is from a medium-sized species, about the size of the extant Water Rail (*Rallus aquaticus*), and closely resembles extant rails in its skeletal morphology. As in most extant Rallidae, the toes are greatly elongated.

Fig. 5.8 Fossils of the gruiform Songziidae. (**a**) Holotype of *Songzia acutunguis* from the early Eocene Yangxi Formation of Hubei Province, China (Institute of Vertebrate Paleontology and Paleoanthropology, Beijing, China, IVPP 18188). (**b**) A *Songzia*-like bird from the early Eocene of the Driftwood Canyon fossil site in the Okanagan Highlands of British Columbia, Canada (Royal British Columbia Museum, Victoria, British Columbia, RBCM.EH2009.015.0001). (Photo in (**a**) courtesy of Min Wang, (**b**) by the author)

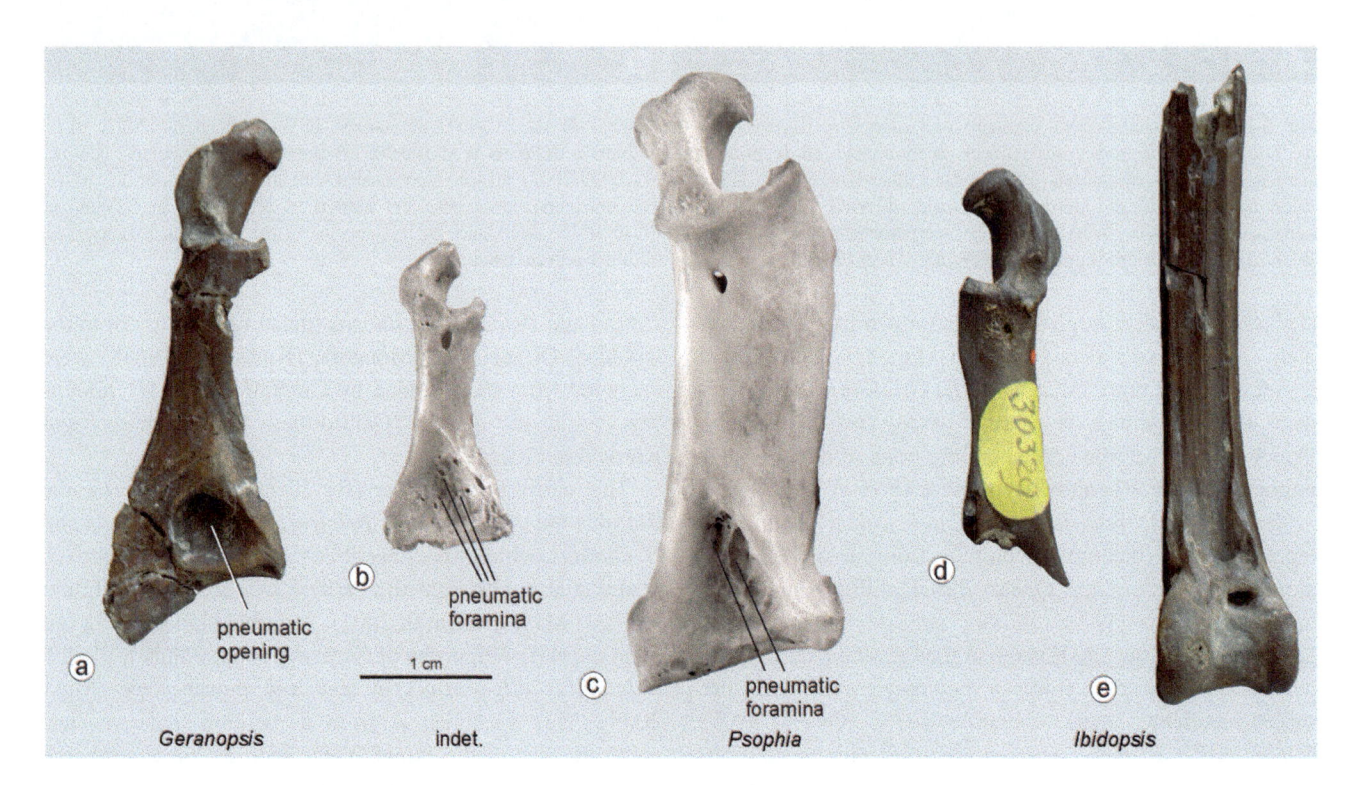

Fig. 5.9 (**a**) Coracoid of *Geranopsis hastingsiae* from the late Eocene of England (Natural History Museum, London, NHMUK A 30331). (**b**) Coracoid of a *Psophia*-like bird from an unknown locality of the Quercy fissure fillings in France (Muséum national d'Histoire naturelle, Paris, MNHN QU 16929); (**c**) Coracoid of the extant Grey-winged Trumpeter, *Psophia crepitans* (Psophiidae). (**d**) Referred coracoid (NHMUK A 30329) and (**e**) distal right tibiotarsus (holotype, NHMUK A 36793) of *Ibidopsis hordwelliensis* from the late Eocene of England. (All photos by Sven Tränkner)

A skeleton of a small rail from the early Oligocene of the Carpathian Basin in Poland was considered to be similar to *Paraortygometra porzanoides* by Mayr and Bochenski (2016); the fossil has a wider sternal extremity of the furcula than most extant rails. Remains of *P. porzanoides* were also reported from the late Oligocene of France by Mourer-Chauviré et al. (2004), and the species shows a resemblance to extant Sarothruridae (De Pietri and Mayr 2014).

Most other Oligocene Rallidae are known from just a few bones, which tell us little apart from the occurrence of rails in European and African deposits of that geological epoch. This applies to a distal tarsometatarsus of an unnamed rail from the early Oligocene of the Jebel Qatrani Formation (Fayum) in Egypt (Rasmussen et al. 1987) and to *Rallicrex kolozsvarensis* from the "middle" Oligocene of Romania, of which a distal tarsometatarsus has been found (Lambrecht 1933).

5.3.4 Parvigruidae

The Parvigruidae comprise medium-sized gruiform birds from the early Oligocene of Europe with moderately long leg bones, which exhibit a mosaic distribution of plesiomorphic rail-like traits and derived features of the Gruoidea. The best-represented species of the taxon is *Parvigrus pohli*, the holotype of which is a complete skeleton from the Luberon area in southern France (Fig. 5.10a; Mayr 2005). *Rupelrallus saxoniensis* from the early Oligocene of Germany is another representative of the Parvigruidae (Fig. 5.10). This species is known from several bones of a single individual, including a fragmentary furcula and coracoid, a complete carpometacarpus, and distal ends of the tibiotarsus and tarsometatarsus. *R. saxoniensis* was assigned to the Rallidae in the original description (Fischer 1997), but the species differs from rails in the deeper sternal extremity of the furcula and the longer lateral process of the coracoid. *R. saxoniensis* is also larger than most extant species of rails and closely resembles the coeval but slightly smaller *Parvigrus pohli* in the preserved skeletal elements (Mayr 2013). Two differently-sized species of the Parvigruidae were furthermore reported from the early Oligocene Boom Formation in Belgium (Mayr 2013). One of these was tentatively assigned to *Parvigrus pohli*, whereas the other represents a different species described as *Rupelrallus belgicus*. One of the fossils exhibits medullary bone, which indicates a breeding female.

An assignment of parvigruids to the Gruoidea is supported by the elongated and narrow sternum, which lacks deep incisions in its caudal margin, and by a medial projection formed by the proximal end of the first phalanx of the fourth toe (Mayr 2005). As in the Rallidae, Gruidae, and Aramidae, the caudal end of the mandible bears a hook-like retroarticular process. The beak is of similar proportions to that of the Psophiidae but has longer nostrils. The sternal extremity of the coracoid bears a marked depression, but unlike in the Psophiidae and Gruidae it lacks pneumatic foramina. The length proportions of the limb bones are "rail-like," that is, the ulna is shorter than the humerus and the femur as long as the humerus. The toes are proportionally longer than those of the Gruidae but less elongated than in the Aramidae. As in all Gruoidea and a few species of the Ralloidea, there are ossified tendons along the hindlimbs.

Initially (Mayr 2005), *P. pohli* was considered to be the sister taxon of the clade (Aramidae + Gruidae), because it agrees with the latter and differs from the Psophiidae in the long nostrils and a plantarly deflected trochlea for the second toe. An analysis by Musser et al. (2019) also supported a sister group relationship between parvigruids and the clade (Aramidae + Gruidae). However, the tibiotarsus of parvigruids lacks a tubercle near the supratendinal bridge, the presence of which is a derived characteristic of the Gruoidea, and the distal end of the bone is also more similar to the distal tibiotarsus of the Rallidae in the proportions of the condyles. The plesiomorphic features of the tibiotarsus, which for reasons of preservation cannot be discerned in the holotype of *P. pohli*, distinguish the Parvigruidae from extant Gruoidea and indicate that these birds are the sister taxon of the Gruoidea (Mayr 2013). This revised phylogenetic position is also in better agreement with the fact that the caudal margin of the sternum of *P. pohli* exhibits a pair of shallow incisions, whereas it is entire in extant Gruoidea.

Another gruiform bird from the Luberon area was described as *Palaeogeranos tourmenti* by Louchart and Duhamel (2021), who considered the fossil to be a possible stem group representative of the Gruidae. This species is solely represented by a coracoid, and even though this bone is somewhat larger than the coracoid of *P. pohli*, it differs only slightly from the latter species in morphological details. Close affinities between *P. tourmenti* and the Gruidae are poorly established, and here it is considered more likely that the species is another representative of the Parvigruidae.

5.3.5 Gruoidea (Trumpeters, Limpkins, and Cranes)

The Psophiidae (trumpeters), Aramidae (limpkins), and Gruidae (cranes) share pneumatic openings in the sternal extremity of the coracoid, a notarium, and a prominent tubercle near the supratendinal bridge of the tibiotarsus. As detailed in Sect. 2.3, a clade including these three taxa is congruently obtained in current phylogenetic analyses of molecular sequence data. Within this clade, the Aramidae and Gruidae are sister groups.

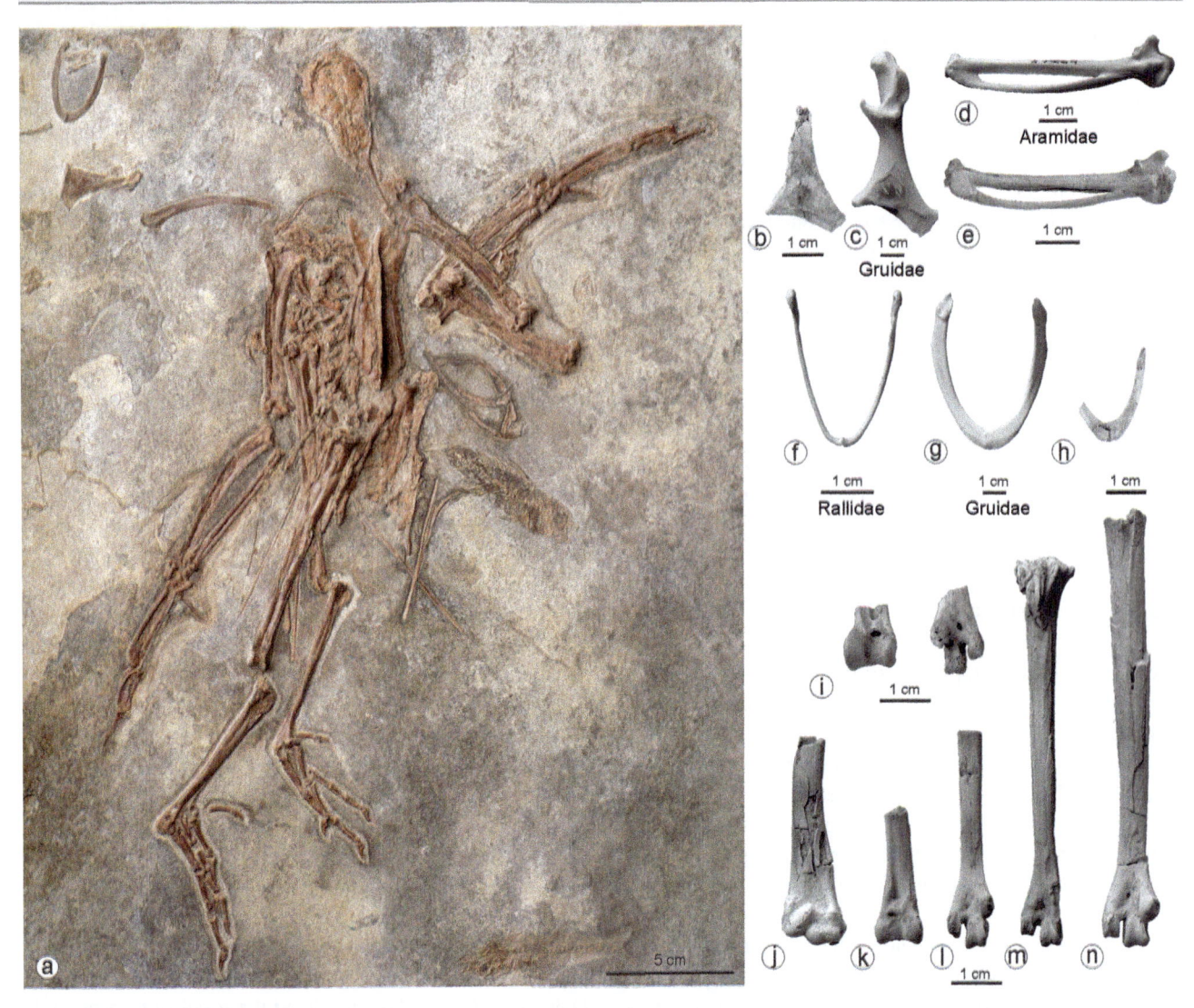

Fig. 5.10 Fossils of the gruiform Parvigruidae. (**a**) Holotype of *Parvigrus pohli* from the early Oligocene of the Luberon area in France (holotype, Wyoming Dinosaur Center, Thermopolis, USA, WDC-CCF 02). (**b**) Sternal end of the coracoid of *Rupelrallus saxoniensis* from the early Oligocene of Germany (holotype, Naturkundemuseum Leipzig, Germany, NML Pa. 3251). (**c**) Coracoid of the extant *Balearica regulorum* (Gruidae). (**d**) Carpometacarpus of the extant *Aramus guarauna* (Aramidae). (**e**) Left carpometacarpus (mirrored) of *R. saxoniensis* (NML Pa. 3251). (**f**)–(**h**) Furcula of (**f**) the extant *Fulica atra* (Rallidae), (**g**) the extant *B. regulorum* (Gruidae), and (**h**) *R. saxoniensis* (NML Pa. 3251). (**i**) Distal ends of the tibiotarsus and tarsometatarsus of *R. saxoniensis* (NML Pa. 3251). (**j**) Distal end of the humerus of *Parvigrus pohli* from the early Oligocene of Belgium (Royal Belgian Institute of Natural Sciences, Brussels, Belgium, IRSNB Av 98). (**k**) Distal tibiotarsus of *P. pohli* from the early Oligocene of Belgium (IRSNB Av 97b). (**l**)–(**n**) Distal tarsometatarsi of (**l**), (**m**) *P. pohli* from the early Oligocene of Belgium (l: IRSNB Av 96a; m: IRSNB Av 96a) and (**n**) *Rupelrallus belgicus* from the early Oligocene of Belgium (holotype, IRSNB Av 100a). (All photos by Sven Tränkner)

A fragmentary distal tarsometatarsus from the late Eocene Submeseta Formation of Seymour Island (Antarctica) was tentatively assigned to the Gruoidea by Davis et al. (2020). However, the specimen lacks most of the trochleae, so that a well-founded classification is impossible.

A partial tarsometatarsus from the early Eocene of the Tsagaan-Khushuu locality of Mongolia was described as *Bumbanipes aramoides* by Zelenkov (2021b), who noted similarities to the tarsometatarsus of extant Gruiformes. The fossil is, however, distinguished from most extant gruiform birds in a mediolaterally narrow trochlea for the fourth toe, and a robust phylogenetic placement requires the discovery of additional material.

The South American Psophiidae do not have an unambiguously identified Paleogene fossil record. Mayr and Mourer-Chauviré (2006) described a trumpeter-like coracoid from an unknown stratigraphic horizon of the Quercy fissure fillings, which agrees with the corresponding bone of extant Psophiidae in the presence of a marked crest along the medial side of the shaft and pneumatic openings in the sternal extremity (Fig. 5.9b). The specimen is, however, from a species much smaller than any extant trumpeter and cannot be referred to this group of birds with confidence.

The extant representatives of the Aramidae and Gruidae have a complementary distribution, with the Aramidae occurring in South and Central America, whereas the Gruidae are found on all other continents except South America and Antarctica. A putative member of the Aramidae from the early Oligocene (Brule Formation) of South Dakota was described as *Badistornis aramus* by Wetmore (1940). This species is known from a nearly complete tarsometatarsus, which resembles that of extant Aramidae in the strongly plantarly deflected trochlea for the second toe. Its classification into the Aramidae was also accepted by Cracraft (1973), but more fossil material of *B. aramus* is desirable to be confident about its phylogenetic affinities. Chandler and Wall (2001) described three eggs from the early Oligocene of the Brule Formation of South Dakota, which they assigned to *B. aramus*.

Loncornis erectus from the late Oligocene (Deseadan) of Argentina is just represented by a fragmentary distal femur. The species was assigned to the Aramidae by Brodkorb (1967), but its affinities are indeterminable on the basis of the known material (Cracraft 1973; Agnolin 2004).

The holotype of "*Palaeogrus*" *hordwelliensis* from the late Eocene (MP 17; Mlíkovský 2002) of England is a distal tibiotarsus (Harrison and Walker 1976a). Because the middle Eocene *Palaeogrus princeps*, the type species of the genus *Palaeogrus*, is now assigned to the palaeognathous Palaeotididae (Sect. 3.2.1), "*P.*" *hordwelliensis* has to be assigned to a new genus. A smaller species, *Geranopsis hastingsiae*, occurs in the late Eocene and early Oligocene of England and is known from coracoids (Fig. 5.9a) and a referred distal tibiotarsus (Lydekker 1891; Harrison and Walker 1976a, 1979a). As noted above (Sect. 4.5.3), the coracoid of this species closely resembles that of the putative anseranatid *Anserpica* from the late Oligocene of France. Mourer-Chauviré (2006) furthermore noted the presence of undescribed Gruidae in the Quercy fissure fillings.

The fossil record of cranes outside Europe is equally sparse. The description of *Eobalearica tugarinovi* from the middle Eocene Ferghana Basin of Uzbekistan was based on a distal tibiotarsus. The holotype and only known specimen is, however, poorly preserved, and its identification as a gruid was doubted by Cracraft (1973). Mayr and Zvonok (2011) hypothesized that the species may be a bony-toothed bird, but close affinities to pelagornithids were contested by Zelenkov and Kurochkin (2015), who maintained a classification in the Gruiformes. Rasmussen et al. (1987) assigned a distal tarsometatarsus from the early Oligocene Jebel Qatrani Formation of Egypt to the Gruidae. According to Olson (1985: 164, 1989: 2024), there are further fossil cranes from the late Eocene (Chadronian) of North America.

The extant species of the Gruidae occur in open grasslands and savannah-like biotopes. The evolution of cranes is therefore likely to have been correlated to the origin of extensive grassland biotas toward the late Paleogene and early Neogene, which may account for the absence of unambiguous fossils of these birds in early Paleogene fossil sites.

References

Agnolin FL (2004) Revisión systemática de algunas aves deseadenses (Oligoceno Medio) descriptas por Ameghino en 1899. Rev Mus Argent Cienc Nat, n s 6:239–244

Alvarenga HMF (1990) Flamingos fósseis da Bacia de Taubaté, estado de São Paulo, Brasil: descrição de nova espécie. An Acad Bras Cienc 62:335–345

Baird RF, Vickers-Rich P (1998) *Palaelodus* (Aves: Palaelodidae) from the Middle to Late Cainozoic of Australia. Alcheringa 22:135–151

Benson RD (1999) *Presbyornis isoni* and other late Paleocene birds from North Dakota. In: Olson SL (ed) Avian paleontology at the close of the 20th century: Proceedings of the 4th international meeting of the Society of Avian Paleontology and Evolution, Washington, DC, 4–7 June 1996. Smithson Contrib Paleobiol 89:253–259

Bertelli S, Chiappe LM, Mayr G (2011) A new Messel rail from the Early Eocene Fur Formation of Denmark (Aves, Messelornithidae). J Syst Palaeontol 9:551–562

Bertelli S, Lindow BEK, Dyke GJ, Mayr G (2013) Another charadriiform-like bird from the Lower Eocene of Denmark. Paleontol J 47:1282–1301

Bessonat G, Michaut A (1973) Découverte d'un squelette complet d'échassier dans le Stampien provençal. Bull Mus Hist Nat Marseille 33:143–145

Boles WE (1999) Early Eocene shorebirds (Aves: Charadriiformes) from the Tingamarra Local Fauna, Murgon, Queensland, Australia. Rec West Austral Mus Suppl 57:229–238

Boles WE (2001) A new emu (Dromaiinae) from the Late Oligocene Etadunna Formation. Emu 101:317–321

Boles WE (2005) A new flightless gallinule (Aves: Rallidae: *Gallinula*) from the Oligo-Miocene of Riversleigh, northwestern Queensland, Australia. Rec Austral Mus 57:179–190

Brodkorb P (1967) Catalogue of fossil birds. Part 3 (Ralliformes, Ichthyornithiformes, Charadriiformes). Bull Fla state Mus. Biol Sci 11:99–220

Brunet J (1970) Oiseaux de l'Éocène supérieur du bassin de Paris. Ann Paléontol 56:3–57

Chandler RM, Parmley D (2003) The earliest North American record of auk (Aves: Alcidae) from the Late Eocene of Central Georgia. Oriole 68:7–9

Chandler RM, Wall WP (2001) The first record of bird eggs from the early Oligocene (Orellan) of North America. Proceedings of the 6th Fossil Resource Conference. Geol Resour Div Techn Rep NPS/NRGRD/GRDTR-01:23–26

Cheneval J, Escuillié F (1992) New data concerning *Palaelodus ambiguus* (Aves: Phoenicopteriformes: Palaelodidae): ecological and evolutionary interpretations. In: Campbell KE (ed) Papers in avian paleontology honoring Pierce Brodkorb. Nat Hist Mus Los Angeles Cty Sci Ser 36:208–224

Cope ED (1880) On a wading bird from the Amyzon shales. Bull U S Geol Surv 6(3):83–85

Cracraft J (1973) Systematics and evolution of the Gruiformes (Class Aves). 3. Phylogeny of the suborder Grues. Bull Am Mus Nat Hist 151:1–127

Dames W (1890) Über Vogelreste aus dem Saltholmskalk von Limhamn bei Malmö. Kungl Svenska Vetenskapsakad Handl 16:3–11

Davis SN, Torres CR, Musser GM, Proffitt JV, Crouch NM, Lundelius EL, Lamanna MC, Clarke JA (2020) New mammalian and avian records from the late Eocene La Meseta and Submeseta formations of Seymour Island, Antarctica. PeerJ 8:e8268

De Pietri VL, Mayr G (2014) Reappraisal of early Miocene rails (Aves, Rallidae) from Central France: diversity and character evolution. J Zool Syst Evol Res 52:312–322

De Pietri VL, Scofield RP (2014) The earliest European record of a stone-curlew (Charadriiformes, Burhinidae) from the late Oligocene of France. J Ornithol 155:421–426

De Pietri VL, Costeur L, Güntert M, Mayr G (2011) A revision of the Lari (Aves, Charadriiformes) from the early Miocene of Saint-Gérand-le-Puy (Allier, France). J Vertebr Paleontol 31:812–828

De Pietri VL, Camens AB, Worthy TH (2015) A plains-wanderer (Pedionomidae) that did not wander plains: a new species from the Oligocene of South Australia. Ibis 157:68–74

De Pietri VL, Scofield RP, Hand SJ, Tennyson AJD, Worthy TH (2016) Sheathbill-like birds (Charadriiformes: Chionoidea) from the Oligocene and Miocene of Australasia. J Roy Soc New Zeal 46:181–199

De Pietri VL, Mayr G, Costeur L, Scofield RP (in press) New records of buttonquails (Aves, Charadriiformes, Turnicidae) from the Oligocene and Miocene of Europe. C R Palevol

Dyke GJ (2001) The fossil waterfowl (Aves: Anseriformes) from the Eocene of England. Am Mus Novit 3354:1–15

Ericson PGP (1999) New material of *Juncitarsus* (Phoenicopteriformes), with a guide for differentiating that genus from the Presbyornithidae (Anseriformes). In: Olson SL (ed) Avian paleontology at the close of the 20th century: Proceedings of the 4th international meeting of the Society of Avian Paleontology and Evolution, Washington, DC, 4–7 June 1996. Smithson Contrib Paleobiol 89:245–251

Ericson PGP, Anderson CL, Britton T, Elzanowski A, Johansson US, Källersjö M, Ohlson JI, Parsons TJ, Zuccon D, Mayr G (2006) Diversification of Neoaves: integration of molecular sequence data and fossils. Biol Lett 2:543–547

Fain MG, Houde P (2007) Multilocus perspectives on the monophyly and phylogeny of the order Charadriiformes (Aves). BMC Evol Biol 7:35

Fain MG, Krajewski C, Houde P (2007) Phylogeny of "core Gruiformes" (Aves: Grues) and resolution of the Limpkin-Sungrebe problem. Mol Phylogenet Evol 43:515–529

Feduccia A (1999) The origin and evolution of birds, 2nd edn. Yale University Press, New Haven

Fischer K (1997) Neue Vogelfunde aus dem mittleren Oligozän des Weißelsterbeckens bei Leipzig (Sachsen). Mauritiana 16:271–288

Gaillard C (1908) Les oiseaux des Phosphorites du Quercy. Ann Univ Lyon (Nouv Sér) 23:1–178

Göhlich UB, Mayr G (2018) The alleged early Miocene auk *Petralca austriaca* is a loon (Aves, Gaviiformes): restudy of a controversial fossil bird. Hist Biol 30:1076–1083

Grande L (2013) The lost world of Fossil Lake. Snapshots from deep time. University of Chicago Press, Chicago

Hackett SJ, Kimball RT, Reddy S, Bowie RCK, Braun EL, Braun MJ, Chojnowski JL, Cox WA, Han K-L, Harshman J, Huddleston CJ, Marks BD, Miglia KJ, Moore WS, Sheldon FH, Steadman DW, Witt CC, Yuri T (2008) A phylogenomic study of birds reveals their evolutionary history. Science 320:1763–1767

Harrison CJO (1979) The Upper Eocene birds of the Paris basin: a brief re-appraisal. Tertiary Res 2:105–109

Harrison CJO (1983) A new wader, Recurvirostridae (Charadriiformes), from the early Eocene of Portugal. Cięnc Terra 7:9–16

Harrison CJO (1984) Further additions to the fossil birds of Sheppey: a new Falconid and three small rails. Tert Res 5:179–187

Harrison CJO, Walker CA (1976a) Birds of the British Upper Eocene. Zool J Linnean Soc 59:323–351

Harrison CJO, Walker CA (1976b) Cranial material of Oligocene and Miocene flamingos: with a description of a new species from Africa. Bull Brit Mus (Nat Hist) 27:305–312

Harrison CJO, Walker CA (1977) Birds of the British Lower Eocene. Tert Res Spec Pap 3:1–52

Harrison CJO, Walker CA (1979a) Birds of the British Lower Oligocene. In: Harrison CJO, Walker CA (eds) studies in Tertiary avian paleontology. Tert Res Spec Pap 5:29–43

Harrison CJO, Walker CA (1979b) Birds of the British Middle Eocene. In: Harrison CJO, Walker CA (eds) Studies in Tertiary avian paleontology. Tert Res Spec Pap 5:19–27

Hesse A (1988a) Die Messelornithidae—eine neue Familie der Kranichartigen (Aves: Gruiformes: Rhynocheti) aus dem Tertiär Europas und Nordamerikas. J Ornithol 129:83–95

Hesse A (1988b) Taxonomie der Ordnung Gruiformes (Aves) nach osteologischen morphologischen Kriterien unter besonderer Berücksichtigung der Messelornithidae Hesse 1988. Cour Forsch-Inst Senckenberg 107:235–247

Hesse A (1990) Die Beschreibung der Messelornithidae (Aves: Gruiformes: Rhynocheti) aus dem Alttertiär Europas und Nordamerikas. Cour Forsch-Inst Senckenberg 128:1–176

Hesse A (1992) A new species of *Messelornis* (Aves: Gruiformes: Messelornithidae) from the Middle Eocene Green River Formation. In: Campbell KE (ed) Papers in avian paleontology honoring Pierce Brodkorb. Nat Hist Mus Los Angeles Cty Sci Ser 36:171–178

Hesse A, Habersetzer J (1993) Infraspecific development of various foot- and wing-proportions of *Messelornis cristata* (Aves: Gruiformes: Messelornithidae). Kaupia 3:41–53

Hood SC, Torres CR, Norell MA, Clarke JA (2019) New fossil birds from the earliest Eocene of Mongolia. Am Mus Novit 3934:1–24

Hou L-H (1990) An Eocene bird from Songzi, Hubei Province. Vertebr PalAsiat 28:34–42. [in Chinese]

Hou L-H (1994) A new Paleocene bird from Anhui, China. Vertebr PalAsiat 32:60–65. [in Chinese]

Hou L-H (2003) Fossil birds of China. Yunnan Science and Technology Press, Kunming

Hou L-H, Ericson PGP (2002) A Middle Eocene shorebird from China. Condor 104:896–899

Hugueney M, Berthet D, Bodergat A-M, Escuillié F, Mourer-Chauviré C, Wattinne A (2003) La limite Oligocène-Miocène en Limagne: changements fauniques chez les mammifères, oiseaux et ostracodes des différents niveaux de Billy-Créchy (Allier, France). Geobios 36:719–731

Kitto GB, Wilson AC (1966) Evolution of malate dehydrogenase in birds. Science 153:1408–1410

Kuhl H, Frankl-Vilches C, Bakker A, Mayr G, Nikolaus G, Boerno ST, Klages S, Timmermann B, Gahr M (2021) An unbiased molecular approach using 3'UTRs resolves the avian family-level tree of life. Mol Biol Evol 38:108–121

Kurochkin EN (1968a) New Oligocene birds from Kazakhstan. Paleontol Ž 1:92–101. [in Russian]

Kurochkin EN (1968b) Fossil remains of birds from Mongolia. Ornitologija 9:323–330. [in Russian]

Kurochkin EN (1976) A survey of the Paleogene birds of Asia. Smithson Contrib Paleobiol 27:5–86

Lambrecht K (1933) Handbuch der Palaeornithologie. Gebrüder Borntraeger, Berlin

Livezey BC (1998) A phylogenetic analysis of the Gruiformes (Aves) based on morphological characters, with an emphasis on the rails (Rallidae). Phil Trans Roy Soc London, Ser B 353:2077–2151

Louchart A, Duhamel A (2021) A new fossil from the early Oligocene of Provence (France) increases the diversity of early Gruoidea and adds constraint on the origin of cranes (Gruidae) and limpkin (Aramidae). J Ornithol 162:977–986

References

Lydekker R (1891) Catalogue of the fossil birds in the British Museum (Natural History). British Museum (Natural History), London

Mayr G (2000) Charadriiform birds from the early Oligocene of Céreste (France) and the Middle Eocene of Messel (Hessen, Germany). Geobios 33:625–636

Mayr G (2004a) Morphological evidence for sister group relationship between flamingos (Aves: Phoenicopteridae) and grebes (Podicipedidae). Zool J Linnean Soc 140:157–169

Mayr G (2004b) Phylogenetic relationships of the early Tertiary Messel rails (Aves, Messelornithidae). Senckenb Lethaea 84:317–322

Mayr G (2005) A chicken-sized crane precursor from the early Oligocene of France. Naturwiss 92:389–393

Mayr G (2006) A rail (Aves, Rallidae) from the early Oligocene of Germany. Ardea 94:23–31

Mayr G (2007) The birds from the Paleocene fissure filling of Walbeck (Germany). J Vertebr Paleontol 27:394–408

Mayr G (2008) Avian higher-level phylogeny: well-supported clades and what we can learn from a phylogenetic analysis of 2954 morphological characters. J Zool Syst Evol Res 46:63–72

Mayr G (2009) Paleogene fossil birds, 1st edn. Springer, Heidelberg

Mayr G (2013) Parvigruidae (Aves, core-Gruiformes) from the early Oligocene of Belgium. Palaeobiodiv Palaeoenv 93:77–89

Mayr G (2014a) The Eocene *Juncitarsus*—its phylogenetic position and significance for the evolution and higher-level affinities of flamingos and grebes. C R Palevol 13:9–18

Mayr G (2014b) On the middle Miocene avifauna of Maboko Island, Kenya. Geobios 47:133–146

Mayr G (2015) Cranial and vertebral morphology of the straight-billed Miocene phoenicopteriform bird *Palaelodus* and its evolutionary significance. Zool Anz 254:18–26

Mayr G (2016a) The world's smallest owl, the earliest unambiguous charadriiform bird, and other avian remains from the early Eocene Nanjemoy Formation of Virginia (USA). Paläontol Z 90:747–763

Mayr G (2016b) Variations in the hypotarsus morphology of birds and their evolutionary significance. Acta Zool 97:196–210

Mayr G (2017) A small, "wader-like" bird from the early Eocene of Messel (Germany). Ann Paléontol 103:141–147

Mayr G (2019) Hypotarsus morphology of the Ralloidea supports a clade comprising *Sarothrura* and *Mentocrex* to the exclusion of *Canirallus*. Acta Ornithol 54:51–58

Mayr G, Bochenski ZM (2016) A skeleton of a small rail from the Rupelian of Poland adds to the diversity of early Oligocene Rallidae. N Jb Geol Paläontol, Abh 282(2):125–134

Mayr G, Knopf C (2007) A stem lineage representative of buttonquails from the Lower Oligocene of Germany—fossil evidence for a charadriiform origin of the Turnicidae. Ibis 149:774–782

Mayr G, Mourer-Chauviré C (2006) An unusual avian coracoid from the Paleogene Quercy fissure fillings in France. Strata, ser 1(13):129–133

Mayr G, Smith R (2001) Ducks, rails, and limicoline waders (Aves: Anseriformes, Gruiformes, Charadriiformes) from the lowermost Oligocene of Belgium. Geobios 34:547–561

Mayr G, Smith R (2002) Avian remains from the lowermost Oligocene of Hoogbutsel (Belgium). Bull Inst Roy Sci Nat Belg 72:139–150

Mayr G, Smith T (2019a) A diverse bird assemblage from the Ypresian of Belgium furthers knowledge of early Eocene avifaunas of the North Sea Basin. N Jb Geol Paläontol, Abh 291:253–281

Mayr G, Smith T (2019b) New Paleocene bird fossils from the North Sea Basin in Belgium and France. Geol Belg 22:35–46

Mayr G, Zvonok E (2011) Middle Eocene Pelagornithidae and Gaviiformes (Aves) from the Ukrainian Paratethys. Palaeontology 54:1347–1359

Mayr G, Archibald SB, Kaiser GW, Mathewes RW (2019) Early Eocene (Ypresian) birds from the Okanagan Highlands, British Columbia (Canada) and Washington State (USA). Can J Earth Sci 56:803–813

Mayr G, Hervet S, Buffetaut E (2020) On the diverse and widely ignored Paleocene avifauna of Menat (Puy-de-Dôme, France): new taxonomic records and unusual soft tissue preservation. Geol Mag 156:572–584

Mayr G, De Pietri V, Scofield RP (2022) New bird remains from the early Eocene Nanjemoy Formation of Virginia (USA), including the first records of the Messelasturidae, Psittacopedidae, and Zygodactylidae from the Fisher/Sullivan site. Hist Biol 34:322–334

Miller AH (1931) An auklet from the Eocene of Oregon. Univ California Publ, Bull Dept Geol 20:23–26

Miller AH (1963) The fossil flamingos of Australia. Condor 65:289–299

Mlíkovský J (2002) Cenozoic birds of the world. Part 1: Europe. Ninox Press, Praha

Mlíkovský J, Kovar J (1987) Eine neue Alkenart (Aves: Alcidae) aus dem Ober-Oligozän Österreichs. Ann Naturhist Mus Wien, Ser A 88:131–147

Mlíkovský J, Švec P (1986) Review of the Tertiary waterfowl (Aves: Anseridae) of Asia. Věstn Československ Spol Zool 50:259–272

Mori H, Miyata K (2021) Early Plotopteridae specimens (Aves) from the Itanoura and Kakinoura Formations (latest Eocene to early Oligocene), Saikai, Nagasaki Prefecture, western Japan. Paleontol Res 25:145–159

Morlo M (2004) Diet of *Messelornis* (Aves: Gruiformes), an Eocene bird from Germany. Cour Forsch-Instit Senckenberg 252:29–33

Mourer-Chauviré C (1978) La poche à phosphate de Ste. Néboule (Lot) et sa faune de vertébrés du Ludien Supérieur. 6—Oiseaux. Palaeovertebrata 8:217–229

Mourer-Chauviré C (1992) The Galliformes (Aves) from the Phosphorites du Quercy (France): systematics and biostratigraphy. In: Campbell KE (ed) Papers in avian paleontology honoring Pierce Brodkorb. Nat Hist Mus Los Angeles Cty Sci Ser 36:67–95

Mourer-Chauviré C (1995) The Messelornithidae (Aves: Gruiformes) from the Paleogene of France. In: Peters DS (ed) Acta palaeornithologica. Cour Forsch-Inst Senckenberg 181:95–105

Mourer-Chauviré C, Berthet D, Hugueney M (2004) The late Oligocene birds of the Créchy quarry (Allier, France), with a description of two new genera (Aves: Pelecaniformes: Phalacrocoracidae, and Anseriformes: Anseranatidae). Scnckenb Lethaea 84:303–315

Mourer-Chauviré C (2006) The avifauna of the Eocene and Oligocene Phosphorites du Quercy (France): an updated list. Strata, sér 1(13):135–149

Mourer-Chauviré C, Pickford M, Senut B (2017) New data on stem group Galliformes, Charadriiformes, and Psittaciformes from the middle Eocene of Namibia. Contrib MACN 7:99–131

Musser G, Clarke JA (2020) An exceptionally preserved specimen from the Green River Formation elucidates complex phenotypic evolution in Gruiformes and Charadriiformes. Front Ecol Evol 8:559929

Musser G, Ksepka DT, Field DJ (2019) New material of Paleocene-Eocene *Pellornis* (Aves: Gruiformes) clarifies the pattern and timing of the extant gruiform radiation. Diversity 11(7):102

Olson SL (1977) A synopsis of the fossil Rallidae. In: Ripley DS (ed) Rails of the world: a monograph of the family Rallidae. Godine, Boston, pp 339–379

Olson SL (1985) The fossil record of birds. In: Farner DS, King JR, Parkes KC (eds) Avian biology, vol 8. Academic Press, New York, pp 79–238

Olson SL (1989 ["1988"]) Aspects of global avifaunal dynamics during the Cenozoic. In: Ouellet H (ed) Acta XIX Congressus Internationalis Ornithologici. University of Ottawa Press, Ottawa, p 2023–2029

Olson SL (1999a) Early Eocene birds from eastern North America: a faunule from the Nanjemoy Formation of Virginia. In: Weems RE, Grimsley GJ (eds) Early Eocene vertebrates and plants from the Fisher/Sullivan site (Nanjemoy Formation) Stafford County, Virginia. Virginia Div Mineral Resour Publ 152:123–132

Olson SL (2003) First fossil record of a finfoot (Aves: Heliornithidae) and its biogeographical significance. Proc Biol Soc Wash 116:732–736

Olson SL (2014) Tracks of a stilt-like bird from the early Eocene Green River Formation of Utah: possible earliest evidence of the Recurvirostridae (Charadriiformes). Waterbirds 37:340–345

Olson SL, Feduccia A (1980) Relationships and evolution of flamingos (Aves: Phoenicopteridae). Smithson Contrib Zool 316:1–73

Ono K, Hasegawa Y (1991) Vertebrate fossils of the Iwaki Formation, III-1; avian fossils. In: Koda Y (ed) The excavation research report of the animal fossils of the Iwaki Formation, Iwaki City, pp 6–17. [in Japanese]

Paton TA, Baker AJ (2006) Sequences from 14 mitochondrial genes provide a well-supported phylogeny of the charadriiform birds congruent with the nuclear RAG-1 tree. Mol Phylogenet Evol 39:657–667

Peters DS (1987) *Juncitarsus merkeli* n. sp. stützt die Ableitung der Flamingos von Regenpfeifervögeln (Aves: Charadriiformes: Phoenicopteridae). Cour Forsch-Inst Senckenberg 97:141–155

Prum RO, Berv JS, Dornburg A, Field DJ, Townsend JP, Lemmon EM, Lemmon AR (2015) A comprehensive phylogeny of birds (Aves) using targeted next-generation DNA sequencing. Nature 526:569–573

Rasmussen DT, Olson SL, Simons EL (1987) Fossil birds from the Oligocene Jebel Qatrani Formation, Fayum Province, Egypt. Smithson Contrib Paleobiol 62:1–20

Rasmussen DT, Simons EL, Hertel F, Judd A (2001) Hindlimb of a giant terrestrial bird from the Upper Eocene, Fayum, Egypt. Palaeontology 44:325–337

Roux T (2002) Deux fossiles d'oiseaux de l'Oligocène inférieur du Luberon. Courr sci Parc nat rég Luberon 6:38–57

Sangster G (2005) A name for the flamingo-grebe clade. Ibis 147:612–615

Švec P (1982) Two new species of diving birds from the lower Miocene of Czechoslovakia. Čas mineral geol 27:243–260

Švec P (1984) Further finds of grebe *Miobaptus walteri* in the Miocene of Bohemia. Čas mineral geol 29:167–170

Torres CR, De Pietri VL, Louchart A, Van Tuinen M (2015) New cranial material of the earliest filter feeding flamingo *Harrisonavis croizeti* (Aves, Phoenicopteridae) informs the evolution of the highly specialized filter feeding apparatus. Org Divers Evol 15:609–618

Vickers-Rich PV (1991) The Mesozoic and Tertiary history of birds on the Australian plate. In: Vickers-Rich P, Monaghan TM, Baird RF, Rich TH (eds) Vertebrate Palaeontology of Australasia. Pioneer Design Studio and Monash University Publications Committee, Melbourne, pp 721–808

Wang M, Mayr G, Zhang J, Zhou Z (2012) Two new skeletons of the enigmatic, rail-like avian taxon *Songzia* Hou, 1990 (Songziidae) from the early Eocene of China. Alcheringa 36:487–499

Warheit KI (2002) The seabird fossil record and the role of paleontology in understanding seabird community structure. In: Schreiber EA, Burger J (eds) Biology of marine birds. CRC Marine Biology Series, Boca Raton, FL, pp 17–55

Weidig I (2010) New birds from the lower Eocene Green River Formation, North America. Rec Austral Mus 62:29–44

Weigel RD (1963) Oligocene birds from Saskatchewan. Q J Fla Acad Sci 26:257–262

Wetmore A (1940) Fossil bird remains from Tertiary deposits in the United States. J Morphol 66:25–37

Worthy TH, Boles WE (2011) *Australlus*, a new genus for *Gallinula disneyi* (Aves: Rallidae) and a description of a new species from Oligo-Miocene deposits at Riversleigh, northwestern Queensland, Australia. Rec Austral Mus 63:61–77

Zelenkov N (2021a) A revision of the Palaeocene-Eocene Mongolian Presbyornithidae (Aves: Anseriformes). Paleontol J 55:323–330

Zelenkov N (2021b) New bird taxa (Aves: Galliformes, Gruiformes) from the early Eocene of Mongolia. Paleontol J 55:438–446

Zelenkov NV, Kurochkin EN (2015) Class Aves. In: Kurochkin EN, Lopatin AV, Zelenkov NV (eds) Fossil vertebrates of Russia and neighbouring countries. Fossil reptiles and birds. Part 2. GEOS, Moscow, pp 86–290. [in Russian]

6 Opisthocomiformes (Hoatzins), "Columbaves" (Doves, Cuckoos, Bustards, and Allies), and Strisores (Nightjars, Swifts, Hummingbirds, and Allies)

The affinities of the taxa included in the present chapter have long been—and for most parts still are—difficult to resolve. The Opisthocomiformes (hoatzins) remain a phylogenetic enigma and are recovered in disparate positions in different analyses of molecular data (Hackett et al. 2008; Prum et al. 2015; Kuhl et al. 2021). Several other notoriously difficult-to-place taxa, such as the Columbiformes (doves), Musophagiformes (turacos), Cuculiformes (cuckoos), and Otidiformes (bustards), form a clade in analyses of molecular sequences, for which the name Columbaves was proposed (Prum et al. 2015). Opisthocomiformes and the taxa of the Columbaves resulted in close proximity to the Strisores (nightjars, swifts, hummingbirds, and allies) in some molecular analyses (Hackett et al. 2008; Prum et al. 2015; Kuhl et al. 2021). The exact interrelationships between these birds are, however, controversially resolved in different analyses, none of which recovered a clade including all of the above taxa. Based on studies of anatomical data, many earlier authors already assumed close affinities between the Columbiformes and Pterocliformes and between the Cuculiformes and Musophagiformes, respectively. Current molecular analyses, by contrast, resulted in a clade (Columbiformes + (Pterocliformes + Mesitornithiformes)) (Prum et al. 2015) or supported a clade ((Cuculiformes + Columbiformes) + (Pterocliformes + Mesitornithiformes)) (Kuhl et al. 2021). With regard to these novel molecular phylogenies, it is worth mentioning that morphological similarities between the Cuculiformes and the Mesitornithiformes (mesites) have also been noted (Mayr and Ericson 2004).

Whereas the fossil record of the Opisthocomiformes and the taxa of the "Columbaves" is very sparse, the Strisores are abundantly represented in various Paleogene fossil sites in Europe and North America. These fossils not only shed light on the evolutionary history of the higher-level taxa they belong to but also show that various groups of the Strisores with a geographically restricted extant range had a much wider distribution in the past.

6.1 Opisthocomiformes (Hoatzin)

The Opisthocomiformes include a single extant species, the Hoatzin (*Opisthocomus hoazin*), which occurs in the Amazon and Orinoco basins in South America. *Onychopteryx simpsoni* from the Eocene (Casamayoran) of Argentina was described by Cracraft (1971), who assigned the species to the new taxon Onychopterygidae, which was considered to be most closely related to the Opisthocomiformes. However, the holotype specimen, a small tarsometatarsus fragment, has rightly been considered to be of indeterminate affinities by Brodkorb (1978).

A species known from more diagnostic material is *Protoazin parisiensis* from the late Eocene of Romainville near Paris in France, of which a partial coracoid and an associated scapula were found (Fig. 6.1c; Mayr and De Pietri 2014). Even though the fossil record is therefore very scanty, the coracoid shows a distinctive morphology in that the sternal end bears a large opening, which characterizes the coracoid of extant Opisthocomiformes. In overall morphology, the bone also corresponds well to the distinctive coracoid of the extant Hoatzin.

The earliest South American record of the Opisthocomiformes is *Hoazinavis lacustris* from the Oligo-Miocene (22–24 Ma) Tremembé Formation of the Taubaté Basin in southeastern Brazil, which was found in an area outside the extant range of hoatzins (Mayr et al. 2011). The species is somewhat smaller than the extant Hoatzin, but the known bones (humerus, coracoid, and scapula; Fig. 6.1b) are otherwise very similar to those of extant Opisthocomiformes.

A stem group representative of the Opisthocomiformes, *Namibiavis senutae*, was also found in the middle Miocene of Namibia (Mayr et al. 2011). *Protoazin* and *Namibiavis* are successive sister taxa of a clade including *Hoazinavis* and crown group Opisthocomiformes (Mayr and De Pietri 2014). This suggests that hoatzins evolved outside South America, and it was hypothesized that stem group Opisthocomiformes dispersed from Africa to South America by rafting across the Atlantic (Mayr et al. 2011).

6.2 Otididae (Bustards)

The 26 extant species of bustards are long-legged denizens of dry grasslands and steppe habitats. Bustards solely occur on the continents of the Old World and are characterized by a foot with just three short toes, with the hallux having been lost. Bustards have no published Paleogene fossil record, but according to Kurochkin (1976) as-yet undescribed remains were found in late Oligocene deposits of Kazakhstan. Mourer-Chauviré (2006) also lists fossils of the Otididae from an unknown stratigraphic horizon of the Quercy fissure fillings. None of these has yet been figured and an assessment of their affinities has to await their formal description.

6.3 Foratidae

This taxon comprises a single named species, *Foro panarium*, the holotype of which is a complete and well-preserved skeleton from the early Eocene Green River Formation (Fig. 6.2a; Olson 1992). *F. panarium* was about the size of a medium-sized phasianid, with rather short wings, and long legs. As shown by Olson (1992), the skull is similar to that of *Opisthocomus hoazin* in its overall proportions but has larger nostrils and deeper mandibular rami. As in the hoatzin, the mandible exhibits short retroarticular processes. *F. panarium* is, however, distinguished from extant Opisthocomiformes in most aspects of its postcranial skeleton, including the absence of a notarium, the much longer legs, and the morphology of the sternum and pectoral girdle. The pelvis bears large preacetabular tubercles, which among extant neognathous birds are equally well-developed only in the Galliformes, Musophagiformes, and some Cuculiformes. A skull from Messel, which closely resembles that of *F. panarium* but is from a smaller species, was reported by Mayr (2016a) (Fig. 6.2d, e).

Olson (1992) considered *F. panarium* to be most similar to extant Musophagiformes and Opisthocomiformes. However, the phylogenetic affinities of the species were "refined mainly by the absence of derived characters" (Olson 1992: 129), which is a problematic approach in the case of early Paleogene fossils, because it is to be expected that archaic stem group representatives of extant taxa lack derived features of their living relatives.

Field and Hsiang (2018) reanalyzed the affinities of *F. panarium* and hypothesized that it is a stem group representative of the Musophagiformes. The main supporting synapomorphy listed by these authors is a presumed lack of fusion of the shafts of the furcula at the sternal extremity of the bone. However, this trait is likely to have been incorrectly identified in the *F. panarium* holotype, in which the furcular shafts appear to be thoroughly co-ossified at their sternal end (as already noted by Olson 1992). The fact that one shaft is broken and somewhat displaced in the fossil (Fig. 6.2b) also suggests the presence of a co-ossified sternal extremity, because it is unlikely from a mechanical point of view that the furcula breaks just above the sternal extremity, if the shafts were not co-ossified. Moreover, the sternal extremity of the furcula invariably is very narrow in taxa, in which the shafts of the furcula are not co-ossified, whereas it is wide in *F. panarium*.

Even if *F. panarium* can be shown to be an archaic stem group representative of the Musophagiformes, it is very different from extant turacos in its skeletal morphology and indicates a bird with disparate ecological preferences. Accordingly, it is not quite appropriate to cite the species as evidence for a relict distribution of the Musophagiformes (contra Field and Hsiang 2018).

Some molecular analyses supported a sister group relationship between the Musophagiformes and the Otidiformes (Kuhl et al. 2021) or between the Musophagiformes and a clade (Otidiformes + Cuculiformes) (Prum et al. 2015). Bustards are long-legged ground-dwelling birds, and Field and Hsiang (2018) hypothesized that long legs are plesiomorphic for the Musophagiformes and that stem group representatives of the clade were more terrestrial than extant turacos. However, it should be noted that the position of turacos is not congruently resolved in molecular analyses and there also is some support for a sister group relationship between the Musophagiformes and the Opisthocomiformes (Braun and Kimball 2021: fig. 4D), which would actually conform better to the mosaic character distribution seen in *Foro panarium*.

6.4 Musophagiformes (Turacos)

The extant species of this taxon are arboreal, frugivorous birds with a semi-zygodactyl foot and only occur in Africa south of the Sahara. The earliest fossils are from the early Oligocene of Egypt (Rasmussen et al. 1987). These specimens, the distal ends of a tarsometatarsus and a humerus, are very similar to the corresponding bones of the extant musophagiform taxon *Crinifer*. The as-yet unnamed species to which they belong was, however, larger than any extant turaco except for the largest species, *Corythaeola cristata*.

A proximal end of a humerus from the late Oligocene (Chattian) of southern Germany was assigned to the Musophagiformes by Ballmann (1970). This fossil comes from a bird larger than any extant species of turacos, from which it also differs in several morphological features (Ballmann 1970). As noted by Mayr (2009), it is more likely

6.4 Musophagiformes (Turacos)

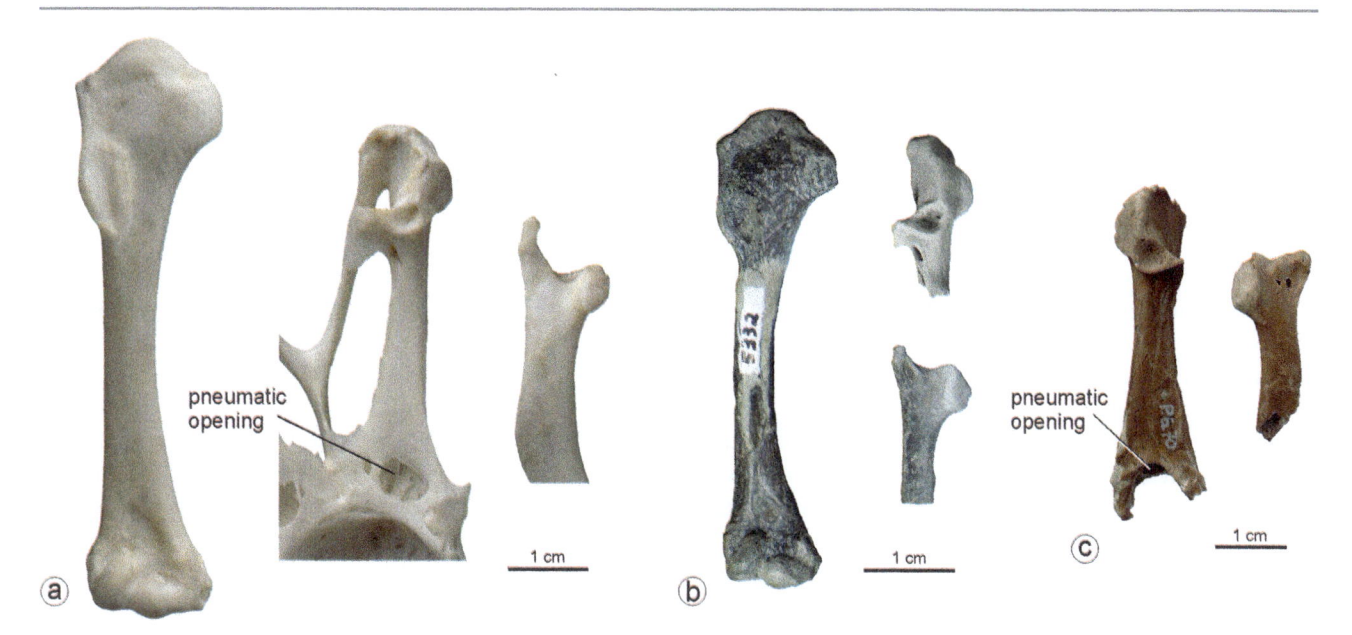

Fig. 6.1 Paleogene fossils of the Opisthocomiformes. (**a**) Humerus, coracoid, and scapula of the extant Hoatzin, *Opisthocomus hoazin*. (**b**) Holotype of *Hoazinavis lacustris* from the late Oligocene to early Miocene Tremembé Formation of the Taubaté Basin in Brazil (humerus, partial coracoid and scapula; Museu de História Natural de Taubaté, Brazil, MHNT-VT 5332). (**c**) Holotype of *Protoazin parisiensis* from the late Eocene of France (partial coracoid and scapula; Naturhistorisches Museum Basel, Switzerland, NMB PG.70). (Photos in (**a**) and (**c**) by Sven Tränkner, (**b**) by the author)

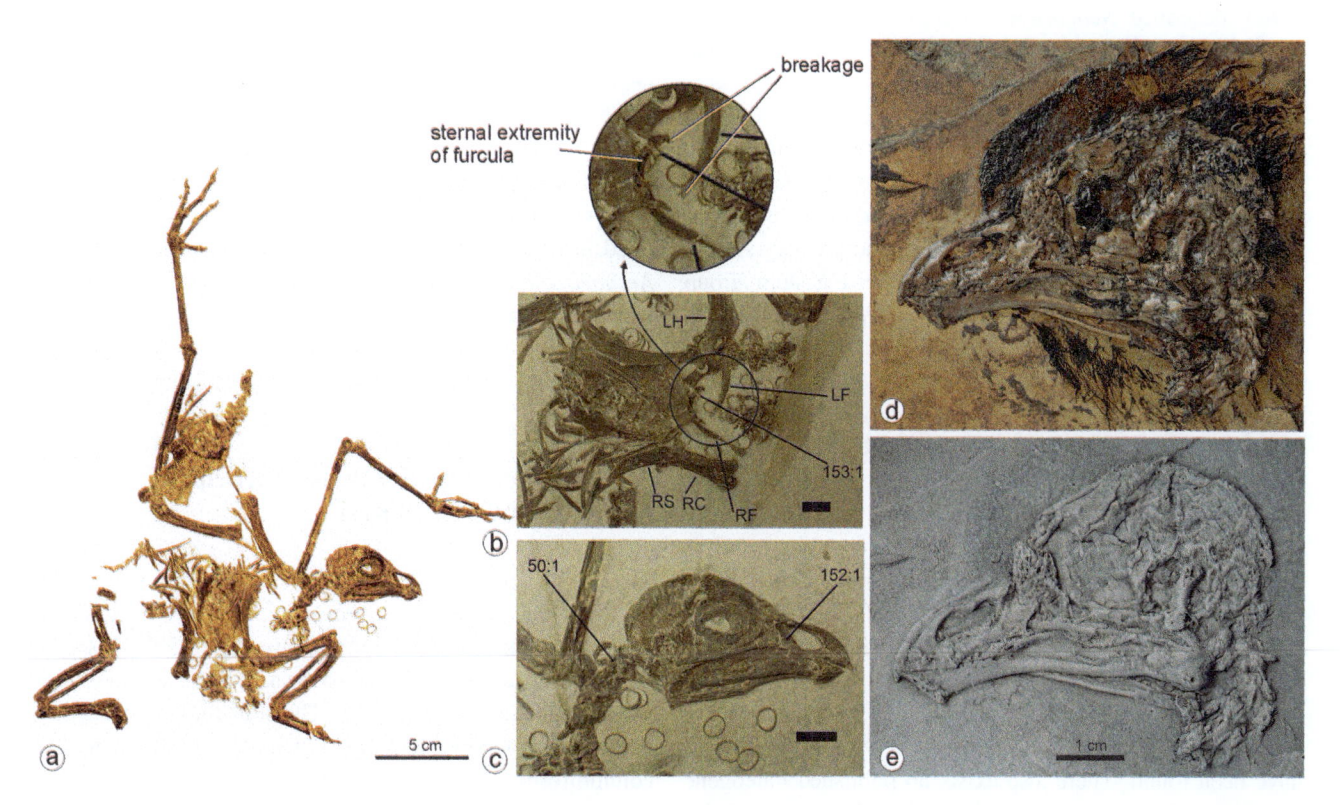

Fig. 6.2 (**a**) Holotype of *Foro panarium* (matrix digitally removed) with details of (**b**) the furcula and (**c**) the skull (from Field and Hsiang 2018: Fig. 1, modified; published under a Creative Commons CC BY 4.0 license). (**d**), (**e**) A *Foro*-like skull from the latest early or earliest middle Eocene of Messel (Hessisches Landesmuseum Darmstadt, Germany, HLMD Me 14,972). (The specimen in (**e**) was coated with ammonium chloride, photos in (**d**) and (**e**) by Sven Tränkner)

that the fossil is from a representative of the Idiornithidae (Sect. 8.4.2), and probably it belongs to the taxon *Dynamopterus* ("*Idiornis*").

6.5 Cuculiformes (Cuckoos)

Although the zygodactyl Cuculiformes have a worldwide distribution today, their fossil record is very poor. The earliest fossil species assigned to the taxon is *Chambicuculus pusillus* from the late early or early middle Eocene of Tunisia. This very small bird is known from several isolated tarsometatarsi as well as a referred coracoid and femur (Mourer-Chauviré et al. 2013, 2016). These bones show a close resemblance to the corresponding elements of extant Cuculiformes, with the tarsometatarsus bearing a large accessory trochlea for the fourth toe. Unlike in extant cuckoos, however, there is a dorsally roofed canal between the trochleae for the third and fourth toes. The coracoid has a very long procoracoid process, which is also present in extant Cuculiformes. In its size, the tiny *C. pusillus* may have corresponded to the smallest extant Cuculiformes, if it was not even smaller.

From the late Eocene (Chadronian) of Canada, Weigel (1963) described *Neococcyx mccorquodalei*. The holotype of this species is a distal humerus of a bird the size of the extant Yellow-billed Cuckoo (*Coccyzus americanus*).

Both *Chambicuculus* and *Neococcyx* are known from just a few isolated bones, so that their assignment to the Cuculiformes needs further substantiation by more fossil material. Especially in the case of *Chambicuculus*, however, the bones exhibit a characteristic derived morphology similar to that found in extant cuckoos. If indeed a stem group representative of the Cuculiformes, *Chambicuculus* may indicate an African origin of the clade.

6.6 Columbiformes (Doves), Pterocliformes (Sandgrouse), and Mesitornithiformes (Mesites)

The Pterocliformes solely occur in Africa and Eurasia, whereas the Columbiformes are globally distributed. In spite of their more restricted extant distribution, however, sandgrouse has a more comprehensive Paleogene fossil record than doves and pigeons.

As yet no fossils of the Madagascan Mesitornithiformes have been found. There also exists no published Paleogene fossil record of the Columbiformes, even though Boles (2001a) noted their presence in the late Oligocene Etadunna Formation in Australia. A "columbid-like" foot from the early Oligocene of the Carpathian Basin in Poland (Bochenski et al. 2010) is too poorly preserved for an unambiguous identification. The earliest formally described columbiform species is from the early Miocene of Florida and was classified into the extant taxon *Columbina* (Becker and Brodkorb 1992). Based on the fossil record, it was hypothesized that the Columbiformes evolved in the Southern Hemisphere and did not arrive in the Northern Hemisphere before the Neogene (Olson 1989).

The description of the alleged columbiform *Microena goodwini* from the early Eocene London Clay was based on a tarsometatarsus, which lacks the trochlea for the fourth toe (Harrison and Walker 1977). Without additional material, the affinities of this species are indeterminable, but apart from being somewhat less stout, the tarsometatarsus actually closely resembles that of *Parvicuculus* (Sect. 6.8.6); affinities to the Strisores are therefore most likely. A distal tarsometatarsus from the early Eocene Nanjemoy Formation in Virginia, USA was tentatively assigned to *Microena* by Mayr (2016b).

Stem group representatives of the Pterocliformes were found in the Quercy fissure fillings (Mourer-Chauviré 1992, 1993). The specimens of the three species of *Archaeoganga*, *A. larvatus*, *A. validus*, and *A. pinguis*, lack stratigraphic data, whereas *Leptoganga sepultus* occurs in late Oligocene (MP 28) deposits. The *Archaeoganga* species are represented by humeri, coracoids, and tarsometatarsi, which closely resemble the corresponding bones of extant Pterocliformes (especially *Pterocles* spp., whereas the species of *Syrrhaptes* exhibit a more derived morphology). Of *Leptoganga* an intertarsal sesamoid bone was described, which is a derived feature of the Pterocliformes (Mourer-Chauviré 1993). Like their extant counterparts, these Paleogene Pterocliformes lived in an open and arid habitat (Mourer-Chauviré 1993). *Archaeoganga larvatus* was the size of the extant Black-bellied Sandgrouse (*Pterocles orientalis*), whereas *A. pinguis* was about twice as large and thus distinctly exceeded any extant species of the Pterocliformes in size.

6.7 Eopachypterygidae, *Eocuculus*, and *Carpathiavis*

The taxon Eopachypterygidae was established by Mayr (2015a) for *Eopachypteryx praeterita*, a species from Messel with a distinctive skeletal morphology (Fig. 6.3b). This small bird exhibits a short and robust beak, a stout humerus, and short legs with anisodactyl feet. The ulna does not exceed the humerus in length, and the coracoid resembles that of columbiform birds. In addition to *E. praeterita*, the Eopachypterygidae include a further unnamed species from Messel. Even though the known material of *E. praeterita* includes a nearly complete skeleton, all specimens are poorly preserved and do not allow the recognition of many skeletal details. The affinities of *Eopachypteryx* are therefore

Fig. 6.3 (**a**) Legs and pelvic girdle of *Eocuculus* cf. *cherpinae* from the early Oligocene of the Luberon area in France (Senckenberg Research Institute, Frankfurt, SMF Av 425). (**b**) *Eopachypteryx praeterita* from the latest early or earliest middle Eocene of Messel (SMF-ME 2426). (**c**) Holotype of *Carpathiavis meniliticus* from the early Oligocene of the Polish Carpathian Basin (SMF Av 649a); specimen coated with ammonium chloride. (Photos in (**a**) and (**c**) by Sven Tränkner, in (**b**) by Anika Vogel)

unresolved, but in the original description, it was noted that it shows a resemblance to *Eocuculus*, an equally enigmatic early Oligocene taxon.

Eocuculus cherpinae was reported by Chandler (1999) from the late Eocene of the Florissant Fossil Beds of Colorado, and very similar fossils from the early Oligocene of the Luberon area in France (Fig. 6.3a) were classified as *E.* cf. *cherpinae* by Mayr (2006, 2008). All records consist of postcranial skeletons. In the initial description, *Eocuculus* was assigned to the Cuculiformes with which it shares, in addition to an at least superficially similar overall skeletal morphology, a sternal keel with a strongly projecting tip and a markedly convex cranial margin. As in the Cuculiformes, the ulna is as long as the humerus, whereas it is usually longer than this bone in extant representatives of the Telluraves (the clade including most arboreal birds). Furthermore, as in extant Cuculiformes, the furcula has a well-developed apophysis (Mayr 2008). However, and in addition to other differences (Mayr 2006), the tarsometatarsal trochlea for the fourth toe does not bear a large accessory trochlea, and in contrast to extant Cuculiformes, *Eocuculus* therefore probably only had semi-zygodactyl feet (Mayr 2006). Again, however, a detailed analysis of the affinities of the taxon is impeded by an insufficient understanding of its skeletal morphology. Mayr (2008) noted similarities to the early Eocene taxon *Pumiliornis*, which has meanwhile been identified as a zygodactyl stem group representative of passerines (Sect. 9.5.1). Analyses by Mayr (2015b, 2020a) did not conclusively resolve the affinities of *Eocuculus*, whereas the taxon was recovered within the Psittacopasseres in an analysis by Ksepka et al. (2019). Still, no conclusive evidence exists for this placement, and in many osteological details *Eocuculus* is clearly distinguished from zygodactyl near-passerines. In light of the fact that molecular analyses suggest close affinities (Prum et al. 2015) or even a sister group relationship (Kuhl et al. 2021) between the Columbiformes and Cuculiformes, the mosaic distribution of columbiform and cuculiform characters in *Eopachypteryx* and *Eocuculus* is notable, but a well-founded placement of either taxon is not possible on the basis of our scant current knowledge of their skeletal morphology.

The holotype and only known specimen of *Carpathiavis meniliticus* from the early Oligocene of the Carpathian Basin in Poland is a partial skeleton lacking the feet (Fig. 6.3c; Mayr 2019). The preserved portions of the hindlimbs indicate a long-legged bird with short wings. This very small bird has an unusually robust furcula with a very long apophysis, and in the original description similarities to *Eocuculus* were noted.

6.8 Strisores (Nightjars, Swifts, Hummingbirds, and Allies)

As detailed in Sect. 2.3, molecular data congruently supported a clade including the paraphyletic "Caprimulgiformes" and the Apodiformes, with these birds sharing an elongated leg (crus longum) of the ulnar carpal bone. This clade was termed Strisores (Mayr 2010a) and is recovered in all current analyses of molecular data. Likewise, a sister group relationship between the Aegotheliformes and the Apodiformes is congruently obtained in all analyses of morphological and molecular data (Mayr 2002, 2010a; Mayr et al. 2003; Ericson et al. 2006; Prum et al. 2015; Chen et al. 2019; Kuhl et al. 2021).

Otherwise, however, there is some conflict between tree topologies derived from morphological and molecular data (Mayr 2010a; Chen et al. 2019). The Steatornithiformes, for example, exhibit a number of plesiomorphic morphological features, which indicate a sister group relationship to all other extant taxa of the Strisores (e.g., quadrate with long orbital process, palatine bone without caudolateral process, and 19 instead of 18 or 17 presacral vertebrae). Morphological characters also support a clade including the Caprimulgiformes, Nyctibiiformes, Aegotheliformes, and Apodiformes, for which the name Cypselomorphae was introduced (Mayr 2002). The Nyctibiiformes and Caprimulgiformes furthermore share derived skull features, including a unique cone-like bony protrusion caudal of the foramen for the optic nerve and strongly widened palatine bones (Fig. 6.4; Mayr 2002).

Most molecular analyses, by contrast, supported a sister group relationship between the Nyctibiiformes and Steatornithiformes, with either this latter clade (Hackett et al. 2008) or the Caprimulgiformes resulting as the sister taxon of all other Strisores (Prum et al. 2015; Kuhl et al. 2021). There is little morphological support for these sequence-based phylogenies, but a sister group relationship between the Caprimulgiformes and all other Strisores may be supported by the plesiomorphic presence of a well-developed lacrimal bone in the former, which is greatly reduced or absent in other Strisores (Chen et al. 2019).

The taxa of the traditional "Caprimulgiformes" are crepuscular or nocturnal birds. Based on the phylogeny in Fig. 6.5, it is most parsimonious to assume that there was a single origin of dark activity in the stem lineage of the Strisores, and a reversal to a diurnal way of living in the stem lineage of the Apodiformes (Mayr 2002). A less parsimonious but possibly more plausible alternative explanation is an independent origin of dark activity in the stem lineages of the Steatornithiformes, Caprimulgiformes, Nyctibiiformes, Podargiformes, and Aegotheliformes (Mayr 2010a). The origin of dark activity in these aerial insectivores may well have been related to the evolution of nocturnal insects in the early Paleogene, such as the lepidopteran Noctuoidea ("owl"-moths), the earliest unambiguous fossil records of which come from that period (Kristensen and Skalski 1999; Sohn et al. 2015).

6.8.1 Archaeotrogonidae

These short-legged aerial insectivores are the most abundant small birds in the Oligocene deposits of the Quercy fissure fillings. Their taxonomy was revised by Mourer-Chauviré (1980), who recognized four species: *Archaeotrogon venustus* (late Eocene to late Oligocene), *A. zitteli* (early to late Oligocene), *A. cayluxensis* (late Oligocene), and *A. hoffstetteri* (from the old collections of unknown exact age).

A. venustus (Fig. 6.6d) is the best-known archaeotrogon and also has the largest temporal range, existing over a period of 15 million years. All major postcranial elements of this species have been found (Mourer-Chauviré 1980, 1995). The humerus is fairly stocky and has a wide proximal end. As far as this can be inferred from the isolated bones and in contrast to most other taxa of the Strisores, the humerus is only slightly shorter than the ulna. The extensor process of the short carpometacarpus forms a pointed spur (Fig. 6.6d), which may have served for defense or intraspecific combats.

The earliest representative of the Archaeotrogonidae is *Archaeodromus anglicus* from the early Eocene of the London Clay of Walton-on-the-Naze (Mayr 2021). The quadrate of this species is distinctive in that the articular facet for the jugal bar directs dorsally and not laterally as it does in most other birds. This characteristic derived morphology is otherwise only found in the Caprimulgiformes, Nyctibiiformes, Aegotheliformes, and swifts (Apodidae and Hemiprocnidae). The holotype of *A. anglicus* is a partial skeleton that includes few of the major limb bones, but additional specimens were collected by Michael Daniels (Fig. 6.6c; Feduccia 1999: table 4.1; Mourer-Chauviré 1995). In one of these fossils, substantial portions of a

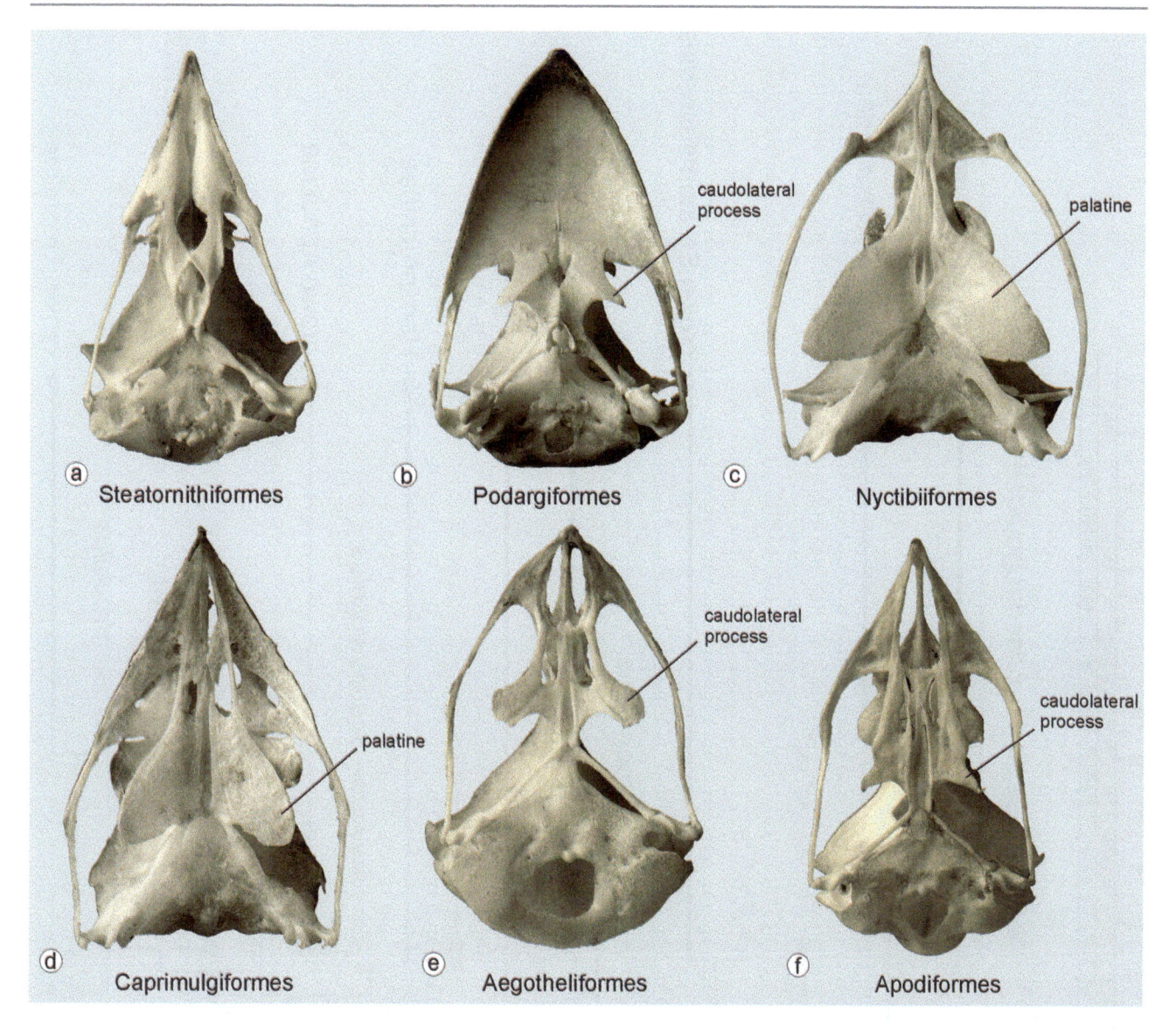

6.8 Strisores (Nightjars, Swifts, Hummingbirds, and Allies)

Fig. 6.4 Skull (ventral view) of taxa of the Strisores. (**a**) Oilbird, *Steatornis caripensis* (Steatornithiformes; right quadrate detached). (**b**) Tawny Frogmouth, *Podargus strigoides* (Podargiformes). (**c**) Common Potoo, *Nyctibius griseus* (Nyctibiiformes). (**d**) White-throated Nightjar, *Eurostopodus mystacalis* (Caprimulgiformes). (**e**) Australian Owlet-nightjar, *Aegotheles cristatus* (Aegotheliformes). (**f**) Alpine Swift, *Apus melba* (Apodiformes). Not to scale. (All photos by Sven Tränkner)

swift- or nightjar-like beak are preserved, and as in *Archaeotrogon* the carpometacarpus bears a spur-like extensor process. As shown by the fossils in the Daniels collection, the hypotarsus of *Archaeodromus* is block-like and does not form bony sulci or canals, which distinguishes the taxon from all other Strisores.

Archaeotrogons are also known from Messel (Mayr 1998, 2004a, 2021). The species from this locality, *Hassiavis laticauda* (Fig. 6.6a, b), was assigned to the Archaeotrogonidae because of a similarly-shaped humerus and tarsometatarsus as well as similar wing-proportions. *H. laticauda* has an owlet-nightjar-like beak and differs from *Archaeotrogon* in that the omal extremity of the coracoid has a hook-shaped outline, whereas it is more rounded in *Archaeotrogon*. The carpometacarpus does not bear a spur.

Some specimens exhibit well-preserved feather remains, and the tail resembles that of nightjars (Caprimulgiformes) in its shape; in two fossils the tail feathers show a distinct barring (Fig. 6.6a; Mayr 1998, 2004a). Within the Strisores, this barring only occurs in nocturnal species, which may indicate the activity pattern of *Hassiavis*.

As indicated by their name, archaeotrogons were originally considered to be representatives of the Trogoniformes, but Mourer-Chauviré (1980) already noted similarities to the Caprimulgiformes. A position of archaeotrogons within the Strisores is supported by the derived morphology of the quadrate of *Archaeodromus* and by the nightjar-like beak of this taxon. *Hassiavis* resulted as a stem group representative of the Aegotheliformes in an analysis by Chen et al. (2019), but the character evidence for this placement is not very

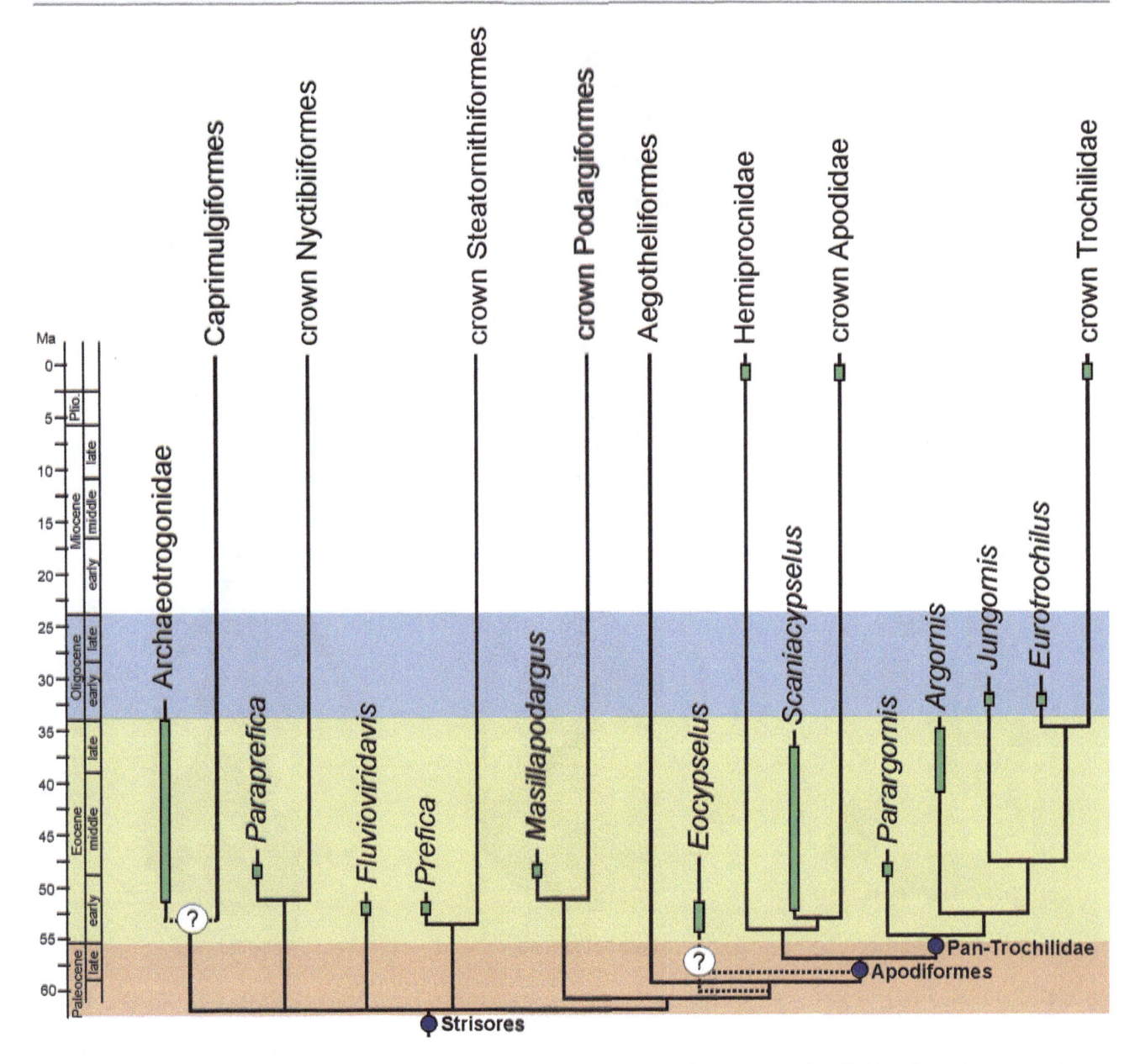

Fig. 6.5 Phylogenetic interrelationships and stratigraphic occurrences of selected taxa of the Strisores. The affinities of taxa shown in a polytomy are controversial, divergence dates are hypothetical; see text for further details

strong, and unlike in the Aegotheliformes and apodiform birds, the coracoid of archaeotrogons does not exhibit a foramen for the supracoracoideus nerve (the absence of this foramen is likely to be plesiomorphic for the Strisores). A phylogenetic analysis by Mayr (2021) supported a sister group relationship between archaeotrogons and the Caprimulgiformes, with which at least *Archaeodromus* shares a characteristic derived morphology of the proximal end of the ulna. Again, no strong evidence exists for this phylogenetic hypothesis, even though recognition of archaeotrogons as stem group representatives of the Caprimulgiformes would fill a conspicuous gap in the fossil record of the Strisores (see next section).

6.8.2 Caprimulgiformes (Nightjars)

The extant representatives of this taxon have a worldwide distribution and include more than 80 species, but its Paleogene fossil record is very sparse. Olson (1999) tentatively assigned to the Caprimulgiformes some isolated bones (proximal humerus, carpometacarpus, and distal tarsometatarsus) from the early Eocene Nanjemoy Formation of Virginia, which belong to at least two different small species. The only other Paleogene specimens assigned to the taxon are two coracoids from the late Eocene of the Quercy fissure fillings, which were described as *Ventivorus ragei* by Mourer-Chauviré (1988). The identification of these fossils

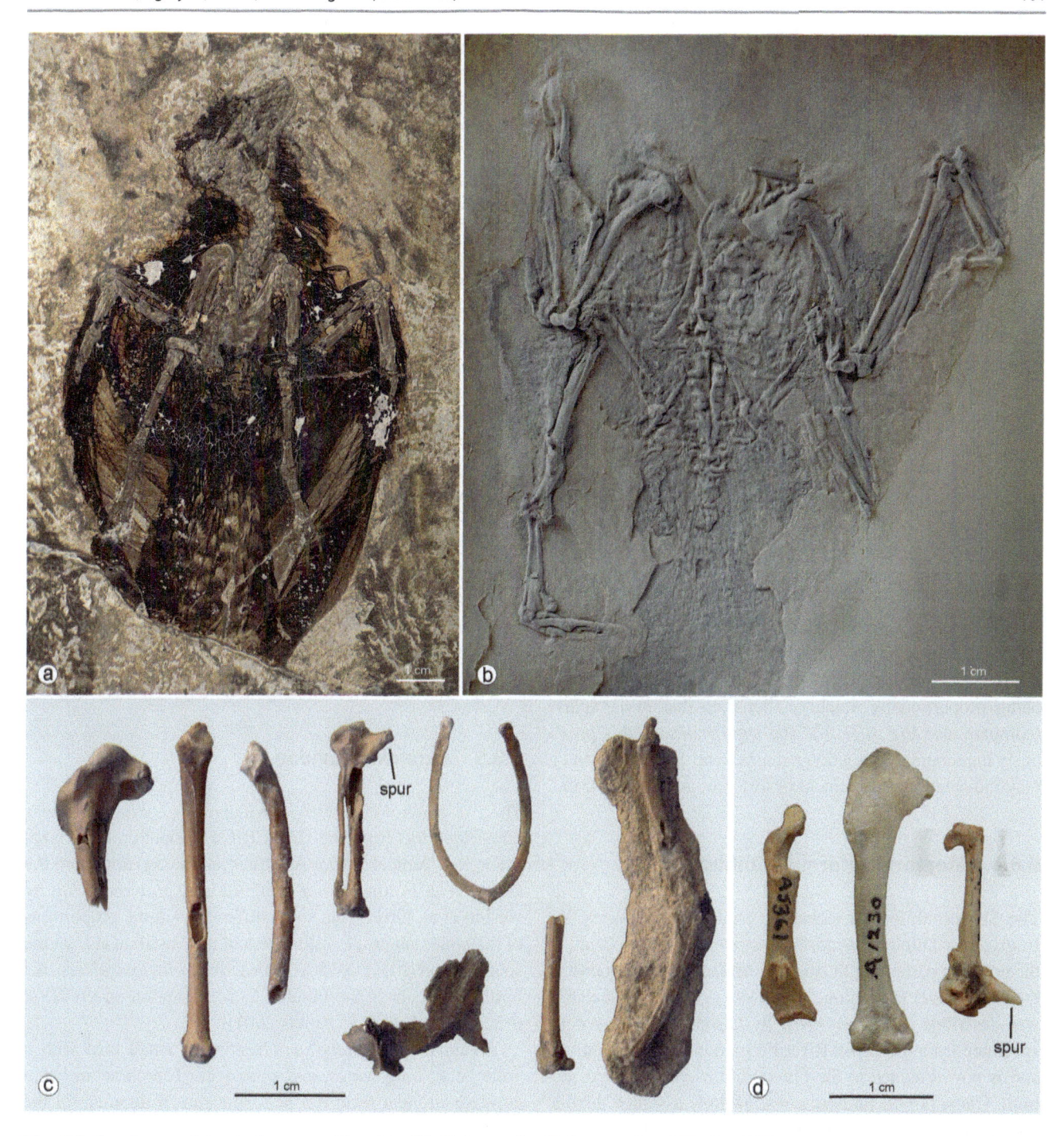

Fig. 6.6 Specimens of the Archaeotrogonidae. (**a**), (**b**) Skeleton of *Hassiavis laticauda* from the latest early or earliest middle Eocene of Messel (Senckenberg Research Institute, Frankfurt, **a**: SMF-ME 3545, **b**: SMF-ME SMF-ME 11702A, coated with ammonium chloride). (**c**) Partial skeleton of *Archaeodromus anglicus* from the early Eocene London Clay of Walton-on-the-Naze (collection of Michael Daniels, Holland-on-Sea, UK, WN 85500); note the mandible in the block of matrix on the right. (**d**) Bones of *Archaeotrogon* cf. *venustus* (Archaeotrogonidae) from an unknown stratigraphic horizon of the Quercy fissure fillings (Natural History Museum, London; coracoid: NHMUK A 5361, humerus: NHMUK A 1230, carpometacarpus: NHMUK A 1228). (All photos by Sven Tränkner)

needs to be corroborated by further material, with the coracoid of *Ventivorus* distinctly differing from that of the Archaeotrogonidae.

6.8.3 Nyctibiiformes (Potoos)

The seven extant species of the Nyctibiiformes occur in forests of tropical Central and South America. In their skeletal morphology, these birds are well-characterized by a derived skull morphology with a very short beak and strongly bowed jugal bars and by a grotesquely abbreviated tarsometatarsus (Fig. 6.7c, d). With the removal of *Euronyctibius* from the Nyctibiiformes (Sect 6.8.4; Mourer-Chauviré 2013), the only Paleogene fossils of potoos belong to *Paraprefica*. This taxon is represented by complete skeletons from Messel (Mayr 1999, 2001a, 2005a), where two species, *Paraprefica major* and *P. kelleri*, have been recognized (Fig. 6.7a, b). As in extant Nyctibiiformes, the palatine bones of *Paraprefica* are greatly enlarged, the tibiotarsus lacks an ossified supratendinal bridge, the tarsometatarsus is greatly abbreviated, and the tail feathers are fairly long. The carpometacarpus is, however, proportionally longer and more slender than that of extant potoos.

In the original description (Mayr 1999), *Paraprefica* was erroneously assigned to the steatornithiform Preficinae (see next section). The similarities between *Prefica* and *Paraprefica* were considered to be plesiomorphic for the Strisores by (Mayr 2005a), but if molecular analyses correctly identify the Nyctibiiformes as the closest extant relatives of the Steatornithiformes, they are more likely to be indicative of close affinities between the two fossil taxa.

The *Paraprefica* fossils from Messel distinctly differ in the relative sizes of their skulls, with the skull of *P. major* being proportionally much smaller than that of *P. kelleri* (compare also Fig. 6.7a, b). The significance of this previously unrecognized variation remains to be determined, but it may indicate the involvement of different genus-level taxa.

6.8.4 Steatornithiformes (Oilbirds)

The Steatornithiformes include a single extant species, the frugivorous Oilbird, *Steatornis caripensis*, which occurs in the northern part of South America. Based on a skeleton from the Green River Formation, a fossil stem group representative was described as *Prefica nivea* by Olson (1987); a second specimen from the Green River Formation was misidentified and is now assigned to the Fluvioviridavidae (see next section). Olson (1999) furthermore tentatively assigned the distal and proximal ends of a humerus from the early Eocene of the Nanjemoy Formation in Virginia to *Prefica*.

The skull of *P. nivea* is unknown, but the shape of its mandible is very similar to that of the extant *S. caripensis*. *P. nivea* also shares an extremely abbreviated tarsometatarsus and well-developed temporal fossae with the extant oilbird; as in extant Steatornithiformes, the tibiotarsus is just about as long as the carpometacarpus (Olson 1987; Mayr 2005a). *P. nivea* is smaller than *S. capensis* and differs in a number of presumably plesiomorphic features, including two pairs of notches in the caudal margin of the sternum and the lack of a fusion between the ilium and the synsacrum. The fossil was classified in a monotypic taxon Preficinae by Olson (1987), who hypothesized that it already had a frugivorous diet similar to that of extant Steatornithiformes.

A putative European stem group representative of the Steatornithiformes is *Euronyctibius kurochkini*. This species was initially established for a proximal humerus from an unknown stratigraphic horizon of the Quercy fissure fillings, which was assigned to the Nyctibiiformes by Mourer-Chauviré (1989). Mourer-Chauviré (2013) identified new material of *E. kurochkini* from a late Eocene (37 Ma) locality and transferred the species to the Steatornithiformes. In light of a possible sister group relationship between the Steatornithiformes and Nyctibiiformes, this taxonomic history is of particular interest, because it demonstrates some similarity of the humeri of both taxa. Mayr (2017) raised the possibility that tarsometatarsi described as *Querycypodargus olsoni* and referred to the Podargiformes by Mourer-Chauviré (1989) may belong to *E. kurochkini*.

An earlier tentative identification of the Steatornithiformes in the Quercy fissure fillings (Mourer-Chauviré 1982) was based on the cranial portion of a sternum from a late Oligocene (26 Ma) deposit. Mayr (2009) proposed that this specimen belongs to a species of the Archaeotrogonidae, but this hypothesis was refuted by Mourer-Chauviré (2013), who considered affinities to *Euronyctibius* more likely.

6.8.5 Fluvioviridavidae

The Fluvioviridavidae include *Fluvioviridavis platyrhamphus* from the Green River Formation (Fig. 6.8a; Mayr and Daniels 2001). A further partial skeleton from the Green River Formation was assigned to *F. platyrhamphus* by Nesbitt et al. (2011), but some differences in the proportions of the bones suggest a distinctness of this fossil on at least the species level (Mayr 2015c). Undescribed *Fluvioviridavis*-like birds also occur in the London Clay of Walton-on-the-Naze (Fig. 6.8b; Mayr and Daniels 2001).

Fluvioviridavis platyrhamphus was a small bird with a wide beak, long wings, and short legs. The coracoid has a cup-like articular facet for the scapula and a foramen for the supracoracoideus nerve, and both the alular and major wing digits bear ungual phalanges. The caudal margin of the sternum exhibits two pairs of shallow incisions. The long hallux reaches nearly the length of the tarsometatarsus.

Fluvioviridavis was assigned to the Podargiformes by Nesbitt et al. (2011), but as detailed by Mayr (2015c) its skeletal morphology is very different from that of the coeval *Masillapodargus* and extant Podargiformes. Mayr (2015c) noted that the humerus and coracoid of *Fluvioviridavis* are actually more similar to the Steatornithiformes. Analyses by Chen et al. (2019) resulted in a sister group between *Fluvioviridavis* and either the Podargiformes (morphological

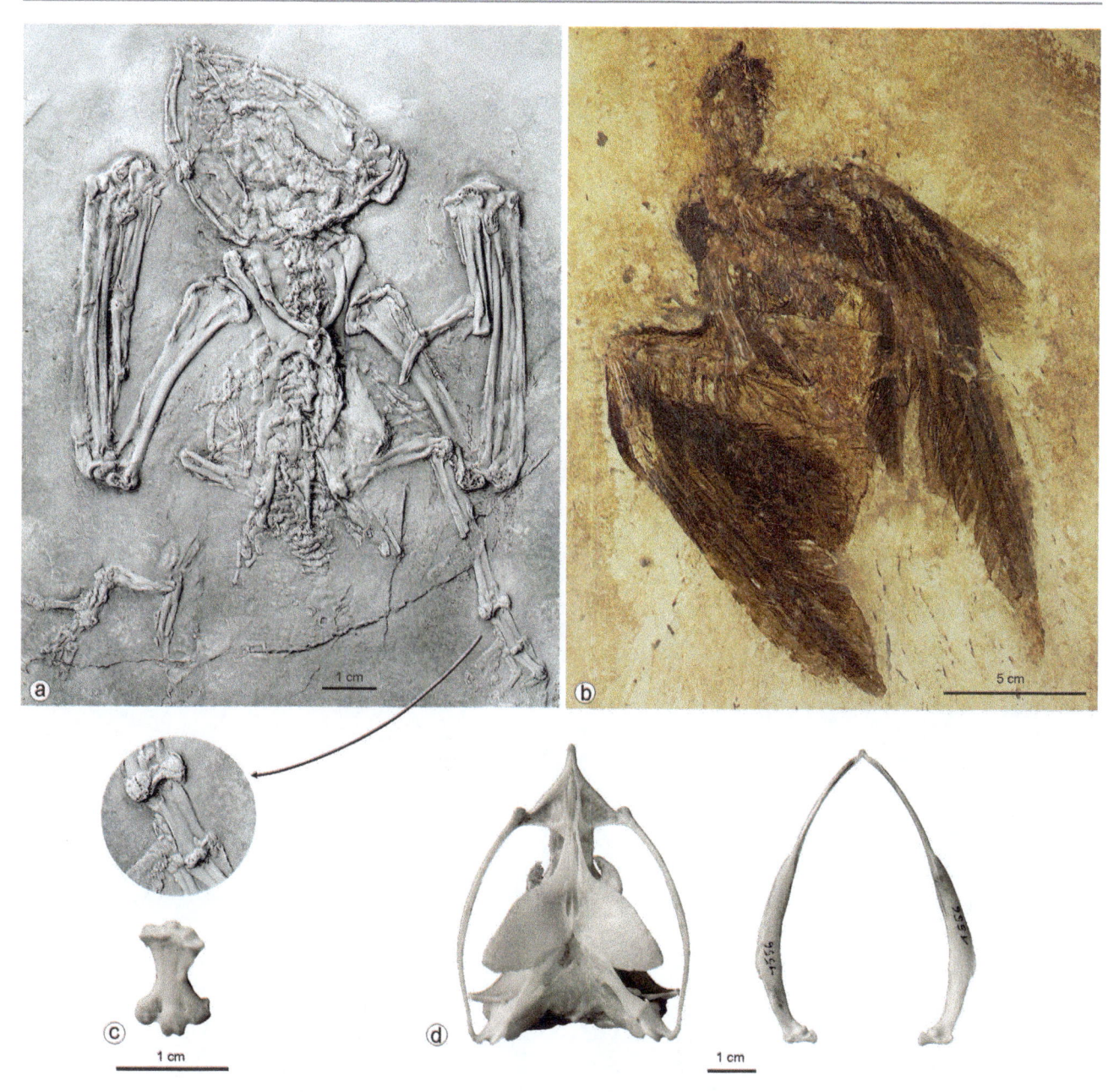

6.8 Strisores (Nightjars, Swifts, Hummingbirds, and Allies)

Fig. 6.7 Fossils of Eocene potoos (Nyctibiiformes). (a) Skeleton of *Paraprefica kelleri* from the latest early or earliest middle Eocene of Messel (Senckenberg Research Institute, Frankfurt, SMF-ME 3727A); specimen coated with ammonium chloride, the arrow indicates an enlarged detail of the tarsometatarsus. (b) Skeleton of *Paraprefica* sp. from Messel (Hessisches Landesmuseum, Darmstadt, Germany, HLMD Be 164); note that this fossil has a proportionally much smaller skull than *P. kelleri*. (c) Tarsometatarsus of the extant *Nyctibius griseus* (Nyctibiidae). (d) Skull and mandible of *N. griseus*. (Photo in (b) by Wolfgang Fuhrmannek, all others by Sven Tränkner)

data alone) or the clade (*Prefica* + *Steatornis*) (combined morphological and molecular data).

Fluvioviridavis resembles *Palaeopsittacus georgei* from the London Clay of Walton-on-the-Naze, which was described as a parrot (Psittaciformes) by Harrison (1982a). The type specimen consists of a coracoid and several fragmentary bones (Fig. 6.8d). Mayr and Daniels (1998) described additional specimens of *P. georgei* from Walton-on-the-Naze (Fig. 6.8c), and Mayr (2003a) tentatively referred to a postcranial skeleton from Messel to this species (Fig. 6.8e, f). The assignment of *P. georgei* to the Psittaciformes is disproved by these specimens, which show that the species lacks a zygodactyl foot. Although the new fossils add to a better understanding of the skeletal morphology of *P. georgei*, the phylogenetic affinities of this species remain uncertain. As in *Fluvioviridavis*, the coracoid of *Palaeopsittacus* exhibits a foramen for the supracoracoideus nerve.

Fig. 6.8 (**a**) Skeleton of *Fluvioviridavis platyrhamphus* (Fluvioviridavidae) from the early Eocene Green River Formation (holotype, Staatliches Museum für Naturkunde Karlsruhe, Germany, SMNK-PAL 2368a). (**b**) *Fluvioviridavis* sp. from the early Eocene London Clay of Walton-on-the-Naze (collection of Michael Daniels, Holland-on-Sea, UK, WN 88588). (**c**) Referred specimen of *Palaeopsittacus georgei* from Walton-on-the-Naze (WN 91682). (**d**) Holotype of *P. georgei* from Walton-on-the-Naze (Natural History Museum, London, NHMUK A 5163). (**e**), (**f**) Postcranial skeleton of *Palaeopsittacus* cf. *georgei* from the latest early or earliest middle Eocene of Messel (SMNK-PAL 3834a); the fossil in (**f**) was coated with ammonium chloride. (All photos by Sven Tränkner)

The recently described putative stem group roller *Ueekenkcoracias tambussiae* from the early Eocene of Argentina (Degrange et al. 2021) is likely to have been misidentified and more closely resembles *Palaeopsittacus*.

However, a definitive classification of this species requires the discovery of more diagnostic material (the holotype just consists of a poorly preserved leg).

6.8.6 Parvicuculidae

Parvicuculus minor was described by Harrison and Walker (1977) on the basis of a tarsometatarsus from the London Clay (Fig. 6.9g). The short and stout bone exhibits a distinctive morphology in that it has a well-developed plantar crest, a large distal vascular foramen, and a plantarly projecting wing-like flange on the trochlea for the fourth toe. Further specimens of *Parvicuculus* are known from the early Eocene of France and the North American Nanjemoy Formation (Fig. 6.9a–f; Mayr and Mourer-Chauviré 2005; Mayr et al. 2022), but these also consist of tarsometatarsi only.

Parvicuculus was considered to be a representative of the Cuculiformes by Harrison and Walker (1977) and Harrison (1982b), but this poorly established classification was disputed by multiple authors (Olson and Feduccia 1979; Martin and Mengel 1984; Baird and Vickers-Rich 1997; Mlíkovský 2002; Mayr and Mourer-Chauviré 2005). Actually, the tarsometatarsus of *Parvicuculus* shows a strong resemblance in its proportions to that of the coeval *Fluvioviridavis*, in which it is equally short and stout and exhibits a plantar crest (Fig. 6.9h; Mayr et al. 2022). As noted in the previous section, *Fluvioviridavis*-like fossils were reported from the London Clay of Walton-on-the-Naze and it is well possible that the tarsometatarsus assigned to *Parvicuculus minor* belongs to these or closely related species. The great similarity of the tarsometatarsi of *Fluvioviridavis* and *Parvicuculus* actually raises the possibility that the taxon Fluvioviridavidae is a junior synonym of the taxon Parvicuculidae.

6.8.7 Podargiformes (Frogmouths)

Paleogene fossils of this group, which is today confined to Australasia, were reported from the late Eocene (MP 16) of the Quercy fissure fillings and from Messel. The French species, *Quercypodargus olsoni*, is known from tarsometatarsi and distal tibiotarsi (Mourer-Chauviré 1989). As noted above (Sect. 6.8.4), there is a possibility that this species is closely related to the putative steatornithiform *Euronyctibius*, which was found in coeval deposits and is only known from wing bones.

Of the species from Messel, *Masillapodargus longipes*, several skeletons have been found (Fig. 6.10). The beak of this species exhibits the characteristic shape of extant Podargiformes, in which it is wide, dorsoventrally flattened, and with a broadly rounded tip (Mayr 1999, 2001a, 2015c). Furthermore, as in extant frogmouths, the ventral surface of the upper beak seems to have been completely ossified. However, the beak of *Masillapodargus longipes* is somewhat narrower than that of crown group Podargiformes and, among others (Mayr 1999, 2001a), the fossil species also differs from extant frogmouths in the morphology of the coracoid and sternum. The legs are proportionally longer than in extant Podargiformes, and the tarsometatarsus is also more slender than that of *Quercypodargus*.

6.8.8 Protocypselomorphus

Protocypselomorphus manfredkelleri is represented by a single skeleton from Messel (Fig. 6.11a; Mayr 2005b). The species has a short, swift-like beak and is of similar size to the extant Common Swift (*Apus apus*). The distal section of the wing (carpometacarpus and distal phalanges) is very long as in apodiform birds, but unlike in the latter, the humerus is not abbreviated. The elongate sternum has a cranially projecting keel. The short tarsometatarsus measures just one-third of the length of the ulna. *P. manfredkelleri* agrees with the Steatornithiformes but differs from other taxa of the Strisores in the presence of 19 or 20 presacral vertebrae. The combination of a short, swift-like beak with long wings and short legs indicates that the species was an aerial insectivore.

A phylogenetic analysis by Mayr (2005b) resulted in a sister group relationship between *Protocypselomorphus* and a clade including the Caprimulgiformes, Nyctibiiformes, Aegotheliformes, and Apodiformes. This clade is, however, not recovered in current molecular analyses. A more recent analysis by Chen et al. (2019) identified *Protocypselomorphus* as a stem group representative of the Steatornithiformes, but the character evidence for this placement is limited to the high number of presacral vertebrae. The count of 19 or 20 presacral vertebrae in *Protocypselomorphus* corresponds to that of most other neognathous birds and is therefore likely to be plesiomorphic for the Strisores, all non-steatornithiform representatives of which have only 18 or 17 presacral vertebrae (Mayr 2005b). *P. manfredkelleri* is much smaller than the coeval *Prefica nivea* and the extant *Steatornis caripensis*, and its swift-like beak is clearly distinguished from the more robust beak of the Oilbird. The beak of *Protocypselomorphus* is also much shorter than that of *Fluvioviridavis*, which likewise resulted as a stem group representative of the Steatornithiformes in some analyses of Chen et al. (2019). The hand section of the wing of *Protocypselomorphus* is furthermore proportionally longer than it is in *Fluvioviridavis*, *Prefica*, and *Steatornis*. If indeed a representative of the Steatornithiformes, *Protocypselomorphus* would document an unexpectedly high diversity of these birds. Even though the affinities of *Protocypselomorphus* are certainly in need of a revision, the holotype and only known specimen of the taxon does

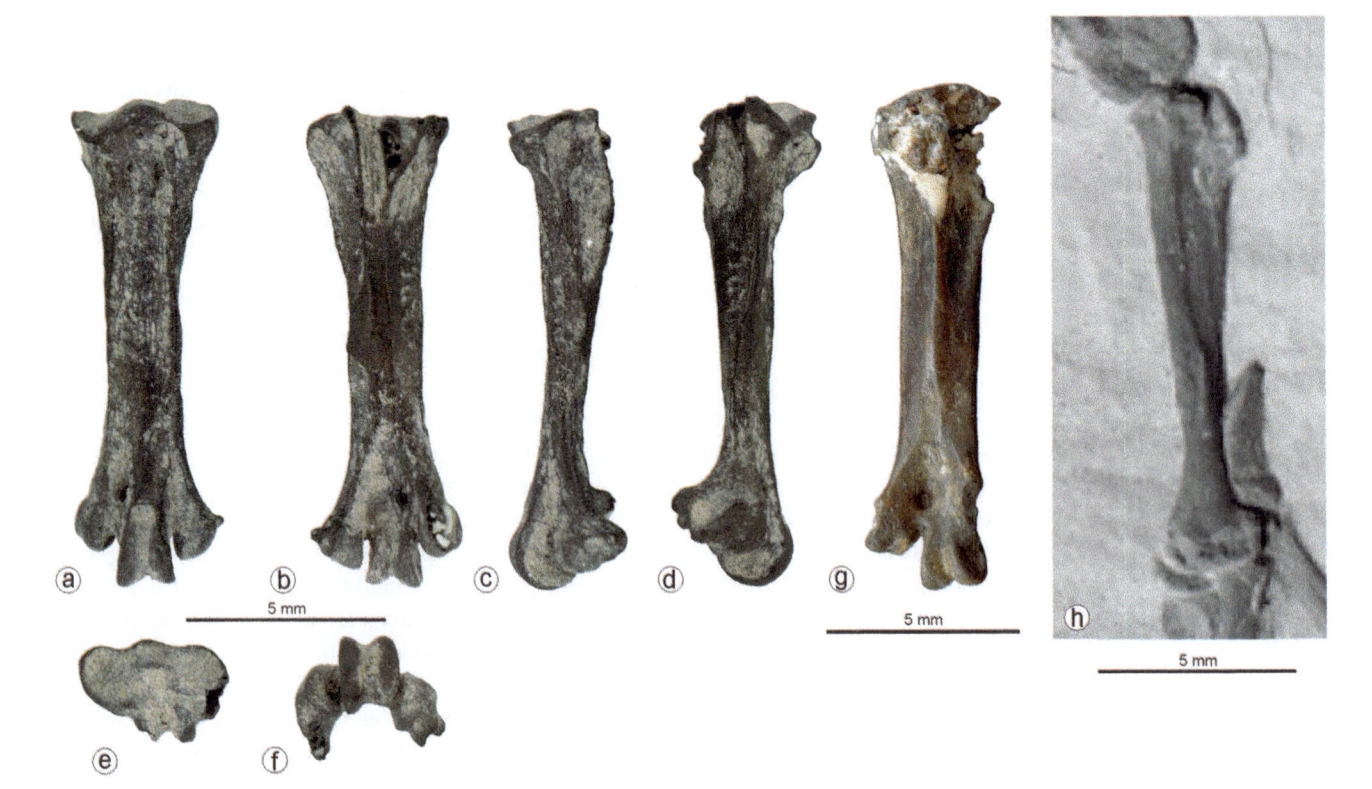

Fig. 6.9 (**a**)–(**f**) Right tarsometatarsus of *Parvicuculus* sp. (Parvicuculidae) from the early Eocene Nanjemoy Formation, Virginia, USA (National Museum of Natural History, Smithsonian Institution, Washington D.C., USA, USNM PAL 496384). (**g**) Right tarsometatarsus of *Parvicuculus minor* (holotype; Natural History Museum, London, NHMUK A 4919). (**h**) Left tarsometatarsus (lateral view) of a fossil from the early Eocene Green River Formation, which was referred to *Fluvioviridavis platyrhamphus* by Nesbitt et al. (2011). From Mayr et al. (2022), published under a Creative Commons CC BY 4.0 license; image in (**h**) from Nesbitt et al. (2011), published under a Creative Commons CC BY 4.0 license

not allow the recognition of many osteological details, and steatornithiform affinities are here considered weakly supported and rather unlikely.

6.8.9 Aegotheliformes (Owlet Nightjars) and Apodiformes (Swifts and Hummingbirds)

There is congruent evidence from multiple analyses of different data that the Australasian Aegotheliformes are the sister taxon of apodiform birds (e.g., Mayr 2002; Mayr et al. 2003; Ericson et al. 2006). Extant Apodiformes include the Southeast Asian Hemiprocnidae (tree swifts), the globally distributed Apodidae (true swifts), and the Trochilidae (hummingbirds) which today occur only in the New World. All representatives of the Apodiformes are small to very small birds that exhibit specialized flight techniques. In all species, the proximal wing bones (humerus and ulna) are strongly abbreviated, and the hand section is very long, with the carpometacarpus being at least as long as the humerus.

The Aegotheliformes have no Paleogene fossil record. An earlier tentative identification of owlet-nightjars in the Quercy fissure fillings (Mourer-Chauviré 1982) was based on a fragmentary sternum, which more likely belongs to a species of the Aegialornithidae (own observation and Mourer-Chauviré 2006). Apodiform birds were diversified in the Paleogene of the Northern Hemisphere, and their fossil record dates back to the early Eocene.

The alleged swift *Laputavis robusta* from the London Clay of Walton-on-the-Naze is certainly no apodiform bird (Mayr 2001b; contra Dyke 2001a, b). This species differs from all representatives of the clade formed by the Aegotheliformes and Apodiformes in the absence of a coracoidal foramen for the supracoracoideus nerve.

Another poorly known bird that was likened to the Apodiformes is *Cypseloramphus dimidius*, which is just known from a partial skeleton from Messel (Fig. 6.11b; Mayr 2016a). This species has a short and wide, "swift-like" beak and an abbreviated humerus, which indicate apodiform affinities.

6.8.9.1 Eocypselidae

This taxon of stem group Apodiformes (Mayr 2003b) was established by Harrison (1984) for *Eocypselus vincenti* from the early Eocene London Clay, the holotype of which consists of wing and pectoral girdle bones (Fig. 6.12g).

Fig. 6.10 Fossils of Eocene frogmouths (Podargiformes). (**a**) Skeleton of *Masillapodargus longipes* from the latest early or earliest middle Eocene of Messel (Staatliches Museum für Naturkunde Karlsruhe, Germany, SMNK-PAL 1083). (**b**) Isolated beak of *M. longipes* from Messel (Senckenberg Research Institute, Frankfurt, SMF-ME 3404B). (**c**) Skull of the extant *Batrachostomus javensis*. (**d**) Partial skeleton of *M. longipes* from Messel (Hessisches Landesmuseum Darmstadt, Germany, HLMD Me 7627a). (All photos by Sven Tränkner)

Two postcranial skeletons from the Danish Fur Formation were assigned to *E. vincenti* by Dyke et al. (2004); Mayr (2010b) described a further complete skeleton from this locality (Fig. 6.12a, b). A North American species of *Eocypselus*, *E. rowei*, was reported from the Green River Formation and is represented by a well-preserved skeleton with feather remains (Ksepka et al. 2013). Tentatively identified fragmentary bones of *Eocypselus* are furthermore known from the early Eocene Nanjemoy Formation (Mayr 2016b).

Eocypselus was initially considered to be closely related to the Hemiprocnidae (Harrison 1984; Dyke et al. 2004), but Mayr (2003b, 2010b) detailed that the taxon is a stem group apodiform. This hypothesis was confirmed by Ksepka et al. (2013). Even though the humerus of *Eocypselus* is stouter than that of non-apodiform birds (including the Aegotheliformes), it is less abbreviated than the humerus of crown group Apodiformes. Together with the more slender ulna, the humerus proportions suggest a position of *E. vincenti* outside crown group Apodiformes. The coracoid resembles the corresponding bone of the Aegotheliformes and is distinguished from that of crown group Apodiformes in the proportionally smaller acrocoracoid process. The caudal margin of the sternum exhibits four incisions, deep lateral and shallow medial ones, whereas the sternum lacks incisions in crown group Apodiformes. Unlike in extant swifts and hummingbirds, the proximal phalanx of the major wing digit of *Eocypselus* bears just a small, distally projecting internal index process (processus internus indicis). The legs of *Eocypselus* are very long compared to those of most extant Apodiformes (Mayr 2010b).

Eocypselus resulted as a stem group representative of the Aegotheliformes in one of the analyses by Chen et al. (2019), whereas an analysis constrained to a molecular scaffold supported a position as a stem group apodiform, as suggested by Mayr (2003b, 2010b). However, *Eocypselus* is not only clearly outside crown group Apodiformes, but the taxon may actually be a stem group representative of the clade including

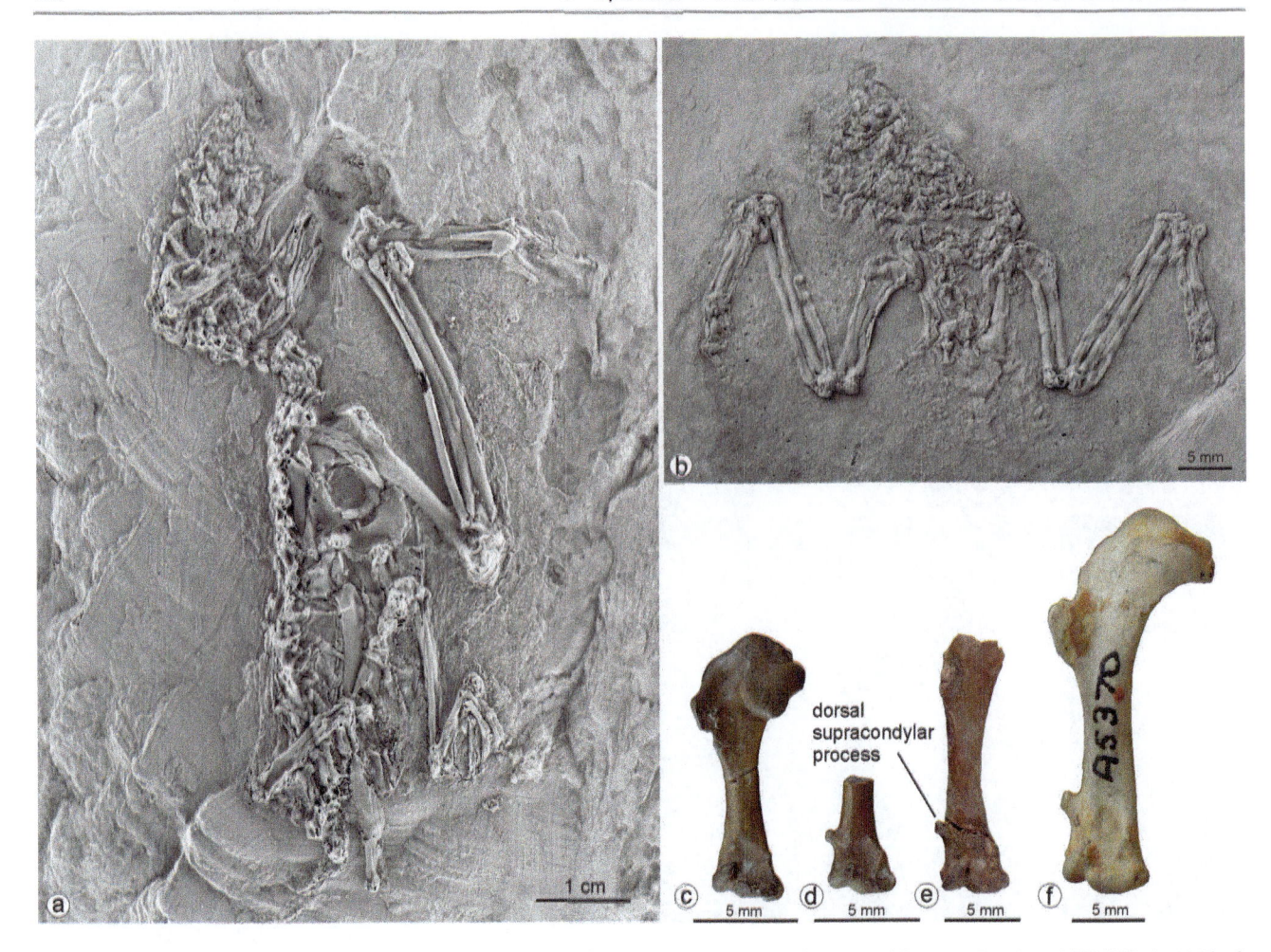

Fig. 6.11 (a) Holotype of *Protocypselomorphus manfredkelleri* from the latest early or earliest middle Eocene of Messel (Senckenberg Research Institute, Frankfurt, SMF-ME 11043); specimen coated with ammonium chloride. (b) Holotype of *Cypseloramphus dimidius* from Messel (SMF-ME 11081); specimen coated with ammonium chloride. (c), (d) Humeri of *Primapus lacki* (Aegialornithidae) from the early Eocene London Clay of Bognor Regis and the Isle of Sheppey (c: holotype, Natural History Museum, London, NHMUK A 2166; d: referred specimen NHMUK A 3710). (e) Humerus of *Aegialornis broweri* (Aegialornithidae) from the middle Eocene of the Geisel Valley (Geiseltalsammlung, Martin- Luther Universität of Halle-Wittenberg, Germany, GMH L-9-1969). (f) Humerus of *Aegialornis gallicus* (Aegialornithidae) from an unknown stratigraphic horizon of the Quercy fissure fillings (NHMUK A 5370). (All photos by Sven Tränkner)

the Aegotheliformes and apodiform birds. This is suggested by an undescribed *Eocypselus* specimen from the London Clay of Walton-on-the-Naze in the Daniels collection, which includes an owlet nightjar- or swift-like upper beak and mandible as well as the axis (second cervical vertebra) and substantial portions of the sternum (Fig. 6.12c). The latter is distinguished from the sternum of crown group Apodiformes by the presumably plesiomorphic presence of a slightly bifurcated spina externa and long craniolateral processes. The morphology of the axis is remarkable in that the spinosus process is well-delimited from the caudal zygapophyses, whereas these structures are connected by a ridge for the attachment of the splenius capitis muscle (Fig. 6.12d–f). The latter morphology is due to an unusual cruciform development of this neck muscle, with contralateral portions of the muscle overlapping and interdigitating with each other, which is one of the hallmark synapomorphies of the Aegotheliformes and Apodiformes (Mayr 2002). The plesiomorphic absence of a ridge formed by the spinosus process and the caudal zygapophyses of the axis indicates that *Eocypselus* lacked a cruciform splenius capitis muscle and may suggest a position of this early Eocene taxon outside the clade formed by the Aegotheliformes and Apodiformes.

6.8.9.2 Aegialornithidae

The Aegialornithidae are among the most abundant small birds in late Eocene sites of the Quercy fissure fillings, where they disappear toward the early Oligocene (Mourer-Chauviré 1978, 1980, 1988). Four species were recognized in these deposits by Mourer-Chauviré (1988), that is, *Aegialornis gallicus*, *Ae. broweri*, *Ae. wetmorei*, and *Ae. leehnardti*. A humerus from the middle Eocene of the Geisel Valley (MP 12) was assigned to *Ae. broweri* by Peters (1998) (this specimen was assigned to a new species *Ae. germanicus*

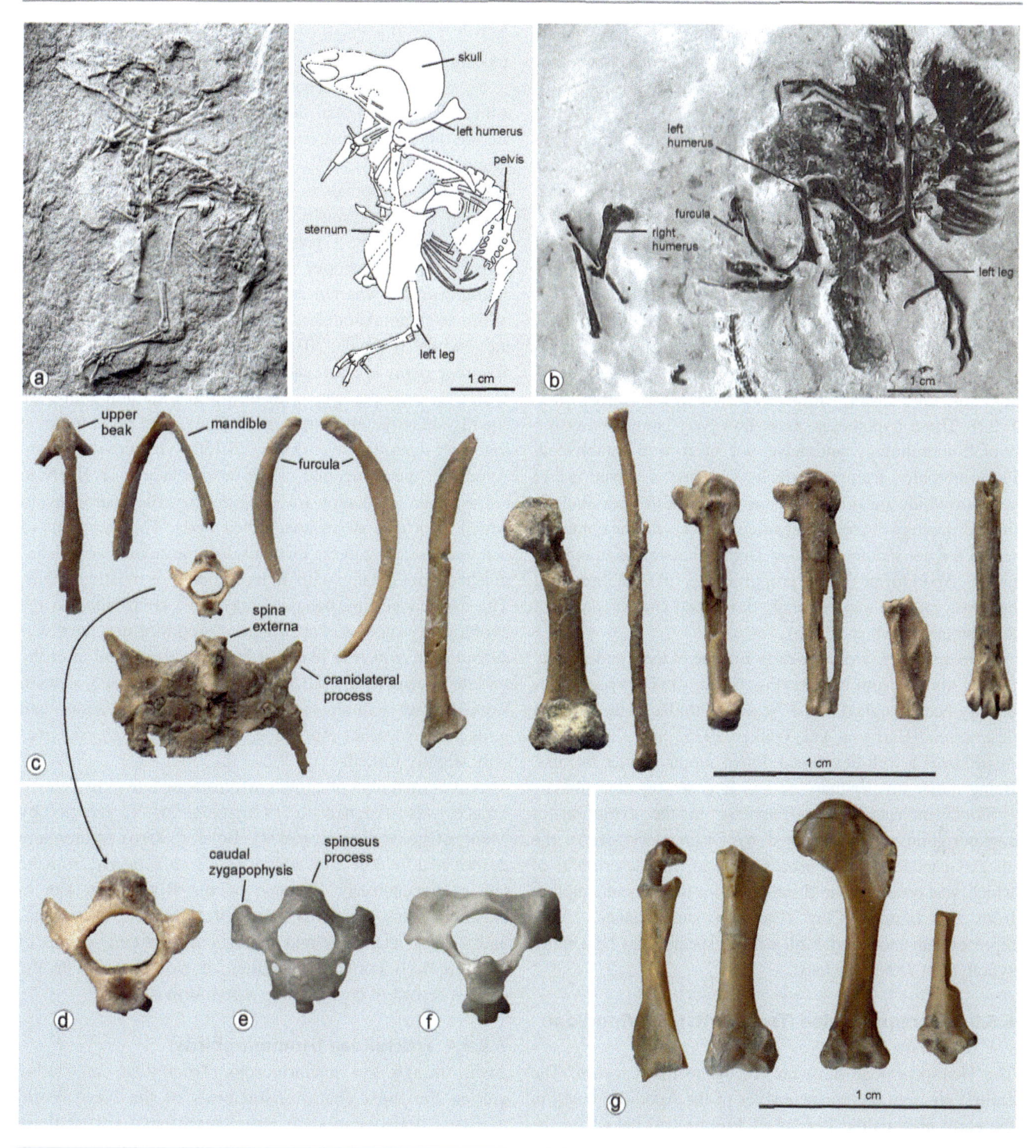

6.8 Strisores (Nightjars, Swifts, Hummingbirds, and Allies)

Fig. 6.12 Fossils of the Eocypselidae. (**a**), (**b**) Skeletons of *Eocypselus vincenti* from the early Eocene Fur Formation in Denmark (Geological Museum, Copenhagen, Denmark, **a**: MGUH 29278 with interpretive drawing, **b**: MGUH 26729). (**c**) Specimen of *Eocypselus* sp. from the early Eocene London Clay of Walton-on-the-Naze (collection of Michael Daniels, Holland-on-Sea, UK, WN 91680), with (**d**) an enlarged view of the axis. (**e**) Axis of the extant *Chordeiles minor* (Caprimulgidae). (**f**) Axis of the extant *Apus melba* (Apodidae). (**g**) Holotype of *Eocypselus vincenti* from Walton-on-the-Naze (Natural History Museum, London, NHMUK A 5429). (**e**) and (**f**) are not to scale. (Photo in (**b**) courtesy of Bent Lindow, all others by Sven Tränkner)

by Mlíkovský 2002, who did, however, not convincingly justify this taxonomic action; Mayr 2020b).

Olson (1999) assigned a tarsometatarsus from the early Eocene of the Nanjemoy Formation in Virginia to the

Aegialornithidae. This specimen is stouter than the tarsometatarsus of *Aegialornis* and is from a representative of the Parvicuculidae (Sect. 6.8.6; Mayr and Mourer-Chauviré 2005; Mayr et al. 2022).

All major limb bones of aegialornithids are known. The humerus is less robust than in crown group Apodiformes, and there is a well-developed dorsal supracondylar process on its distal end (Fig. 6.11e, f). The ulna is short and stout but also less so than that of extant Apodiformes. The fenestrated proximal phalanx of the major wing digit bears a well-developed internal index process.

Earlier authors considered the Aegialornithidae to either be closely related to the Hemiprocnidae or to be a transitional taxon linking the Hemiprocnidae and Apodidae (e.g., Harrison 1975, 1984; Mourer-Chauviré 1988; Karhu 1988, 1992). These hypotheses were, however, largely based on overall morphology and have not been well established. Plesiomorphic features which support a position of aegialornithids outside crown group Apodiformes include the morphology of the sternal articular facets for the coracoid, which are not saddle-shaped or slightly convex as in crown-group Apodiformes, the proportionally less abbreviated humerus, and the more strongly developed cnemial crests of the tibiotarsus (Mayr 2003b).

Primapus lacki from the early Eocene of the London Clay is only known from humeri (Fig. 6.11c, d). The species has already been assigned to the Aegialornithidae in the original description (Harrison and Walker 1975) and shares with *Aegialornis* a well-developed dorsal supracondylar process. Being much smaller than the late Eocene Aegialornithidae, *P. lacki* indicates a size increase in the evolution of aegialornithids. As first noted by Olson (1985: 135), the alleged cuculiform *Procuculus minutus*, the description of which was based on the distal end of a tiny tarsometatarsus from the London Clay (Harrison and Walker 1977), represents an apodiform bird, and this species may be a junior synonym of *Primapus lacki*.

6.8.9.3 Hemiprocnidae (Tree Swifts) and Apodidae (True Swifts)

The Hemiprocnidae have no Paleogene fossil record. The earliest stem group representatives of the Apodidae occur in the early and middle Eocene of Europe and belong to the taxon *Scaniacypselus*. *Scaniacypselus wardi* from the early Eocene (MP 8; Mlíkovský 2002) of Denmark was described by Harrison (1984) and is known from the wing bones of a single individual (Fig. 6.13d). A second species, *S. szarskii*, is represented by several skeletons from Messel (Fig. 6.13a–c) and a humerus from the middle Eocene (MP 10/11) of the Quercy fissure fillings (Peters 1985; Mayr and Peters 1999; Mourer-Chauviré 2006; Mayr 2015d). A record of *Scaniacypselus* sp. also comes from the late Eocene (MP 17) of the Quercy fissure fillings (Mourer-Chauviré and Sigé 2006).

S. szarskii was erroneously assigned to the Aegialornithidae in the original description, from which it differs, among others, in the strongly abbreviated humerus and ulna (Mayr and Peters 1999). With regard to the latter traits, *S. szarskii* corresponds to extant Apodidae even though the sternum is proportionally shorter than that of extant true swifts, the ulna proportionally longer, and the carpometacarpus shorter. These differences indicate that *Scaniacypselus* was not as well adapted to an aerial mode of life as extant Apodidae, which spend most of their life on the wing (Mayr 2015d). The tarsometatarsus of *Scaniacypselus* is much shorter than that of most extant true swifts. Because the legs of tree swifts and hummingbirds are equally abbreviated, short legs are likely to be plesiomorphic for the Apodiformes (Mayr 2015d). In crown group Apodidae, the elongation of the tarsometatarsus is likely to represent an adaptation for clinging to vertical surfaces, to which most true swifts attach their nests. The short legs of *Scaniacypselus* suggest that this Eocene taxon was distinguished from extant swifts in its breeding or roosting habits. The derived nesting behavior may have contributed to the evolutionary success of crown group Apodiformes, the diversification of which is likely to have occurred well after the middle Eocene (Mayr 2015d). Many specimens of *S. szarskii* from Messel exhibit excellent feather preservation, and unlike many extant Apodidae, the tail of *S. szarskii* was only slightly forked.

A proximal ulna of a true swift from the late Oligocene of France was assigned to *Procypseloides* cf. *ignotus* by Mourer-Chauviré et al. (2004). Boles (2001b) furthermore described a swiftlet (Collocaliini), *Collocalia buday*, from the late Oligocene/early Miocene of the Riversleigh site in Australia. This species is known from humeri, a coracoid, and a tarsometatarsus, and is larger than extant species of *Collocalia*. If correctly identified, it would constitute the earliest record of crown group Apodiformes.

6.8.9.4 Trochilidae (Hummingbirds)

Even though the nectarivorous Trochilidae are today among the most characteristic birds of the New World avifauna, Paleogene stem group representatives of these birds were only found in fossil sites in Europe. Wing bones of a small, hummingbird-like apodiform were first reported by Karhu (1988) from the early Oligocene of the Caucasus, who assigned these fossils to the species *Jungornis tesselatus*, within the new taxon Jungornithidae. An almost complete humerus of another species of *Jungornis*, *J. geraldmayri*, was reported by Mourer-Chauviré and Sigé (2006) from the late Eocene (MP 17b) of the Quercy fissure fillings.

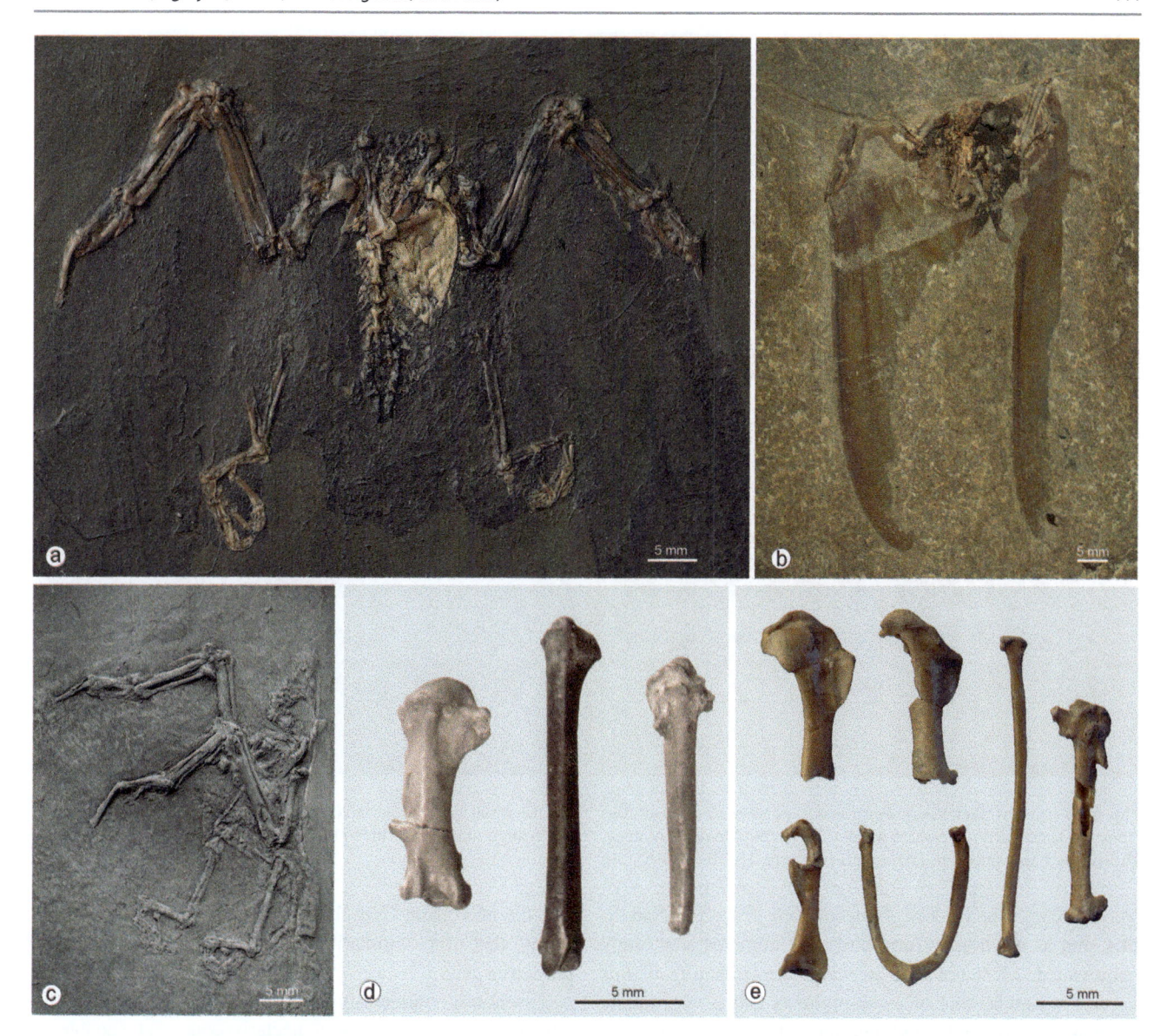

6.8 Strisores (Nightjars, Swifts, Hummingbirds, and Allies)

Fig. 6.13 (a)–(c) Skeletons of *Scaniacypselus szarskii* (Apodidae) from the latest early or earliest middle Eocene of Messel (**a**: Staatliches Museum für Naturkunde Karlsruhe, Germany, SMNK-PAL 301; **b**: Senckenberg Research Institute, Frankfurt, SMF-ME 11345A, photo taken before the preparation of the specimen; **c**: SMF-ME 3409A, coated with ammonium chloride). (**d**) Holotype of *Scaniacypselus wardi* (Apodidae) from the early Eocene Røsnæs Clay in Denmark (holotype, Natural History Museum, London, NHMUK A 5430). (**e**) Undescribed *Scaniacypselus*-like apodiform bird from the early Eocene London Clay of Walton-on-the-Naze (collection of Michael Daniels, Holland-on-Sea, UK, WN 93766). (Photo in (**b**) by Michael Ackermann, all others by Sven Tränkner)

Karhu (1999) also assigned a slightly older and morphologically less advanced species from late Eocene deposits of the Caucasus to the Jungornithidae. This bird was described as *Argornis caucasicus* and is likewise just known from wing and pectoral girdle elements. Karhu (1988, 1992, 1999) noted that *Jungornis* and *Argornis* share characteristic features with extant hummingbirds, including a derived morphology of the proximal ulna and, in *Jungornis*, a protrusion on the humeral head, which is solely found in hummingbirds and allows a rotation of the humerus during hovering flight. He attributed these similarities to convergence and considered *Argornis* and *Jungornis* to be aberrant swifts with hummingbird-like specializations. However, there are no derived morphological traits that are exclusively shared by *Argornis*, *Jungornis*, and swifts which are not also found in hummingbirds. Not surprisingly, therefore, an analysis of the interrelationships of extant and fossil Apodiformes showed the "Jungornithidae" sensu Karhu (1999) to be paraphyletic, with *Argornis* being the sister taxon of a clade including *Jungornis* and crown-group Trochilidae (Mayr 2003b).

Argornis and *Jungornis* are only known from wing elements, but an *Argornis*-like bird, *Parargornis messelensis*, was described from Messel and is known from a complete skeleton with well-preserved feather remains (Fig. 6.14;

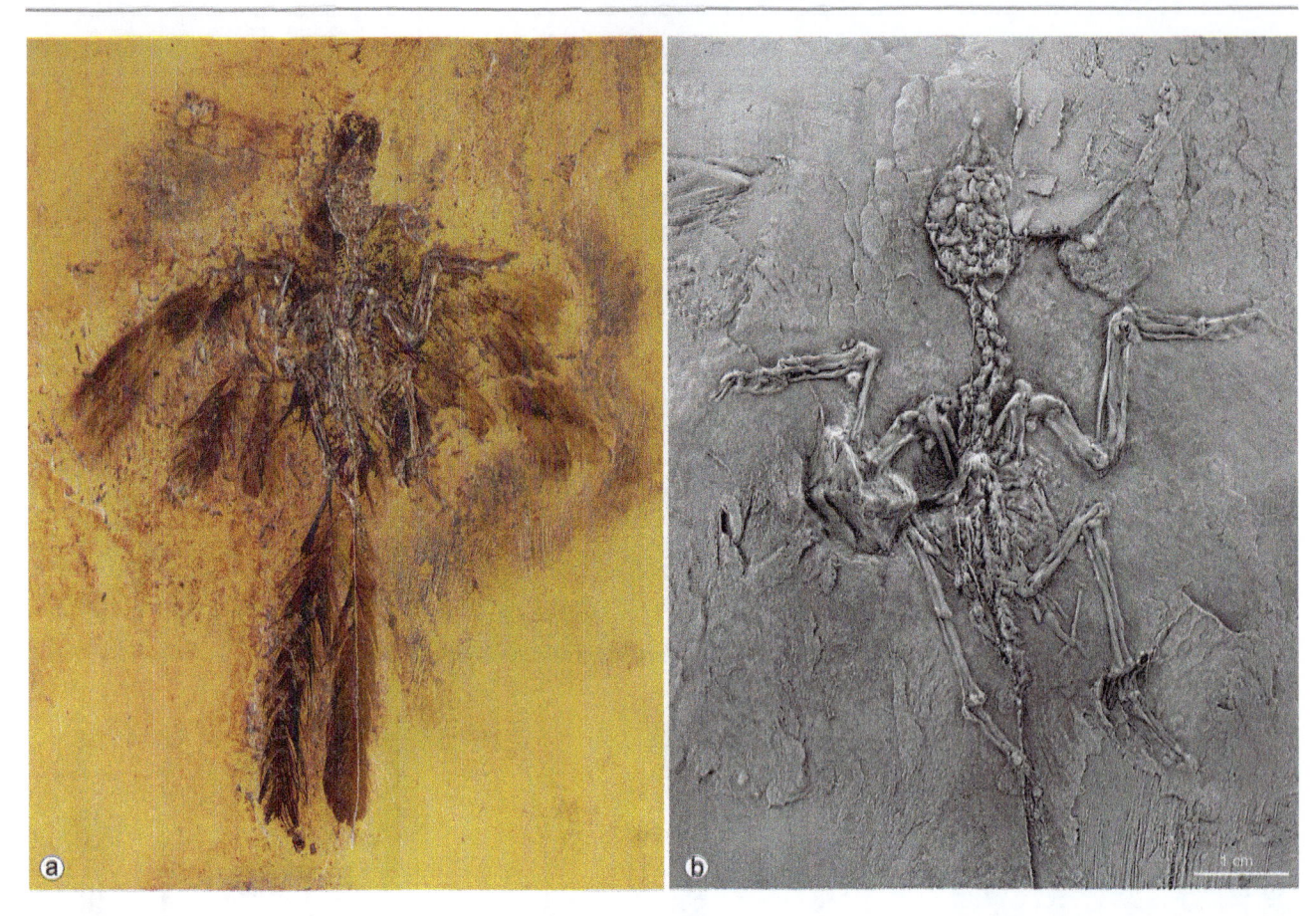

Fig. 6.14 (**a**), (**b**) Skeleton of *Parargornis messelensis* (Trochilidae) from the latest early or earliest middle Eocene of Messel (holotype, Hessisches Landesmuseum Darmstadt, Germany, HLMD Be 163). In (**b**) the fossil was coated with ammonium chloride. (Photo in (**a**) by Wolfgang Fuhrmannek, in (**b**) by Sven Tränkner)

Mayr 2003c). The beak of *P. messelensis* is short and swift-like, and as in all apodiform birds, the humerus is strongly abbreviated. Most notably, however, the feathering of this species resembles that of owlet-nightjars and is very unlike the feathering of swifts and hummingbirds in that the wings are short and broad and the tail is long. The peculiar wing morphology of *Parargornis*, that is, the combination of a greatly abbreviated humerus with a short and broad wing, has no counterpart among extant birds and may represent an early stage in the evolution of hovering flight.

Another archaic stem group representative of the Trochilidae is *Cypselavus gallicus* from the late Eocene and early Oligocene (MP 16–23) of the Quercy fissure fillings. This species was classified into the Hemiprocnidae by earlier authors (Harrison 1984; Peters 1985; Mourer-Chauviré 1988). The coracoid assigned to *C. gallicus* by Mourer-Chauviré (1978) was transferred to the Jungornithidae and described as a new species, *Palescyvus escampensis*, by Karhu (1988). More likely, however, all skeletal elements of *Cypselavus* belong to a single species, which is also a stem group representative of the Trochilidae. Mourer-Chauviré (2006) classified *Parargornis*, *Argornis*, and *Cypselavus* into the new taxon Cypselavidae, but there remains a possibility that the similarities between these birds are plesiomorphic.

Essentially modern-type stem group hummingbirds, which appear to have been nectarivorous and capable of sustained hovering flight, occur in the early Oligocene of Europe. The first described species is *Eurotrochilus inexpectatus* from Wiesloch-Frauenweiler in Germany, which is known from partial skeletons of four individuals (Fig. 6.15a, b; Mayr 2004b, 2007; Mayr and Micklich 2010). Another species of *Eurotrochilus*, *E. noniewiczi*, was described on the basis of a partial skeleton from the early Oligocene of the Carpathian Basin in Poland (Bochenski and Bochenski 2008; an isolated foot described by Bochenski et al. 2016 is likely to be from the same or a closely related species). A complete skeleton of *Eurotrochilus* sp. was found in the early Oligocene of the Luberon area in France (Fig. 6.15c; Louchart et al. 2008).

In addition to a similar overall skeletal morphology, *E. inexpectatus* shares with extant Trochilidae a tiny size and a greatly elongated beak, as well as a marked depression on the caudal surface of the proximal end of the ulna and a

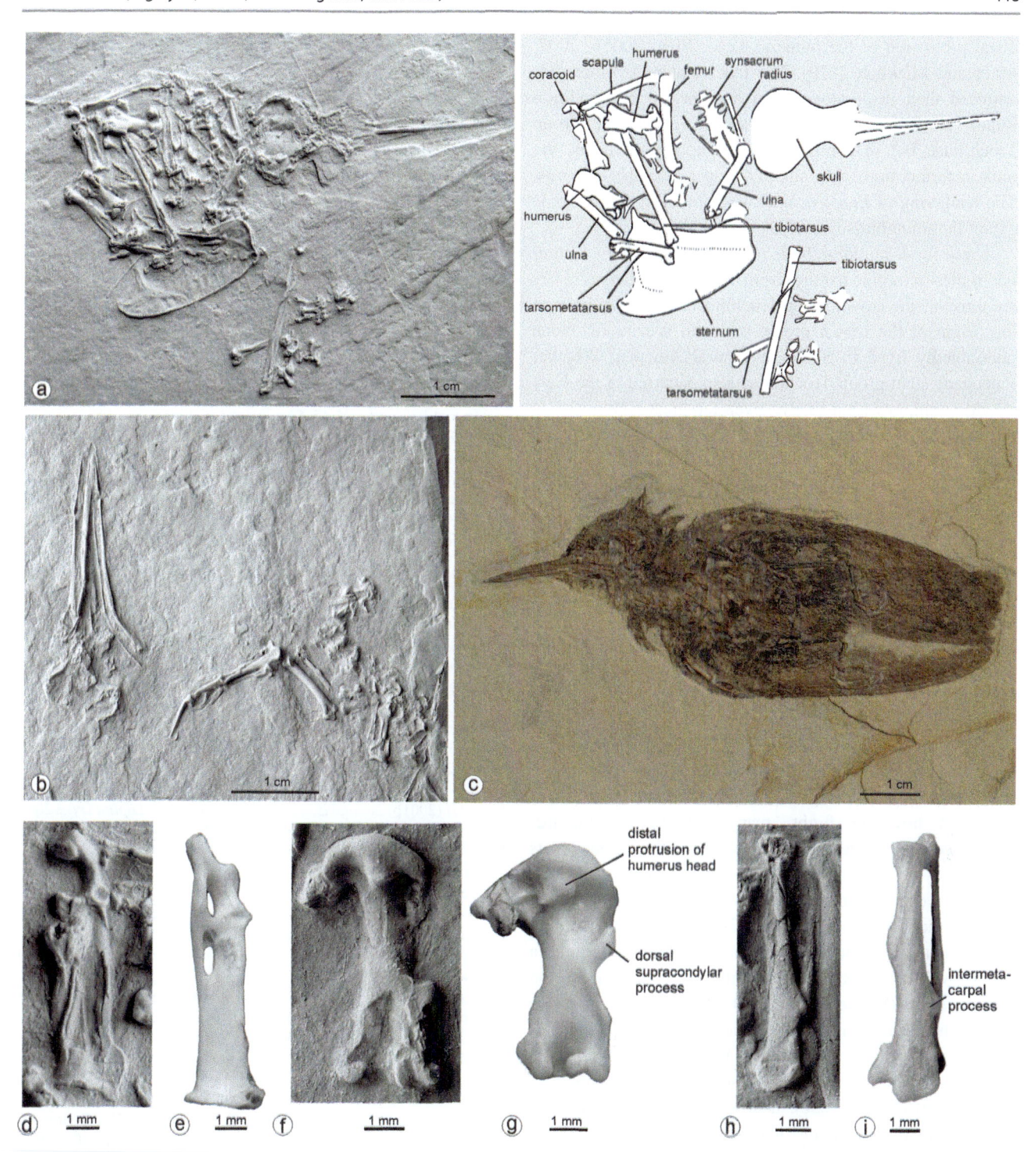

6.8 Strisores (Nightjars, Swifts, Hummingbirds, and Allies)

Fig. 6.15 Fossils of Oligocene hummingbirds (Trochilidae). (**a**) Skeleton of *Eurotrochilus inexpectatus* (Trochilidae) from the early Oligocene of Wiesloch-Frauenweiler in Germany with interpretive drawing (counter slab of holotype, Staatliches Museum für Naturkunde Karlsruhe, Germany, SMNK-PAL 5591). (**b**) Another *E. inexpectatus* specimen from Wiesloch-Frauenweiler (SMNK-PAL 4410a). (**c**) Skeleton *Eurotrochilus* sp. from the early Oligocene of the Luberon area in France (private collection of Nicholas Tourment, Marseille, NT-LBR-040). (**d**), (**e**) Coracoid of *E. inexpectatus* (left, SMNK-PAL 5591) and the extant Great-billed Hermit, *Phaethornis malaris* (right). (**f**), (**g**) Humerus of *E. inexpectatus* (left, SMNK-PAL 6599) and the extant *Ph. malaris* (right). (**h**), (**i**) Carpometacarpus of *E. inexpectatus* (left, SMNK-PAL 6599) and the extant Hairy Hermit, *Glaucis hirsuta* (right). All fossil specimens except (**c**) coated with ammonium chloride. (All photos except (**c**) by Sven Tränkner, photo in (**c**) by the author)

distal protrusion of the humerus head (Mayr 2004b, 2007; Mayr and Micklich 2010). The latter feature is more pronounced than in *Jungornis* and approaches the condition found in extant hummingbirds. As in crown group Trochilidae but in contrast to other apodiform birds, the ulna is furthermore only slightly longer than the humerus. The feathering of *Eurotrochilus* was similar to that of some extant hummingbirds (Louchart et al. 2008).

These fossil hummingbirds nevertheless differ in a number of plesiomorphic features from their living relatives. They are thus outside crown group Trochilidae and do not bear on the origin of the crown group, the stem species of which undoubtedly lived in South or Central America. Whether Paleogene stem group Trochilidae also occurred in the New World is unknown but likely, given the great similarities in the early Oligocene avifaunas of North America and Europe (Sect. 11.1.1).

The long beak of *E. inexpectatus* indicates that the species was nectarivorous, whereas the swift-like beak of *Parargornis* suggests that the stem species of Pan-Trochilidae was insectivorous. If the above interpretation of the wing morphology of *Parargornis* is correct, the capability for hovering flight evolved in the stem lineage of the Trochilidae before the origin of a nectarivorous feeding ecology and may have enabled *Parargornis* to glean insects from the underside of leaves or around flowers (Mayr and Manegold 2002; Mayr 2005c).

The *Eurotrochilus* fossils are also of interest with regard to the evolution of ornithophilous (bird-pollinated) flowers in the Old World. Because hummingbirds are capable of sustained hovering flight, many extant hummingbird-pollinated plants differ from Old World ornithophilous plants in that they do not provide perches near the flowers. Surprisingly, however, some extant Old World plants that occur in areas without hovering avian pollinators exhibit a similar flower morphology to those pollinated by hummingbirds in the New World. Examples, therefore, are the Himalayan *Agapetes* spp. (Ericaceae), the East African *Canarina eminii* (Campanulaceae), the South African *Tecomaria capensis* (Bignoniaceae), as well as the West African *Impatiens sakeriana* (Balsaminaceae) and *Sabicea speciosa* (Rubiaceae) (Mayr 2005c). It now seems possible that the flower traits of these plants may indeed primarily constitute an adaptation to pollination by hummingbirds, which, after the disappearance of hummingbirds from the Old World, may have been adopted by long-tongued bees and other insects (Mayr 2004b, 2005c).

References

Baird RF, Vickers-Rich P (1997) *Eutreptodactylus itaboraiensis* gen. et. sp. nov., an early cuckoo (Aves: Cuculiformes) from the Late Paleocene of Brazil. Alcheringa 21:123–127

Ballmann P (1970) Ein neuer Vertreter der Musophagidae (Aves) aus dem Chattium von Gaimersheim bei Ingolstadt (Bayern). Mitt Bayer Staatssamml Paläontol Hist Geol 10:271–275

Becker JJ, Brodkorb P (1992) An early Miocene ground-dove (Aves: Columbidae) from Florida. In: Campbell KE (ed) Papers in avian paleontology honoring Pierce Brodkorb. Nat Hist Mus Los Angeles Cty Sci Ser 36:189–193

Bochenski Z, Bochenski ZM (2008) An Old World hummingbird from the Oligocene: a new fossil from Polish Carpathians. J Ornithol 149:211–216

Bochenski ZM, Tomek T, Swidnicka E (2010) A columbid-like avian foot from the Oligocene of Poland. Acta Ornithol 45:233–236

Bochenski ZM, Tomek T, Swidnicka E (2016) A tiny short-legged bird from the early Oligocene of Poland. Geolo Carpath 67:463–469

Boles WE (2001a) A new emu (Dromaiinae) from the late Oligocene Etadunna Formation. Emu 101:317–321

Boles WE (2001b) A swiftlet (Apodidae: Collocaliini) from the Oligo-Miocene of Riversleigh, northwestern Queensland. Mem Assoc Austral Palaeontol 25:45–52

Braun EL, Kimball RT (2021) Data types and the phylogeny of Neoaves. Birds 2:1–22

Brodkorb P (1978) Catalogue of fossil birds. Part 5 (Passeriformes). Bull Fla state Mus. Biol Sci 23:139–228

Chandler RM (1999) Fossil birds of Florissant, Colorado: with a description of a new genus and species of cuckoo. Geol Resour Div Techn Rep NPS/NRGRD/GRDTR-99:49–53

Chen A, White ND, Benson RB, Braun MJ, Field DJ (2019) Total-evidence framework reveals complex morphological evolution in nightbirds (Strisores). Diversity 11(9):143

Cracraft J (1971) A new family of hoatzin-like birds (Order Opisthocomiformes) from the Eocene of South America. Ibis 113:229–233

Degrange FJ, Pol D, Puerta P, Wilf P (2021) Unexpected larger distribution of Paleogene stem-rollers (Aves, Coracii): new evidence from the Eocene of Patagonia, Argentina. Sci Rep 11:1363

Dyke GJ (2001) A primitive swift from the London Clay and the relationships of fossil apodiform birds. J Vertebr Paleontol 21:195–200

Dyke GJ (2001b) *Laputavis*, a replacement name for *Laputa* Dyke 2001 (preoccupied name). J Vertebr Paleontol 21:401

Dyke GJ, Waterhouse DM, Kristoffersen AV (2004) Three new fossil landbirds from the early Paleogene of Denmark. Bull Geol Soc Denmark 51:47–56

Ericson PGP, Anderson CL, Britton T, Elzanowski A, Johansson US, Källersjö M, Ohlson JI, Parsons TJ, Zuccon D, Mayr G (2006) Diversification of Neoaves: integration of molecular sequence data and fossils. Biol Lett 2:543–547

Feduccia A (1999) The origin and evolution of birds, 2nd edn. Yale University Press, New Haven

Field DJ, Hsiang AY (2018) A North American stem turaco, and the complex biogeographic history of modern birds. BMC Evol Biol 18:102

Hackett SJ, Kimball RT, Reddy S, Bowie RCK, Braun EL, Braun MJ, Chojnowski JL, Cox WA, Han K-L, Harshman J, Huddleston CJ, Marks BD, Miglia KJ, Moore WS, Sheldon FH, Steadman DW, Witt CC, Yuri T (2008) A phylogenomic study of birds reveals their evolutionary history. Science 320:1763–1767

Harrison CJO (1975) Ordinal affinities of the Aegialornithidae. Ibis 117:164–170

Harrison CJO (1982a) The earliest parrot: a new species from the British Eocene. Ibis 124:203–210

Harrison CJO (1982b) Cuculiform, piciform and passeriform birds in the Lower Eocene of England. Tertiary Res 4:71–81

Harrison CJO (1984) A revision of the fossil swifts (vertebrata, Aves, suborder Apodi), with descriptions of three new genera and two new species. Meded Werkgr Tert Kwart Geol 21:157–177

Harrison CJO, Walker CA (1975) A new swift from the Lower Eocene of Britain. Ibis 117:162–164

References

Harrison CJO, Walker CA (1977b) Birds of the British Lower Eocene. Tert Res Spec Pap 3:1–52

Karhu A (1988) A new family of swift-like birds from the Paleogene of Europe. Paleontol J 3:78–88. [in Russian]

Karhu A (1992) Morphological divergence within the order Apodiformes as revealed by the structure of the humerus. In: Campbell KE (ed) Papers in avian paleontology honoring Pierce Brodkorb. Nat Hist Mus Los Angeles Cty Sci Ser 36:379–384

Karhu A (1999) A new genus and species of the family Jungornithidae (Apodiformes) from the Late Eocene of the Northern Caucasus, with comments on the ancestry of hummingbirds. In: Olson SL (ed) Avian paleontology at the close of the 20th century: Proceedings of the 4th international meeting of the Society of Avian Paleontology and Evolution, Washington, D.C., 4-7 June 1996. Smithson Contrib Paleobiol 89:207–216

Kristensen NP, Skalski AW (1999) Phylogeny and palaeontology. In: Kristensen NP (ed) Lepidoptera, moth and butterflies, Evolution, systematics, and biogeography, vol 1. de Gruyter, Berlin, pp 7–25

Ksepka DT, Clarke JA, Nesbitt SJ, Kulp FB, Grande L (2013) Fossil evidence of wing shape in a stem relative of swifts and hummingbirds (Aves, Pan-Apodiformes). Proc R Soc Lond Ser B 280:20130580

Ksepka DT, Grande L, Mayr G (2019) Oldest finch-beaked birds reveal parallel ecological radiations in the earliest evolution of passerines. Curr Biol 29:657–663

Kuhl H, Frankl-Vilches C, Bakker A, Mayr G, Nikolaus G, Boerno ST, Klages S, Timmermann B, Gahr M (2021) An unbiased molecular approach using 3'UTRs resolves the avian family-level tree of life. Mol Biol Evol 38:108–121

Kurochkin EN (1976) A survey of the Paleogene birds of Asia. Smithson Contrib Paleobiol 27:5–86

Louchart A, Tourment N, Carrier J, Roux T, Mourer-Chauviré C (2008) Hummingbird with modern feathering: an exceptionally well-preserved Oligocene fossil from southern France. Naturwiss 95:171–175

Martin LD, Mengel RM (1984) A new cuckoo and a chachalaca from the early Miocene of. Colorado. Carnegie Mus Nat Hist Spec Publ 9:171–177

Mayr G (1998) Ein Archaeotrogon (Aves: Archaeotrogonidae) aus dem Mittel-Eozän der Grube Messel (Hessen, Deutschland)? J Ornithol 139:121–129

Mayr G (1999) Caprimulgiform birds from the Middle Eocene of Messel (Hessen, Germany). J Vertebr Paleontol 19:521–532

Mayr G (2001a) Comments on the osteology of *Masillapodargus longipes* Mayr 1999 and *Paraprefica major* Mayr 1999, caprimulgiform birds from the Middle Eocene of Messel (Hessen, Germany). Neues Jahrb Geol Paläontol Mh 2001:65–76

Mayr G (2001b) The relationships of fossil apodiform birds—a comment on Dyke (2001). Senckenb Lethaea 81:1–2

Mayr G (2002) Osteological evidence for paraphyly of the avian order Caprimulgiformes (nightjars and allies). J Ornithol 143:82–97

Mayr G (2003a) A postcranial skeleton of *Palaeopsittacus* Harrison, 1982 (Aves incertae sedis) from the Middle Eocene of Messel (Germany). Oryctos 4:75–82

Mayr G (2003b) Phylogeny of early Tertiary swifts and hummingbirds (Aves: Apodiformes). Auk 120:145–151

Mayr G (2003c) A new Eocene swift-like bird with a peculiar feathering. Ibis 145:382–391

Mayr G (2004a) New specimens of *Hassiavis laticauda* (Aves: Cypselomorphae) and *Quasisyndactylus longibrachis* (Aves: Alcediniformes) from the Middle Eocene of Messel, Germany. Cour Forsch-Inst Senckenberg 252:23–28

Mayr G (2004b) Old World fossil record of modern-type hummingbirds. Science 304(5672):861–864

Mayr G (2005a) The Palaeogene Old World potoo *Paraprefica* Mayr, 1999 (Aves, Nyctibiidae): its osteology and affinities to the New World Preficinae Olson, 1987. J Syst Palaeontol 3:359–370

Mayr G (2005b) A new cypselomorph bird from the Middle Eocene of Germany and the early diversification of avian aerial insectivores. Condor 107:342–352

Mayr G (2005c) Fossil hummingbirds in the Old World. Biologist 52:12–16

Mayr G (2006) A specimen of *Eocuculus* Chandler, 1999 (Aves, ? Cuculidae) from the early Oligocene of France. Geobios 39:865–872

Mayr G (2007) New specimens of the early Oligocene Old World hummingbird *Eurotrochilus inexpectatus*. J Ornithol 148:105–111

Mayr G (2008) *Pumiliornis tessellatus* Mayr, 1999 revisited—new data on the osteology and possible phylogenetic affinities of an enigmatic Middle Eocene bird. Paläontol Z 82:247–253

Mayr G (2009) Paleogene fossil birds, 1st edn. Springer, Heidelberg

Mayr G (2010a) Phylogenetic relationships of the paraphyletic "caprimulgiform" birds (nightjars and allies). J Zool Syst Evol Res 48:126–137

Mayr G (2010b) Reappraisal of *Eocypselus*—a stem group representative of apodiform birds from the early Eocene of Northern Europe. Palaeobiodiv Palaeoenv 90:395–403

Mayr G (2015a) Towards completion of the early Eocene aviary: a new bird group from the Messel oil shale (Aves, Eopachypterygidae, fam. nov.). Zootaxa 4013:252–264

Mayr G (2015b) A reassessment of Eocene parrotlike fossils indicates a previously undetected radiation of zygodactyl stem group representatives of passerines (Passeriformes). Zool Scr 44:587–602

Mayr G (2015c) Eocene fossils and the early evolution of frogmouths (Podargiformes): further specimens of *Masillapodargus* and a comparison with *Fluvioviridavis*. Palaeobiodiv Palaeoenv 95:587–596

Mayr G (2015d) Skeletal morphology of the middle Eocene swift *Scaniacypselus* and the evolutionary history of true swifts (Apodidae). J Ornithol 156:441–450

Mayr G (2016a) Fragmentary but distinctive: three new avian species from the early Eocene of Messel, with the earliest record of medullary bone in a Cenozoic bird. N Jb Geol Paläontol, Abh 279:273–286

Mayr G (2016b) The world's smallest owl, the earliest unambiguous charadriiform bird, and other avian remains from the early Eocene Nanjemoy Formation of Virginia (USA). Paläontol Z 90:747–763

Mayr G (2017) Avian evolution: the fossil record of birds and its paleobiological significance. Wiley-Blackwell, Chichester

Mayr G (2019) A skeleton of a small bird with a distinctive furcula morphology, from the Rupelian of Poland, adds a new taxon to early Oligocene avifaunas. Palaeodiv 12:113–123

Mayr G (2020a) A remarkably complete skeleton from the London Clay provides insights into the morphology and diversity of early Eocene zygodactyl near-passerine birds. J Syst Palaeontol 18:1891–1906

Mayr G (2020b) An updated review of the middle Eocene avifauna from the Geiseltal (Germany), with comments on the unusual taphonomy of some bird remains. Geobios 62:45–59

Mayr G (2021) An early Eocene fossil from the British London Clay elucidates the evolutionary history of the enigmatic Archaeotrogonidae (Aves, Strisores). Pap Palaeontol 7:2049–2064

Mayr G, Daniels M (2001) A new short-legged landbird from the early Eocene of Wyoming and contemporaneous European sites. Acta Palaeontol Pol 46:393–402

Mayr G, De Pietri VL (2014) Earliest and first Northern Hemispheric hoatzin fossils substantiate Old World origin of a "Neotropic endemic". Naturwissenschaften 101:143–148

Mayr G, Ericson PGP (2004) Evidence for a sister group relationship between the Madagascan mesites (Mesitornithidae) and cuckoos (Cuculidae). Senck Biol 84:119–135

Mayr G, Manegold A (2002) Eozäne Stammlinienvertreter von Schwalmvögeln und Seglern aus der Grube Messel bei Darmstadt. Sitzungsber Ges Naturforsch Freunde Berl (N F) 41:21–35

Mayr G, Micklich N (2010) New specimens of the avian taxa *Eurotrochilus* (Trochilidae) and *Palaeotodus* (Todidae) from the early Oligocene of Germany. Paläontol Z 84:387–395

Mayr G, Mourer-Chauviré C (2005) A specimen of *Parvicuculus* Harrison & Walker 1977 (Aves: Parvicuculidae) from the early Eocene of France. Bull Brit Ornithol Club 125:299–304

Mayr G, Peters DS (1999) On the systematic position of the Middle Eocene swift *Aegialornis szarskii* Peters 1985 with description of a new swift-like bird from Messel (Aves, Apodiformes). Neues Jahrb Geol Paläontol Mh 1999:312–320

Mayr G, Manegold A, Johansson U (2003) Monophyletic groups within "higher land birds"—comparison of morphological and molecular data. J Zool Syst Evol Res 41:233–248

Mayr G, Alvarenga HMF, Mourer-Chauviré C (2011) Out of Africa: fossils shed light on the origin of the hoatzin, an iconic Neotropic bird. Naturwiss 98:961–966

Mayr G, De Pietri V, Scofield RP (2022) New bird remains from the early Eocene Nanjemoy Formation of Virginia (USA), including the first records of the Messelasturidae, Psittacopedidae, and Zygodactylidae from the Fisher/Sullivan site. Hist Biol 34:322–334

Mlíkovský J (2002) Cenozoic birds of the world. Part 1: Europe. Ninox press, Praha

Mourer-Chauviré C (1978) La poche à phosphate de Ste. Néboule (Lot) et sa faune de vertébrés du Ludien Supérieur. 6—Oiseaux. Palaeovertebrata 8:217–229

Mourer-Chauviré C (1980) The Archaeotrogonidae from the Eocene and Oligocene deposits of "Phosphorites du Quercy", France. In: Campbell KE (ed) Papers in avian paleontology honoring Hildegarde Howard. Nat Hist Mus Los Angeles Cty Contrib Sci 330:17–31

Mourer-Chauviré C (1982) Les oiseaux fossiles des Phosphorites du Quercy (Eocène supérieur à Oligocène supérieur): implications paléobiogéographiques. Geobios, mém spéc 6:413–426

Mourer-Chauviré C (1988) Les Aegialornithidae (Aves: Apodiformes) des Phosphorites du Quercy. Comparaison avec la forme de Messel. Cour Forsch-Inst Senckenberg 107:369–381

Mourer-Chauviré C (1989 [1988]) Les Caprimulgiformes et les Coraciiformes de l'Éocène et de l'Oligocène des Phosphorites du Quercy et description de deux genres nouveaux de Podargidae et Nyctibiidae. In: Ouellet H (ed) Acta XIX Congressus Internationalis Ornithologici. University of Ottawa Press, Ottawa, pp 2047–2055

Mourer-Chauviré C (1992) Un ganga primitif (Aves, Columbiformes, Pteroclidae) de très grande taille de Paléogène des Phosphorites du Quercy (France). C R Acad Sci Sér II 314:229–235

Mourer-Chauviré C (1993) Les gangas (Aves, Columbiformes, Pteroclidae) du Paléogène et du Miocène inférieur de France. Palaeovertebrata 22:73–98

Mourer-Chauviré C (1995) Les Garouillas et les sites contemporains (Oligocène, MP 25) des Phosphorites du Quercy (Lot, Tarn-et-Garonne, France) et leurs faunes de Vertébrés. 3. Oiseaux. Palaeontographica (A) 236:33–38

Mourer-Chauviré C (2006) The avifauna of the Eocene and Oligocene Phosphorites du Quercy (France): an updated list. Strata, sér 1 13:135–149

Mourer-Chauviré C (2013) New data concerning the familial position of the genus *Euronyctibius* (Aves, Caprimulgiformes) from the Paleogene of the Phosphorites du Quercy, France. Paleontol J 47:1315–1322

Mourer-Chauviré C, Sigé B (2006) Une nouvelle espèce de *Jungornis* (Aves, Apodiformes) et de nouvelles formes de Coraciiformes s.s. dans l'Éocène supérieur du Quercy. Strata, ser 1 13:151-159

Mourer-Chauviré C, Berthet D, Hugueney M (2004) The late Oligocene birds of the Créchy quarry (Allier, France), with a description of two new genera (Aves: Pelecaniformes: Phalacrocoracidae, and Anseriformes: Anseranatidae). Senckenb Lethaea 84:303–315

Mourer-Chauviré C, Tabuce R, El Mabrouk E, Marivaux L, Khayati H, Vianey-Liaud M, Ben Haj Ali M (2013) A new taxon of stem group Galliformes and the earliest record for stem group Cuculidae from the Eocene of Djebel Chambi, Tunisia. In: Göhlich UB, Kroh A (eds) Paleornithological research 2013—Proceedings of the 8th international meeting of the Society of Avian Paleontology and Evolution. Natural History Museum Vienna, Vienna, pp 1–15

Mourer-Chauviré C, Ammar HK, Marivaux L, Marzougui W, Temani R, Vianey-Liaud M, Tabuce R (2016) New remains of the very small cuckoo, *Chambicuculus pusillus* (Aves, Cuculiformes, Cuculidae) from the late Early or early Middle Eocene of Djebel Chambi, Tunisia. Palaeovertebr 40(1) -e2:1–4

Nesbitt SJ, Ksepka DT, Clarke JA (2011) Podargiform affinities of the enigmatic *Fluvioviridavis platyrhamphus* and the early diversification of Strisores ("Caprimulgiformes" + Apodiformes). PLoS One 6: e26350

Olson SL (1985) The fossil record of birds. In: Farner DS, King JR, Parkes KC (eds) Avian biology, vol 8. Academic Press, New York, pp 79–238

Olson SL (1987) An early Eocene oilbird from the Green River Formation of Wyoming (Caprimulgiformes: Steatornithidae). Doc Lab Géol Lyon 99:57–69

Olson SL (1989) Aspects of global avifaunal dynamics during the Cenozoic. In: Ouellet H (ed) Acta XIX Congressus Internationalis Ornithologici. University of Ottawa Press, Ottawa, pp 2023–2029

Olson SL (1992) A new family of primitive landbirds from the Lower Eocene Green River Formation of Wyoming. In: Campbell KE (ed) Papers in avian paleontology honoring Pierce Brodkorb. Nat Hist Mus Los Angeles Cty Sci Ser 36:137–160

Olson SL (1999) Early Eocene birds from eastern North America: a faunule from the Nanjemoy Formation of Virginia. In: Weems RE, Grimsley GJ (eds) Early Eocene vertebrates and plants from the Fisher/Sullivan site (Nanjemoy Formation) Stafford County, Virginia. Virginia Div Mineral Resour Publ 152:123–132

Olson SL, Feduccia A (1979) An Old-World occurrence of the Eocene avian family Primobucconidae. Proc Biol Soc Wash 92:494–497

Peters DS (1985) Ein neuer Segler aus der Grube Messel und seine Bedeutung für den Status der Aegialornithidae (Aves: Apodiformes). Senckenb Lethaea 66:143–164

Peters DS (1998) Erstnachweis eines Seglers aus dem Geiseltal (Aves: Apodiformes). Senckenb Lethaea 78:211–212

Prum RO, Berv JS, Dornburg A, Field DJ, Townsend JP, Lemmon EM, Lemmon AR (2015) A comprehensive phylogeny of birds (Aves) using targeted next-generation DNA sequencing. Nature 526:569–573

Rasmussen DT, Olson SL, Simons EL (1987) Fossil birds from the Oligocene Jebel Qatrani Formation, Fayum Province, Egypt. Smithson Contrib Paleobiol 62:1–20

Sohn JC, Labandeira CC, Davis DR (2015) The fossil record and taphonomy of butterflies and moths (Insecta, Lepidoptera): implications for evolutionary diversity and divergence-time estimates. BMC Evol Biol 15:12

Weigel RD (1963) Oligocene birds from Saskatchewan. Q J Fla Acad Sci 26:257–262

7. Phaethontiformes and Aequornithes: The Aquatic and Semi-aquatic Neoavian Taxa

As outlined in Sect. 2.3, there is strong molecular support for a clade including the non-monophyletic "Pelecaniformes" except the Phaethontiformes (tropicbirds), as well as the Sphenisciformes (penguins), Gaviiformes (loons), Procellariiformes (tubenoses), and the taxa of the non-monophyletic "Ciconiiformes". This "waterbird clade" was termed Aequornithes (Mayr 2011), and the Phaethontiformes result as close relatives of it in current analyses.

The fossil record shows that the Aequornithes have a very long evolutionary history (Fig. 7.1). Stem group representatives of the Sphenisciformes date back to the mid-Paleocene and stem group Procellariiformes may have already existed in the Late Cretaceous (Sect. 2.4). The early evolution of the Aequornithes remains, however, poorly understood. The skeletal morphology of early Eocene stem group representatives of the Threskiornithidae, for example, is markedly different from that of coeval Sphenisciformes, which indicates a long evolutionary separation of the lineages leading to these two taxa already by that time. It is also elusive where the stem species of the Aequornithes occurred, although this may well have been on one of the southern continents.

7.1 Phaethontiformes (Tropicbirds)

Extant tropicbirds (Phaethontidae) are highly pelagic and predominantly piscivorous birds, which inhabit tropical and subtropical seas and capture prey with plunge-dives into the water. The three living species are medium-sized birds with wingspans of about one meter, very short legs, and a pair of long, streamer-like tail feathers. Tropicbirds were traditionally assigned to the "Pelecaniformes." However, current analyses of molecular data congruently support a sister group relationship between the Phaethontiformes and the Eurypygiformes, which include the South American Sunbittern (*Eurypyga helias*) and the New Caledonian Kagu (*Rhynochetos jubatus*) (Jarvis et al. 2014; Prum et al. 2015; Kuhl et al. 2021). The Eurypygiformes were traditionally considered to be "gruiform" birds and morphological data do not indicate a sister group relationship to the Phaethontiformes (Mayr 2015a).

Tropicbirds have a long fossil record and as noted in Sect. 2.4, the taxon *Novacaesareala* from the Cretaceous/Paleocene boundary of the North American Hornerstown Formation, shows a resemblance to Paleogene Phaethontiformes (Mayr and Scofield 2016). The oldest Paleogene specimens are fragmentary wing and pectoral girdle bones from the mid-Paleocene of the Waipara Greensand in New Zealand (Mayr and Scofield 2016). These fossils belong to an unnamed species and resemble phaethontiform specimens from the Paleocene and Eocene of the Northern Hemisphere.

Phaethontiform fossils were found in late Paleocene to middle Eocene marine localities of Europe, Asia, North Africa, and eastern North America. Most belong to the Prophaethontidae, a taxon originally established for *Prophaethon shrubsolei* from the early Eocene London Clay of the Isle of Sheppey. The holotype of this species is a partial skeleton, which includes a nearly complete skull (Fig. 7.2), a pelvis, as well as pectoral girdle and leg bones (Andrews 1899; Harrison and Walker 1976a). Olson (1981) noted that a humerus fragment from the Isle of Sheppey, which was described as the putative ibis *Proplegadis fisheri* by Harrison and Walker (1971), more closely resembles the Phaethontiformes. Mayr (2009a) considered the bone too small to be from *P. shrubsolei*, but this assumption was erroneous, and fossils of *Prophaethon* from the London Clay of Walton-on-the-Naze as well as putative prophaethontiform from the early Eocene of Egem in Belgium confirmed the synonymy of *Proplegadis* and *Prophaethon* (Mayr 2015a; Mayr and Smith 2019). Virtual endocasts of the skull of *P. shrubsolei* were examined by Milner and Walsh (2009).

A further species of the Prophaethontidae, *Lithoptila abdounensis*, was reported by Bourdon et al. (2005) from late Paleocene and early Eocene deposits of the Ouled

Abdoun Basin in Morocco (Bourdon et al. 2005, 2008a). Bourdon et al. (2008a) furthermore detailed that the humerus of *Lithoptila* is very similar to that of *Zhylgaia aestiflua* from the late Paleocene of Kazakhstan. Hence, this species, which was originally identified as a charadriiform bird (Nessov 1988, 1992), was also assigned to the Prophaethontidae by Bourdon et al. (2008a). Mayr and Scofield (2016) showed that tarsometatarsi from the type locality of *Z. aestiflua*, which were described as *Tshulia litorea*, also belong to *Z. aestiflua*. *L. abdounensis*, *P. shrubsolei*, and *Z. aestiflua* were of similar size and had wingspans of about one meter (Bourdon et al. 2008a).

Bourdon et al. (2008b) tentatively assigned a proximal humerus from the early Eocene of the Ouled Abdoun Basin in Morocco to the Phaethontidae. The specimen was assigned to a new species, *Phaethusavis pelagicus*, but its actual distinctness from the Prophaethontidae still needs to be verified by additional material. The humerus of *Phaethusavis* closely resembles that of the above-mentioned mid-Paleocene phaethontiform species from the Waipara Greensand in New Zealand (Mayr and Scofield 2016).

A distal end of a humerus and an incomplete coracoid from the Paleocene of Maryland (USA) were assigned to the Prophaethontidae by Olson (1994), and a coracoid of a prophaethontid was reported from the middle Eocene of Belgium by Mayr and Smith (2002a). This latter fossil is the as-yet youngest fossil record of the Prophaethontidae.

An analysis by Bourdon et al. (2005) resulted in a sister group relationship between the Prophaethontidae and Phaethontidae, which, among others, is supported by a very large dorsal tympanic recess in the otic region of the skull. Like extant Phaethontidae and other representatives of the polyphyletic traditional "Pelecaniformes," *Prophaethon* furthermore exhibits an articular facet for the furcula on the tip of the sternal keel and a distinct nasofrontal hinge. The skull of *Prophaethon* is similar to that of extant tropicbirds in overall morphology but differs in that the nostrils are very long and slit-like, whereas they are greatly reduced in extant tropicbirds (Fig. 7.2).

The long-winged prophaethontids were probably offshore or pelagic feeders (Bourdon et al. 2008a), and these birds seem to have been more aquatic than their extant relatives. In contrast to the wide pelvis of extant Phaethontidae, the pelvis of *Prophaethon* is elongated and narrow, and the tibiotarsus bears strongly projecting cnemial crests (Harrison and Walker 1976a; Bourdon et al. 2008a; Mayr 2015a). The tarsometatarsus is proportionally longer and less dorsoventrally flattened than that of extant tropicbirds. These differences indicate that propaethontids were better adapted to swimming than extant Phaethontidae, and the close resemblance to the pelvis and leg bones of albatrosses (Fig. 7.2) suggests that prophaethontids were surface-feeding and foraged while swimming. This habit may be plesiomorphic for the Phaethontiformes, the extant representatives of which perform plunge-dives to capture prey (Mayr 2015a). Different morphologies of the caudal vertebrae indicate that prophaethontids lacked long tail streamers, and in contrast to their living relatives they were probably no aerial acrobates (Mayr 2015a).

Extant tropicbirds are highly pelagic birds, which breed on remote oceanic islands and usually forage far offshore. The occurrence of prophaethontid fossils in near-shore deposits of the London Clay and the Ouled Abdoun Basin in Morocco suggests that these birds visited littoral habitats on a more regular basis than do their extant relatives (Bourdon et al. 2008a; Mayr 2015a). Increasing competition and predation at breeding sites during Cenozoic turnovers in coastal ecosystems may have confined the distribution of tropicbirds to the less productive seas of the tropical and subtropical regions, where these birds today occur (Mayr 2015a).

In contrast to the fairly comprehensive early Paleogene fossil record, there is just a single Oligocene fossil of a putative phaethontiform. This specimen, a femur from the early Oligocene Boom Formation in Belgium, was described by Mayr and Smith (2013). The scant late Paleogene and Neogene fossil record of tropicbirds may indicate that these birds already assumed a predominantly pelagic way of living by that time.

Analyses of nuclear sequence data supported a sister group relationship between the clade (Phaethontiformes + Eurypygiformes) and the Aequornithes, which include other taxa of the traditional "Pelecaniformes" (Prum et al. 2015; Kuhl et al. 2021). Even though "pelecaniform" characteristics, such as throat pouch and totipalmate feet, are only weakly developed in tropicbirds, there is no morphological evidence in support of a sister group relationship to the Eurypygiformes.

The Eurypygiformes include the sister taxa Eurypygidae (sunbittern) and Rhynochetidae (kagu), which were traditionally assigned to the "Gruiformes." The New Caledonian Rhynochetidae have no Paleogene fossil record. Olson (1989) considered a skeleton from the Green River Formation, which was figured by Grande (2013: fig. 124), to be from a representative of the Eurypygidae, the single extant species of which is restricted to South America. This fossil bears no resemblance to the Prophaethontidae or extant tropicbirds.

It is highly improbable that the gruiform-like Eurypygiformes evolved from an ancestor with a seabird habitus. If the Phaethontiformes and Eurypygiformes form a clade, the seabird traits of tropicbirds most likely evolved convergently to those of other taxa of the Aequornithes. This hypothesis is, however, challenged by the fossil record, which shows that stem group Phaethontiformes were even more aquatic (in terms of their presumed swimming capabilities) than the extant species of tropicbirds. Of all

Fig. 7.1 Interrelationships and known stratigraphic occurrences of the taxa of the Aequornithes. The divergence dates are hypothetical

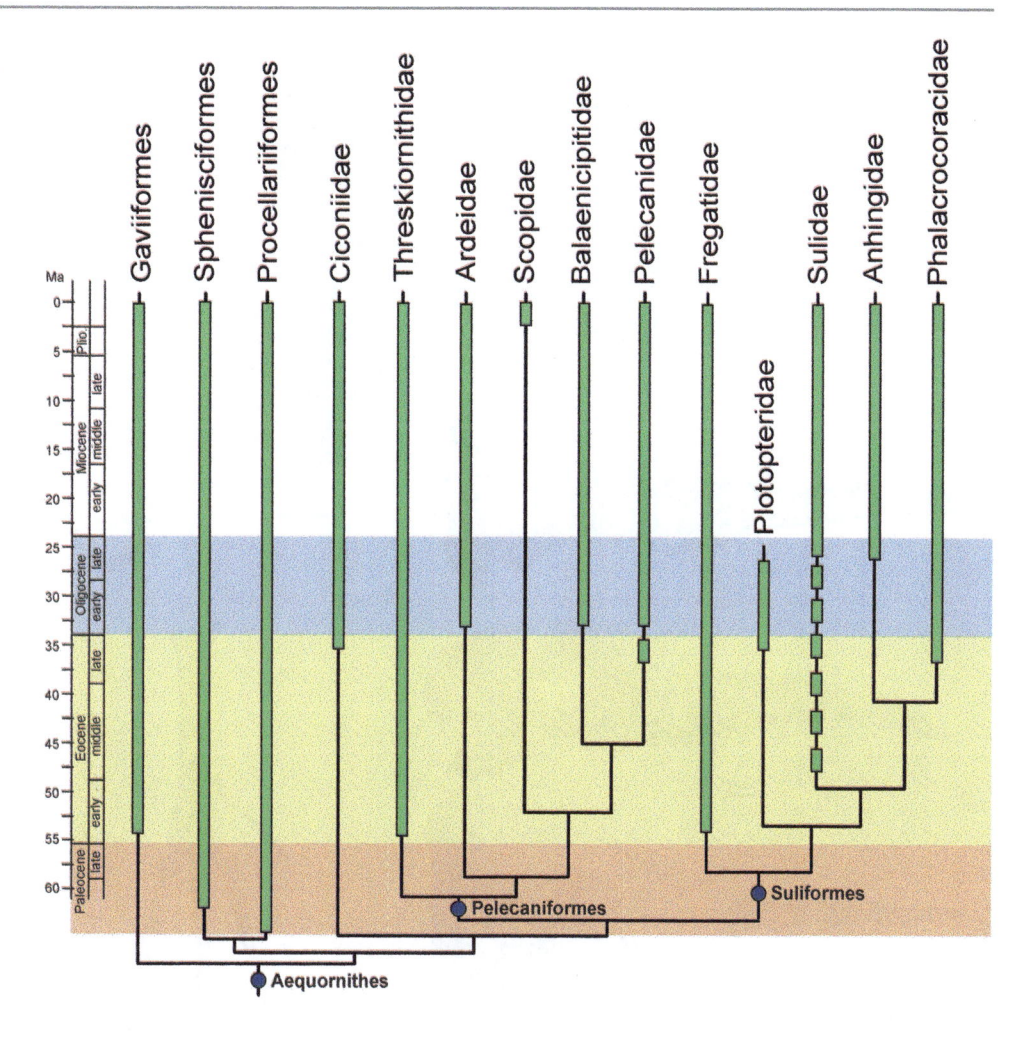

novel phylogenetic proposals based on molecular analyses, a sister group relationship between the Phaethontiformes and Eurypygiformes is therefore among those that show the greatest conflict to morphological and fossil data.

7.2 Gaviiformes (Loons)

The five extant species of the Gaviiformes are foot-propelled diving birds of the Northern Hemisphere, which breed in northern freshwater sites but winter in coastal areas at lower latitudes. According to analyses of nuclear gene sequences, the Gaviiformes are the sister taxon of all other Aequornithes (Fig. 7.1). Because stem group representatives of the Sphenisciformes, which branch next in the phylogeny of the Aequornithes, are known from the early Paleocene (Sect. 7.4), the origin of loons must also date back to the earliest Paleocene or, more likely, Late Cretaceous. However, even though fossils from the Late Cretaceous of Antarctica and South America were assigned to the Gaviiformes (Sect. 2.4), their classification is disputed and all unambiguously identified specimens of fossil loons are from Cenozoic fossil sites in Europe and, possibly (see below), North America.

Michael Daniels collected a loon from the early Eocene of Walton-on-the-Naze. This fossil is as yet undescribed and consists of a partial skeleton (Fig. 7.3a). It is identified as a diver by the morphology of its mediolaterally compressed tarsometatarsus, which has a short, plantarly deflected trochlea for the second toe and is very similar to the tarsometatarsus of the middle Eocene *Colymbiculus* (see below). Among others, the London Clay fossil differs from geologically younger loons in the craniocaudally shorter extensor process of the carpometacarpus and the proportionally longer femur. The humerus has an unusual shape in that its shaft is sigmoidally curved. This fossil shows that early Eocene Gaviiformes were less adapted to foot-propelled diving, and these birds also appear to have differed significantly from modern loons in aspects of their flight apparatus.

The earliest formally described fossil that was tentatively assigned to the Gaviiformes consists of a pair of wings from an unnamed species from the early Eocene (53 Ma) McAbee site of the Okanagan Highlands in Canada (Mayr et al. 2019a). This specimen exhibits feather remains, but the

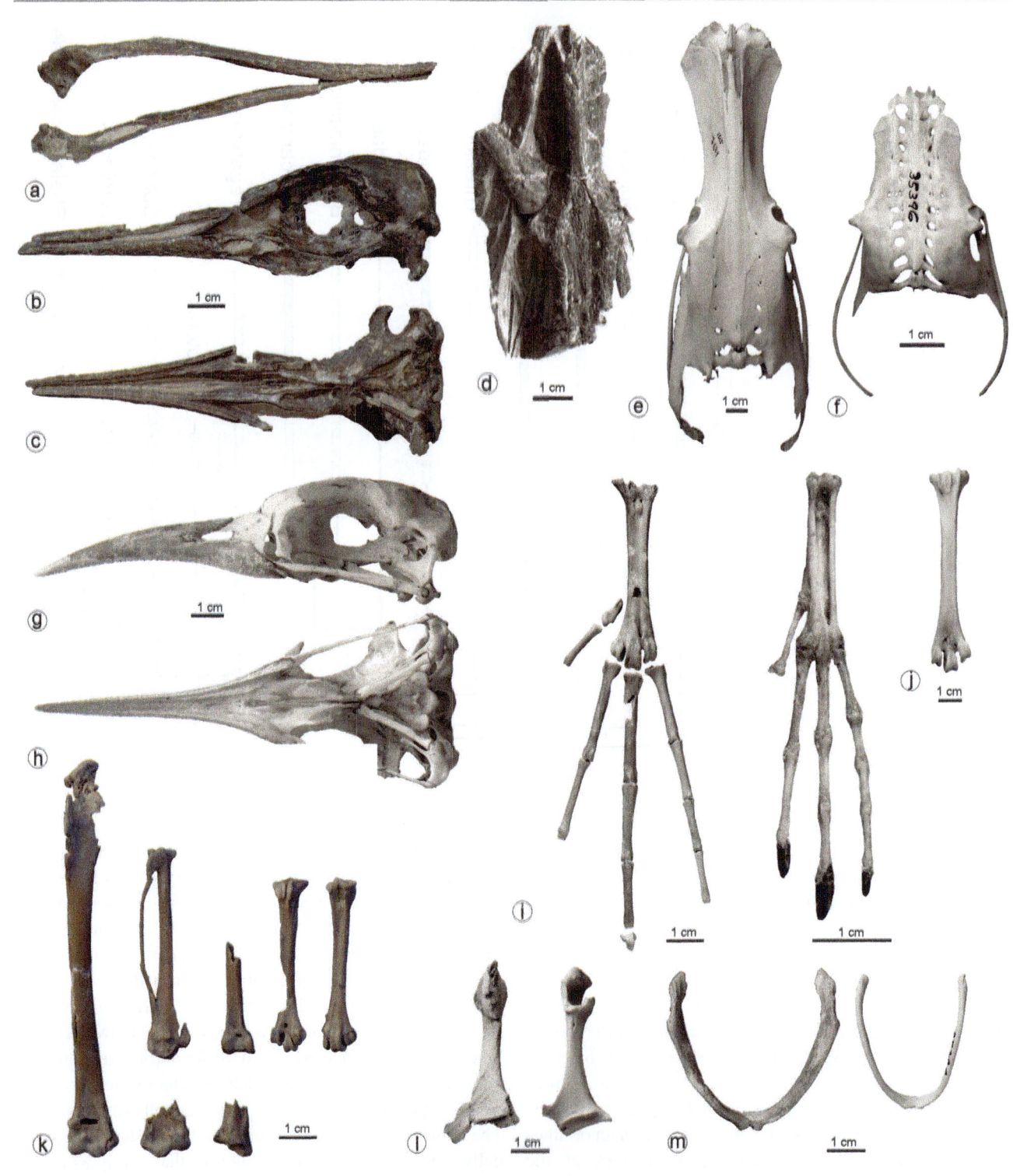

Fig. 7.2 Fossils of the phaethontiform Prophaethontidae. (**a**) Mandible, (**b**), (**c**) skull, and (**d**) pelvis of *Prophaethon shrubsolei* (Prophaethontidae) from the early Eocene London Clay of the Isle of Sheppey (holotype, Natural History Museum, London, NHMUK A 683). (**e**), (**f**) Pelvis of (**e**) *Diomedea antipodensis* (Diomedeidae) and (**f**) *Phaethon lepturus* (Phaethontidae). (**g**), (**h**) Skull of the extant Red-tailed Tropicbird, *Phaethon aethereus* (Phaethontidae). (**i**) Foot of *P. shrubsolei* from the London Clay of Walton-on-the-Naze (left, Senckenberg Research Institute, Frankfurt, SMF Av 602) and the extant *P. lepturus* (right) for comparison. (**j**) Tarsometatarsus of *Thalassarche melanophris* (Diomedeidae). (**k**) Partial skeleton of *P. shrubsolei* from Walton-on-the-Naze (collection of Michael Daniels, Holland-on-Sea, UK, WN 86536). (**l**) Coracoid of *P. shrubsolei* from Walton-on-the-Naze (left, SMF Av 602) and the extant *P. lepturus* (right). (**m**) Furcula of *P. shrubsolei* from Walton-on-the-Naze (left, SMF Av 602) and the extant *P. lepturus* (right). (All photos by Sven Tränkner)

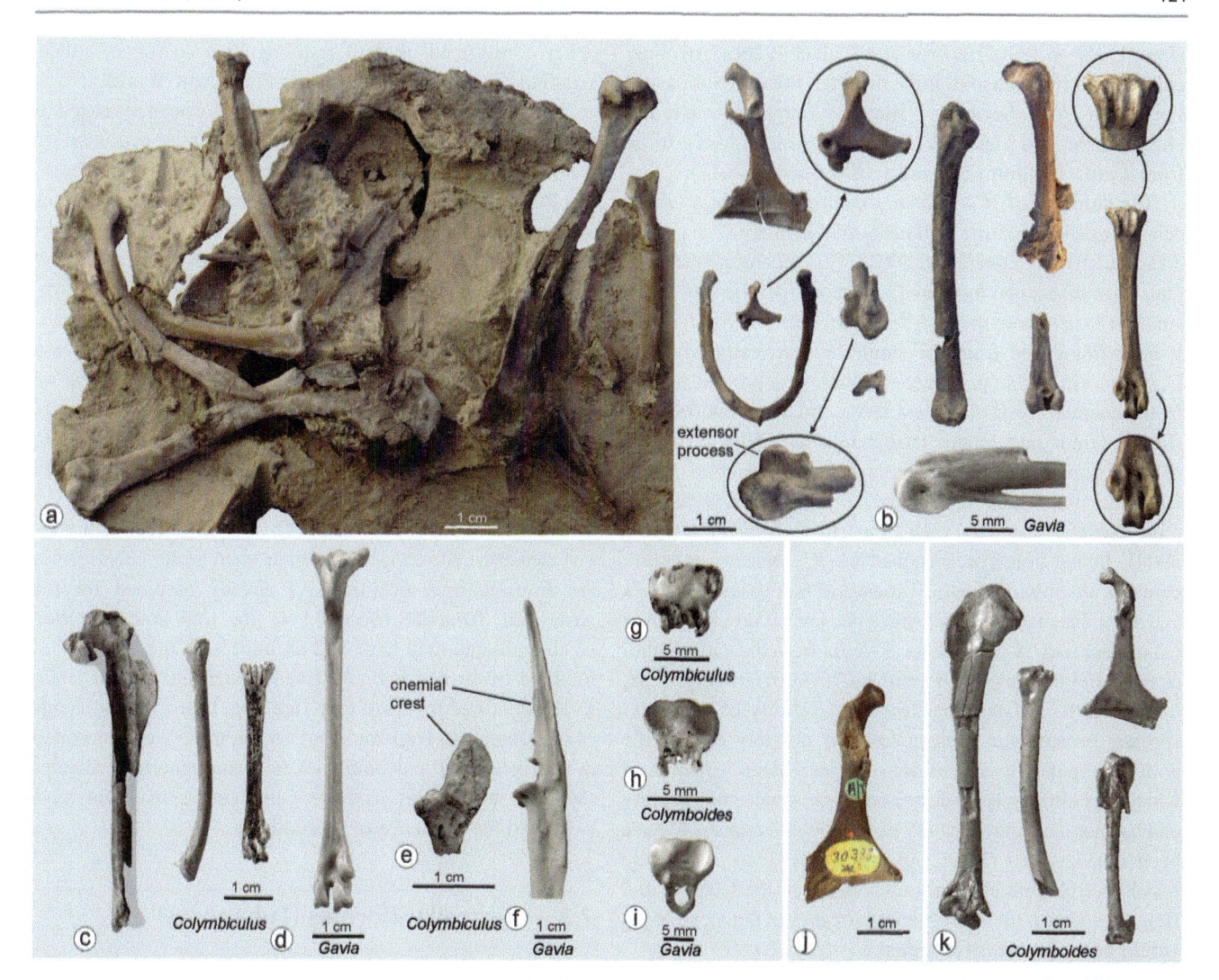

Fig. 7.3 Paleogene loons (Gaviiformes). (**a**) Partial skeleton of an undescribed species from the early Eocene London Clay of Walton-on-the-Naze (collection of Michael Daniels, Holland-on-Sea, UK, WN 92720); the block of matrix on the left includes the sternum, both humeri, and other bones; the arrows denote enlarged details of the quadrate, carpometacarpus, and tarsometatarsus. (**b**) Proximal end of the carpometacarpus of the extant *Gavia stellata*. (**c**) Humerus, ulna, and tarsometatarsus of *Colymbiculus udovichenkoi* from the middle Eocene of the Ukraine (Senckenberg Research Institute Frankfurt, humerus: SMF Av 548, ulna: SMF Av 548, tarsometatarsus: SMF Av 592). (**e**) Proximal end of the tibiotarsus of *C. udovichenkoi* (SMF Av 588). (**f**) Proximal tibiotarsus of the extant *G. stellata*. (**d**) Left tarsometatarsus of the extant *Gavia stellata* to show the much larger size of extant loons. (**g**)–(**i**) proximal end of the tarsometatarsus (proximal view) of (**g**) *C. udovichenkoi* (SMF Av 592), (**h**) the early Miocene *Colymboides minutus* (Naturhistorisches Museum Basel, Switzerland, NMB S.G. 20829), and (**i**) extant *G. stellata*. (**j**) left coracoid of *Colymboides anglicus* from the late Eocene of England (holotype, Natural History Museum, London; NHMUK A 30330). (**k**) Humerus, ulna, coracoid, and carpometacarpus of ?*Colymboides metzleri* from the early Oligocene of Belgium (Royal Belgian Institute of Natural Sciences, Brussels, Belgium, IRSNB Av 85). (All photos by Sven Tränkner)

bones do not allow an assessment of osteological details. In overall shape, however, they resemble the corresponding elements of early Paleogene stem group representatives of the Gaviiformes, even though the humerus shaft is straighter than in the London Clay fossil.

Of *Colymbiculus udovichenkoi* from the middle Eocene of Ukraine, several isolated and well-preserved limb bones are known (Fig. 7.3c; Mayr and Zvonok 2011, 2012; Mayr et al. 2013). As in the above fossil from the London Clay, the proximal end of the humerus is proportionally wider than in geologically younger Gaviiformes. The cnemial crests of the tibiotarsus are furthermore not as strongly proximally projecting as in post-Eocene Gaviiformes, which indicates that *C. udovichenkoi* was less well adapted to diving than more advanced loons. Unlike in extant loons, the hypotarsus of the tarsometatarsus exhibits well-developed tendinal furrows, the presence of which is a plesiomorphic trait of the Gaviiformes.

A late Eocene species of the Gaviiformes is *Colymboides anglicus* from the Priabonian (MP 17; Mlíkovský 2002) of England, which is represented by a coracoid (Fig. 7.3j), a referred humerus, and a portion of the frontal bones of the skull (Harrison 1976; Harrison and Walker 1976b). An allegedly gaviiform coracoid from the late Eocene Submeseta Formation of Seymour Island, Antarctica (Acosta Hospitaleche and Gelfo 2015), by contrast, is likely to be from a procellariiform bird (Mayr and Goedert 2017).

The holotype of ?*Colymboides metzleri* from early Oligocene (Rupelian) marine sediments of Germany is a partial, disarticulated skeleton (Mayr 2004). This species was smaller than *C. anglicus* and measured just about half the size of the smallest extant loon, the Red-throated Diver (*Gavia stellata*). It is distinguished from the similarly-sized early Miocene *C. minutus* in a more prominent and narrower process formed by the cnemial crests of the tibiotarsus and in a proportionally shorter femur and longer tarsometatarsus. With regard to these features, ?*C. metzleri* more closely resembles extant Gaviidae, for which reason the species may actually be more closely related to *Gavia* than to *Colymboides* (Mayr 2004). In the holotype specimen of ?*C. metzleri*, stomach contents are preserved, which consist of numerous fish bones and show Paleogene loons to have been piscivorous like their extant relatives. A partial loon skeleton from the early Oligocene of Belgium closely resembles ?*C. metzleri* (Fig. 7.3k; Mayr 2009b) and may challenge the validity of this latter species, because the Belgian fossil is possibly conspecific with the putative anseriform species "*Anas*" *benedeni*, which has already been described in the nineteenth century and has nomenclatural priority even though it appears to have been misidentified (Mayr and Smith 2013). In case future fossils corroborate close affinities between the German and Belgian species, the correct name for these Oligocene loons would be *Colymboides benedeni*.

Another species of *Colymboides*, *C. belgicus*, was described from lacustrine sediments from the earliest Oligocene of Belgium on the basis of a proximal carpometacarpus and a distal ulna (Mayr and Smith 2002b). *C. belgicus* was much larger than the aforementioned species and only slightly smaller than the extant *Gavia stellata*.

Kurochkin (1976) noted the presence of *Colymboides* in the late Oligocene of Kazakhstan. A leg of a small gaviiform bird was also found in the late Oligocene (MP 28) Enspel maar lake in Germany (Mayr and Poschmann 2009). This fossil is remarkable, because it is preserved in association with a crocodilian tooth, which appears to have been lost during an attack on the living bird or the floating carcass. Today, loons coexist with crocodilians in only a very small part of their range in southeastern Asia.

An allegedly Paleogene North American species assigned to the Gaviiformes is *Gaviella pusilla*, the holotype of which is a proximal carpometacarpus from an unknown locality and stratigraphic horizon in Wyoming. Wetmore (1940) assumed that the specimen comes from Oligocene deposits and, because of its morphological distinctness, classified it into a monotypic taxon Gaviellinae. Unless the age of this fossil is better constrained, it is of little significance for an understanding of the evolution of loons in the New World.

The distribution of loons appears to have been restricted to the Northern Hemisphere throughout the entire evolutionary history of these birds. The fossil record furthermore shows that Paleogene stem group representatives of the Gaviiformes occurred in marine and freshwater environments with a subtropical palaeoclimate. The factors which restricted the inland distribution of loons to the northern latitudes are unknown, even though increased predation or competition at the breeding sites may have played a role. The significant size increase in the lineage leading to the extant taxa is likely to be an example of Bergmann's rule, which postulates that in closely related endothermic species those in cooler climates tend to have a larger body size owing to a more favorable—with regard to heat loss—volume to surface ratio.

Whereas the Neogene taxon *Colymboides* occurs in various lacustrine sites, most Paleogene stem group Gaviiformes are from marine localities. As already assumed for the anseriform Anatidae (Sect. 4.5.4), the Cenozoic evolution of crocodilians may have had an impact on the diversity of loons in freshwater habitats. Even though the earliest tentatively identified fossils of loons are from lacustrine sediments of the Okanagan Highlands in Canada, these were deposited at relatively high paleoaltitudes with microthermal climates (Mayr et al. 2019a), and the cold winters may not have permitted the existence of crocodilians.

7.3 Procellariiformes (Tubenoses)

All Procellariiformes are pelagic birds, which mainly feed on squid or fish and are characterized by the possession of tubular nostrils and a peculiar morphology of the hind toe, the first phalanx of which is lacking so that the ungual phalanx articulates with the metatarsal. The extant species are classified into four taxa, the Diomedeidae (albatrosses), Oceanitidae (southern storm petrels), Hydrobatidae (northern storm-petrels), and Procellariidae (fulmars, petrels, shearwaters). Current molecular analyses indicate that the Diomedeidae are the sister taxon of the remaining groups of extant Procellariiformes (Prum et al. 2015; Kuhl et al. 2021), but an earlier study by Hackett et al. (2008) identified the Oceanitidae as the earliest diverging clade. The distal end of the humerus of the Diomedeidae and Procellariidae bears a prominent dorsal supracondylar process; this process is much smaller in the Oceanitidae and in the procellariid taxon *Pelecanoides* (the humerus of which is modified owing to its use for underwater propulsion).

Even though the main distribution of extant Procellariiformes is in the Southern Hemisphere, many Paleogene taxa were found in European and North American fossil sites. This is likely to be an artifact of the more complete fossil record of the Northern Hemisphere, and in recent years a few specimens from Southern Hemispheric fossil localities were reported.

An incomplete humerus from the latest Cretaceous or earliest Paleocene Hornerstown Formation of New Jersey was described as *Tytthostonyx glauconiticus* by Olson and Parris (1987), who tentatively assigned the species to the Procellariiformes, within a monotypic taxon Tytthostonychidae. The fossil resembles the humerus of extant Procellariiformes in most features but has a very small dorsal supracondylar process. Even though Bourdon et al. (2008a) considered the specimen to be from a representative of the Phaethontiformes, its morphology conforms better to that of procellariiform birds (Mayr and Scofield 2016).

Records of putative Procellariiformes from early Paleogene deposits are likewise based on fragmentary remains. *Eopuffinus kazachstanensis* from the late Paleocene of Kazakhstan was described on the basis of a cranium fragment, which bears distinct, *Puffinus*-like fossae for nasal glands (see Nessov 1992). *Marinavis longirostris* is a fairly large species from the London Clay, the holotype of which is a partial beak (Harrison and Walker 1977). A partial carpometacarpus referred to the species in the original description is likely to be from a pelagornithid species, from which the aforementioned beak is, however, clearly distinguished in the lack of pseudo-teeth. On the basis of a distal humerus end from the London Clay, another much smaller species the size of extant Audubon's Shearwater (*Puffinus lherminieri*) was described as *Primodroma bournei* (Harrison and Walker 1977). The distal portion of a tarsometatarsus of a similar-sized procellariiform species was reported from the early Eocene Nanjemoy Formation in Virginia (USA), which also yielded further remains of procellariiform birds (Mayr 2016; Mayr et al. 2022).

An equally small species is *Kievornis rogovitshi* from the late Eocene of Ukraine, which is known from humeri and ulnae. Originally, this species was assigned to the "Graculavidae" (Averianov et al. 1990), but procellariiform affinities were suggested by Mayr (2009a), and these were subsequently confirmed by Zvonok et al. (2015).

Feduccia and McPherson (1993) described a distal tibiotarsus from the late Eocene of Louisiana (USA), which was considered to be most similar to that of the extant taxon *Pterodroma* (Procellariidae). However, these similarities are likely to be plesiomorphic, and the exact affinities of the fossil within the Procellariiformes are indeterminable without further material. Another procellariiform species from the Paleogene of North America is *Makahala mirae* from the latest Eocene or earliest Oligocene Makah Formation of Washington State, USA (Fig. 7.4; Mayr 2015b). The holotype and only known specimen of this species consists of wing bones. Unlike extant Diomedeidae and Procellariidae, *M. mirae* lacks an enlarged dorsal supracondylar process, and the radius has an unusually large distal end with a strongly convex margin. The deep brachial fossa (fossa musculi brachialis) of the humerus suggests that *M. mirae* is within a clade formed by the Oceanitidae, Hydrobatidae, and Procellariidae. Its position within this clade is, however, unresolved.

The description of *Argyrodyptes microtarsus* from the late Eocene or early Oligocene of Argentina was based on the distal portions of a tibiotarsus and a femur. The species was assigned to the Procellariidae by Agnolin (2007), who therefore confirmed earlier identifications of this fossil (e.g., Tonni 1980). "*Larus*" *raemdonckii* from the early Oligocene (Rupelian) of Belgium is based on an incomplete humerus that bears a large dorsal supracondylar process. The species was transferred from the charadriiform Laridae to the Procellariiformes by Brodkorb (1962), who proposed a classification into the extant taxon *Puffinus*; procellariiform affinities of "*L.*" *raemdonckii* were also assumed by Miller and Sibley (1941). Unfortunately, the holotype appears to be lost, and the affinities of the species are poorly resolved. As yet undescribed fossils of procellariiform birds from the late Oligocene of North America were mentioned by Olson (1985).

7.3.1 The Early Oligocene Diomedeoididae

Arguably the group of Paleogene procellariiform birds with the most substantial fossil record are the Diomedeoididae, which are represented by isolated bones and articulated skeletons from various Oligocene sites in Germany, France, Belgium, Poland, the Czech Republic, Switzerland, and Iran (Cheneval 1995; Fischer 1983, 1985, 1997, 2003; Peters and Hamedani 2000; Mayr et al. 2002; Gregorová 2006; Mayr 2009c; De Pietri et al. 2010; Mayr and Smith 2012a; Elzanowski et al. 2012). All fossils were found in marine sediments, and diomedeoidids are actually the most common seabirds in the Rupelian of Europe.

Specimens of these birds were previously assigned to the taxa *Diomedeoides* and *Frigidafons*. These have now been synonymized with *Rupelornis*, which has nomenclatural priority and was established by van Beneden (1871) for *Rupelornis definitus*, a species described on the basis of a distal tibiotarsus from the Rupelian of Belgium (Mayr and Smith 2012a). Various further names proposed by van Beneden (1871) and other authors for misidentified representatives of the Diomedeoididae likewise fall into synonymy with *R. definitus* (Mayr and Smith 2012a).

Three species of *Rupelornis* are well established: *R. definitus*, which includes *Diomedoides* ("*Frigidafons*") *brodkorbi* (Fig. 7.5a, b), as well as the larger *R. lipsiensis* from the Oligocene of Europe and *R. babaheydariensis* from the Rupelian of Iran. A record of the Diomedeoididae from

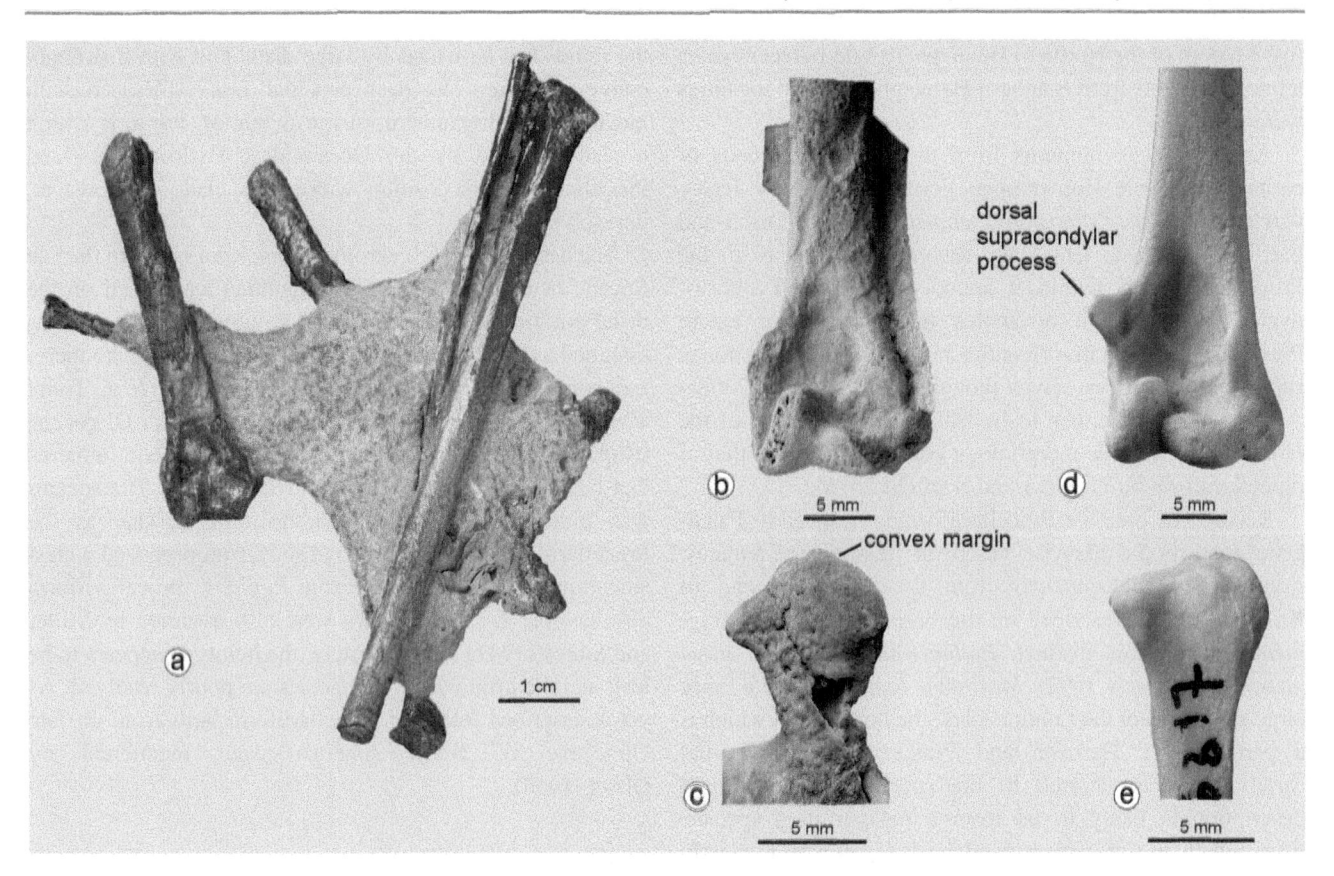

Fig. 7.4 (a) Holotype of the procellariiform bird *Makahala mirae* from the late Eocene/early Oligocene of Washington State, USA (Senckenberg Research Institute, Frankfurt, SMF Av 603), with details of the distal ends of (b) the humerus and (c) the radius (coated with ammonium chloride). (d), (e) Distal ends of the humerus and radius of the extant Northern Fulmar, *Fulmarus glacialis* (Procellariidae). (All photos by Sven Tränkner)

the early Miocene of Germany (Cheneval 1995) is of doubtful stratigraphic provenance (Mayr and Smith 2012a). Taxonomic validity and phylogenetic affinities of a large-sized species from the early Oligocene of Hungary, which was described as *Diomedeoides harmathi* by Kessler (2009), is likewise in need of further substantiation, and the holotype only consists of a poorly preserved humerus and a few associated wing elements (Mayr and Smith 2012a).

The Diomedeoididae are best characterized by their peculiar feet, which have greatly widened and flattened pedal phalanges with "nail-like" ungual phalanges that are strikingly similar to those of the extant Polynesian Storm-petrel *Nesofregetta fuliginosa* (Oceanitidae) (Fig. 7.5c; Mayr et al. 2002: fig. 5). Among extant Oceanitidae, such widened phalanges also occur in *Fregetta* and *Pelagodroma*, but they are absent in the oceanitid taxa *Garrodia* and *Oceanites*. This indicates that the distinctive morphology of the phalanges evolved convergently in the Diomedeoididae and Oceanitidae.

Also as in the Oceanitidae and unlike in other crown group Procellariiformes, the coracoid of diomedeoidids has a cup-like articular facet for the scapula (Fig. 7.5d; De Pietri et al. 2010; Mayr and Smith 2012a). The legs are furthermore very long as in extant Oceanitidae. In other skeletal features, however, diomedeoidids are clearly distinguished from southern storm-petrels. The wing proportions are like those of extant Procellariidae, whereas the wing of the Oceanitidae is proportionally much shorter. In addition to being considerably larger, *Rupelornis* furthermore differs from all extant Oceanitidae in its longer and more slender beak, the much more marked temporal fossae of the skull, the presence of incisions in the caudal margin of the sternum (which is entire in all extant Oceanitidae), and in the morphology of the hypotarsus. Of these features, at least the entire caudal margin of the sternum is a derived character, which is shared by *Nesofregetta* and other Oceanitidae but absent in *Rupelornis*. The tip of the beak of diomedeoidids is furthermore less deeply hooked than that of extant Procellariiformes.

An analysis by Mayr and Smith (2012a) recovered diomedeoidids as the sister taxon of either all crown group Procellariiformes or the Diomedeidae. A position outside crown group Procellariiformes is supported by the cup-like articular facet of the coracoid for the scapula. On the other hand, diomedeoidids and diomedeids share a derived morphology of the pelvis, in which the preacetabular blades of the ilium are fused along their midlines (Fig. 7.5k, l; Mayr and Smith 2012a).

7.3 Procellariiformes (Tubenoses)

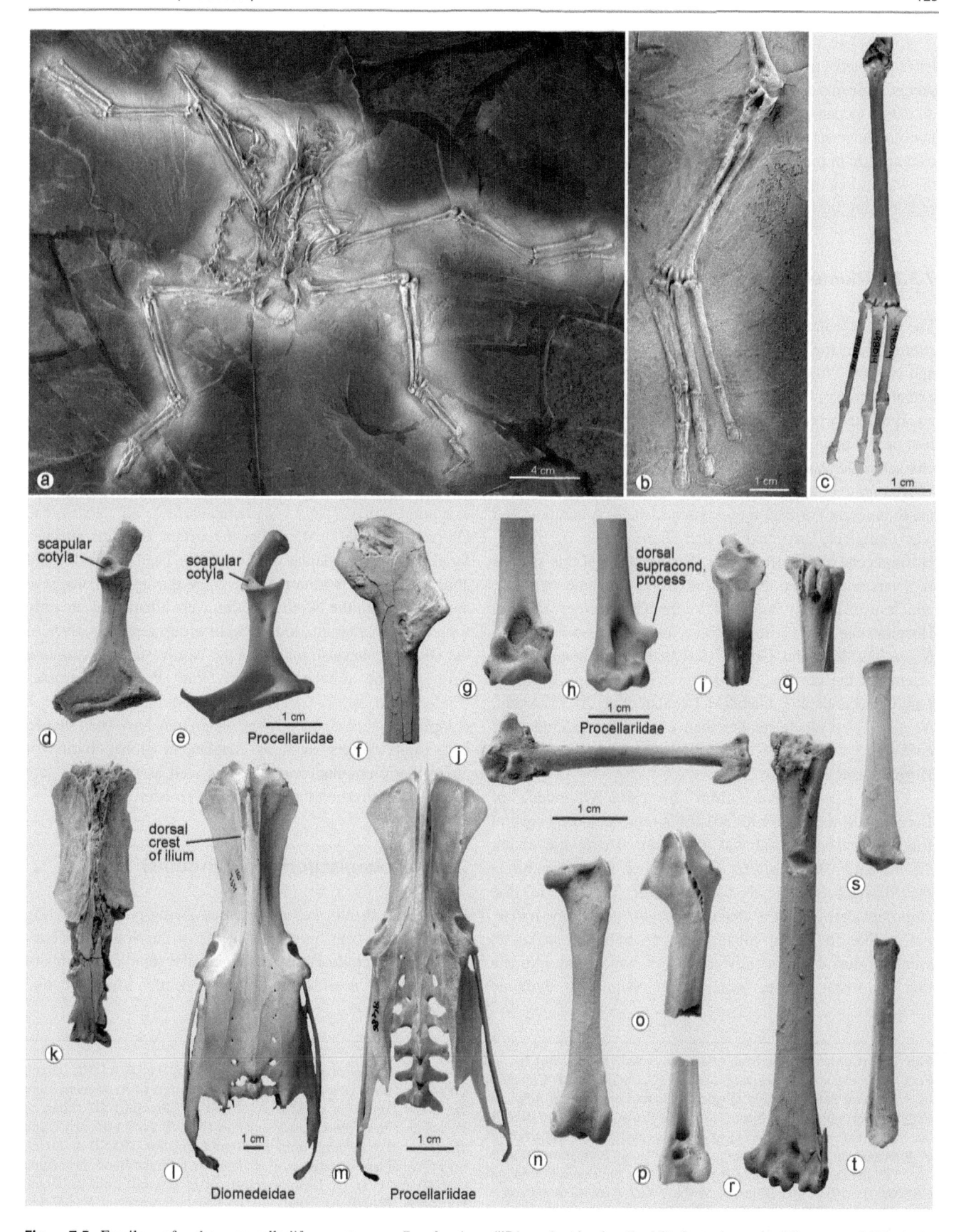

Fig. 7.5 Fossils of the procellariiform taxon *Rupelornis* (Diomedeoididae). (**a**), (**b**) Skeleton and right foot of *R. definitus* ("*Diomedeoides brodkorbi*") from the early Oligocene of Wiesloch-Frauenweiler in Germany (Staatliches Museum für Naturkunde

The species of the Diomedeoididae appear to have been less adapted to sustained gliding than most extant procellariiforms and more likely employed flap-gliding, as do extant Oceanitidae (e.g., Pennycuick 1982). The feet may not only have served as a brake for rapid stops when the bird caught sight of prey near the water surface but, immersed into the water, they may have also facilitated a stationary position of the flying bird at open sea (Mayr 2009c).

7.3.2 Diomedeidae (Albatrosses)

The earliest tentative record of the Diomedeidae is *Murunkus subitus* from the middle Eocene of Uzbekistan (Panteleyev and Nessov 1993). This species was much smaller than any extant albatross, but the only known fossil material consist of a carpometacarpus. Even though an unambiguous identification of this bone is not possible, the extensor process has the characteristic shape found in albatrosses.

Notoleptos giglii was described from the late Eocene of the Submeseta Formation of Seymour Island, Antarctica, and was tentatively assigned to the Diomedeidae by Acosta Hospitaleche and Gelfo (2017). The holotype of this species is a tarsometatarsus. A coracoid from the same deposits, which was initially assigned to the Gaviiformes (Acosta Hospitaleche and Gelfo 2015), is likely to also belong to *N. giglii* (Mayr and Goedert 2017), and the bone closely resembles the coracoid of *Maaqwi casacadensis* from the Late Cretaceous of Canada (McLachlan et al. 2017). Again, *N. giglii* was much smaller than extant Diomedeidae and coexisted with other procellariiform species (Acosta Hospitaleche and Gelfo 2017).

Tydea septentrionalis from the early Oligocene of Belgium (Mayr and Smith 2012b) is represented by several fragmentary bones and was about the size of the extant *Thalassarche melanophris*. The species likewise exhibits the derived shape of the extensor process of the carpometacarpus that is characteristic for the Diomedeidae, but unlike in extant albatrosses, its humerus lacks an enlarged dorsal supracondylar process, which may indicate that the fossil species was not yet adapted to dynamic soaring to the same degree as extant albatrosses (Mayr and Smith 2012b).

Another stem group representative of the Diomedeidae is *Diomedavus knapptonensis* from the late Oligocene (26–28 Ma) Lincoln Creek Formation of Washington State, USA (Fig. 7.6; Mayr and Goedert 2017). *D. knapptonensis* is the as-yet oldest record of an albatross from the North Pacific Basin and is smaller than *T. septentrionalis* and extant Diomedeidae. Olson (1985) mentioned an as-yet undescribed specimen of the Diomedeidae from the late Oligocene of South Carolina, which needs to be compared with *D. knapptonensis*.

The fossil material assigned to *Manu antiquus* from the late Oligocene of New Zealand consists of a furcula and tentatively referred incomplete femora. This large-sized species was assigned to the Diomedeidae by Marples (1946), from which the furcula is, however, clearly distinguished in the much more expanded shafts.

The fossil record of modern-type albatrosses is restricted to the Neogene period (Mayr and Goedert 2017), and with the exception of *Tydea septentrionalis*, all Paleogene Diomedeidae are smaller than the extant albatross species. Stem group Diomedeidae appear to have been diversified and abundant on the Northern Hemisphere during the Paleogene and occurred in the North Atlantic, were albatrosses are only found as rare vagrants today. Stem group representatives of the Diomedeidae also inhabited the North Pacific Basin long before the occurrence of extant North Pacific albatrosses (*Phoebastria* spp.). Clearly, albatrosses therefore have a complex biogeographic history, which may have been shaped by various factors, such as changes in global marine or atmospheric circulation systems as well as the loss of safe breeding grounds on land (Mayr and Goedert 2017).

7.4 Sphenisciformes (Penguins)

Extant Sphenisciformes are flightless, wing-propelled diving birds, which occur on all continents of the Southern Hemisphere and include 18 or 19 species, the standing heights of which range from about 40 cm (Little Blue Penguin,

Fig. 7.5 (Continued) Karlsruhe, Germany, SMNK-PAL 3812). (**c**) Foot of the extant *Nesofregetta fuliginosa* (Oceanitidae). (**d**) Coracoid of *R. definitus* from the early Oligocene Boom Formation in Belgium (Royal Belgian Institute of Natural Sciences, Brussels, Belgium, IRSNB Av 112a). (**e**) Coracoid of the extant *Fulmarus glacialis* (Procellariidae). (**f**) Proximal humerus of *R. definitus* from the Boom Formation (IRSNB Av 111). (**g**) Distal humerus of *R. definitus* from the Boom Formation (IRSNB Av 102a). (**h**) Humerus of the extant *Lugensa brevirostris* (Procellariidae). (**i**) Proximal ulna of *R. definitus* from the Boom Formation (IRSNB Av 109f). (**j**) Carpometacarpus of *R. definitus* from the Boom Formation (IRSNB Av 105). (**k**) Partial pelvis of *R. definitus* from the Boom Formation (IRSNB Av 109l). (**l**) Pelvis of the extant *Diomedea antipodensis* (Diomedeidae). (**m**) Pelvis of the extant *Calonectris diomedea* (Procellariidae). (**n**) Femur of *R. definitus* from the Boom Formation (IRSNB Av 112e). (**o**) Proximal tibiotarsus of *R. definitus* from the Boom Formation (IRSNB Av 114b). (**p**) Distal tibiotarsus of *R. definitus* from the Boom Formation (IRSNB Av 103c). (**q**) Proximal tarsometatarsus of *R. definitus* from the Boom Formation (IRSNB Av 113). (**r**) Tarsometatarsus of *R. definitus* from the Boom Formation (IRSNB Av 108d). (**s**, **t**) Pedal phalanges of *R. definitus* from the Boom Formation (s: IRSNB Av 112f, t: IRSNB Av 104g). All fossil specimens coated with ammonium chloride, same scale for all fossils in (**d**)–(**t**). (All photos by Sven Tränkner)

7.4 Sphenisciformes (Penguins)

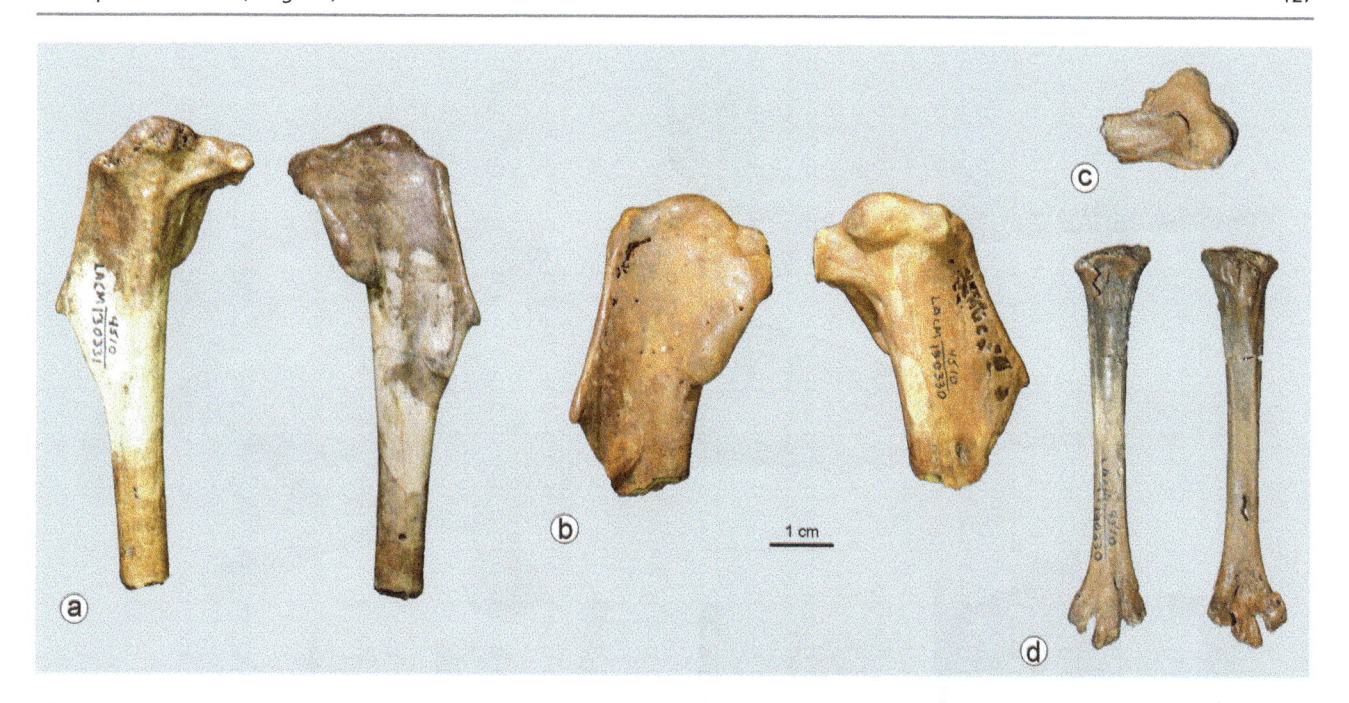

Fig. 7.6 (**a**), (**b**) Humeri (cranial and caudal view), (**c**) proximal carpometacarpus and (**d**) tarsometatarsus (dorsal and plantar view) of the stem group albatross *Diomedavus knapptonensis* (Diomedeidae) from the late Oligocene of Washington State, USA (**a**: Natural History Museum of Los Angeles County, USA, LACM 130331; **b**-**d**: holotype, LACM 130330)

Eudyptula minor) to slightly more than one meter (Emperor Penguin, *Aptenodytes forsteri*). Owing to the fact that penguin bones are fairly robust, and because these birds live in an aquatic environment that offers favorable conditions for fossilization, their fossil record is very comprehensive and dates back into the Paleocene (Jadwiszczak 2009; Ksepka and Ando 2011; Mayr 2017).

The evolutionary origins of penguins, however, remain elusive and all unambiguously identified fossils already exhibit the characteristic skeletal morphology of these flightless, wing-propelled seabirds. Stem group Sphenisciformes must have lost their flight capabilities very early, around the K/Pg boundary, but as yet no undisputed volant ancestor of penguins has been described. The sole candidate species is *Australornis lovei* from the mid-Paleocene Waipara Greensand in New Zealand, which is known from a few wing bones of a single individual (Mayr and Scofield 2014). The humerus of *A. lovei* has unusually thick bone walls, which may indicate a diving bird. The craniocaudally flattened shaft of the bone and the very long deltopectoral crest also conform to the morphologies found in penguins, but otherwise, there is no strong evidence for sphenisciform affinities of *Australornis*. Alternatively, the taxon may be closely related to an unnamed species represented by a long and narrow—and thereby clearly non-sphenisciform—tarsometatarsus from the latest Cretaceous or early Paleocene of New Zealand (Ksepka and Cracraft 2008). Agnolín et al. (2017) assigned *Australornis* to the Vegaviidae, but this was shown to have been erroneous (Mayr et al. 2018a).

All Paleogene penguin fossils are from the Southern Hemisphere and were found within the extant range of the Sphenisciformes, in Antarctica, New Zealand, Australia, and South America. No Paleogene fossils were described from South Africa, where the earliest penguins come from late Miocene/early Pliocene deposits (Acosta Hospitaleche 2006; Ksepka and Ando 2011; Mayr 2017).

The oldest penguin fossils are from the New Zealand region, where the last common ancestor of flightless penguins may have lived. One of the earliest species is *Kupoupou stilwelli* from the late early to middle Paleocene (62.5–60 Ma) of Chatham Island (Blokland et al. 2019). *K. stilwelli* was slightly smaller than the extant *Aptenodytes patagonicus* and is known from skeletal elements of multiple individuals; the type locality also yielded specimens of a larger, unnamed sphenisciform species (Blokland et al. 2019). Coeval fossils are known from 60.5–61.6 million-years-old deposits of the lower sections of the Waipara Greensand in New Zealand. One of these, *Waimanu manneringi*, was among the first described Paleocene penguins but is as yet only known from hindlimb and pelvis bones (Fig. 7.7b), which indicate a bird about the size of the Emperor Penguin. Much better represented is another species from the lower sections of the Waipara Greensand, *Sequiwaimanu rosiae*, of which a partial skeleton was described (Fig. 7.7a; Mayr et al. 2018b). Another well-represented species from the Waipara Greensand is *Muriwaimanu* ("*Waimanu*") *tuatahi*, which is from geologically younger (58–60 Ma) strata and somewhat smaller than

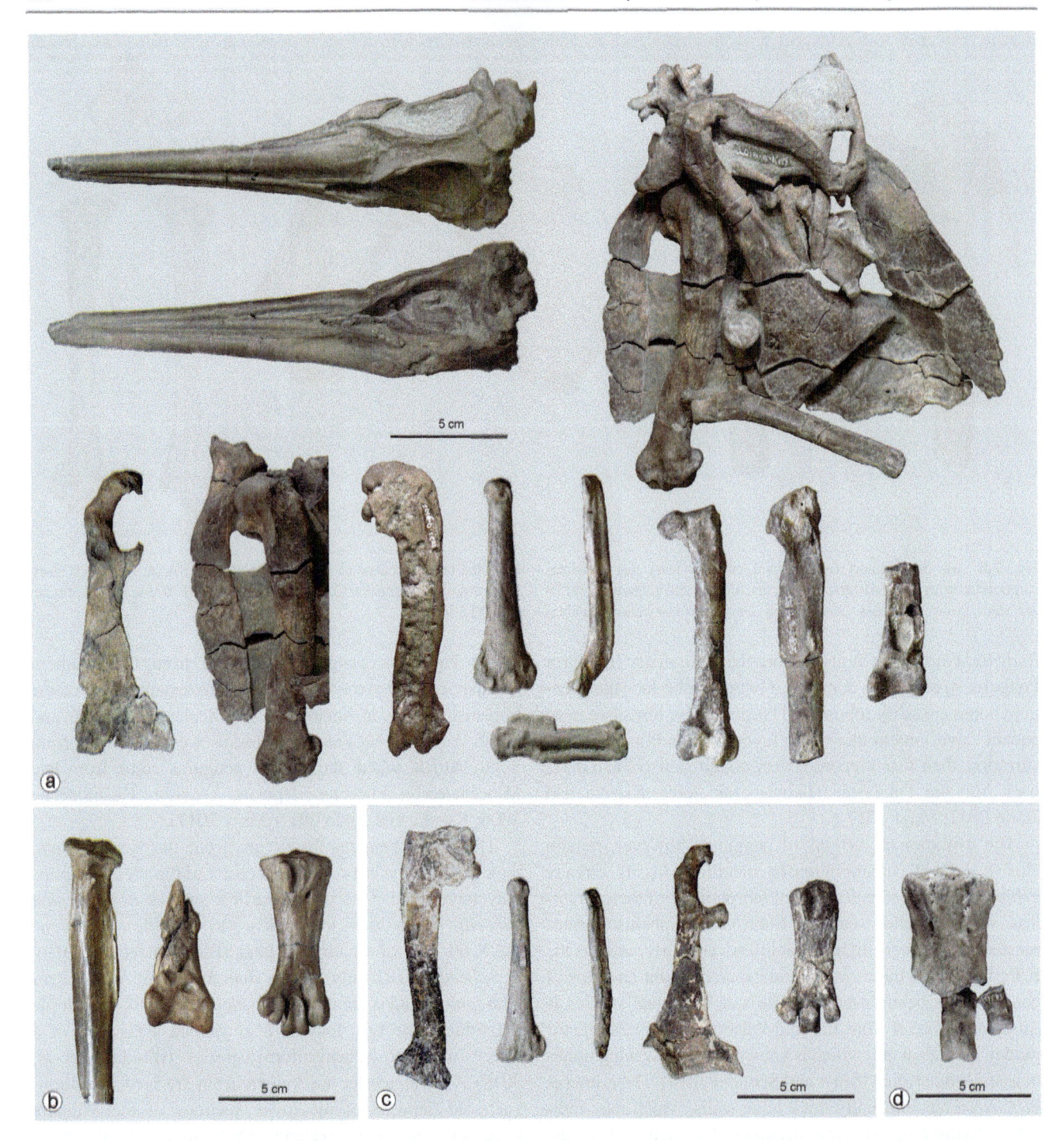

Fig. 7.7 Fossils of Paleocene penguins (Sphenisciformes) from the Waipara Greensand in New Zealand. (**a**) Selected bones (skull and major pectoral girdle, wing, and leg bones) of the holotype of *Sequiwaimanu rosieae* (Sphenisciformes) (Canterbury Museum, Christchurch, New Zealand, CM 2016.6.1); shown are two views of the skull as well as major limb and pectoral girdle bones. (**b**) Partial tibiotarsus and tarsometatarsus of the holotype of *Waimanu manneringi* (CM zfa 35). (**c**) Specimens referred to *Muriwaimanu tuatahi* (humerus and coracoid: CM zfa 34, ulna and radius: CM 2009.99.1). (**d**) Tarsometatarsus of an unnamed very large species (CM 2016.158.1). (All photos by the author)

W. manneringi (Fig. 7.7c; Slack et al. 2006; Mayr et al. 2018b). Of *M. tuatahi* several partial skeletons have been found and all major bones are known (Slack et al. 2006; Mayr et al. 2018b, 2020a).

The above species had standing heights of 80–100 cm and were already flightless, wing-propelled divers with an upright stance (Slack et al. 2006). In osteological details, however, they distinctly differ from crown group Sphenisciformes. As

7.4 Sphenisciformes (Penguins)

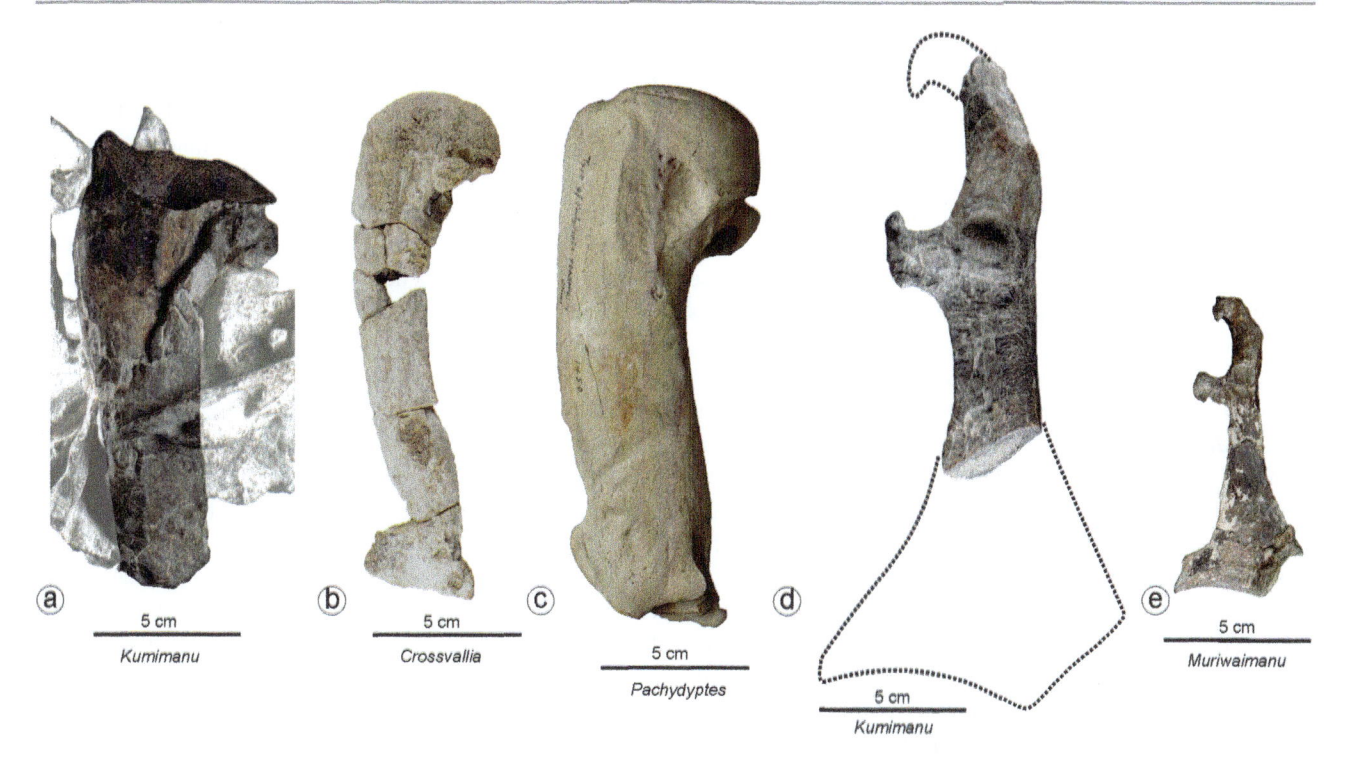

Fig. 7.8 Bones of giant Paleogene penguins (from Mayr et al. 2017b, modified; published under a Creative Commons CC BY 4.0 license). (**a**) Incomplete left humerus of *Kumimanu biceae* from the late Paleocene of New Zealand (holotype, Museum of New Zealand Te Papa Tongarewa, Wellington, New Zealand, NMNZ S.45877); the specimen is still partly embedded in matrix and the surrounding bones and matrix were digitally brightened. (**b**) Humerus of *Crossvallia unienwillia* from the late Paleocene of Antarctica (holotype, Museum de La Plata, Argentina, MLP 00-I-10-1). (**c**) Humerus of *Pachydyptes ponderous* from the late Eocene of New Zealand (holotype, NMNZ OR.001450). (**d**) Partial coracoid of *K. biceae* (holotype); the dotted lines indicate the reconstructed outline of the bone. (**e**) Coracoid of *Muriwaimanu tuatahi* from the late Paleocene of New Zealand (Canterbury Museum, Christchurch, New Zealand, CM zfa 34; left coracoid, mirrored to ease comparisons). (Photos in (**a**), (**c**), and (**d**) by Claude Stahl; (**b**) by P. Scofield, and (**e**) by the author)

in other Paleogene penguins, the beak of *Muriwaimanu* is long and dagger-like (the beaks of *Kupoupou* and *Waimanu* are unknown). Unlike in extant penguins, the wings of *Muriwaimanu* did not form stiff flippers (see below), with the scapula not being as greatly widened as in crown group Sphenisciformes and the distal portions of the ulna and radius not being flattened. The tarsometatarsi of *Muriwaimanu* and *Waimanu* are not as stout as those of extant Sphenisciformes.

In addition to the above taxa, the Waipara Greensand yielded further fossils of Paleocene stem group Sphenisciformes. Most notable are those of two very large species, which distinctly exceed the size of the Emperor Penguin. One of these is solely represented by a tarsometatarsus (Fig. 7.7d) and a few other bones; this unnamed species is notable, because its tarsometatarsus differs from other Paleocene stem group Sphenisciformes and more closely resembles the tarsometatarsus of some early Eocene species (Mayr et al. 2017a). A further very large species from the Waipara Greensand was described as *Crossvallia waiparensis* on the basis of leg bones from a single individual (Mayr et al. 2020b). The taxon *Crossvallia* was originally established for a stem group sphenisciform from the late Paleocene Cross Valley Formation of Seymour Island, Antarctica. This species, *Crossvallia unienwillia*, is known from various poorly preserved bones of a single individual, including an incomplete humerus, femur, and tibiotarsus (Fig. 7.8b; Tambussi et al. 2005; Jadwiszczak et al. 2013).

Another species of giant late Paleocene penguin, *Kumimanu biceae*, comes from the Moeraki Formation at Hampden Beach in Otago, New Zealand (Mayr et al. 2017b). *K. biceae* is one of the largest fossil penguins described so far and has an unusually stout humerus (Fig. 7.8a). The holotype consists of bones of a single individual, which were found in a very hard sedimentary concretion.

Although the record of Paleocene penguins has been significantly improved in the past years, early Eocene fossils of stem group Sphenisciformes are still fairly scant. The holotype of *Kaiika maxwelli* from the Kauru Formation of New Zealand is a humerus, and the species was about the size of the extant Emperor Penguin (Fordyce and Thomas 2011). Jadwiszczak (2006b) commented on some penguin specimens from the early Eocene of the La Meseta Formation of Seymour Island, Antarctica, and Jadwiszczak and Chapman (2011) identified remains of an unnamed medium-sized species from these deposits; tarsometatarsi of the very large-sized *Anthropornis* from early Eocene strata of the La Meseta Formation were described by Jadwiszczak (2013).

The fossil record of middle and late Eocene stem group Sphenisciformes, by contrast, is very extensive. Several thousand bones, which accumulated in the vicinity of former rookeries, were collected from the deposits of the La Meseta and Submeseta formations of Seymour Island. Myrcha et al. (2002) briefly reviewed the collection history of these specimens and distinguished the following taxa on the basis of their tarsometatarsi: *Anthropornis nordenskjoeldi*, *A. grandis*, *Palaeeudyptes gunnari*, *P. klekowskii*, *Delphinornis larseni*, *D. gracilis*, *D. arctowskii*, *Mesetaornis polari*, *Marambiornis exilis*, and *Archaeospheniscus wimani*.

Much research has been devoted to the penguins from Seymour Island during the past years and substantial new material was reported. Jadwiszczak (2006a) identified other postcranial elements of some of the above species, and Ksepka and Bertelli (2006) and Haidr and Acosta Hospitaleche (2017) described cranial remains, which could, however, not be assigned to particular taxa. Well-identified skull material is known from *Anthropornis grandis* (Acosta Hospitaleche et al. 2019a), and partial skeletons have been reported for *Anthropornis* sp. (Jadwiszczak 2012), *Delphinornis larseni* (Jadwiszczak and Mörs 2019), *Palaeeudyptes gunnari* (Acosta Hospitaleche and Reguero 2010; Acosta Hospitaleche et al. 2020), and *P. klekowskii* (Acosta Hospitaleche and Reguero 2014). Tambussi et al. (2006) furthermore described humeri from the Submeseta Formation as *Tonniornis mesetaensis* and *T. minimum*, but these identifications were disputed by Jadwiszczak (2006b). Tambussi et al. (2006) also assumed the presence of *Archaeospheniscus lopdelli* and *Palaeeudyptes antarcticus* in the Eocene of Seymour Island. These two species are otherwise only known from the Oligocene of New Zealand. Their identification in the fossil record of Seymour Island was, however, likewise questioned by Jadwiszczak (2006b). Jadwiszczak (2008) described a new but as yet not named small sphenisciform species from the late Eocene of the Submeseta Formation, which is represented by a tarsometatarsus fragment. Another very small species, *Aprosdokitos mikrotero*, was described on the basis of a humerus (Acosta Hospitaleche et al. 2017). Further fossils of unnamed sphenisciforms indicate an even higher taxonomic diversity (Jadwiszczak and Mörs 2017). Most recently, Jadwiszczak et al. (2021) described a tarsometatarsus from the late Eocene of the Submeseta Formation as *Marambiornopsis sobrali*; this bone closely resembles the tarsometatarsus of *Marambiornis exilis*.

Other sphenisciform taxa were reported from the late Eocene and Oligocene of New Zealand, and the following species are currently considered valid (Simpson 1971, 1975; Worthy and Holdaway 2002; Acosta Hospitaleche 2006; Ksepka and Ando 2011; Ksepka et al. 2012; Giovanardi et al. 2021): *Palaeeudyptes marplesi* (late Eocene), *P. antarcticus* (probably early Oligocene [Worthy and Holdaway 2002]), *Pachydyptes ponderosus* (late Eocene), *Kairuku grebneffi*, *K. waitaki*, and *K. waewaeroa* (early Oligocene), *Archaeospheniscus lopdelli* (late Oligocene), *A. lowei* (late Oligocene), *Duntroonornis parvus* (late Oligocene), *Platydyptes novaezealandiae* (late Oligocene), *P. amiesi* (late Oligocene), ?*P. marplesi* (late Oligocene), and *Korora oliveri* (late Oligocene). The standing heights of these species varied from about 60 cm in the smallest (*Duntroonornis parvus*) to about 150 cm in the largest (*Pachydyptes ponderosus*; Fig. 7.8c), which reached twice the size of the largest extant penguin and had an estimated weight of about 100 kilograms (Simpson 1971: 376).

By contrast, only a few sphenisciform fossils are known from the Paleogene of Australia, with the most substantial record being a partial skeleton from the late Eocene, which constitutes the holotype of *Pachydyptes simpsoni*. A few further remains were found in late Eocene deposits; some of these may belong to *Palaeeudyptes*, whereas other late Eocene and Oligocene fossils are indeterminable (Park and Fitzgerald 2012).

The earliest well-dated South American sphenisciform species is *Perudyptes devriesi* from the middle Eocene (42 Ma) of Peru (Clarke et al. 2007; Ksepka and Clarke 2010). This species was about the size of the extant King Penguin (*Aptenodytes patagonicus*) and is known from a partial skeleton including substantial parts of the skull. Roughly coeval to *P. devriesi* is a partial skeleton from the late middle Eocene of Tierra del Fuego (Argentina), which consists of portions of the pelvis and leg bones, including a complete femur, and was first studied by Clarke et al. (2003). These authors refrained from a classification of the fossil, but more recently it was assigned to *Palaeeudyptes gunnari* (Acosta Hospitaleche and Olivero 2016). This species was initially described from the Eocene of Antarctica (see above) and, if correctly identified, the South American fossil would indicate a biogeographic corridor for the dispersal of Eocene penguins from Antarctica. Detailed comparisons of the specimen from Tierra del Fuego and *Perudyptes devriesi* are, however, desirable before a definitive classification of the former is established. A putative record of *Palaeeudyptes* was also reported by Sallaberry et al. (2010) from the middle to late Eocene of Chile.

The holotype of *Icadyptes salasi* from the late Eocene of Peru consists of a nearly complete skull and associated wing elements (Clarke et al. 2007; Ksepka et al. 2008). This species was very large, with a minimum standing height of 1.5 meters (Clarke et al. 2007); as in *Muriwaimanu* but in contrast to extant penguins, it seems to have had a freely moveable alular wing phalanx (Ksepka et al. 2008). Of *Inkayacu paracasensis*, also from the late Eocene of Peru, a largely complete skeleton was found, in which wing feathers and the skin of the toes are preserved (Clarke et al. 2010). The skeleton of *Inkayacu* resembles that of *Palaeeudyptes*, and in both taxa the humerus exhibits a marked sulcus for the coracobrachialis nerve. Other Paleogene South American Sphenisciformes were discovered in the Argentinean part of Patagonia and, according to Acosta Hospitaleche (2006) and Acosta Hospitaleche et al. (2004), include the late Eocene/early Oligocene *Arthrodytes andrewsi* and *Paraptenodytes robustus*, as well as the late Oligocene/early Miocene *Eretiscus tonnii*, which is among the smallest species of Paleogene Sphenisciformes.

As already noted by Simpson (1975), Paleogene penguin faunas were much more diversified than the extant ones, and 10–15 sympatric species coexisted in the late Eocene of Seymour Island alone (Tambussi and Acosta Hospitaleche 2007; Ksepka and Ando 2011; Jadwiszczak 2013; Acosta Hospitaleche et al. 2019b). The Eocene penguins from Seymour Island show a remarkable variation in size, from small species the size of the extant Macaroni Penguin, *Eudyptes chrysolophus*, to the giant *Anthropornis nordenskjoeldi*, which reached an estimated standing height of 1.5–1.7 meters and body masses up to 80 kilograms (Jadwiszczak 2001; Ksepka and Ando 2011). *A. grandis* and the species of *Palaeeudyptes* were also much larger than extant penguins, and the large size range spanned by these species may have facilitated the coexistence of so many taxa in a geographically restricted area (Ksepka et al. 2006). Niche partitioning is also indicated by different bill morphologies and cranial specializations (Haidr and Acosta Hospitaleche 2012).

An remarkable feature of Paleogene penguins is the very large size of many species, with representatives of the taxa *Crossvallia*, *Kumimanu*, *Anthropornis*, *Pachydyptes*, *Icadyptes*, *Palaeeudyptes*, and *Archaeospheniscus* having been distinctly larger than the extant Emperor Penguin. The largest species of these taxa had standing heights of 1.5–1.7 m and may have weighed up to 80 kg (Jadwiszczak 2001; Ksepka and Ando 2011). For *Kumimanu biceae*, a body length of 1.8 m and a weight of 100 kg was estimated (Mayr et al. 2017b). Exceptionally large remains of *Palaeeudyptes klekowskii* suggest that this late Eocene species may have even reached had a standing height of two meters (Acosta Hospitaleche 2014).

Clarke et al. (2007) hypothesized that a giant size, that is, a standing height of 1.5 meters or more, evolved just once within Sphenisciformes, in the early Eocene, and persisted into the Oligocene, across the Eocene/Oligocene climate transition which marks the onset of global cooling. New fossil finds have, however, shown that even some Paleocene taxa, such as *Kumimanu* and *Crossvallia*, attained a very large size and that a very large size evolved multiple times independently in penguin evolution (Mayr et al. 2017b, 2020a). Ksepka et al. (2006) also noted that a large body size must have been attained very rapidly in the early evolution of penguins, and even the early Paleocene species *Waimanu manneringi* and *Sequiwaimanu rosieae* were as large or even larger than the Emperor Penguin. However, these giant early Paleogene species coexisted with smaller ones, and the volant stem species of the Sphenisciformes was certainly a small to midsize bird. A very large size could only have been positively selected after aerodynamic constraints ceased to exist.

Regarding extant penguins, the large *Aptenodytes* species occur in the extreme south, whereas the near-equatorial Galapagos Penguin is one of the smallest species. It has therefore been suggested by earlier authors that there is a correlation between size and latitude in penguins, in accordance with Bergmann's rule, which postulates that within clades of endothermic animals the species in colder climates are larger than closely related ones that live in warmer areas. However, the validity of this rule in the case of Sphenisciformes was already questioned by Simpson (1971). In fact, the average size of penguins seems to have declined over the Cenozoic global cooling (Clarke et al. 2007), and fossils of the giant *Pachydyptes ponderosus* were associated with warm water foraminiferans (Clarke et al. 2007; see also Tambussi et al. 2006).

From the occurrence of the large *Icadyptes* in near-equatorial fossil deposits, Clarke et al. (2007) concluded that giant size in Paleogene penguins was not related to either cooler temperatures or higher latitudes but may have been due to increased ocean productivity. However, to explain the body size of these early Sphenisciformes, it may be more appropriate not to ask why there were such large species in the past, but why there are none today (see Sect. 11.3.3.2).

All Paleogene penguins are stem group representatives of the Sphenisciformes, and the earliest crown group species occurred in the Miocene (Ksepka and Ando 2011; Mayr 2017). In the past years, several analyses have been performed, which shed light on the interrelationships of fossil Sphenisciformes (Fig. 7.9). In general, geologically younger taxa are more closely related to the crown group, which indicates a high selective pressure on a refinement of the penguin body plan. The tarsometatarsi of *Kupoupou*, *Waimanu*, *Muriwaimanu*, *Crossvallia*, *Delphinornis*, *Marambiornis*, and *Mesetaornis* still exhibit a distal vascular foramen, which is lost in other sphenisciform taxa closer to the crown group (Chávez-Hoffmeister 2014; Mayr et al. 2018b; Blokland et al. 2019). The earliest known sphenisciform taxa, that is, the Paleocene *Kupoupou*, *Waimanu*, *Muriwaimanu*, *Sequiwaimanu*, and *Crossvallia*, are outside a clade including all other Sphenisciformes, and the early Eocene taxa *Delphinornis*, *Marambiornis*, and *Mesetaornis* are branching next (Ksepka and Ando 2011; Mayr et al. 2017b; Blokland et al. 2019).

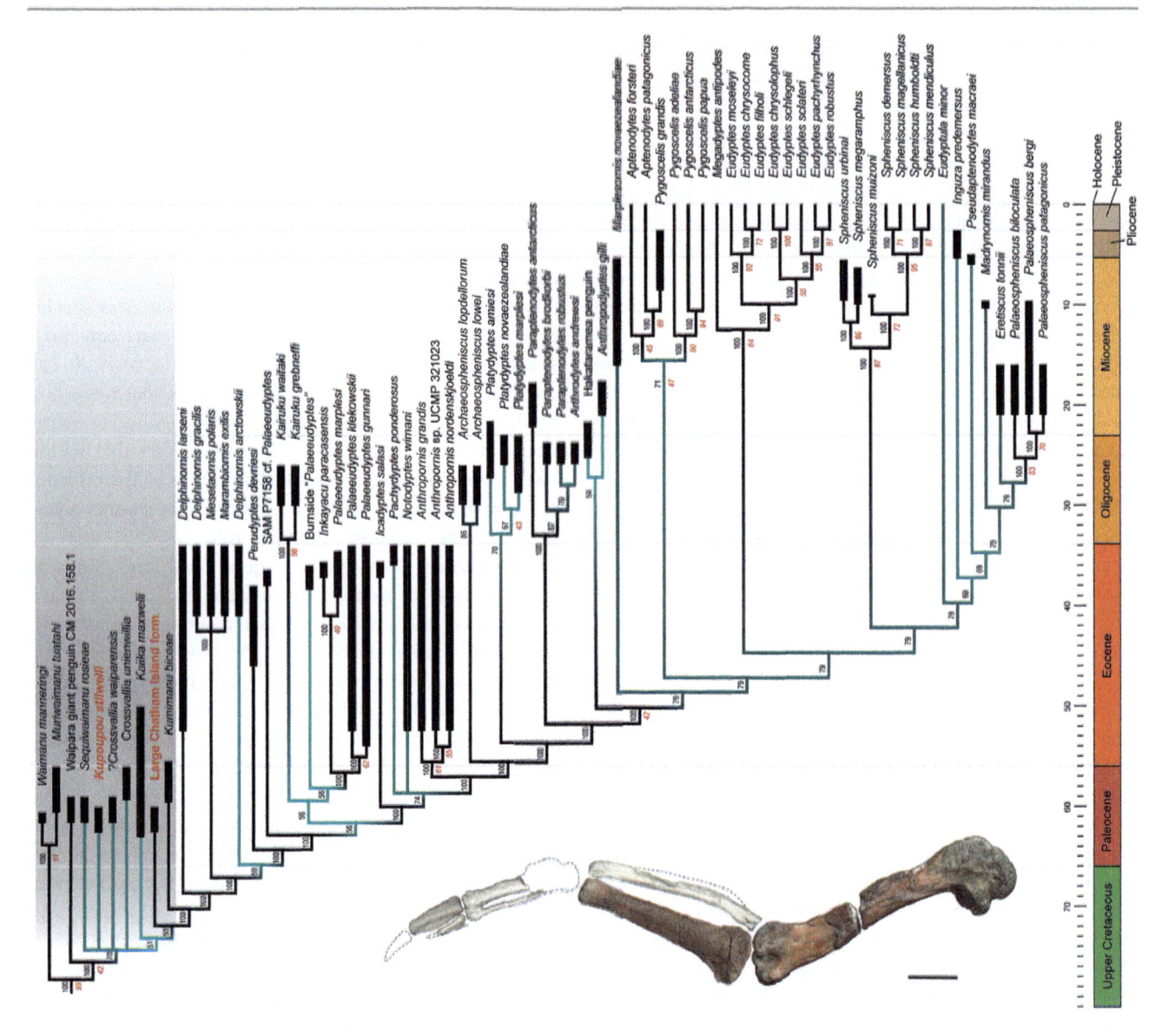

Fig. 7.9 Phylogenetic interrelationships of fossil and extant Sphenisciformes (from Blokland et al. 2019: fig. 13), with a reconstruction of the skeleton of the left wing of *Kupoupou stilwelli* from the late early or early middle Paleocene of Chatham Island (from Blokland et al. 2019: fig. 6). Both images were published under a Creative Commons CC BY 4.0 license

The giant Eocene and Oligocene species were classified into the Palaeeudyptinae by Simpson (1971), but a monophyly of this taxon is not supported by current analyses (Ksepka et al. 2006, 2012; Clarke et al. 2007; Chávez-Hoffmeister 2014; Mayr et al. 2018b; Blokland et al. 2019). Among these large species, *Inkayacu* (late Eocene of Peru), *Kairuku* (early Oligocene of New Zealand), and *Palaeeudyptes* (late Eocene of Seymour Island) share an unusually short and stout femur (Mayr 2017: fig. 10.3), whereas *Icadyptes* (late Eocene of Peru) as well as *Palaeeudyptes*, *Pachydyptes*, and *Anthropornis* (late Eocene of Seymour Island) have a very stout and robust humerus in common.

Extant Sphenisciformes exhibit a highly derived skeletal morphology, and the osteological characteristics of the crown group were gradually acquired in the long evolutionary history of penguins. Even though all of the known Paleogene penguins already attained the fundamental characteristics of the sphenisciform body plan, some of the earliest taxa show distinct differences to the modern species. The wings of extant penguins are strongly modified for underwater propulsion and form stiff paddles, whereas those of early Paleogene stem group representatives of the Sphenisciformes exhibit a much less specialized morphology.

A nearly complete wing of the mid-Paleocene *Muriwaimanu tuatahi* (Fig. 7.10) shows that this species

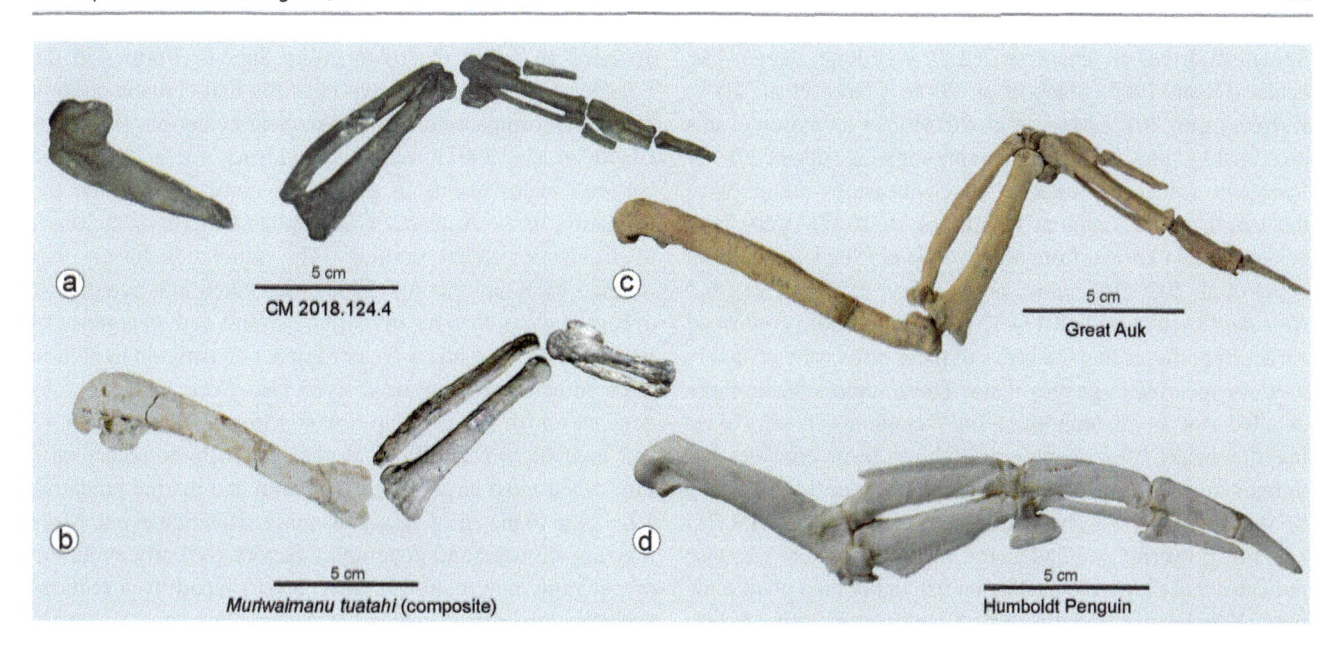

Fig. 7.10 (a) Wing bones of a Paleocene stem group sphenisciform from the Paleocene Waipara Greensand in New Zealand (cf. *Muriwaimanu tuatahi*; Canterbury Museum, Christchurch, New Zealand, CM 2018.124.4); the surrounding matrix was digitally removed and the bones were reassembled for the figure. (b) Composite wing skeleton of *Muriwaimanu tuatahi* from the Waipara Greensand (humerus: CM 2010.108.3; ulna and radius: CM 2009.99.1, both left side, mirrored; carpometacarpus: CM zfa 34; left side, mirrored). (c) Wing of the flightless Great Auk, *Alca impennis* (Alcidae); the specimen lacks the phalanx of the minor digit. (d) Wing of the extant Humboldt Penguin, *Spheniscus humboldti*. From Mayr et al. (2020a), modified; published under a Creative Commons CC BY 4.0 license

did not have stiff flippers, with the proportions of the major wing bones being more similar to those of the charadriiform auks (Alcidae; Mayr et al. 2020a). In further contrast to crown group Sphenisciformes, the phalanx of the alular digit was not co-ossified with the carpometacarpus, which suggests the presence of a functional alula ("bastard wing") and possibly also implies less reduced wing feathers than in extant penguins (Mayr et al. 2020a). Possibly, the wing feathers of this Paleocene taxon were therefore not as short and scale-like as they are in extant penguins but remained differentiated into functional categories (Mayr et al. 2020a). A similar morphology of the wing skeleton is also known from the slightly older *Kupoupou stilwelli* (Fig. 7.9). Even in the late Eocene *Inkayacu*, at least a rudimentary alula appears to have been present (Ksepka et al. 2008).

In most Paleogene Sphenisciformes the pneumotricipital fossa of the humerus is simple and not bipartite as it is in extant penguins, the tarsometatarsus is more elongated, the metatarsals are more strongly co-ossified (e.g., Myrcha et al. 2002), and the hypotarsal crests are better developed (Jadwiszczak 2006b). In many Paleogene Sphenisciformes the sternal margin of the coracoid is furthermore convex rather than concave as in extant penguins (Simpson 1975; Ksepka et al. 2006: fig. 7).

On land, penguins are characterized by an unusual upright stance. In the extant species, the stiff tail feathers support the weight of the standing bird, and the pygostyle, to which the tail feathers are attached, is elongated and unusually narrow. In this regard, it is notable that the pygostyle has a more plesiomorphic shape in the early Oligocene *Kairuku*, which suggests that unlike in extant penguins the tail feathers did not brace the standing bird (Ksepka et al. 2012).

The humerus of *Muriwaimanu* lacks a humeral plexus, a vascular system that in extant penguins serves a heat retention structure and enables foraging in cold offshore waters over extended periods of time. However, this trait first evolved in stem group Sphenisciformes that lived in tropical climates and it is one of the preadaptations, which allowed penguins to disperse into areas with very cold climates (Thomas et al. 2011).

In the *Inkayacu* holotype, feather remains are preserved that allow an examination of the melanosomes. As shown by Clarke et al. (2010), these cell organelles did not have the unique shape found in extant penguins, in which they are unusually large and ellipsoidal, and it was concluded that the plumage of *Inkayacu* was grayish or brownish and not black and white as in extant penguins.

The beaks of extant penguins show different shapes, which reflect the nature of their main prey items. The planktonivorous *Eudyptes* species have wide beaks and deep mandibles, whereas the beaks of the piscivorous *Aptenodytes* species are long and pointed. Many Paleogene

Sphenisciformes are characterized by very long, dagger-like beaks (Olson 1985; Slack et al. 2006; Clarke et al. 2007; Ksepka et al. 2012; Mayr et al. 2018b). In its extreme, this presumably plesiomorphic morphology is displayed by *Icadyptes salasi*, the beak of which measures about twice the length of the cranium (Clarke et al. 2007). Very long beaks are also known from *Muriwaimanu* (Slack et al. 2006; Mayr et al. 2020a), *Sequiwaimanu* (Mayr et al. 2018b), and *Kairuku* (Ksepka et al. 2012). Ksepka et al. (2008) confirmed earlier hypotheses that the beaks of these birds were probably used for spearing large prey items. These authors furthermore detailed that the rhamphotheca (the keratinous sheath covering the beak) of *Icadyptes* must have been considerably thinner than in extant penguins. Paleogene stem group Sphenisciformes also show strongly developed temporal fossae for powerful jaw adductor muscles, which likewise indicates that they were feeding on fish rather than planktonic prey (Ksepka and Bertelli 2006; Ksepka et al. 2006; Acosta Hospitaleche and Haidr 2011). Different foraging strategies of early Paleogene stem group Sphenisciformes to those of the extant species are also suggested by the caudal ends of the mandible of *Muriwaimanu*, which lack derived characteristics of crown group Sphenisciformes, such as long retroarticular processes (Mayr et al. 2020a). The feeding ecology of penguins may have been impacted by the formation of the Antarctic Circumpolar Current, which was critical for the evolution of notothenioid fishes and krill, both of which today constitute a major part of the diet of Antarctic and Subantarctic penguins (Mayr et al. 2020a). However, if the identification of some fragmentary fossils by Jadwiszczak (2006b) is correct, not all early Paleogene taxa had dagger-like beaks, and some diversity in the feeding adaptations of Paleocene penguins was also noted by Haidr and Acosta Hospitaleche (2012).

Current fossil evidence suggests that penguins originated in the New Zealand region in the latest Cretaceous or earliest Cenozoic, from where they dispersed into Antarctica and South America (Ksepka and Thomas 2012). It is notable that all fossil penguins are from the Southern Hemisphere and occurred within the geographic range of extant penguins. Today, oceanic circulation systems and the distribution of productive marine upwellings constitute geographic barriers (Ksepka and Thomas 2012). However, penguins evolved in geological periods with very different climatic regimes, and it remains elusive why these birds did not disperse into the Northern Hemisphere.

Glaciation of the Antarctic continent resulted from the formation of the Antarctic Circumpolar Current after the late Eocene (about 36 Ma) opening of the Drake Passage, which separates South America from Antarctica (e.g., Woodburne and Case 1996). The origin of Pan-Sphenisciformes clearly predates this event (Ksepka and Ando 2011; Mayr et al. 2017b). Baker et al. (2006) assumed an origin of crown group Sphenisciformes in the Eocene of Antarctica and hypothesized that climatic cooling had a major impact on their subsequent evolution. However, Clarke et al. (2007) found no evidence for a correlation between major events in penguin evolution and either the formation of the Antarctic Circumpolar Current or the onset of Cenozoic global cooling. The same conclusion was reached by Ksepka et al. (2006) who noted that geographic events, such as changes in marine currents and emergence of islands, may have had a greater impact on penguin evolution than climatic cooling itself (see also Sect. 11.3.3.2). As detailed above, stem group Sphenisciformes appear to have lost their flight capabilities around the K/Pg boundary, at a time when most larger-sized terrestrial and marine predators fell victim to the end-Cretaceous mass extinction event. More than any climatic and geographic factors, the early evolution of penguins may therefore have been shaped by a reduced predation pressure at sea and at the breeding sites.

7.5 Ciconiidae (Storks)

Paleogene fossils of the Ciconiidae are only known from the Old World, where also most extant species of storks occur. The putative ciconiid *Eociconia sangequanensis* from the middle Eocene of China was described on the basis of a fragmentary distal tarsometatarsus (Hou 1989, 2003). Although the specimen is of similar size and proportions to the tarsometatarsus of the extant *Ciconia*, its unambiguous identification requires the discovery of additional fossil material. Wang et al. (2012) reported a leg from the middle Eocene of China, the distal tarsometatarsus of which shows a similarity to that of *Eociconia*. This leg belongs, however, to a much smaller species described as *Sanshuiornis zhangi*. The hypotarsus of *S. zhangi* exhibits distinct crests and sulci and resembles that of the threskiornithid taxon *Rynchaeites* (Sect. 7.7), but the affinities of the species likewise cannot be determined without further material.

The earliest well-represented and unambiguously identified stork is *Palaeoephippiorhynchus dietrichi* from the early Oligocene of the Jebel Qatrani Formation of the Fayum in Egypt. This species was described by Lambrecht (1930) on the basis of a three-dimensionally preserved skull (Fig. 7.11a) and was about the size of the extant Saddlebill (*Ephippiorhynchus senegalensis*). As in the extant taxa *Ephippiorhynchus* and *Jabiru*, the tip of the beak is slightly upturned. Rasmussen et al. (1987) tentatively referred a distal tibiotarsus from the early Oligocene of the Jebel Qatrani Formation to *P. dietrichi*. Another distal tibiotarsus of a stork was reported from late Eocene deposits of this locality by Miller et al. (1997), who refrained from assigning it to *P. dietrichi* because of slight differences to the early Oligocene tibiotarsus.

Fig. 7.11 (a) Skull of *Palaeoephippiorhynchus dietrichi* (Ciconiidae) from the early Oligocene of the Jebel Qatrani Formation in Egypt (Staatliches Museum für Naturkunde Stuttgart, Germany, SMNS 12653); from Lambrecht (1930). (b), (c) Holotype skeleton of *Protoplotus beauforti* from Paleocene or Eocene of Sumatra; from Lambrecht (1931) with labels added, (c) is an x-ray photograph of the specimen

Boles (2005) described *Ciconia louisebolesae* from the late Oligocene/early Miocene of the Riversleigh site in Australia. The fossil material of this species, which was about the size of the extant White Stork (*C. ciconia*), consists of a partial skull and incomplete limb bones. The taxon *Ciconia* was also described from the Neogene of Australia, and it is unknown why it became extinct on the continent, where the sole extant species of stork belongs to *Ephippiorhynchus* (Boles 2005).

7.6 Suliformes: Fregatidae (Frigatebirds) and Suloidea (Gannets, Boobies, Cormorants, and Anhingas)

Phylogenetic analyses congruently support a sister group relationship between the Fregatidae and the Suloidea, that is, a clade including the Sulidae, Phalacrocoracidae, and Anhingidae (e.g., Ericson et al. 2006; Mayr 2008; Prum et al. 2015; Kuhl et al. 2021). Among others, these birds share a greatly abbreviated tarsometatarsus, which measures just about half of the length of the carpometacarpus, a distally protruding trochlea for the second toe, and a pectinate claw of the third toe (e.g., Mayr 2003, 2008). Within the Phalacrocoracidae, the tarsometatarsal features are only present in the small *Microcarbo* species, which are the sister taxon of a clade formed by all other extant cormorants (Mayr 2007).

Mangystania humilicristata was described on the basis of a proximal humerus from the middle Eocene of Kazakhstan (Zvonok et al. 2016), which shows a resemblance to the proximal humerus of extant Suloidea and to that of the Sulidae in particular. The fossil exhibits, however, an unusual combination of presumably plesiomorphic and autapomorphic features which impedes a straightforward classification.

7.6.1 Protoplotidae

One of the earliest representatives of the Suliformes is *Protoplotus beauforti*, the holotype of which is a nearly complete skeleton from lacustrine sediments of Sumatra (Fig. 7.11b, c; Lambrecht 1931). The age of these deposits is debated, but they probably date from the Eocene or even Paleocene (van Tets et al. 1989; Stidham et al. 2005). *P. beauforti* is a small species, of similar size to the extant Little Pied Cormorant, *Microcarbo melanoleucos*. The tarsometatarsus is very short, and the trochlea for the second toe protrudes distally beyond the trochleae for the other toes. As in the stem group frigatebird *Limnofregata* (Sect. 7.6.2), the nostrils of *P. beauforti* are long and slit-like, whereas they are greatly reduced in extant Suliformes. *Protoplotus* differs from *Limnofregata* in, e.g., the shape of the mandible and the proportionally longer legs.

P. beauforti was originally described as a representative of the Anhingidae (Lambrecht 1931). Van Tets et al. (1989) classified the species into a new taxon Protoplotidae and detailed that its anhingid affinities have not been well established. *P. beauforti* differs in many osteological features from crown group Anhingidae, e.g., the mandibular rami are closely adjacent in their rostral half (as in the stem group tropicbird *Prophaethon*; Sect. 7.1), the cervical vertebrae are proportionally shorter, and the limb elements have different proportions, with ulna and tarsometatarsus being relatively longer. In its overall skeletal morphology, the species is closer to the Anhingidae and Phalacrocoracidae than to the Fregatidae and Sulidae, but affinities to the former two taxa have not yet been established with derived characters.

The holotype specimen of *P. beauforti* is remarkable, because it exhibits a compact mass of large gastroliths, which led Lambrecht (1931) to hypothesize that the species was granivorous. By contrast, van Tets et al. (1989) assumed that it was foraging under water and fed on small fish and invertebrates. These authors furthermore noted that the gastroliths may have served as ballast to aid diving, whereas Zhou et al. (2004) argued that *Protoplotus* could have employed seasonal diet switching from fish to plant material.

7.6.2 Fregatidae (Frigatebirds)

The Fregatidae include five extant species, which occur in all tropical and subtropical oceans. Frigatebirds are highly aerial birds and can neither walk on land nor are they able to swim. All Paleogene fossils of these birds are from the early Eocene of North America. In addition to three species of the taxon *Limnofregata* (Fig. 7.12a) from the Green River and Wasatch formations of Wyoming (Olson 1977; Olson and Matsuoka 2005; Stidham 2015), a proximal carpometacarpus from the Nanjemoy Formation in Virginia was tentatively referred to the Fregatidae (Mayr 2016).

Limnofregata hutchisoni from the early Eocene (53–54 Ma) Wasatch Formation of Wyoming, USA is only represented by a humerus and a coracoid, which were found in close association (Stidham 2015). Together with the tentatively referred carpometacarpus from the Nanjemoy Formation (Mayr 2016), *L. hutchinsoni* is the earliest record of the Fregatidae. The two species from the Green River Formation, *Limnofregata azygosternon* and *L. hasegawai*, mainly differ in size and the proportions of their beaks. Both are known from several articulated skeletons, some of which exhibit feather remains. The smaller species, *L. azygosternon*, was about the size of the males of the Lesser Frigatebird (*Fregata ariel*), which is the smallest extant species of frigatebird (Olson 1977).

Limnofregata shares with crown group Fregatidae a greatly abbreviated tarsometatarsus, a subtriangular-shaped deltopectoral crest (humerus), and a short and wide pelvis. However, despite a similar overall morphology, it distinctly differs in several presumably plesiomorphic features. Among others, the beak of the fossil taxon is shorter and its tip less strongly hooked, and the nostrils are long and slit-like, whereas they are greatly reduced in extant Fregatidae (juvenile frigatebirds, however, still have slit-like nostrils; Olson 1977: fig. 9). The hindlimbs of *Limnofregata* are proportionally longer than those of extant Fregatidae, and unlike in the latter, the coracoid is not co-ossified with the sternum and furcula. Olson and Matsuoka (2005) furthermore assumed that, in contrast to extant frigatebirds, *Limnofregata* was not sexually dimorphic in size.

Whereas extant frigatebirds are pelagic birds, all *Limnofregata* fossils were found in lacustrine deposits. However, this fact does not necessarily indicate a strictly non-marine way of living, and Olson and Matsuoka (2005) hypothesized that *Limnofregata* may have visited the lakes that formed the sediments of the Green River Formation

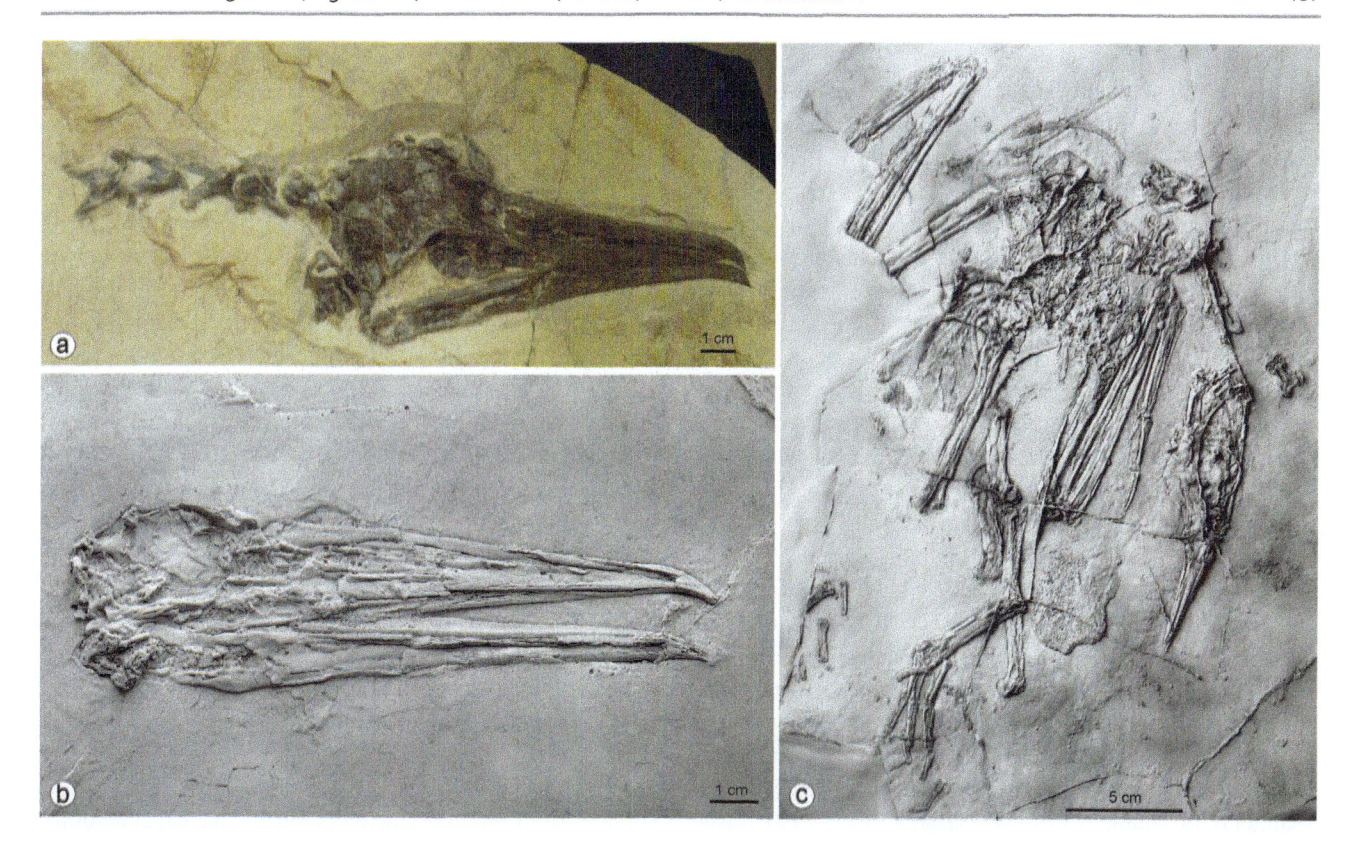

Fig. 7.12 (a) Skull of the early Eocene frigatebird *Limnofregata hasegawai* (Fregatidae) from the Green River Formation of Wyoming, USA (holotype; Gunma Museum of Natural History, Gunma, Japan, GMNH PV 170). (b) Skull of *Masillastega rectirostris* (?Sulidae) from the latest early or earliest middle Eocene of Messel (holotype, Palaeontological Institute of the University of Bonn, Germany, IPB 140b); coated with ammonium chloride. (c) Skeleton of ?*Oligocorax stoeffelensis* (Phalacrocoracidae) from the late Oligocene of Enspel in Germany (holotype, Landesamt für Denkmalpflege Rheinland-Pfalz, Mainz, Germany, PW 2005/5022-LS); coated with ammonium chloride. Photo in (a) by the author, in (b) and (c) by Sven Tränkner

during periodic dieoffs of fishes. The proportionally longer legs suggest that this Eocene taxon of frigatebirds was better adapted to swimming than its extant relatives and may have occupied an ecological niche similar to extant gulls (Olson 1977 and Olson and Matsuoka 2005). The virtual absence of fossil frigatebirds in post-Eocene sites indicates that, as in the case of tropicbirds (Sect. 7.1), these birds adopted a strictly pelagic way of living after the early Eocene (Stidham 2015).

7.6.3 Sulidae (Gannets and Boobies)

The extant species of the Sulidae are exclusively marine birds, which plunge-dive to capture fish or squid. All Paleogene fossils of the taxon were reported from European fossil sites and mainly consist of fragmentary remains of uncertain affinities.

Masillastega rectirostris from Messel is known from an isolated skull and was tentatively identified as a stem group representative of the Sulidae (Mayr 2002a). The species has a long and straight beak, which is deep in its proximal part and, unlike the tapering beak of extant Sulidae, has a slightly hooked tip (Fig. 7.12b). As in extant Sulidae, the nostrils are greatly reduced and the cutting edges of the mandibular rami are very wide. *M. rectirostris* furthermore shares with extant Sulidae a deep upper beak and mandible, which both bear numerous impressions of blood vessels, as well as a deeply excavated dorsal tympanic recess in the otic region of the skull (Mayr 2002a). If its assignment to the Sulidae is confirmed by future postcranial remains, the species indicates that stem group representatives of the taxon also occurred in freshwater environments.

Eostega lebedinskyi was described on the basis of an incomplete mandible from marine and putatively middle Eocene deposits of Romania and was considered to be closely related to extant Suloidea by Lambrecht (1929). It was referred to the Sulidae by Mlíkovský (2002, 2007), but this assignment was based on overall resemblance rather than shared derived characteristics. Mlíkovský (2007) furthermore synonymized *Masillastega* with *Eostega*, an action, the validity of which needs to be assessed by a direct comparison of the specimens.

Prophalacrocorax ronzoni from early Oligocene (MP 21; Mlíkovský 2002) lacustrine deposits of Ronzon in France is known from a fragmentary pelvis. The species was originally described as a merganser ("*Mergus ronzoni*"), that is, a representative of the Anseriformes, but it was transferred to the Sulidae by Milne-Edwards (1867–1871), as "*Sula*" *ronzoni*. Harrison (1975) assigned it to the taxon *Prophalacrocorax* and considered *P. ronzoni* to be a species of the Phalacrocoracidae. Without more material, the affinities of this species probably cannot be determined (see also Olson 1985). Putative Sulidae were also reported from the late Eocene or early Oligocene of Japan (Ono and Hasegawa 1991; Mori and Miyata 2021), but these fossils have not yet been formally described.

Empheresula arvernensis from the late Oligocene (MP 30; Mlíkovský 2002) lacustrine deposits of Gannat in France is also solely known from an incomplete pelvis and a referred sternum (Milne-Edwards 1867–1871). As first noted by Olson (1985), the sternum of this species differs from that of crown group Sulidae in the presence of four incisions in the caudal margin. Its assignment to the Sulidae also needs to be established with additional material. Remains of sulids were furthermore reported from late Oligocene marine sediments of southern Germany (Darga et al. 1999: distal end of a humerus).

7.6.4 Phalacrocoracidae (Cormorants) and Anhingidae (Anhingas)

The piscivorous Phalacrocoracidae live in freshwater and coastal marine habitats and are today globally distributed, being particularly species-rich in northern and southern latitudes. They are the sister taxon of the Anhingidae, which are specialized divers and occur in tropical regions worldwide.

One of the earliest records of a cormorant-like bird is an incomplete upper beak from the late Eocene (MP 17; Mlíkovský 2002) of England, which was described as *Piscator tenuirostris* by Harrison and Walker (1976b). There are undescribed skeletons of small cormorants from the early Oligocene of the Luberon area in France in private collections (Roux 2002 and own observation), and as yet undescribed fossils of cormorants from an unknown stratigraphic horizon of the Quercy fissure fillings were listed by Mourer-Chauviré (1995, 2006). A cormorant-like sternum was found in marine deposits of the latest Eocene or earliest Oligocene Makah Formation in Washington State, USA (Mayr and Goedert 2018). Undescribed fossils of the Phalacrocoracidae were also reported from the late Eocene or early Oligocene of Japan (Ono and Hasegawa 1991; Mori and Miyata 2021). Rasmussen et al. (1987) assigned a partial beak from the early Oligocene of the Jebel Qatrani Formation of the Fayum in Egypt to the Phalacrocoracidae.

Borvocarbo guilloti was established on the basis of a coracoid from the late Oligocene (MP 30) of France, which differs from the corresponding bone of extant Phalacrocoracidae and Anhingidae in the plesiomorphic presence of a concave articular facet for the scapula (Mourer-Chauviré et al. 2004). This character may support a position of *B. guilloti* outside the clade including the Anhingidae and Phalacrocoracidae. Mayr (2010) tentatively assigned a distal humerus end from the late Oligocene or early Miocene of the Mainz Basin in Germany to *Borvocarbo* cf. *guilloti*.

Another late Oligocene (MP 28) species, ?*Oligocorax stoeffelensis* from the Enspel maar lake in Germany, was initially assigned to *Borvocarbo* and is known from a foot and two skeletons (Fig. 7.12c; Mayr 2001, 2007, 2015c). This species was slightly larger than the extant Pygmy Cormorant (*Microcarbo pygmeus*) and has a similar overall morphology to extant cormorants. The Enspel cormorant exhibits derived characters of the Phalacrocoracidae that are absent in the Anhingidae, including a prominent median crest in the caudal portion of the skull (crista nuchalis sagittalis), a very large patella, and a long fourth toe (Mayr 2007, 2015c; Smith 2010). Plesiomorphic traits, such as the larger acromial process of the furcula and the less strongly protruding medial hypotarsal crest, support a position outside crown group Phalacrocoracidae. A phylogenetic assignment of ?*O. stoeffelensis* to the stem group of the Phalacrocoracidae is, however, not straightforward, and at least two presumably plesiomorphic features, that is, less prominent paroccipital processes (skull) and a shorter bicipital crest (humerus), even distinguish the fossil species from crown group representatives of the clade (Phalacrocoracidae + Anhingidae) (Mayr 2007). Unlike in extant Phalacrocoracidae, the sternal extremity of the furcula of ?*O. stoeffelensis* is co-ossified with the tip of the sternal keel.

The description of *Limicorallus saiensis* was based on a distal humerus from the late Oligocene (*Indricotherium* Beds) of Kazakhstan. The species was originally described as a rail, but it was identified as a representative of the Phalacrocoracidae by Mlíkovský and Švec (1986). *L. saiensis* was slightly smaller than the Pygmy Cormorant, which is the smallest extant cormorant.

Two species from the late Oligocene or early Miocene Etadunna and Namba formations of Australia, *Nambashag billerooensis* and *N. microglaucus*, are well represented by multiple isolated bones (Worthy 2011). A phylogenetic analysis recovered *Nambashag* as the sister taxon of crown group Phalacrocoracidae. *Oligocorax* was not included in the analysis, and the exact interrelationships of the late Oligocene taxa *Nambashag* and *Oligocorax* still have to be resolved.

The earliest fossil record of the Anhingidae is "*Anhinga*" *walterbolesi* from the late Oligocene (26–25 Ma) Etadunna

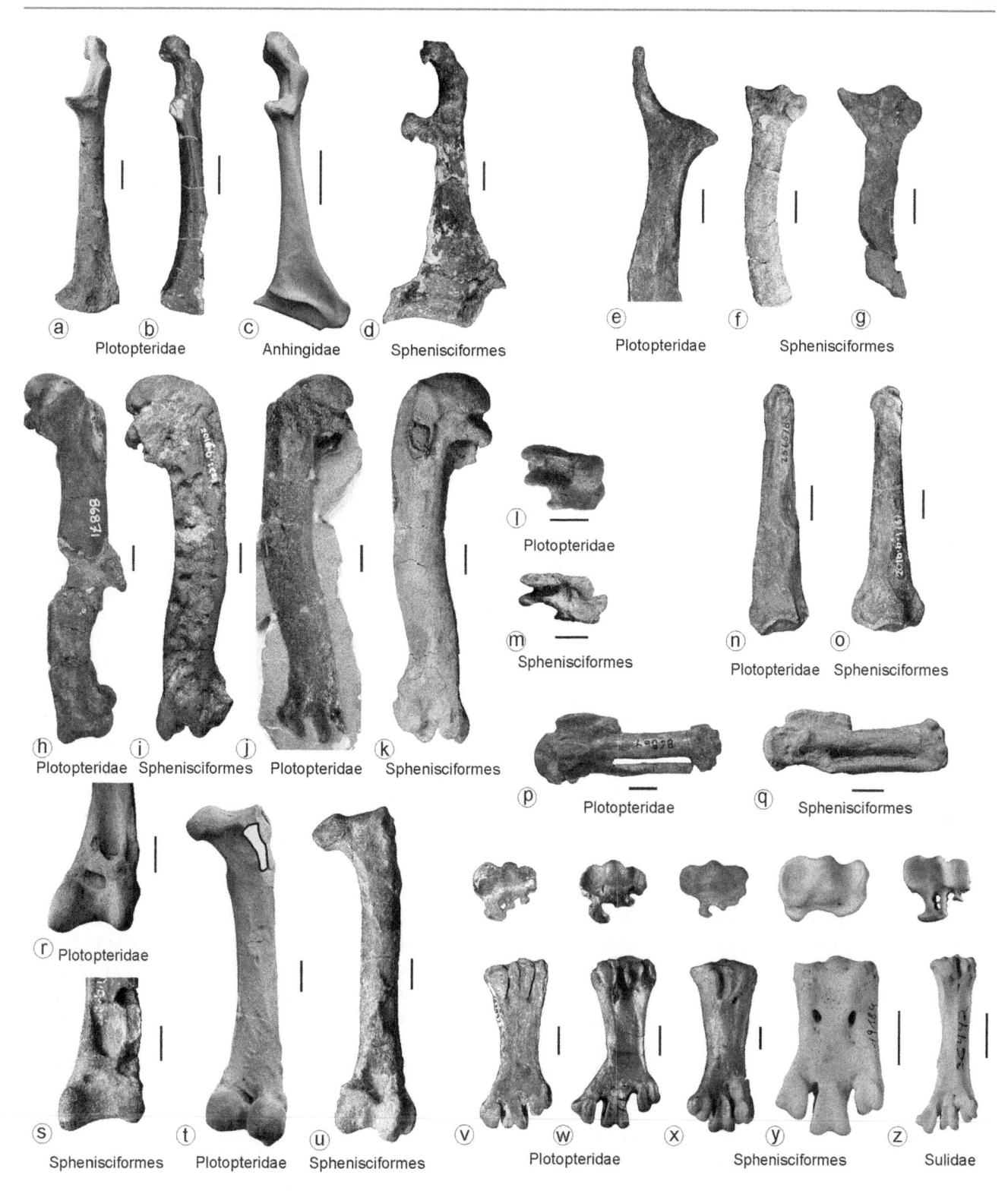

7.6 Suliformes: Fregatidae (Frigatebirds) and Suloidea (Gannets, Boobies, Cormorants... 139

Fig. 7.13 Selected skeletal elements of the Plotopteridae in comparison to Paleogene Sphenisciformes (after modified figures in Mayr et al. 2021; published under a Creative Commons CC BY 4.0 license). (**a**)–(**d**) Coracoids of (**a**) the plotopterid *Klallamornis abyssa* (Burke Museum of Natural History and Culture, Seattle, USA, UWBM 108400), (**b**) the plotopterid *Stemec suntokum* (Royal British Columbia Museum, Victoria, Canada, RBCM RBCM.EH2014.032.0001.001), (**c**) the extant *Anhinga anhinga* (Suliformes, Anhingidae), (**d**) the stem sphenisciform *Muriwaimanu tuatahi* (Canterbury Museum, Christchurch, New Zealand, CM zfa 34, mirrored). (**e**)–(**g**) Cranial ends of the scapulae of (**e**) the plotopterid *Tonsala hildegardae* (cast of holotype in the Natural History Museum of Los Angeles County, USA, LACM 256518), (**f**) *M. tuatahi* (CM zfa 34, mirrored), and (**g**) an undescribed Paleocene stem group sphenisciform from the Waipara Greensand (CM 2009.99.1). (**h**)–(**m**) Humeri of (**h**) a plotopterid referred to "*Tonsala*" *buchanani* by Dyke et al. (2011) (UWBM

Formation of South Australia (Worthy 2012). The holotype of this species is a tarsometatarsus, which differs from that of extant Anhingidae in some osteological features. All extant species of darters are classified in the taxon *Anhinga*. However, the Anhingidae include various extinct Neogene genus-level taxa, which may be more closely related to the crown group than is "*A*". *walterbolesi* (e.g., *Macranhinga* and *Giganhinga*), in which case the assignment of the Australian taxon to the genus *Anhinga* needs to be revised.

7.6.5 Plotopteridae

The Plotopteridae are flightless, wing-propelled seabirds, the skeleton of which shows a striking resemblance to that of penguins (Fig. 7.13). Remains of these birds were found in late Eocene to early Miocene deposits of the north Pacific coast of western North America and Japan. Many fossils are from deep-water sediments, which were deposited offshore (Goedert and Cornish 2002; see, however, Sakurai et al. 2008).

The taxon Plotopteridae was originally established for *Plotopterum joaquinensis* from the early Miocene of California (Howard 1969). *P. joaquinensis* is one of the smallest and geologically youngest plotopterids and is only known from the holotype, a partial coracoid. A very similar species, *Stemec suntokum*, was reported from the late Oligocene (24–25 Ma) of British Columbia, Canada, and is likewise just represented by a coracoid (Fig. 7.13b; Kaiser et al. 2015). Both *Plotopterum* and *Stemec* were smaller than earlier, late Eocene and early Oligocene, plotopterids and exhibit a less derived and more "cormorant-like" coracoid morphology than geologically older representatives of the Plotopteridae.

The earliest plotopterid specimen is a partial skeleton of an as-yet unnamed large-sized species from the late Eocene (~35 Ma) of the Olympic Peninsula (Washington State, USA), which was referred to as the "Whiskey Creek specimen" by Goedert and Cornish (2002) and found in association with three small pebbles that were interpreted as gastroliths. The description of the first species to be scientifically named, *Phocavis maritimus*, was based on a tarsometatarsus from the latest Eocene (33.8 Ma) of Oregon (Goedert 1988). The tarsometatarsus of *P. maritimus* is less robust than that of geologically younger species and still exhibits a well-developed distal vascular foramen, which is reduced in later species (Fig. 7.13v). This plesiomorphic feature suggests a sister group relationship between *P. maritimus* and the other plotopterid taxa, of which the tarsometatarsus is known.

Goedert and Cornish (2002) already noted the presence of at least five different taxa of plotopterids in the latest Eocene and early Oligocene of the Olympic Peninsula of Washington State (USA), which demonstrates a high diversity of these birds in the Northeastern Pacific during the Paleogene. The first named species from these deposits is the early Oligocene *Tonsala hildegardae* (Figs. 7.13e; 7.14a), the original description of which (Olson 1980) was complemented by Goedert and Cornish (2002), Mayr et al. (2015), and Mayr and Goedert (2018). A second, larger species described as *Tonsala buchanani* (Fig. 7.14c; Dyke et al. 2011) shows some distinct differences to *T. hildegardae* and is likely to belong to a different genus-level taxon (Mayr et al. 2015; Mayr and Goedert in press). Three further species from the Olympic Peninsula are *Klallamornis abyssa* and ?*K. clarki* from the Makah and Pysht formations (Fig. 7.14f), as well as *Olympidytes thieli* from the late Eocene/early Oligocene Lincoln Creek Formation (Mayr et al. 2015; Mayr and Goedert 2016). The species of *Klallamornis* were medium-sized to very large plotopterids, whereas *O. thieli* represents a smaller species.

The earliest plotopterids from Japan are likewise from late Eocene deposits (Sakurai et al. 2008), but most published fossils come from Oligocene sites (Olson and Hasegawa 1979, 1996; Kimura et al. 1998; Sakurai et al. 2008). From the Nishisonogi Group of Kyushu, a femur from latest Eocene/earliest Oligocene strata and a distal tibiotarsus from early Oligocene sediments were reported; the latter was tentatively assigned to *Olympidytes* (Mori and Miyata 2021). The majority of published Japanese plotopterids, however, are from the early Oligocene Ashiya Group of Kyushu Island. The first formally published fossils from these deposits are two very large species, which were described as *Copepteryx hexeris* and *C. titan* (Olson and Hasegawa 1996; these species were originally considered to be of late Oligocene age, but see Kaiser et al. 2015 for the revised age of the fossil sites). *C. hexeris* is known from partial skeletons;

Fig. 7.13 (Continued) 86871), (**i**) the stem sphenisciform *Sequiwaimanu rosieae* (CM 2016.6.1), (**j**) cf. "*Tonsala*" *buchanani* (SMF Av 601, mirrored), and (**k**) *S. rosieae* (CM 2016.6.1). (**l**), (**m**) Distal ends of humeri (distal view) of (**l**) a plotopterid referred to "*T.*" *buchanani* (UWBM 86871) and (**m**) *S. rosieae* (CM 2016.6.1, mirrored). (**n**), (**o**) Ulnae of (**n**) *T. hildegardae* (cast of holotype, LACM 256518) and (**o**) *S. rosieae* (CM 2016.6.1). (**p**), (**q**) Carpometacarpi of (**p**) an unnamed, large late Eocene plotopterid (cf. *Klallamornis*), from Whiskey Creek, Washington State (UWBM 86869) and (**q**) *S. rosieae* (CM 2016.6.1). (**r**), (**s**) Distal tibiotarsi of (**r**) the plotopterid *Olympidytes thieli* (SMF Av 608) and (**s**) *S. rosieae* (CM 2016.6.1). (**t**), (**u**) Femora of (**t**) *O. thieli* (SMF Av 608, cast, mirrored) and (**u**) *S. rosieae* (CM 2016.6.1). (**v**)–(**z**) tarsometatarsi of (**v**) the plotopterid *Phocavis maritimus* (LACM 123897), (**w**) an undescribed plotopterid from the early Oligocene of Kitakyushu, Japan (CM Ffa 18, cast), (**x**) the stem sphenisciform *Waimanu manneringi* (CM zfa 35), (**y**) the extant *Pygoscelis adeliae* (Spheniscidae), (**z**) the extant *A. anhinga*; the upper row shows the proximal end of the tarsometatarsus in proximal view. The scale bars represent 1 cm. (All photos by Sven Tränkner)

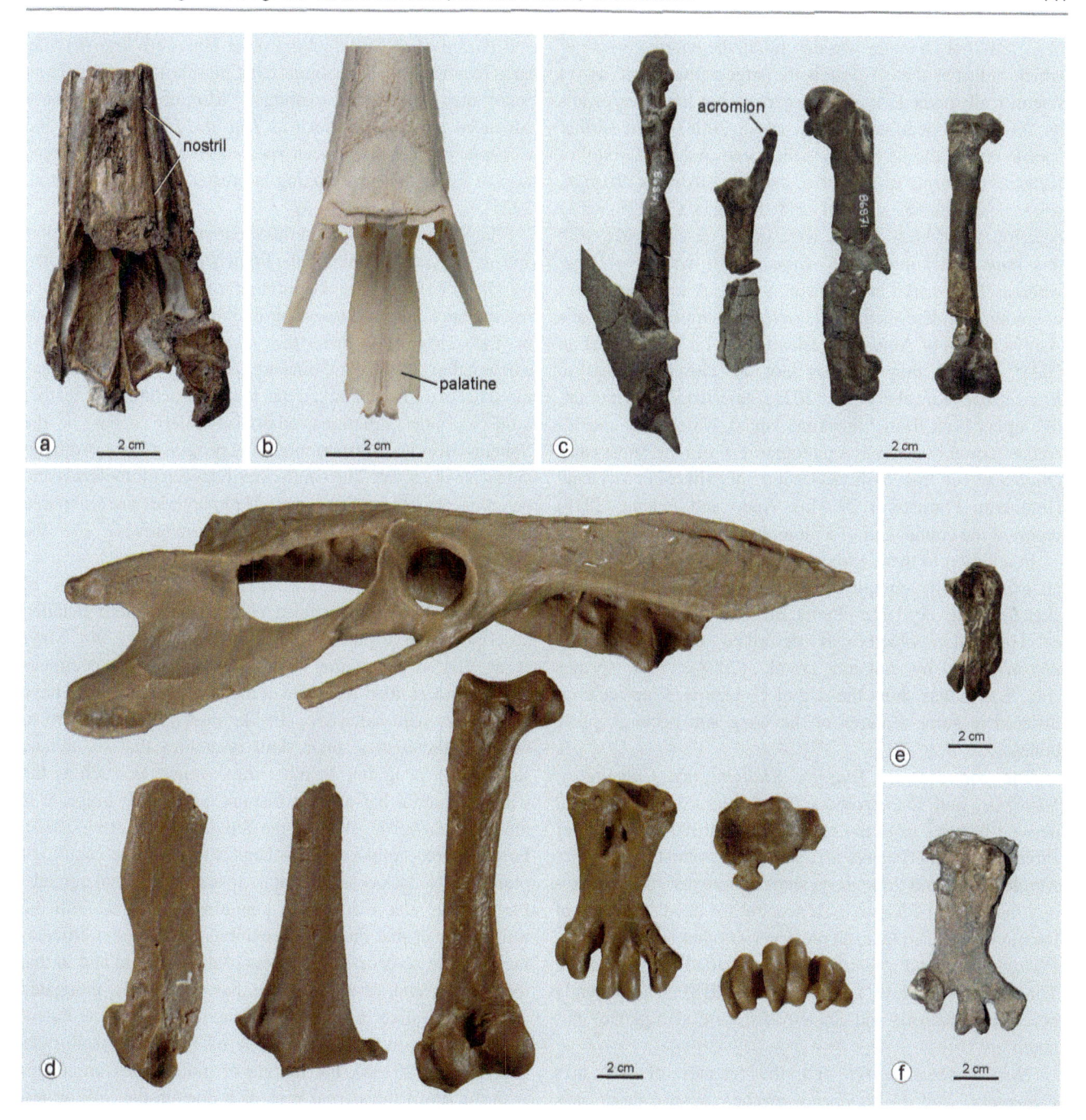

Fig. 7.14 Differently sized representatives of late Eocene and Oligocene plotopterids from Japan and North America. (**a**) Partial skull (palatal elements and proximal portion of beak in dorsal view) from the Pysht Formation of the Olympic Peninsula, Washington State, USA, which was referred to *Tonsala hildegardae* (Senckenberg Research Institute, SMF Av 599). (**b**) Partial skull of the extant *Morus bassanus* (Sulidae) to illustrate the different morphology of the palatine bones. (**c**) Selected skeletal elements (coracoid, partial scapula, humerus, and femur) of a plotopterid from the late Eocene or early Oligocene of the Makah Formation of the Olympic Peninsula, which was assigned to "*Tonsala*" *buchanani* by Dyke et al. (2011) (Burke Museum of Natural History and Culture, Seattle, USA, UWBM 86871). (**d**) Pelvis, partial humerus and coracoid, femur, and tarsometatarsus (dorsal, proximal, and distal view) of the very large *Hokkaidornis abashiriensis* from the late Oligocene of Japan (uncatalogued cast of the holotype in Senckenberg Research Institute Frankfurt). (**e**) Tarsometatarsus of a plotopterid from the Makah Formation of the Olympic Peninsula, which was assigned to "*T.*" *buchanani* by Dyke et al. (2011) (Burke Museum of Natural History and Culture, Seattle, USA, UWBM 86870). (**f**) Tarsometatarsus of the large-sized ?*Klallamornis clarki* from the Makah Formation of the Olympic Peninsula (holotype, Natural History Museum of Los Angeles County, USA, LACM 129405). (Photos in (**a**), (**b**), and (**d**) by Sven Tränkner, others by the author)

of *C. titan* only a femur was described, the enormous size of which indicates that *C. titan* is the largest plotopterid as yet scientifically named. An even larger species, however, awaits its formal publication (Sakurai et al. 2008). Two further species from the late early Oligocene Ashiya Group of Kyushu, *Stenornis kanmonensis* and *Empeirodytes okazakii*, were described by Ohashi and Hasegawa (2020). The descriptions of both species were based on coracoids, with this bone indicating that *S. kanmonensis* was very large, whereas *E. okazakii* had a medium size. A tarsometatarsus of another undescribed smaller-sized plotopterid from the Ashiya Group of Ainoshima Island (Fig. 7.13w; Mayr et al. 2021: fig. 4j) may or may not be closely related to *E. okazakii*. Kawabe et al. (2021) described a cormorant-like upper beak from Ainoshima Island, which they considered to possibly be from a plotopterid. From the latest early Oligocene of the Ashiya Group of Hikoshima Island (Jinnobaru Formation, 29 Ma), Ando and Fukata (2018) reported the cranial end of a plotopterid scapula.

In addition to these early Oligocene species, there is also an exceptionally complete and well preserved postcranial skeleton of a very large plotopterid from the late Oligocene of Hokkaido, which was described as *Hokkaidornis abashiriensis* by Sakurai et al. (2008). This species (Fig. 7.14d) was about the size of *C. hexeris*, from which it differed in some features of the wing and pectoral girdle bones.

The coracoids of *Tonsala*, *Klallamornis*, *Copepteryx*, *Stenornis*, and *Empeirodytes* are distinguished in several derived features from the more cormorant-like coracoids of *Plotopterum* and *Stemec*, and the tarsometatarsi of *Tonsala*, *Klallamornis*, and *Copepteryx* share a more derived morphology than that of *Phocavis*. To account for the distinctness of the more advanced taxa, all plotopterids other than *Phocavis*, *Plotopterum*, and *Stemec* were classified in the taxon Tonsalinae by Mayr and Goedert (2018). The recently described *Stenornis* and *Empeirodytes* are also part of this taxon.

?*Klallamornis clarki* and the species of the taxa *Copepteryx* and *Hokkaidornis* reached a very large size, with that of the giant *C. titan* rivaling the size of the largest Paleogene penguins. The great morphological similarity of *Copepteryx* and *Hokkaidornis* suggests that these two taxa are closely related, whereas it is less straightforward to constrain the exact affinities of ?*K. clarki*. The cormorant-sized species of *Plotopterum* and *Stemec* are at the other end of the size spectrum of plotopterids, and other taxa have intermediate sizes. Whether a very large size evolved just once in plotopterids or several times independently, as was the case in stem group Sphenisciformes, can only be determined when a well-resolved phylogeny for plotopterids is available.

The disparity in coracoid morphology indicates that plotopterids occupied different ecological niches, and *Stemec* and *Plotopterum* (of which no other bones are known) may have retained a more plesiomorphic morphology owing to an occurrence in littoral habitats. Morphologically more advanced plotopterids, such as *Tonsala*, *Hokkaidornis*, and *Copepteryx*, may have been specialized for a more pelagic way of living and for foraging in greater depths (Mayr et al. 2021).

The forelimbs of plotopterids are transformed into paddles and are strikingly similar to the flippers of penguins. As noted by Olson (1980: 54), the humerus of plotopterids has "a resemblance among known birds only to penguins," and as in the Sphenisciformes, the scapular blade is greatly expanded and radius and ulna are flattened. The ulna furthermore bears a row of marked pits for the attachment of feather quills; similar structures otherwise solely occur in the Sphenisciformes. Furthermore, as in penguins, the tarsometatarsus is very short. The similarities between plotopterids and penguins are especially evident, if plotopterids are compared to Paleocene stem group representatives of the Sphenisciformes (Fig. 7.13; Mayr et al. 2021).

Even though there are anecdotic earlier reports of plotopterid skulls (Hasegawa et al. 1979), more detailed descriptions were published just recently (Fig. 7.14a; Mayr et al. 2015), and some of the best-preserved specimens remain undescribed (Kawabe et al. 2014). Plotopterids have long beaks with extensive, slit-like nostrils. With regard to these characteristics, their skull resembles that of archaic stem group penguins. In other traits, however, such as the wide internarial bar and the distinct nasofrontal hinge, it is more similar to the skull of the Sulidae (Mayr et al. 2015). Furthermore, unlike the Sphenisciformes, the skull of plotopterids lacks supraorbital fossae for nasal glands. Plotopterids also differ from penguins and agree with the Suliformes in that the furcula exhibits a prominent articular facet for the acrocoracoid process of the coracoid and in that the cranial end of the scapula has a strongly elongated acromion. Unlike in extant penguins but as in some Paleocene stem group representatives of the Sphenisciformes (Mayr et al. 2018b), the furcula of plotopterids articulates with the tip of the sternal keel, as it does in the Suliformes. Skull endocasts show the brain shape of plotopterids to have been similar to that of extant penguins, but these endocasts are clearly distinguished from those of Paleocene stem group Sphenisciformes (Kawabe et al. 2014; Proffitt et al. 2016; Tambussi et al. 2015).

Plotopterids share a greatly enlarged patella with extant Anhingidae and Phalacrocoracidae, but unlike in crown group Suliformes (Sulidae, Anhingidae, and Phalacrocoracidae) the palatine bones of plotopterids do not form an essentially flat horizontal platform and the nostrils are not greatly reduced (Fig. 7.14a, b; Mayr et al. 2015). These plesiomorphic traits suggest that the large patella evolved convergently in plotopterids, cormorants, and anhingas.

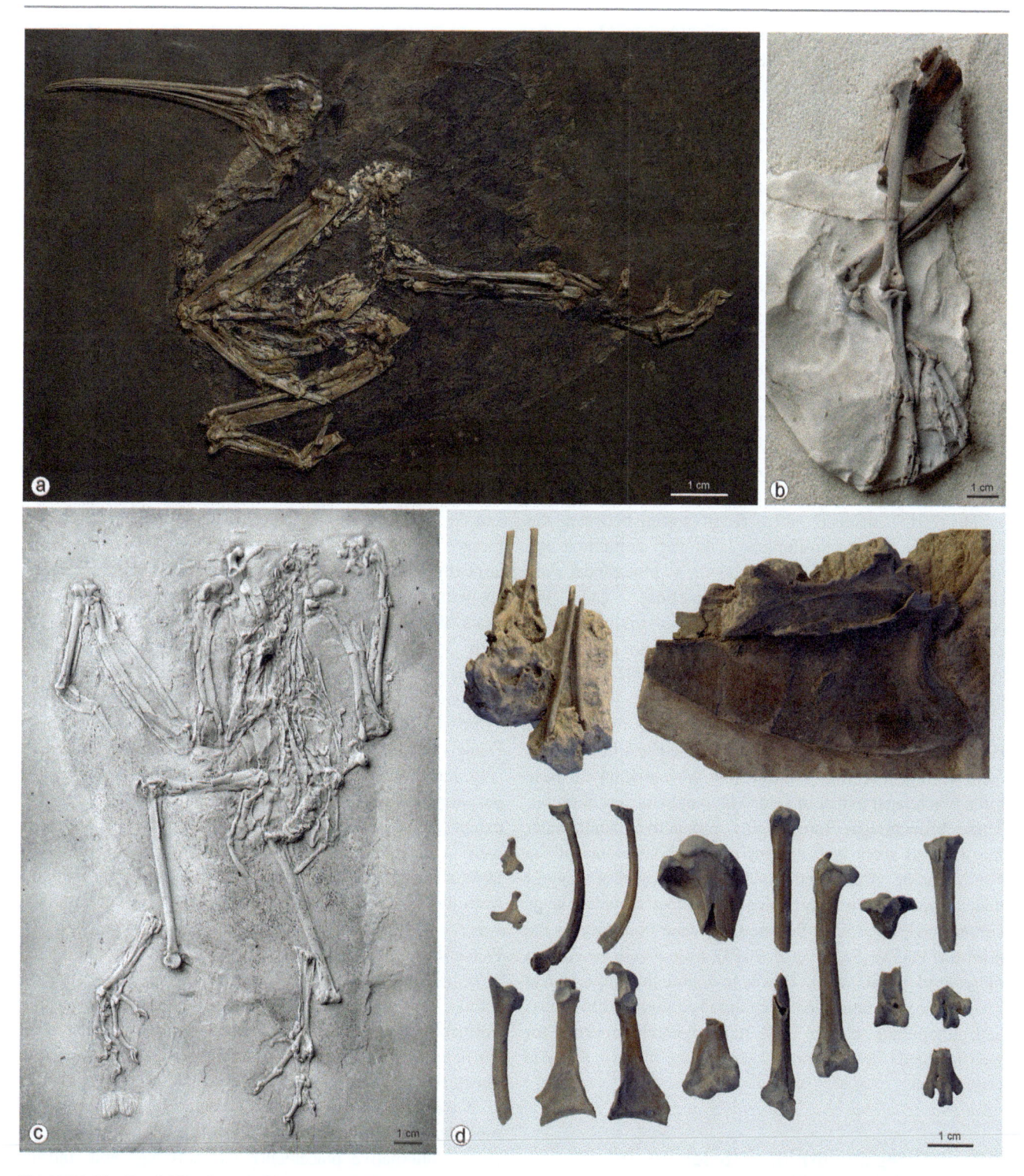

Fig. 7.15 Fossils of *Rhynchaeites* (Threskiornithidae). (**a**) Skeleton of *Rhynchaeites messelensis* from the latest early or earliest middle Eocene of Messel (Staatliches Museum für Naturkunde Karlsruhe, SMNK-PAL 725). (**b**) Legs of *Rhynchaeites* sp. from the early Eocene Fur Formation in Denmark (Geological Museum of the University of Copenhagen, Denmark, MGUH 20288); specimen coated with ammonium chloride (**c**) Postcranial skeleton of *Rhynchaeites messelensis* from Messel (Senckenberg Research Institute, Frankfurt, SMF-ME 3577). (**d**) Selected skeletal elements of a skeleton of *Rhynchaeites* sp. from the early Eocene London Clay of Walton-on-the-Naze (collection of Michael Daniels, Holland-on-Sea, UK, WN 91677); the fossil includes a partial skull (upper left), the sternum (upper right) as well as various limb bones. (All photos by Sven Tränkner)

The similarities between the Plotopteridae and sphenisciform birds were attributed to convergence by Olson (1980) and Olson and Hasegawa (1979, 1996), who considered plotopterids to be most closely related to the suliform Phalacrocoracidae and Anhingidae. Mayr (2005), by contrast, proposed a sister group relationship between the Plotopteridae and Sphenisciformes and hypothesized that paedomorphosis (the retention of juvenile characters in adulthood) may account for the absence of derived characters in penguins, which are shared by plotopterids and suliform birds. In 2005, the higher-level affinities of penguins were poorly resolved, but subsequent molecular analyses congruently refused close affinities to the Suliformes. A more recent analysis of morphological characters by Smith (2010) recovered a sister group relationship between the Plotopteridae and the clade (Anhingidae + Phalacrocoracidae), but a reanalysis of the revised data by Mayr et al. (2015) supported a sister group relationship between plotopterids and sphenisciforms. Analysis of another data set by Mayr et al. (2015) resulted in a sister group between the Plotopteridae and the Suliformes. If the very cormorant-like taxon *Stemec suntokum* is indeed a plotopterid, the similarities between the Plotopteridae and the Sphenisciformes are clearly the result of a striking convergent evolution, and it is to be hoped that more data on the skeletal anatomy of this species will be available in the near future.

The geographic occurrence of plotopterids in the North Pacific and their diversification toward the latest Eocene and earliest Oligocene may indicate that the radiation of these birds was related to that of kelp (Mayr and Goedert in press). These brown algae form highly productive "underwater forests" and are a key element in nearshore biotas of the North Pacific. Calibrated molecular phylogenies suggest that the radiation of morphologically more complex kelps commenced toward the Eocene-Oligocene boundary in the northeast Pacific (Starko et al. 2019), from where the earliest plotopterid fossils stem. However, although temporal and geographic correlations between the evolution of kelp forests and plotopterids seem to exist, more research is needed for well-founded evolutionary scenarios of a possible coevolution.

7.7 Threskiornithidae (Ibises and Spoonbills)

The earliest species that was associated with the Threskiornithidae is *Dakotornis cooperi* from the late Paleocene of North Dakota (Erickson 1975). The holotype of this species is a humerus, which shows some overall resemblance to that of ibises. *D. cooperi* was assigned to the putatively charadriiform "Graculavidae" by Benson (1999), but its actual affinities cannot be resolved without further fossil material.

Much better represented is *Rhynchaeites messelensis* from Messel, of which more than a dozen partial and complete skeletons were found (Fig. 7.15a, c). *R. messelensis* was originally considered closely related to the charadriiform Rostratulidae (painted snipes) by Wittich (1898), and the species was also misidentified as a charadriiform bird by Hoch (1980). Its threskiornithid affinities were recognized by Peters (1983), and *R. messelensis* was classified in the monotypic taxon Rhynchaeitinae by Mayr (2002b). Derived similarities shared by *R. messelensis* and extant Threskiornithidae include a long, decurved, and schizorhinal beak, as well as a notarium consisting of at least three co-ossified thoracic vertebrae (Peters 1983). In many other aspects of its skeletal morphology, *R. messelensis* is, however, very different from extant ibises, which is most evident in the much shorter legs. The tip of the upper beak lacks openings for sensory nerves, which are characteristic for extant ibises, and the Eocene taxon therefore may have been a less tactile forager (Mayr 2002b). The sternum is proportionally much larger than that of extant Threskiornithidae and is very different in its morphology: whereas the caudal margin of this bone bears two pairs of rather shallow incisions in extant ibises, there is just a single pair of very deep incisions in *Rhynchaeites* (Mayr 2002b). The coracoid exhibits a plesiomorphic, cup-like articular facet for the scapula.

Fossils of *Rhynchaeites* were also collected by Michael Daniels in the London Clay of Walton-on-the-Naze (Fig. 7.15d). These as-yet undescribed specimens correspond very well with the Messel fossils and show that, in contrast to extant Threskiornithidae, the maxillary bones of the beak were widely separated in *Rhynchaeites* and the coracoid lacks a foramen for the supracoracoideus nerve. A humerus from the early Eocene of the Danish Fur Formation, which was described as *Mopsitta tanta* and assigned to the Psittaciformes by Waterhouse et al. (2008), was considered to be similar to that of *Rhynchaeites* by Mayr and Bertelli (2011), who reported a *Rhynchaeites*-like leg from the Fur Formation (Fig. 7.15b). A largely complete skeleton of *Rhynchaeites* was furthermore found in the North American Green River Formation (Gary McFadden, pers. comm).

If *Rhynchaeites* is correctly identified as a stem group representative of the Threskiornithidae (and there is no compelling alternative placement), its short legs and unusual morphological features are likely to represent plesiomorphic traits. In traditional classifications, ibises were united with other long-legged "ciconiiform" birds, such as herons and storks, which would complicate an evolutionary interpretation of the morphology and limb bone proportions of *Rhynchaeites*. However, current sequence-based analyses recover ibises in a clade together with pelecaniform birds, and if *Rhynchaeites* is compared with the Pelecanidae

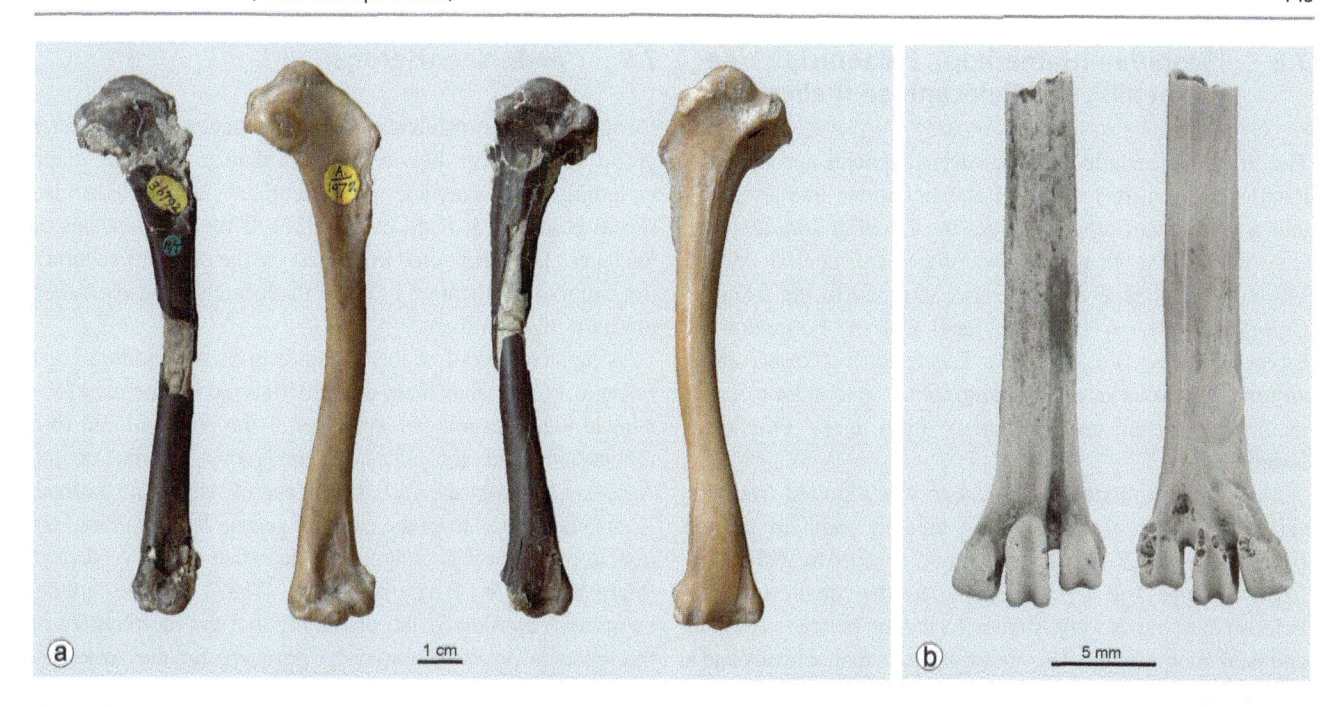

Fig. 7.16 (a) Humerus referred to *Actiornis anglicus* (Threskiornithidae) from the late Eocene of Hampshire in England (Natural History Museum, London, NHMUK A 36792) in comparison to the humerus of *Threskiornis aethiopicus* from the Pleistocene of Madagascar (NHMUK A 1972); cranial view and caudal view (*A. anglicus* on the left). (b) Distal tarsometatarsus of ?*Proardea deschutteri* (Ardeidae) from the earliest Oligocene of Belgium, Royal Belgian Institute of Natural Sciences, Brussels, Belgium, IRSNB Av 129); fossil coated with ammonium chloride. (Photo in (a) by Sven Tränkner, in (b) by Thierry Smith)

(pelicans), some of its distinct morphological features are less unusual (e.g., the short tarsometatarsus and the deep scapular cotyla of the coracoid) and may reflect the morphology of the stem species of the clade including ibises and pelicans.

Fossils of more modern-type Threskiornithidae are known from the late Eocene onward. *Actiornis anglicus* from the late Eocene (MP 17; Mlíkovský 2002) of England was assigned to the Threskiornithidae by Harrison and Walker (1976b), and a humerus referred to this species closely matches the corresponding bone of extant ibises (Fig. 7.16a). A postcranial skeleton of a representative of the Threskiornithidae was described by Roux (2002) from the early Oligocene of Céreste in France. This unnamed species also exhibits a proportionally shorter tarsometatarsus than extant ibises. A partial carpometacarpus from the early Oligocene Boom Formation in Belgium was tentatively assigned to the Threskiornithidae by Mayr and Smith (2013).

Outside Europe, the Paleogene fossil record of the Threskiornithidae is very scant. A threskiornithid-like bird from the Green River Formation was described as *Vadaravis brownae* on the basis of a postcranial skeleton (Smith et al. 2013). The species resulted as the sister taxon the Threskiornithidae in a phylogenetic analysis constrained to a molecular scaffold. The actual morphological character support for this placement is, however, restricted to the presence of a four-notched sternum. As noted above, this morphology is not present in *Rhynchaeites*, which was not included in the analysis by Smith et al. (2013). Judging from the distinctive pelvis morphology, *Vadaravis* appears to be a member of the Aequornithes, but its exact position within the clade is difficult to determine on the basis of the available fossil material. Fragmentary remains of an unnamed threskiornithid-like bird were also reported from the early Eocene Nanjemoy Formation in Virginia, USA (Mayr 2016). Their identification was considered tentative and needs to be revisited once more material becomes available.

Based on a distal tibiotarsus and a referred distal ulna, Hou (1982) described *Minggangia changgouensis* from the late Eocene of China as a member of the Threskiornithidae. Without further material, the exact affinities of this fossil remain indeterminable. Stidham et al. (2005) furthermore reported a distal tibiotarsus of a small ibis-like bird from the middle Eocene of Myanmar. The authors themselves noted, however, that an unambiguous identification of this bone is not possible and that there is a possibility that it belongs to the Geranoididae or Eogruidae (Sects. 3.2.2 and 3.2.3). Here it is noted that the specimen likewise shows a resemblance to the distal tibiotarsus of the Palaeotididae (Sect. 3.2.1). Jadwiszczak et al. (2008) identified a putative partial beak of an ibis-like bird from the late Eocene of Seymour Island (Antarctica), but this fossil was subsequently identified as the dorsal spine of a chondrichthyan fish (Agnolin et al. 2019).

7.8 Scopidae (Hamerkop), Balaenicipitidae (Shoebill), and Pelecanidae (Pelicans)

The Scopidae include a single African species and have no Paleogene fossil record. The Balaenicipitidae likewise contain a sole extant representative, the Shoebill (*Balaeniceps rex*), which lives in remote swamps of east-central Africa. The only Paleogene fossil species assigned to the taxon is *Goliathia andrewsi* from the Jebel Qatrani Formation of Egypt (Rasmussen et al. 1987). This species is known from an ulna, which lacks exact stratigraphic data, and a referred incomplete distal tarsometatarsus from early Oligocene sediments.

A skull of a modern-type pelican was reported from the early Oligocene of the Luberon area in southern France (Louchart et al. 2011). The unnamed species to which this fossil belongs had a similar size to the smallest extant Pelecanidae and is very similar to extant pelicans in skull and beak morphology. This fossil suggests that pelicans had a long evolutionary history, which dates back well before the early Oligocene.

The fossil from the Luberon area may also lend further support to pelecanid affinities of *Protopelicanus cuvierii* from the late Eocene Paris Gypsum, which is just represented by a femur. Brunet (1970) considered this species to be a pelican, whereas it was classified into the Sulidae by Harrison (1979) and compared with the Pelagornithidae by Olson (1985).

Extant Pelecanidae occur in all continents except Antarctica, but their position within a clade that otherwise only includes the Balaenicipitidae and Scopidae may indicate an origin in Africa. Such an assumption would explain the fact that pelicans occur in essentially their modern form in the early Oligocene of Europe.

Most recently, the hypothesis of an African origin of the Pelecanidae gained further support from the description of a pelican from the late Eocene (early Priabonian, ~36 Ma) of Egypt. The holotype of *Eopelecanus aegyptiacus* is a nearly complete tibiotarsus from marine strata (El Adli et al. 2021). The species was about the size of the smallest extant pelican species, *Pelecanus occidentalis*, and was considered to be too large to be conspecific with the roughly coeval *Protopelicanus cuvierii*, the pelecanid affinities of which were contested by El Adli et al. (2021).

A species from the late Oligocene or early Miocene of the Namba Formation in Australia, *Pelecanus tirarensis*, is represented by fragmentary tarsometatarsi (Miller 1966). These fossils document that pelicans already achieved a wide distribution by that time, even though their first occurrence in the New World may not have been before the Neogene.

7.9 Ardeidae (Herons)

Current analyses of nuclear gene sequences support a sister group relationship between the Ardeidae and the clade including the Balaenicipitidae, Scopidae, and Pelecanidae (Prum et al. 2015; Kuhl et al. 2021). If the stem species of the latter clade originated in Africa (see the previous section), the initial diversification of the Ardeidae may have also taken place on this continent.

A putative record of the Ardeidae from the middle to late Eocene of Chile (Sallaberry et al. 2010) is too fragmentary for a reliable identification (the specimen is the proximal end of a tibiotarsus), and the oldest unambiguous records of the Ardeidae are from the early Oligocene of Africa and Europe. A tarsometatarsus from the early Oligocene Jebel Qatrani Formation of Egypt was assigned to the extant taxon *Nycticorax* (night herons) by Rasmussen et al. (1987), but the authors themselves considered the possibility that the morphology of this bone in *Nycticorax* may be primitive for the Ardeidae. Rasmussen et al. (1987) furthermore described a second, slightly larger, and also unnamed species of the Ardeidae from the Jebel Qatrani Formation, which is represented by an incomplete rostrum and a few referred bones.

Another early Oligocene record of the Ardeidae was described as ?*Proardea deschutteri* on the basis of a partial tarsometatarsus from the early Oligocene of Hoogbutsel in Belgium (Fig. 7.16b; Mayr et al. 2019b). A very similar but slightly larger species, *Proardea amissa* from an unknown locality and stratigraphic horizon of the Quercy fissure fillings, is likewise represented by an incomplete tarsometatarsus (Milne-Edwards 1892). As yet undescribed remains of this latter species were found in the late Oligocene (MP 28) Quercy locality Pech Desse (C. Mourer-Chauviré, personal communication, and own observation).

Mlíkovský and Švec (1989) noted that "*Anas*" *basaltica* from the "middle" Oligocene of the Czech Republic is a representative of the Ardeidae. Kurochkin (1976) furthermore mentioned the occurrence of as-yet undescribed Ardeidae in the early Oligocene of Mongolia.

Olson (1985) found the holotypic distal humerus of *Gnotornis aramiellus* from the early Oligocene (Whitneyan) of South Dakota to be very similar to that of herons. If further material confirms ardeid affinities of this species, it would constitute the earliest New World fossil record of the Ardeidae (disregarding the above-mentioned specimen from the Eocene of Chile).

7.10 Xenerodiopidae

This taxon includes a single species, *Xenerodiops mycter* from the early Oligocene of Egypt, which was established on the basis of a rostrum and a tentatively referred humerus

(Rasmussen et al. 1987). Whether these bones indeed belong to the same taxon cannot be definitely shown, but as noted by Rasmussen et al. (1987) there is a possibility that they do. *X. mycter* was slightly smaller than the extant Abdim's Stork, *Ciconia abdimii*. The humerus shows some overall similarity to that of extant Ardeidae, but the pneumotricipital fossa lacks pneumatic openings (Rasmussen et al. 1987). The rostrum exhibits weak lateral furrows, which occur in many of the taxa discussed in the present chapter, as well as a completely ossified ventral surface. Humerus morphology also suggests that *X. mycter* is probably most closely related to the birds united here, but its phylogenetic affinities are best considered unresolved. Undoubtedly, however, the species represents a distinctive taxon, and it is to be hoped that future specimens shed more light on its relationships.

References

Acosta Hospitaleche C (2006) Taxonomic longevity in penguins (Aves, Spheniscidae). Neues Jahrb Geol Palaeontol Abh 241:383–403

Acosta Hospitaleche C (2014) New giant penguin bones from Antarctica: systematic and paleobiological significance. C R Palevol 13:555–560

Acosta Hospitaleche C, Gelfo JN (2015) New Antarctic findings of upper Cretaceous and lower Eocene loons (Aves: Gaviiformes). Ann Paléontol 101:315–324

Acosta Hospitaleche C, Gelfo JN (2017) Procellariiform remains and a new species from the latest Eocene of Antarctica. Hist Biol 29:755–769

Acosta Hospitaleche C, Haidr N (2011) Penguin cranial remains from the Eocene La Meseta Formation, Isla Marambio (Seymour Island), Antarctic Peninsula. Antarct Sci 23:369–378

Acosta Hospitaleche C, Olivero E (2016) Re-evaluation of the fossil penguin *Palaeeudyptes gunnari* from the Eocene Leticia Formation, Argentina: additional material, systematics and palaeobiology. Alcheringa 40:373–382

Acosta Hospitaleche C, Reguero MA (2010) First articulated skeleton of *Palaeeudyptes gunnari* from the late Eocene of Isla Marambio (Seymour Island), Antarctica. Antarct Sci 22:289–298

Acosta Hospitaleche C, Reguero M (2014) *Palaeeudyptes klekowskii*, the best-preserved penguin skeleton from the Eocene-Oligocene of Antarctica: Taxonomic and evolutionary remarks. Geobios 47:77–85.

Acosta Hospitaleche C, Tambussi C, Cozzuol M (2004) *Eretiscus tonnii* (Simpson) (Aves, Sphenisciformes): materiales adicionales, status taxonómico y distribución geográfica. Rev Mus Argent Cienc Nat, n s 6:233–237.

Acosta Hospitaleche C, Reguero M, Santillana S (2017) *Aprosdokitos mikrotero* gen. et sp. nov., the tiniest Sphenisciformes that lived in Antarctica during the Paleogene. N Jb Geol Paläont (Abh) 283:25–34

Acosta Hospitaleche C, Haidr N, Paulina-Carabajal A, Reguero M (2019a) The first skull of *Anthropornis grandis* (Aves, Sphenisciformes) associated with postcranial elements. C R Palevol 18:599–617

Acosta Hospitaleche C, Jadwiszczak P, Clarke JA, Cenizo M (2019b) The fossil record of birds from the James Ross Basin, West Antarctica. Adv Polar Sci 30:251–273

Acosta Hospitaleche C, De Los RM, Santillana S, Reguero M (2020) First fossilized skin of a giant penguin from the Eocene of Antarctica. Lethaia 53:409–420

Agnolin FL (2007) *Argyrodyptes microtarsus* Ameghino, 1905: un petrel (Procellariiformes) del Eoceno-Oligoceno de Argentina. Studia Geol Salmanticensia 43:207–213

Agnolín FL, Egli FB, Chatterjee S, Marsà JAG, Novas FE (2017) Vegaviidae, a new clade of southern diving birds that survived the K/T boundary. Sci Nat 104:87

Agnolín FL, Bogan S, Rozadilla S (2019) Were ibises (Aves, Threskiornithidae) present in Antarctica? Antarct Sci 31:35–36

Ando T, Fukata K (2018) A well-preserved partial scapula from Japan and the reconstruction of the triosseal canal of plotopterids. PeerJ 6: e5391

Andrews CW (1899) On the remains of a new bird from the London Clay of Sheppey. Proc Zool Soc London 1899:776–785

Averianov AO, Potapova OR, Nessov LA (1990) [On original native bone finds of ancient birds]. Proc Zool Inst Leningrad 210:3–9 [in Russian]

Baker AJ, Pereira SL, Haddrath OP, Edge K-E (2006) Multiple gene evidence for expansion of extant penguins out of Antarctica due to global cooling. Proc R Soc Lond Ser B 273:11–17

Benson RD (1999) *Presbyornis isoni* and other late Paleocene birds from North Dakota. In: Olson SL (ed) Avian paleontology at the close of the 20th century: Proceedings of the 4th international meeting of the society of avian paleontology and evolution, Washington, DC, 4–7 June 1996. Smithson Contrib Paleobiol 89:253–259

Blokland JC, Reid CM, Worthy TH, Tennyson AJ, Clarke JA, Scofield RP (2019) Chatham Island Paleocene fossils provide insight into the palaeobiology, evolution, and diversity of early penguins (Aves, Sphenisciformes). Palaeontol Electron 22.3.78:22:1–92

Boles WE (2005) A review of the Australian fossil storks of the genus *Ciconia* (Aves: Ciconiidae), with the description of a new species. Rec Austral Mus 57:165–178

Bourdon E, Bouya B, Iarochène M (2005) Earliest African neornithine bird: a new species of Prophaethontidae (Aves) from the Paleocene of Morocco. J Vertebr Paleontol 25:157–170

Bourdon E, Mourer-Chauviré C, Amaghzaz M, Bouya B (2008a) New specimens of *Lithoptila abdounensis* (Aves, Prophaethontidae) from the Lower Paleogene of Morocco. J Vertebr Paleontol 28:751–761

Bourdon E, Amaghzaz M, Bouya B (2008b) A new seabird (Aves, cf. Phaethontidae) from the Lower Eocene phosphates of Morocco. Geobios 41:455–459

Brodkorb P (1962) The systematic position of two Oligocene birds from Belgium. Auk 79:706–707

Brunet J (1970) Oiseaux de l'Éocène supérieur du bassin de Paris. Ann Paléontol 56:3–57

Chávez-Hoffmeister MC (2014) Phylogenetic characters in the humerus and tarsometatarsus of penguins. Pol Polar Res 35:469–496

Cheneval J (1995) A fossil shearwater (Aves: Procellariiformes) from the Upper Oligocene of France and the Lower Miocene of Germany. In: Peters DS (ed) Acta palaeornithologica. Cour Forsch-Inst Senckenberg 181:187–198

Clarke JA, Olivero EB, Puerta P (2003) Description of the earliest fossil penguin from South America and first Paleogene vertebrate locality of Tierra Del Fuego, Argentina. Am Mus Novit 3423:1–18

Clarke JA, Ksepka DT, Stucchi M, Urbina M, Giannini N, Bertelli S, Narváez Y, Boyd CA (2007) Paleogene equatorial penguins challenge the proposed relationship between penguin biogeography, diversity, and Cenozoic climate change. Proc Natl Acad Sci USA 104:11545–11550

Clarke JA, Ksepka DT, Salas-Gismondi R, Altamirano AJ, Shawkey MD, D'Alba L, Vinther J, DeVries TJ, Baby P (2010) Fossil evidence for evolution of the shape and color of penguin feathers. Science 330:954–957

Darga R, Böhme M, Göhlich UB, Rössner G (1999) Reste höherer Wirbeltiere aus dem Alttertiär des Alpenvorlandes bei Siegsdorf/Oberbayern. Mitt Bayer Staatssamml Paläontol Hist Geol 39:91–114

De Pietri VL, Berger JP, Pirkenseer C, Scherler L, Mayr G (2010) New skeleton from the early Oligocene of Germany indicates a stem-group position of diomedeoidid birds. Acta Palaeontol Pol 55:23–34

Dyke GJ, Wang X, Habib MB (2011) Fossil plotopterid seabirds from the Eo-Oligocene of the Olympic Peninsula (Washington State, USA): descriptions and functional morphology. PLoS One 6(10): e25672

El Adli JJ, Wilson Mantilla JA, Antar MSM, Gingerich PD (2021, in press) The earliest recorded fossil pelican, recovered from the late Eocene of Wadi Al-Hitan, Egypt. J Vertebr Paleontol:e1903910

Elzanowski A, Bieńkowska-Wasiluk M, Chodyń R, Bogdanowicz W (2012) Anatomy of the coracoid and diversity of the Procellariiformes (Aves) in the Oligocene of Europe. Palaeontology 55:1199–1221

Erickson BR (1975) *Dakotornis cooperi*, a new Paleocene bird from North Dakota. Sci Pub Sci Mus Minnesota 3:1–7

Ericson PGP, Anderson CL, Britton T, Elzanowski A, Johansson US, Källersjö M, Ohlson JI, Parsons TJ, Zuccon D, Mayr G (2006) Diversification of Neoaves: integration of molecular sequence data and fossils. Biol Lett 2:543–547

Feduccia A, McPherson AB (1993) A petrel-like bird from the late Eocene of Louisiana: Earliest record for the order Procellariiformes. Proc Biol Soc Wash 106:749–751

Fischer K (1983) Möwenreste (Laridae, Charadriiformes, Aves) aus dem mitteloligozänen Phosphoritknollenhorizont des Weisselsterbeckens bei Leipzig (DDR). Mitt Zool Mus Berlin 59, Suppl: Ann Ornithol 7:151–155

Fischer K (1985) Ein albatrosartiger Vogel (*Diomedeoides minimus* nov. gen., nov. sp., Diomedeoididae nov. fam., Procellariiformes) aus dem Mitteloligozän bei Leipzig (DDR). Mitt Zool Mus Berlin 61, Suppl: Ann Ornithol 9:113–118

Fischer K (1997) Neue Vogelfunde aus dem mittleren Oligozän des Weißelsterbeckens bei Leipzig (Sachsen). Mauritiana 16:271–288

Fischer K (2003) Weitere Vogelknochen von *Diomedeoides* (Diomedeoididae, Procellariiformes) und *Paraortyx* (Paraortygidae, Galliformes) aus dem Unteroligozän des Weißelsterbeckens bei Leipzig (Sachsen). Mauritiana 18:387–395

Fordyce RE, Thomas D (2011) *Kaiika maxwelli*, a new Early Eocene archaic penguin (Sphenisciformes, Aves) from Waihao Valley, South Canterbury, New Zealand. New Zealand J Geol Geophys 54:43–51

Giovanardi S, Ksepka DT, Thomas DB (2021) A giant Oligocene fossil penguin from the North Island of New Zealand. J Vertebr Paleontol: e1953047

Goedert JL (1988) A new late Eocene species of Plotopteridae (Aves: Pelecaniformes) from northwestern Oregon. Proc Calif Acad Sci 45:97–102

Goedert JL, Cornish J (2002) A preliminary report on the diversity and stratigraphic distribution of the Plotopteridae (Pelecaniformes) in Paleogene rocks of Washington State, USA. In: Zhou Z, Zhang F (eds) Proceedings of the 5th symposium of the Society of Avian Paleontology and Evolution, Beijing, 1–4 June 2000. Science Press, Beijing, pp 63–76

Grande L (2013) The lost world of Fossil Lake: Snapshots from deep time. University of Chicago Press, Chicago

Gregorová R (2006) A new discovery of a seabird (Aves: Procellariiformes) in the Oligocene of the "Menilitic Formation" in Moravia (Czech Republic). Hantkeniana 5:90

Hackett SJ, Kimball RT, Reddy S, Bowie RCK, Braun EL, Braun MJ, Chojnowski JL, Cox WA, Han K-L, Harshman J, Huddleston CJ, Marks BD, Miglia KJ, Moore WS, Sheldon FH, Steadman DW, Witt CC, Yuri T (2008) A phylogenomic study of birds reveals their evolutionary history. Science 320:1763–1767

Haidr N, Acosta Hospitaleche C (2012) Feeding habits of Antarctic Eocene penguins from a morphofunctional perspective. N Jb Geol Paläont (Abh) 263:125–131

Haidr N, Acosta Hospitaleche C (2017) A new penguin cranium from Antarctica and its implications for body size diversity during the Eocene. N Jb Geol Paläont (Abh) 286:229–233

Harrison CJO (1975) The taxonomic status of Milne-Edward's [sic] fossil sulids. Bull Br Ornithol Club 95:51–54

Harrison CJO (1976) The wing proportions of the Eocene diver *Colymboides anglicus*. Bull Br Ornithol Club 96:64–65

Harrison CJO (1979) The Upper Eocene birds of the Paris basin: a brief re-appraisal. Tertiary Res 2:105–109

Harrison CJO, Walker CA (1971) A new ibis from the Lower Eocene of Britain. Ibis 113:367–368

Harrison CJO, Walker CA (1976a) A reappraisal of *Prophaethon shrubsolei* Andrews (Aves). Bull Brit Mus (Nat Hist) 27:1–30

Harrison CJO, Walker CA (1976b) Birds of the British Upper Eocene. Zool J Linnean Soc 59:323–351

Harrison CJO, Walker CA (1977) Birds of the British Lower Eocene. Tert Res Spec Pap 3:1–52

Hasegawa Y, Isotani S, Nagai K, Seki K, Suzuki T, Otsuka H, Ota M, Ono K (1979) [Preliminary notes on the Oligo-Miocene penguin-like birds from Japan (Parts I-VII)]. Bull Kitakyushu Mus Nat Hist 1:41–60 [In Japanese]

Hoch E (1980) A new Middle Eocene shorebird (Aves: Charadriiformes, Charadrii) with columboid features. Nat Hist Mus Los Angeles Cty, Contrib Sci 330:33–49

Hou L-H (1982) [New form of the Threskiornithidae from the Upper Eocene of the Minggang, Henan]. Vertebr PalAsiat 20:196–202 [in Chinese]

Hou L-H (1989) [A middle Eocene bird from Sangequan, 'Xinjiang]. Vertebr PalAsiat 27:65–70 [in Chinese]

Hou L-H (2003) Fossil birds of China. Yunnan Science and Technology Press, Kunming

Howard H (1969) A new avian fossil from Kern County, California. Condor 71:68–69

Jadwiszczak P (2001) Body size of Eocene Antarctic penguins. Pol Polar Res 22:147–158

Jadwiszczak P (2006a) Eocene penguins of Seymour Island, Antarctica: Taxonomy. Pol Polar Res 27:3–62

Jadwiszczak P (2006b) Eocene penguins of Seymour Island, Antarctica: The earliest record, taxonomic problems and some evolutionary considerations. Pol Polar Res 27:287–302

Jadwiszczak P (2008) An intriguing penguin bone from the Late Eocene of Seymour Island, Antarctic Peninsula. Antarct Sci 20:589

Jadwiszczak P (2009) Penguin past: The current state of knowledge. Pol Polar Res 30:3–28

Jadwiszczak P (2012) Partial limb skeleton of a "giant penguin" *Anthropornis* from the Eocene of Antarctic Peninsula. Pol Polar Res 33:259–274

Jadwiszczak P (2013) Taxonomic diversity of Eocene Antarctic penguins: a changing picture. Geol Soc Lond Spec Publ 381:129–138

Jadwiszczak P, Chapman SD (2011) The earliest fossil record of a medium-sized penguin. Polish Pol Res 32:269–277

Jadwiszczak P, Mörs T (2017) An enigmatic fossil penguin from the Eocene of Antarctica. Polar Res 36:1291086

Jadwiszczak P, Mörs T (2019) First partial skeleton of *Delphinornis larseni* Wiman, 1905, a slender-footed penguin from the Eocene of Antarctic Peninsula. Palaeontol Electron 22.2(32A):1–31

Jadwiszczak P, Gaździcki A, Tatur A (2008) An ibis-like bird from the Upper La Meseta Formation (Late Eocene) of Seymour Island, Antarctica. Antarct Sci 20:413–414

Jadwiszczak P, Acosta Hospitaleche C, Reguero M (2013) Redescription of *Crossvallia unienviella*: The only Paleocene Antarctic penguin. Ameghiniana 50:545–553

Jadwiszczak P, Reguero M, Mörs T (2021) A new small-sized penguin from the late Eocene of Seymour Island with additional material of *Mesetaornis polaris*. GFF 143:283–291

Jarvis ED, Mirarab S, Aberer AJ, Li B, Houde P, Li C, Ho SYW, Faircloth BC, Nabholz B, Howard JT, Suh A, Weber CC, da Fonseca RR, Li J, Zhang F, Li H, Zhou L, Narula N, Liu L, Ganapathy G, Boussau B, Bayzid MS, Zavidovych V, Subramanian S, Gabaldón T, Capella-Gutiérrez S, Huerta-Cepas J, Rekepalli B, Munch K, Schierup M. et al. (75 further co-authors) (2014) Whole-genome analyses resolve early branches in the tree of life of modern birds. Science 346:1320–1331

Kaiser G, Watanabe J, Johns M (2015) A new member of the family Plotopteridae (Aves) from the late Oligocene of British Columbia, Canada. Palaeontol Electron 18.3(52A):1–18

Kawabe S, Ando T, Endo H (2014) Enigmatic affinity in the brain morphology between plotopterids and penguins, with a comprehensive comparison among water birds. Zool J Linnean Soc 170:467–493

Kawabe S, Ando Y, Kawano S, Matsui K (2021) New record of a rostrum of waterbird (Aves, Suliformes) from the Oligocene of Ashiya Group in Ainoshima Island, Kyushu, Japan. Bull Kitakyushu Mus Nat Hist Hum Hist Ser A 19:35–39

Kessler E (2009) The oldest modern bird (Ornithurinae) remains from the Early Oligocene of Hungary. Fragm Palaeontol Hungar 27:93–96

Kimura M, Sakurai K, Katoh T (1998) An extinct fossil bird (Plotopteridae) from the Tokoro Formation (Late Oligocene) in Abashiri City, northeastern Hokkaido, Japan. J Hokkaido Univ Education (Sect IIB) 48:11–16

Ksepka DT, Ando T (2011) Penguins past, present, and future: Trends in the evolution of the Sphenisciformes. In: Dyke G, Kaiser G (eds) Living dinosaurs: The evolutionary history of modern birds. John Wiley & Sons, Chichester, pp 155–186

Ksepka DT, Bertelli S (2006) Fossil penguin (Aves: Sphenisciformes) cranial material from the Eocene of Seymour Island (Antarctica). Hist Biol 18:389–395

Ksepka DT, Clarke JA (2010) The basal penguin (Aves: Sphenisciformes) *Perudyptes devriesi* and a phylogenetic evaluation of the penguin fossil record. Bull Am Mus Nat Hist 337:1–77

Ksepka DT, Cracraft J (2008) An avian tarsometatarsus from near the K-T boundary of New Zealand. J Vertebr Paleontol 28:1224–1227

Ksepka DT, Thomas DB (2012) Multiple Cenozoic invasions of Africa by penguins (Aves, Sphenisciformes). Proc R Soc B Biol Sci 279:1027–1032

Ksepka DT, Bertelli S, Giannini NP (2006) The phylogeny of the living and fossil Sphenisciformes (penguins). Cladistics 22:412–441

Ksepka DT, Clarke JA, DeVries TJ, Urbina M (2008) Osteology of *Icadyptes salasi*, a giant penguin from the Eocene of Peru. J Anat 213:131–147

Ksepka DT, Fordyce RE, Ando T, Jones CM (2012) New fossil penguins (Aves, Sphenisciformes) from the Oligocene of New Zealand reveal the skeletal plan of stem penguins. J Vertebr Paleontol 32:235–254

Kuhl H, Frankl-Vilches C, Bakker A, Mayr G, Nikolaus G, Boerno ST, Klages S, Timmermann B, Gahr M (2021) An unbiased molecular approach using 3'UTRs resolves the avian family-level tree of life. Mol Biol Evol 38:108–121

Kurochkin EN (1976) A survey of the Paleogene birds of Asia. Smithson Contrib Paleobiol 27:5–86

Lambrecht K (1929) Mesozoische und tertiäre Vogelreste aus Siebenbürgen. In: Csiki E (ed) Xe Congrés International de Zoologie. Stephaneum, Budapest, pp 1262–1275

Lambrecht K (1930) Studien über fossile Riesenvögel. Geol Hung, Ser Palaeontol 7:1–37

Lambrecht K (1931) *Protoplotus Beauforti* n.g. n.sp., ein Schlangen-halsvogel aus dem Tertiär von W.-Sumatra. Wet Meded Dienst Mijnb Nederlandsch-Indie 17:15–24

Louchart A, Tourment N, Carrier J (2011) The earliest known pelican reveals 30 million years of evolutionary stasis in beak morphology. J Ornithol 152:15–20

Marples BJ (1946) Notes on some neognathous bird bones from the early Tertiary of New Zealand. Trans Roy Soc New Zealand 76:132–134

Mayr G (2001) A cormorant from the late Oligocene of Enspel, Germany (Aves, Pelecaniformes, Phalacrocoracidae). Senck leth 81:329–333

Mayr G (2002a) A skull of a new pelecaniform bird from the Middle Eocene of Messel, Germany. Acta Palaeontol Pol 47:507–512

Mayr G (2002b) A contribution to the osteology of the Middle Eocene ibis *Rhynchaeites messelensis* (Aves: Threskiornithidae: Rhynchaeitinae nov. subfam.). Neues Jahrb Geol Paläontol, Mh 2002:501–512

Mayr G (2003) The phylogenetic relationships of the shoebill, *Balaeniceps rex*. J Ornithol 144:157–175

Mayr G (2004) A partial skeleton of a new fossil loon (Aves, Gaviiformes) from the early Oligocene of Germany with preserved stomach content. J Ornithol 145:281–286

Mayr G (2005) Tertiary plotopterids (Aves, Plotopteridae) and a novel hypothesis on the phylogenetic relationships of penguins (Spheniscidae). J Zool Syst Evol Res 43:61–71

Mayr G (2007) A small representative of the Phalacrocoracoidea (cormorants and anhingas) from the late Oligocene of Germany. Condor 109:929–942

Mayr G (2008) Avian higher-level phylogeny: well-supported clades and what we can learn from a phylogenetic analysis of 2954 morphological characters. J Zool Syst Evol Res 46:63–72

Mayr G (2009a) Paleogene fossil birds, 1st edn. Springer, Heidelberg

Mayr G (2009b) A small loon and a new species of large owl from the Rupelian of Belgium (Aves: Gaviiformes, Strigiformes). Paläontol Z 83:247–254

Mayr G (2009c) Notes on the osteology and phylogenetic affinities of the Oligocene Diomedeoididae (Aves, Procellariiformes). Fossil Rec 12:133–140

Mayr G (2010) Mousebirds (Coliiformes), parrots (Psittaciformes), and other small birds from the late Oligocene/early Miocene of the Mainz Basin, Germany. N Jb Geol Paläont (Abh) 258:129–144

Mayr G (2011) Metaves, Mirandornithes, Strisores and other novelties – a critical review of the higher-level phylogeny of neornithine birds. J Zool Syst Evol Res 49:58–76

Mayr G (2015a) New remains of the Eocene *Prophaethon* and the early evolution of tropicbirds (Phaethontiformes). Ibis 157:54–67

Mayr G (2015b) A procellariiform bird from the early Oligocene of North America. N Jb Geol Paläont (Abh) 275:11–17

Mayr G (2015c) A new skeleton of the late Oligocene "Enspel cormorant"—from *Oligocorax* to *Borvocarbo*, and back again. Palaeobiodiv Palaeoenv 95:87–101

Mayr G (2016) The world's smallest owl, the earliest unambiguous charadriiform bird, and other avian remains from the early Eocene Nanjemoy Formation of Virginia (USA). Paläontol Z 90:747–763

Mayr G (2017) Avian Evolution: The fossil record of birds and its paleobiological significance. Wiley-Blackwell, Chichester

Mayr G, Bertelli S (2011) A record of *Rhynchaeites* (Aves, Threskiornithidae) from the early Eocene Fur Formation of Denmark, and the affinities of the alleged parrot *Mopsitta*. Palaeobiodiv Palaeoenv 91:229–236

Mayr G, Goedert JL (2016) New late Eocene and Oligocene remains of the flightless, penguin-like plotopterids (Aves, Plotopteridae) from western Washington State. USA J Vertebr Paleontol 36:e1163573

Mayr G, Goedert JL (2017) Oligocene and Miocene albatross fossils from Washington State (USA) and the evolutionary history of North Pacific Diomedeidae. Auk 134:659–671

Mayr G, Goedert JL (2018) First record of a tarsometatarsus of *Tonsala hildegardae* (Plotopteridae) and other avian remains from the late Eocene/early Oligocene of Washington State (USA). Geobios 51:51–59

Mayr G, Goedert JL (in press) New late Eocene and Oligocene plotopterid fossils from Washington State (USA), with a revision of "*Tonsala*" *buchanani* (Aves, Plotopteridae). J Paleontol

Mayr G, Poschmann M (2009) A loon leg (Aves, Gaviidae) with crocodilian tooth from the late Oligocene of Germany. Waterbirds 32:468–471

Mayr G, Scofield RP (2014) First diagnosable non-sphenisciform bird from the early Paleocene of New Zealand. J Roy Soc New Zealand 44:48–56

Mayr G, Scofield RP (2016) New avian remains from the Paleocene of New Zealand: the first early Cenozoic Phaethontiformes (tropicbirds) from the Southern Hemisphere. J Vertebr Paleontol 36:e1031343

Mayr G, Smith R (2002a) A new record of the Prophaethontidae (Aves: Pelecaniformes) from the Middle Eocene of Belgium. Bull Inst Roy Sci Nat Belg 72:135–138

Mayr G, Smith R (2002b) Avian remains from the lowermost Oligocene of Hoogbutsel (Belgium). Bull Inst Roy Sci Nat Belg 72:139–150

Mayr G, Smith T (2012a) Phylogenetic affinities and taxonomy of the Oligocene Diomedeoididae, and the basal divergences amongst extant procellariiform birds. Zool J Linnean Soc 166:854–875

Mayr G, Smith T (2012b) A fossil albatross from the early Oligocene of the North Sea Basin. Auk 129:87–95

Mayr G, Smith T (2013) Galliformes, Upupiformes, Trogoniformes, and other avian remains (?Phaethontiformes and ?Threskiornithidae) from the Rupelian stratotype in Belgium, with comments on the identity of "*Anas*" *benedeni* Sharpe, 1899. In: Göhlich UB, Kroh A (eds) Paleornithological Research 2013—Proceedings of the 8th international meeting of the Society of Avian Paleontology and Evolution. Natural History Museum Vienna, Vienna, pp 23–35

Mayr G, Smith T (2019) A diverse bird assemblage from the Ypresian of Belgium furthers knowledge of early Eocene avifaunas of the North Sea Basin. N Jb Geol Paläont (Abh) 291:253–281

Mayr G, Zvonok E (2011) Middle Eocene Pelagornithidae and Gaviiformes (Aves) from the Ukrainian Paratethys. Palaeontology 54:1347–1359

Mayr G, Zvonok E (2012) A new genus and species of Pelagornithidae with well-preserved pseudodentition and further avian remains from the middle Eocene of the Ukraine. J Vertebr Paleontol 32:914–925

Mayr G, Peters DS, Rietschel S (2002) Petrel-like birds with a peculiar foot morphology from the Oligocene of Germany and Belgium (Aves: Procellariiformes). J Vertebr Paleontol 22:667–676

Mayr G, Zvonok E, Gorobets L (2013) The tarsometatarsus of the middle Eocene loon *Colymbiculus udovichenkoi*. In: Göhlich UB, Kroh A (eds) Paleornithological Research 2013—Proceedings of the 8th international meeting of the Society of Avian Paleontology and Evolution. Natural History Museum Vienna, Vienna, pp 17–22

Mayr G, Goedert JL, Vogel O (2015) Oligocene plotopterid skulls from western North America and their bearing on the phylogenetic affinities of these penguin-like seabirds. J Vertebr Paleontol 35: e943764

Mayr G, De Pietri VL, Scofield RP (2017a) A new fossil from the mid-Paleocene of New Zealand reveals an unexpected diversity of world's oldest penguins. Sci Nat 104:9

Mayr G, Scofield RP, De Pietri VL, Tennyson AJD (2017b) A Paleocene penguin from New Zealand substantiates multiple origins of gigantism in fossil Sphenisciformes. Nature Comm 8:1927

Mayr G, Scofield P, De Pietri V, Worthy T (2018a) On the taxonomic composition and phylogenetic affinities of the recently proposed clade Vegaviidae Agnolín et al., 2017—neornithine birds from the Upper Cretaceous of the Southern Hemisphere. Cretac Res 86:178–185

Mayr G, De Pietri VL, Love L, Mannering AA, Scofield RP (2018b) A well-preserved new mid-Paleocene penguin (Aves, Sphenisciformes) from the Waipara Greensand in New Zealand. J Vertebr Paleontol 37:e139816915

Mayr G, Archibald SB, Kaiser GW, Mathewes RW (2019a) Early Eocene (Ypresian) birds from the Okanagan Highlands, British Columbia (Canada) and Washington State (USA). Can J Earth Sci 56:803–813

Mayr G, De Pietri VL, Scofield RP, Smith T (2019b) A fossil heron from the early Oligocene of Belgium—the earliest temporally well-constrained record of the Ardeidae. Ibis 161:79–90

Mayr G, De Pietri VL, Love L, Mannering AA, Bevitt JJ, Scofield RP (2020a) First complete wing of a stem group sphenisciform from the Paleocene of New Zealand sheds light on the evolution of the penguin flipper. Diversity 12(2):46

Mayr G, De Pietri VL, Love L, Mannering AA, Scofield RP (2020b) Leg bones of a new penguin species from the Waipara Greensand add to the diversity of very large-sized Sphenisciformes in the Paleocene of New Zealand. Alcheringa 44:194–201

Mayr G, Goedert JL, De Pietri V, Scofield RP (2021) Comparative osteology of the penguin-like mid-Cenozoic Plotopteridae and the earliest true fossil penguins, with comments on the origins of wing-propelled diving. J Zool Syst Evol Res 59:264–276

Mayr G, De Pietri V, Scofield RP (2022) New bird remains from the early Eocene Nanjemoy Formation of Virginia (USA), including the first records of the Messelasturidae, Psittacopedidae, and Zygodactylidae from the Fisher/Sullivan site. Hist Biol 34:322–334

McLachlan SM, Kaiser GW, Longrich NR (2017) *Maaqwi cascadensis*: A large, marine diving bird (Avialae: Ornithurae) from the Upper Cretaceous of British Columbia, Canada. PLoS One 12:e0189473

Miller AH (1966) The fossil pelicans of Australia. Mem Queensl Mus 14:181–190

Miller AH, Sibley CG (1941) A Miocene gull from Nebraska. Auk 58: 563–566

Miller ER, Rasmussen DT, Simons EL (1997) Fossil storks (Ciconiidae) from the Late Eocene and Early Miocene of Egypt. Ostrich 68:23–26

Milne-Edwards A (1867–1871) Recherches anatomiques et paléonto-logiques pour servir à l'histoire des oiseaux fossiles de la France. Victor Masson et fils, Paris

Milne-Edwards A (1892) Sur les oiseaux fossiles des dépots éocènes de phosphate de chaux du Sud de la France. C R Second Congr Ornithol Internat:60–80

Milner AC, Walsh SA (2009) Avian brain evolution: new data from Palaeogene birds (Lower Eocene) from England. Zool J Linnean Soc 155:198–219

Mlíkovský J (2002) Cenozoic birds of the world. Part 1: Europe. Ninox Press, Praha

Mlíkovský J (2007) Taxonomic identity of *Eostega lebedinskyi* Lambrecht, 1929 (Aves) from the middle Eocene of Romania. Ann Naturhist Mus Wien, Ser A 109:19–27

Mlíkovský J, Švec P (1986) Review of the Tertiary waterfowl (Aves: Anseridae) of Asia. Věstn Českoslov Spol Zool 50:259–272

Mlíkovský J, Švec P (1989) Review of the Tertiary waterfowl (Aves: Anseridae) of Czechoslovakia. Čas Mineral Geol 34:199–203

Mori H, Miyata K (2021) Early Plotopteridae specimens (Aves) from the Itanoura and Kakinoura Formations (latest Eocene to early Oligocene), Saikai, Nagasaki Prefecture, western Japan. Paleontol Res 25:145–159

Mourer-Chauviré C (1995) Dynamics of the avifauna during the Paleogene and the Early Neogene of France. Settling of the recent fauna. Acta Zool Cracov 38:325–342

Mourer-Chauviré C (2006) The avifauna of the Eocene and Oligocene Phosphorites du Quercy (France): an updated list. Strata, sér 1 13: 135–149

Mourer-Chauviré C, Berthet D, Hugueney M (2004) The late Oligocene birds of the Créchy quarry (Allier, France), with a description of two new genera (Aves: Pelecaniformes: Phalacrocoracidae, and Anseriformes: Anseranatidae). Senck Leth 84:303–315

Myrcha A, Jadwiszczak P, Tambussi CP, Noriega JI, Gaździcki A, Tatur A, del Valle RA (2002) Taxonomic revision of Eocene

Antarctic penguins based on tarsometatarsal morphology. Pol Polar Res 23:5–46

Nessov LA (1988) [New Cretaceous and Paleocene birds of Soviet Middle Asia and Kazakhstan and their environments]. Proc Zool Inst, Leningrad 182:116–123 [in Russian]

Nessov LA (1992) Mesozoic and Paleogene birds of the USSR and their paleoenvironments. In: Campbell KE (ed) Papers in avian paleontology honoring Pierce Brodkorb. Nat Hist Mus Los Angeles Cty Sci Ser 36:465–478

Ohashi T, Hasegawa Y (2020) New Species of Plotopteridae (Aves) from the Oligocene Ashiya Group of Northern Kyushu, Japan. Paleontol Res 24:285–297

Olson SL (1977) A Lower Eocene frigatebird from the Green River Formation of Wyoming (Pelecaniformes: Fregatidae). Smithson Contrib Paleobiol 35:1–33

Olson SL (1980) A new genus of penguin-like pelecaniform bird from the Oligocene of Washington (Pelecaniformes: Plotopteridae). Nat Hist Mus Los Angeles Cty, Contrib Sci 330:51–57

Olson SL (1981) The generic allocation of *Ibis pagana* Milne-Edwards, with a review of fossil ibises (Aves: Threskiornithidae). J Vertebr Paleontol 1:165–170

Olson SL (1985) The fossil record of birds. In: Farner DS, King JR, Parkes KC (eds) Avian biology, vol 8. Academic Press, New York, pp 79–238

Olson SL (1989 ["1988"]) Aspects of global avifaunal dynamics during the Cenozoic. In: Ouellet H (ed) Acta XIX Congressus Internationalis Ornithologici. University of Ottawa Press, Ottawa, pp 2023–2029

Olson SL (1994) A giant *Presbyornis* (Aves: Anseriformes) and other birds from the Paleocene Aquia Formation of Maryland and Virginia. Proc Biol Soc Wash 107:429–435

Olson SL, Hasegawa Y (1979) Fossil counterparts of giant penguins from the North Pacific. Science 206:688–689

Olson SL, Hasegawa Y (1996) A new genus and two new species of gigantic Plotopteridae from Japan (Aves: Plotopteridae). J Vertebr Paleontol 16:742–751

Olson SL, Matsuoka H (2005) New specimens of the early Eocene frigatebird *Limnofregata* (Pelecaniformes: Fregatidae), with the description of a new species. Zootaxa 1046:1–15

Olson SL, Parris DC (1987) The Cretaceous birds of New Jersey. Smithson Contrib Paleobiol 63:1–22

Ono K, Hasegawa Y (1991) Vertebrate fossils of the Iwaki Formation, III-1; avian fossils. In: Koda Y (ed) The excavation research report of the animal fossils of the Iwaki Formation. Iwaki City, Japan, pp 6–17. [in Japanese]

Panteleyev AV, Nessov LA (1993) A small tubinare (Aves: Procellariiformes) from the Eocene of Middle Asia. Tr zool Inst 252:95–103

Park T, Fitzgerald EMG (2012) A review of Australian fossil penguins (Aves: Sphenisciformes). Mem Mus Victoria 69:309–325

Pennycuick CJ (1982) The flight of petrels and albatrosses (Procellariiformes), observed in South Georgia and its vicinity. Phil Trans Roy Soc London, Ser B 300:75–106

Peters DS (1983) Die „Schnepfenralle" *Rhynchaeites messelensis* Wittich 1898 ist ein Ibis. J Ornithol 124:1–27

Peters DS, Hamedani A (2000) *Frigidafons babaheydariensis* n. sp., ein Sturmvogel aus dem Oligozän des Irans (Aves: Procellariidae). Senck leth 80:29–37

Proffitt JV, Clarke JA, Scofield RP (2016) Novel insights into early neuroanatomical evolution in penguins from the oldest described penguin brain endocast. J Anat 229:228–238

Prum RO, Berv JS, Dornburg A, Field DJ, Townsend JP, Lemmon EM, Lemmon AR (2015) A comprehensive phylogeny of birds (Aves) using targeted next-generation DNA sequencing. Nature 526:569–573

Rasmussen DT, Olson SL, Simons EL (1987) Fossil birds from the Oligocene Jebel Qatrani Formation, Fayum Province, Egypt. Smithson Contrib Paleobiol 62:1–20

Roux T (2002) Deux fossiles d'oiseaux de l'Oligocène inférieur du Luberon. Courr sci Parc nat rég Luberon 6:38–57

Sakurai K, Kimura M, Katoh T (2008) A new penguin-like bird (Pelecaniformes: Plotopteridae) from the Late Oligocene Tokoro Formation, northeastern Hokkaido, Japan. Oryctos 7:83–94

Sallaberry MA, Yury-Yáñez RE, Otero RA, Soto-Acuña S, Torres GT (2010) Eocene birds from the western margin of southernmost South America. J Paleontol 84:1061–1070

Simpson GG (1971) A review of the pre-Pliocene Penguins of New Zealand. Bull Am Mus Nat Hist 144:319–378

Simpson GG (1975) Fossil Penguins. In: Stonehouse B (ed) The Biology of Penguins. Macmillan Press, London, pp 19–41

Slack KE, Jones CM, Ando T, Harrison GL, Fordyce RE, Arnason U, Penny D (2006) Early penguin fossils, plus mitochondrial genomes, calibrate avian evolution. Mol Biol Evol 23:1144–1155

Smith ND (2010) Phylogenetic analysis of Pelecaniformes (Aves) based on osteological data: implications for waterbird phylogeny and fossil calibration studies. PLoS One 5:e13354

Smith ND, Grande L, Clarke JA (2013) A new species of Threskiornithidae-like bird (Aves, Ciconiiformes) from the Green River Formation (Eocene) of Wyoming. J Vertebr Paleontol 33:363–381

Starko S, Gomez MS, Darby H, Demes KW, Kawai H, Yotsukura N, Lindstrom SC, Keeling PJ, Graham SW, Martone PT (2019) A comprehensive kelp phylogeny sheds light on the evolution of an ecosystem. Mol Phylogenet Evol 136:138–150

Stidham TA (2015) A new species of *Limnofregata* (Pelecaniformes: Fregatidae) from the Early Eocene Wasatch Formation of Wyoming: implications for palaeoecology and palaeobiology. Palaeontology 58:239–249

Stidham TA, Holroyd PA, Gunnell GF, Ciochon RL, Tsubamoto T, Egi N, Takai M (2005) A new ibis-like bird (Aves: cf. Threskiornithidae) from the late middle Eocene of Myanmar. Contrib Mus Paleontol Univ Michigan 31:179–184

Tambussi CP, Acosta Hospitaleche C (2007) Antarctic birds (Neornithes) during the Cretaceous-Eocene times. Rev Asoc Geol Argent 62:604–617

Tambussi CP, Reguero MA, Marenssi SA, Santillana SN (2005) *Crossvallia unienwillia*, a new Spheniscidae (Sphenisciformes, Aves) from the Late Paleocene of Antarctica. Geobios 38:667–675

Tambussi CP, Acosta Hospitaleche CI, Reguero MA, Marenssi SA (2006) Late Eocene penguins from West Antarctica: systematics and biostratigraphy. In: Francis JE, Pirrie D, Crame JA (eds) Cretaceous-Tertiary High-Latitude Palaeoenvironments, James Ross Basin, Antarctica. Geol Soc, London, Spec Pub 258:145–161

Tambussi CP, Degrange FJ, Ksepka DT (2015) Endocranial anatomy of Antarctic Eocene stem penguins: implications for sensory system evolution in Sphenisciformes (Aves). J Vertebr Paleontol 35: e981635

Thomas DB, Ksepka DT, Forydce RE (2011) Penguin heat-retention structures evolved in a greenhouse Earth. Biol Lett 7:461–464

Tonni EP (1980) The present state of knowledge of the Cenozoic birds of Argentina. In: Campbell KE (ed) Papers in avian paleontology honoring Hildegarde Howard. Nat Hist Mus Los Angeles Cty Contrib Sci 330:105–114

van Beneden PJ (1871) Les oiseaux de l'argile rupelienne. Bull Acad Roy Belg 2(32):256–261

van Tets GF, Rich PV, Marino-Hadiwardoyo HR (1989) A reappraisal of *Protoplotus beauforti* from the early Tertiary of Sumatra and the basis of a new pelecaniform family. Bull Geol Res Dev Cent, Paleontol Ser 5:57–75

Wang M, Mayr G, Zhang J, Zhou Z (2012) New bird remains from the Middle Eocene of Guangdong, China. Acta Palaeontol Pol 57:519–526

Waterhouse DW, Lindow BEK, Zelenkov N, Dyke GJ (2008) Two new parrots (Psittaciformes) from the Lower Eocene Fur Formation of Denmark. Palaeontology 51:575–582

Wetmore A (1940) Fossil bird remains from Tertiary deposits in the United States. J Morphol 66:25–37

Wittich E (1898) Beiträge zur Kenntnis der Messeler Braunkohle und ihrer Fauna. Abh geol Landesanst 3:79–147

Woodburne MO, Case JA (1996) Dispersal, vicariance, and the late Cretaceous to early Tertiary land mammal biogeography from South America to Australia. J Mamm Evol 3:121–161

Worthy TH (2011) Descriptions and phylogenetic relationships of a new genus and two new species of Oligo-Miocene cormorants (Aves: Phalacrocoracidae) from Australia. Zool J Linnean Soc 163:277–314

Worthy TH (2012) A new species of Oligo-Miocene darter (Aves: Anhingidae) from Australia. Auk 129:96–104

Worthy TH, Holdaway RN (2002) The Lost World of the Moa. Prehistoric Life of New Zealand. Indiana University Press, Bloomington

Zhou Z-H, Clarke JA, Zhang F-C, Wings O (2004) Gastroliths in *Yanornis*: an indication of the earliest radical diet-switching and gizzard plasticity in the lineage leading to living birds? Naturwiss 91:571–574

Zvonok EA, Mayr G, Gorobets L (2015) New material of the Eocene marine bird *Kievornis* Averianov et al., 1990 and a reassessment of the affinities of this taxon. Vertebr PalAsiat 53:238–244

Zvonok EA, Zelenkov NV, Danilov IG (2016) A new unusual waterbird (Aves, ?Suliformes) from the Eocene of Kazakhstan. J Vertebr Paleontol 36:e1035783

8 Accipitriformes (New World Vultures, Hawks, and Allies), Falconiformes (Falcons), and Cariamiformes (Seriemas and Allies)

Sequence-based analyses do not support the monophyly of diurnal birds of prey, and even though the Cathartidae, Sagittariidae, Pandionidae, and Accipitridae form a clade in these studies, the Falconidae are usually united with the Cariamiformes, Psittaciformes, and Passeriformes (Sect. 2.3). The association of the Cariamiformes with gruiform birds by earlier authors has never been convincingly established, but the placement of these birds together with parrots and passerines in sequence-based analyses likewise lacks any support from anatomical or fossil data.

In most molecular analyses, the Cariamiformes and Falconiformes resulted as successive sister groups of the clade (Psittaciformes + Passeriformes). However, a clade including these four taxa is not recovered in all sequence-based analyses, and under some settings, molecular phylogenies supported a sister group relationship between the Cariamiformes and Falconiformes (Kuhl et al. 2021: fig. S4A-C; Braun and Kimball 2021: figs. 4D and 7B), which conforms much better to the fossil record than a successive branching of Cariamiformes and Falconiformes at the base of the (Psittaciformes + Passeriformes) clade. Not only do the Cariamiformes include raptorial stem group representatives, but there are also fossils of long-legged early Eocene falconiform-like birds, which may constitute a morphological link between the Falconiformes and the Cariamiformes.

Early Paleogene fossils of the Falconiformes and Accipitriformes are scarce, with the oldest falconiforms being from the early Eocene of South America, whereas the first records of the Accipitriformes come from the early Eocene of Europe. The Cariamiformes, by contrast, have a fairly extensive fossil record in the Americas and Europe, even though the exact affinities of many taxa remain controversial.

8.1 Accipitriformes (New World Vultures, Hawks, and Allies)

Within the Accipitriformes, the Cathartidae and Sagittariidae are consecutive sister taxa of the clade (Pandionidae + Accipitridae). Extant Cathartidae and Sagittariidae exhibit a plesiomorphic hypotarsus morphology, which is simple and block-like, whereas it forms two well-developed crests, sometimes enclosing a canal, in the Pandionidae and Accipitridae.

8.1.1 Teratornithidae and Cathartidae (New Word Vultures)

The Teratornithidae are a group of New World birds of prey, which includes the largest known flying bird, *Argentavis magnificens* from the late Miocene of Argentina, the wingspan of which was estimated at 6–8 meters (Campbell and Tonni 1983). Teratorns closely resemble the Cathartidae in many osteological features, and some bones were initially indeed mistaken for those of New World vultures (Miller 1910). Although most earlier authors assumed a close relationship between teratorns and New World vultures (e.g., Brodkorb 1964), others emphasized the differences between these taxa and considered teratorns to be more closely related to "pelecaniform" birds (e.g., Campbell and Tonni 1980). A sister group relationship between the Cathartidae and Teratornithidae is, however, well supported by derived skull features, in particular the co-ossification of the lacrimal, ectethmoid, and frontal bones (Miller 1909).

The sole Paleogene fossil record of the Teratornithidae comes from the late Oligocene/early Miocene of the Taubaté Basin in Brazil and was described as *Taubatornis campbelli* by Olson and Alvarenga (2002). The species, which is solely known from the distal end of a tibiotarsus and the proximal end of an ulna, is the smallest representative of the Teratornithidae. As noted by Olson and Alvarenga (2002),

T. campbelli supports the hypothesis that the Teratornithidae are of South American origin and did not reach North America until the late Neogene (Campbell and Tonni 1981).

A South American origin of the Teratornithidae suggests that the stem species of the Cathartidae also lived on this continent. However, two putative stem group representatives of this taxon, *Diatropornis ellioti* and *Parasarcoramphus milneedwardsi*, were reported from the Quercy fissure fillings in France (Cracraft and Rich 1972; Mourer-Chauviré 2002). Whereas several carpometacarpi and tarsometatarsi of *D. ellioti* were identified in middle and late Eocene (MP 16–17) deposits, *P. milneedwardsi* is only known from a single tarsometatarsus of unknown age (Mourer-Chauviré 1988, 2002). The tarsometatarsus of *D. ellioti* closely resembles that of extant Cathartidae, but that of *Parasarcoramphus* has a more dorsally projecting trochlea for the fourth toe, which is more widely separated from the trochlea for the third toe and bears a more plantarly directed wing-like flange. A caveat that has to be placed on the classification of these birds is that comparisons were so far restricted to the Cathartidae. Because the tarsometatarsus and carpometacarpus of teratorns closely resemble the corresponding elements of New World vultures, there remains a possibility that *Diatropornis* and *Parasarcoramphus* are actually stem group representatives of the clade (Teratornithidae + Cathartidae).

A putative New World vulture from the Paleogene of North America, *Phasmagyps patritus*, is represented by a distal tibiotarsus from the late Eocene (Chadronian) of Colorado (Wetmore 1927). Another Paleogene New World record of the Cathartidae is *Brasilogyps faustoi* from the late Oligocene/early Miocene of the Taubaté Basin in Brazil (Alvarenga 1985a). The fossil material assigned to this species consists of a distal tibiotarsus and an associated proximal tarsometatarsus. Olson (1985) furthermore mentioned undescribed Cathartidae from the early Oligocene of Mongolia.

8.1.2 Sagittariidae

The single extant species of this taxon is the long-legged Secretary Bird, *Sagittarius serpentarius*, which occurs in Sub-Saharan Africa and predominantly feeds on snakes. *Pelargopappus schlosseri*, a stem group representative from the early and late Oligocene of the Quercy fissure fillings, is smaller than *S. serpentarius* and known from leg bones and a referred ulna (Gaillard 1908; Mourer-Chauviré and Cheneval 1983). It has greatly elongated legs and also otherwise resembles extant Sagittariidae in the morphology of the known bones. However, the tarsometatarsus has a more strongly developed trochlea for the second toe and a more pronounced medial hypotarsal crest, which by outgroup comparisons with the Pandionidae and Accipitridae probably represents the plesiomorphic condition for the Sagittariidae (Mayr 2005a).

8.1.3 Pandionidae (Ospreys), and Accipitridae (Hawks and Allies)

The earliest records of hawk-like accipitriform birds are a proximal tarsometatarsus and pedal phalanges of an unnamed species from the early Eocene (Ypresian; 50.5–52 Ma) of Egem in Belgium (Mayr and Smith 2019a). This fossil predates, and thereby disproves, a recent molecular age estimate of 50 Ma for the split between the Sagittariidae and Accipitridae, which was derived from calibrated sequence data (Mindell et al. 2018).

Calibrated molecular data suggest a divergence of the Accipitridae and Pandionidae some 42 million years ago (Mindell et al. 2018), and the fossil record likewise indicates that the accipitrid and pandionid lineages diverged before the Oligocene. Extant ospreys have a nearly global distribution, but all Paleogene fossils come from Old World sites. The earliest specimen is an ungual phalanx from the late Eocene of England (Harrison and Walker 1976). Although isolated ungual phalanges of birds seldom can be reliably identified, those of ospreys exhibit a characteristic derived morphology, which is related to their specialized mode of foraging, that is, feet-first plunges into water bodies from considerable heights (Mayr 2006a). Harrison and Walker (1976) referred the above fossil to *Palaeocircus*, following Brunet's (1970) identification of *Palaeocircus cuvieri* from the late Eocene Paris Gypsum as an osprey. However, *P. cuvieri* is solely known from a carpometacarpus, and there are thus no skeletal elements in common with the British specimen. Correct identification of *P. cuvieri* as an osprey was furthermore doubted by Olson (1985). Another ungual phalanx of an osprey was described by Mayr (2006a) from the early Oligocene (Rupelian) of Germany. Rasmussen et al. (1987) reported a distal humerus and a tentatively referred incomplete carpometacarpus of a stem group representative of the Pandionidae from the early Oligocene of the Jebel Qatrani Formation in Egypt.

The oldest representative of the Accipitridae is *Milvoides kempi* from the middle Eocene of England, a small species that is known from an incomplete distal tarsometatarsus (Harrison and Walker 1979a). Mayr and Smith (2002) tentatively assigned another distal tarsometatarsus from the early Oligocene of Belgium to the Accipitridae. Two accipitrid species were furthermore described from unknown localities and stratigraphic horizons of the Quercy fissure fillings: "*Aquila*" ("*Aquilavus*") *hypogaea* was described on the basis of a femur, whereas the holotype of "*A*". *corroyi* is a tarsometatarsus, which resembles the corresponding bone of extant kites (*Milvus* spp.) in its proportions. Because these species are represented by different bones, their interrelationships cannot be assessed and a classification in the same genus-level taxon (either "*Aquila*" as in the original descriptions or "*Aquilavus*" as per Brodkorb 1964) is conjectural and almost certainly incorrect. As first recognized by Mourer-Chauviré (in Olson 1985: 110), the alleged cuculiform "*Dynamopterus*" *boulei* is a species of the Accipitridae and may well be a junior synonym of *A. corroyi*, with which it corresponds in size. *Palaeohierax gervaisii* from the late Oligocene of France is also based on a tarsometatarsus and was nearly twice as large as the kite-sized *A. corroyi*.

One of the few Paleogene accipitrid species for which dietary habits can at least tentatively be inferred is *Aviraptor longicrus* from the early Oligocene of the Carpathian Basin in Poland (Fig. 8.1a; Mayr and Hurum 2020). This species is known from a nearly complete albeit poorly preserved skeleton on a slab and is the size of the smallest extant Accipitridae, such as the Tiny Hawk (*Accipiter superciliosus*). In addition to its small size, *A. longicrus* is characterized by a very long tarsometatarsus. In extant Accipitriformes, the combination of these two traits—a very small size and a long tarsometatarsus—is characteristic for avivorous species, that is, those predominantly preying upon other birds. Extant small avivorous hawks mainly forage on passerines and hummingbirds, and it is therefore notable that the early Oligocene localities of the Polish Carpathian Basin also yielded the earliest European records of the latter two groups (see Sects. 6.8.9.4 and 9.5.3; Mayr and Hurum 2020).

From the "middle" Oligocene of Mongolia, Kurochkin (1968) described two species of putative Buteoninae as *Buteo circoides* (distal ulna) and *Venerator* ("*Tutor*") *dementjevi* (distal humerus). A putative accipitrine species named *Gobihierax edax* by Kurochkin (1968) on the basis of a fragmentary distal humerus was identified as a galliform bird by Zelenkov and Kurochkin (2015). From the early Oligocene of Mongolia, Kurochkin (1976: 78) furthermore mentioned "several rather large femora resembling the recent Aegypiinae (...) in a number of important characteristics." A few fossils of undetermined accipitrids are also known from the early late Oligocene of Mongolia (Daxner-Höck et al. 2019), and Kurochkin (1976) noted the occurrence of an accipitrid taxon ("*Aquilavus*") in the late Oligocene (*Indricotherium* Beds) of Kazakhstan. A distal tarsometatarsus of a large accipitrid from the late Eocene of the Jebel Qatrani Formation (Fayum) in Egypt resembles extant sea eagles of the taxon *Haliaeetus* (Rasmussen et al. 1987).

Several species of the Accipitridae were described from the late Eocene and early Oligocene White River Group of North America. "*Buteo*" *grangeri* and "*B*". *fluviaticus* were found in early Oligocene deposits of the White River Formation of South Dakota and Colorado, respectively (Wetmore and Case 1934; Miller and Sibley 1942). The holotype of "*B*". *fluviaticus* is a distal tarsometatarsus, whereas of "*B*." *grangeri* only the skull is known; Miller and Sibley (1942) already noted that these specimens may be from the same species. "*B*." *antecursor* from the late Eocene or earliest Oligocene (Chadronian) Brule Formation of Wyoming is represented by a tarsometatarsus. The assignment of all of these species to the extant taxon *Buteo* is based on overall similarity and is likely to be incorrect.

Another species from the early Oligocene (Whitneyan Poleslide Member of the Brule Formation; 30–32 Ma) of Wyoming is *Palaeoplancus sternbergi*, the holotype of which consists of a partial skeleton including the skull (Wetmore 1933a). This roughly Osprey-sized species was classified in the monotypic taxon Palaeoplancinae by Wetmore (1933a). It is very close in its skeletal morphology to extant Accipitridae, from which it, however, differs in a long and slender acromial process of the furcula. This presumably plesiomorphic morphology also occurs in the Sagittariidae and Pandionidae and suggests a position of *P. sternbergi* outside crown group Accipitridae. This hypothesis conforms to calibrated molecular data, which date the earliest divergence within the Accipitridae, between elanine kites and the remaining taxa, to the Eocene/Oligocene boundary, some 34 Ma (Mindell et al. 2018).

More recently, a complete tarsometatarsus from the late Eocene (Chadronian) of the White River Group (Fig. 8.1b) was tentatively assigned to *Palaeoplancus* and described as ?*P. dammanni* by Mayr and Perner (2020), who considered it possible that "*Buteo*" *antecursor* is another representative of the taxon *Palaeoplancus*. Mayr and Perner (2020) furthermore detailed that the accipitrid species from the White River Group are distinctly larger than Eocene and Oligocene Accipitridae from Europe, which may reflect different palaeoenvironments: whereas the North American badlands were savannah-like habitats, Europe was more densely forested during the late Paleogene and therefore provided fewer habitats for large diurnal birds of prey. The exact foraging strategies of *Palaeoplancus* and kin, however, remain unknown.

Based on distal tarsometatarsi, three accipitrid species were also described from the late Oligocene (early Arikareean) of Nebraska, namely "*Geranoaetus*" *ales*, *Palaeastur atavus*, and *Promilio efferus* (see Brodkorb 1964). These birds add to the diversity of larger accipitrids in the Paleogene of North America, but their fossil record is too incomplete for a well-founded phylogenetic placement.

There are only a few Paleogene accipitrids from South America, and Eocene or early Oligocene fossils are altogether unknown from this continent. Agnolin (2006a) and earlier authors (e.g., Brodkorb 1964) assigned *Climacarthrus incompletus* from the late Oligocene (Deseadan) of the Santa Cruz Province in Argentina to the Accipitridae. The holotype of the species is a very fragmentary and badly preserved distal tarsometatarsus, and its definite identification

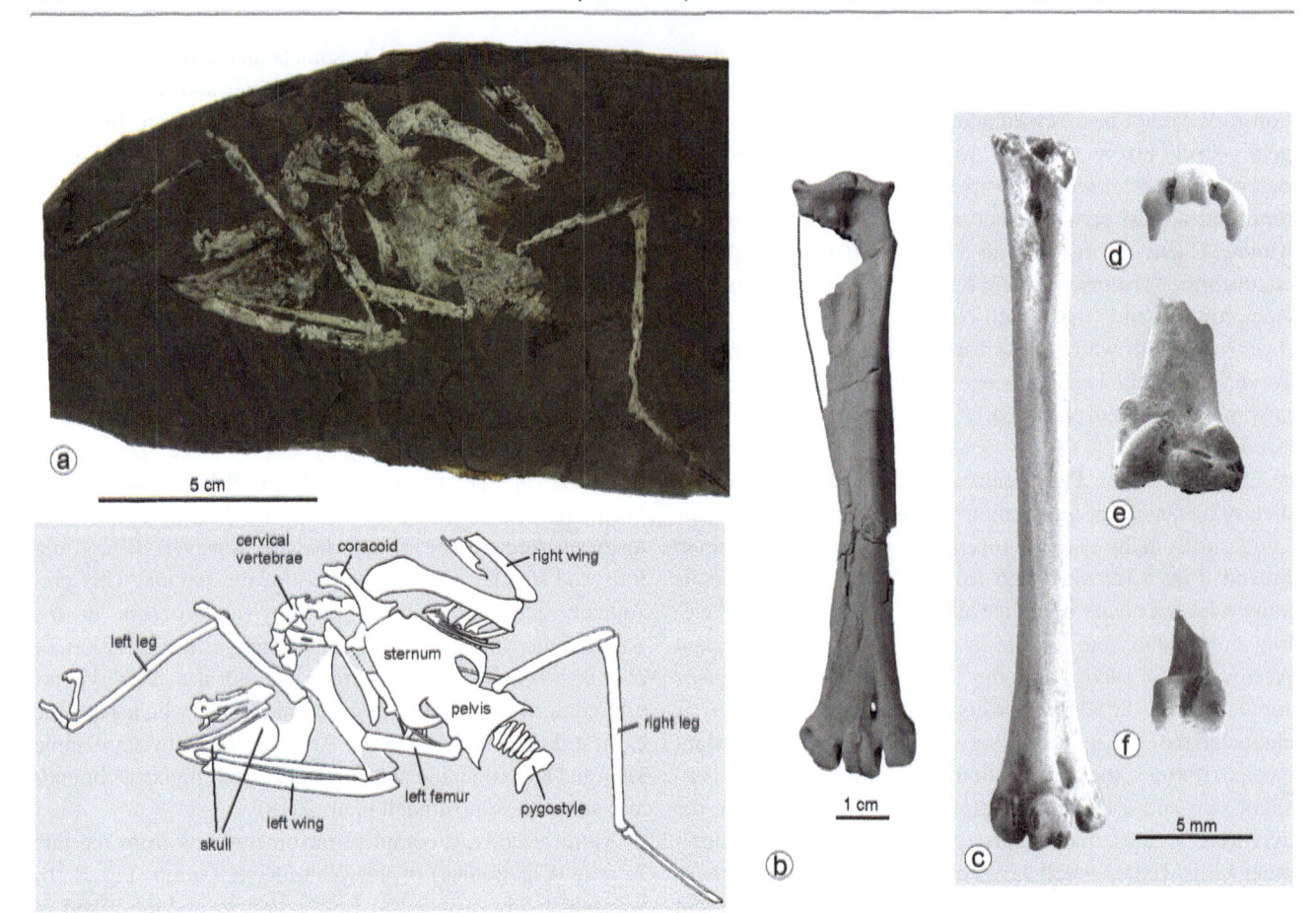

Fig. 8.1 Fossils of accipitrid diurnal birds of prey. (**a**) *Aviraptor longicrus* from the early Oligocene of the Carpathian Basin in Poland (holotype, Natural History Museum, Oslo, Norway, PMO 234.584a), with interpretive drawing. (**b**) Tarsometatarsus of ?*Palaeoplancus dammanni* from the White River Group of Wyoming, USA (holotype, Senckenberg Research Institute, Frankfurt, SMF Av 651); the line denotes the reconstructed medial margin of the damaged proximal end of the bone. (**c**)–(**f**) Selected skeletal elements of *Horusornis vianeyliaudae* (Horusornithidae) from the late Eocene locality La Bouffie of the Quercy fissure fillings in France (**c**, **d**: holotype tarsometatarsus in dorsal and distal view, Université des Sciences et Techniques du Languedoc, Montpellier, France, BFI 1974; **e**: distal end of humerus, BFI 1960; **f**: distal end of tibiotarsus, BFI 1962). (Photo in (**a**) by Jørn Hurum, in (**b**) by Sven Tränkner; photos in (**c**)–(**f**) courtesy of Cécile Mourer-Chauviré)

requires more complete specimens. The same is true for *Cruschedula revola*, also from the Deseadan of Argentina, the holotype of which is the cranial extremity of a scapula. *C. revola* was initially classified into the Cladornithidae (Sect. 9.3) by Ameghino (1899), but it was assigned to the Accipitridae by Brodkorb (1964), Tonni (1980), and Agnolin (2006a). I do not consider it possible to reliably establish the phylogenetic affinities of this fragmentary bone.

Pengana robertbolesi from the late Oligocene or early Miocene of Riversleigh in Australia is represented by a distal tibiotarsus. The specimen exhibits mediolaterally narrow condyles, which permitted a high flexibility of the intertarsal joint (Boles 1993). In extant Accipitridae, a similar morphology is found in the South American *Geranospiza* and the African *Polyboroides*, which are, however, considerably smaller than *Pengana*. The two extant taxa are only distantly related and none is likely to be close to the fossil taxon (Boles 1993); both use their feet to capture prey in hollows inaccessible to other predatory birds. Despite a presumed similar function of the intertarsal joint, the tibiotarsus of *Pengana* is very different from that of the Horusornithidae (see next section) in, e.g., the more accipitrid-like shape of the condyles and the presence of a supratendinal bridge. Another accipitrid from the late Oligocene Namba Formation of Australia was described as *Archaehierax sylvestris* (Mather et al. in press); this species is based on a partial skeleton, but its exact phylogenetic affinities are uncertain.

8.1.4 Horusornithidae

The Horusornithidae (Fig. 8.1c–f) occur in the late Eocene (MP 17) locality La Bouffie of the Quercy fissure fillings and were described by Mourer-Chauviré (1991), who also mentioned as yet undescribed remains from the "early Oligocene" of the USA. The single named species of the taxon,

Horusornis vianeyliaudae, was about the size of the Red-footed Kestrel (*Falco vespertinus*) and is represented by most of the major limb elements. Although the humerus, too, differs from that of extant diurnal birds of prey (Accipitriformes and Falconiformes), the most distinctive traits are found in the hindlimb bones of *H. vianeyliaudae*. The species is distinguished from other diurnal birds of prey in the absence of an ossified supratendinal bridge on the distal tibiotarsus, the condyles of which also exhibit a characteristic morphology in being proximodistally taller and less widely separated than in extant Accipitridae. According to Mourer-Chauviré (1991), this morphology indicates a hyperflexible intertarsal joint. The tarsometatarsus is elongated, and the trochlea for the second toe is more strongly plantarly directed than in most extant Accipitridae.

Among other features (Mourer-Chauviré 1991), horusornithids share with extant Pandionidae, Accipitridae, and Falconidae a derived morphology of the hypotarsus, which exhibits two crests that are separated by a wide sulcus, as well as a derived morphology of the coracoid and ungual phalanges. Their exact relationships to other diurnal birds of prey are, however, unresolved, and the present assignment to the Accipitriformes is provisional.

8.2 Falconiformes (Falcons)

The phylogenetic interrelationships of the extant species indicate that the Falconiformes, which today have a worldwide distribution, originated in South America (Griffiths 1999). Their earliest fossil record is *Antarctoboenus carlinii* from the early Eocene (49–53 Ma) of the La Meseta Formation of Seymour Island, Antarctica (Cenizo et al. 2016; see also Tambussi and Acosta Hospitaleche 2007). This species is represented by the distal end of a tarsometatarsus, which is from a bird the size of the extant Crested Caracara (*Caracara plancus*). The trochlea for the second toe, however, does not reach as far distally as in crown group Falconidae, which identifies the fossil as a stem group falconiform (Mayr 2009a, Cenizo et al. 2016).

Otherwise, the Paleogene fossil record of falcons is very poor and includes no unambiguously identified specimens. Harrison (1982) described a tiny distal tarsometatarsus and associated pedal phalanges from the London Clay of Walton-on-the-Naze as *Parvulivenator watteli* and tentatively assigned the species to the Falconidae. The distal width of the specimen is just about 3 mm, and its morphology does not suggest a closer relationship to falcons or any other diurnal bird of prey. As noted by Mayr (2005a), there is a second distal tarsometatarsus of this species, also from the London Clay, in the collection of the Natural History Museum (London).

The holotype of another putative falconiform from the London Clay of the Isle of Sheppey, *Stintonornis mitchelli*, is likewise the distal end of a tarsometatarsus (Fig. 8.2g; Harrison 1984). This specimen is more similar to extant Falconiformes, but it likewise resembles the distal tarsometatarsus of a *Masillaraptor*-like bird from the London Clay (see next section). Mourer-Chauviré (2006) mentioned the occurrence of still undescribed Falconiformes from late Eocene and early Oligocene deposits of the Quercy fissure fillings.

8.3 Masillaraptoridae

Masillaraptor parvunguis is a species from Messel, which is represented by two largely complete but poorly preserved skeletons, a skull with associated vertebrae, and a tarsometatarsus (Fig. 8.2a, b, f; Mayr 2006b, 2009b, 2020). The long, raptor-like beak of the species is similar to that of the falconiform Caracarinae (caracaras) in its proportions and has an equal depth over most of its length; the straight culmen (dorsal ridge) curves just before the tip. The lacrimal forms a long supraorbital process of similar shape to that of extant Falconiformes (Mayr 2009b). As in the Caracarinae, the legs are fairly long, and as in all Falconiformes and non-cathartid Accipitriformes, the two central phalanges of the fourth toe are shortened. In further concordance with the Falconiformes and the accipitriform Pandionidae and Accipitridae, the first phalanx of the second toe is also abbreviated. The ungual phalanges, however, do not exhibit the characteristic derived shape found in falconiform birds. The phylogenetic affinities of *M. parvunguis* are difficult to establish because of the poor preservation of the specimens. However, before the species was formally described, one of the fossils was already considered to be possibly related to the Falconiformes (Peters 1989: fig. 2), and this classification is supported by the raptor-like beak, the long supraorbital processes, and the abbreviated pedal phalanges.

Tarsometatarsi of *Masillaraptor* were also reported from the middle Eocene deposits of the Geisel Valley (Fig. 8.2e; Mayr 2002a, 2020). These were initially (Mayr 2002a) likened to *Coturnipes cooperi* from the London Clay. However, close similarities between *Masillaraptor* and *Coturnipes* were subsequently disproved by Mayr and Smith (2019a), who tentatively assigned a distal tarsometatarsus and a partial mandible from the early Eocene of Egem in Belgium to *Masillaraptor* (Fig. 8.2d; as noted in Sect. 5.3.1, the taxon *Coturnipes* is more likely to belong to the gruiform Messelornithidae). An as yet undescribed *Masillaraptor*-like skull was furthermore found in the Green River Formation (Grande 2013: fig. 143A).

Fossils of two differently-sized *Masillaraptor*-like birds from the London Clay of Walton-on-the-Naze are in the collection of Michael Daniels (the smaller one of these was erroneously referred to *Coturnipes cooperi* by Olson 1999, which prompted me to liken *Masillaraptor* to *Coturnipes*).

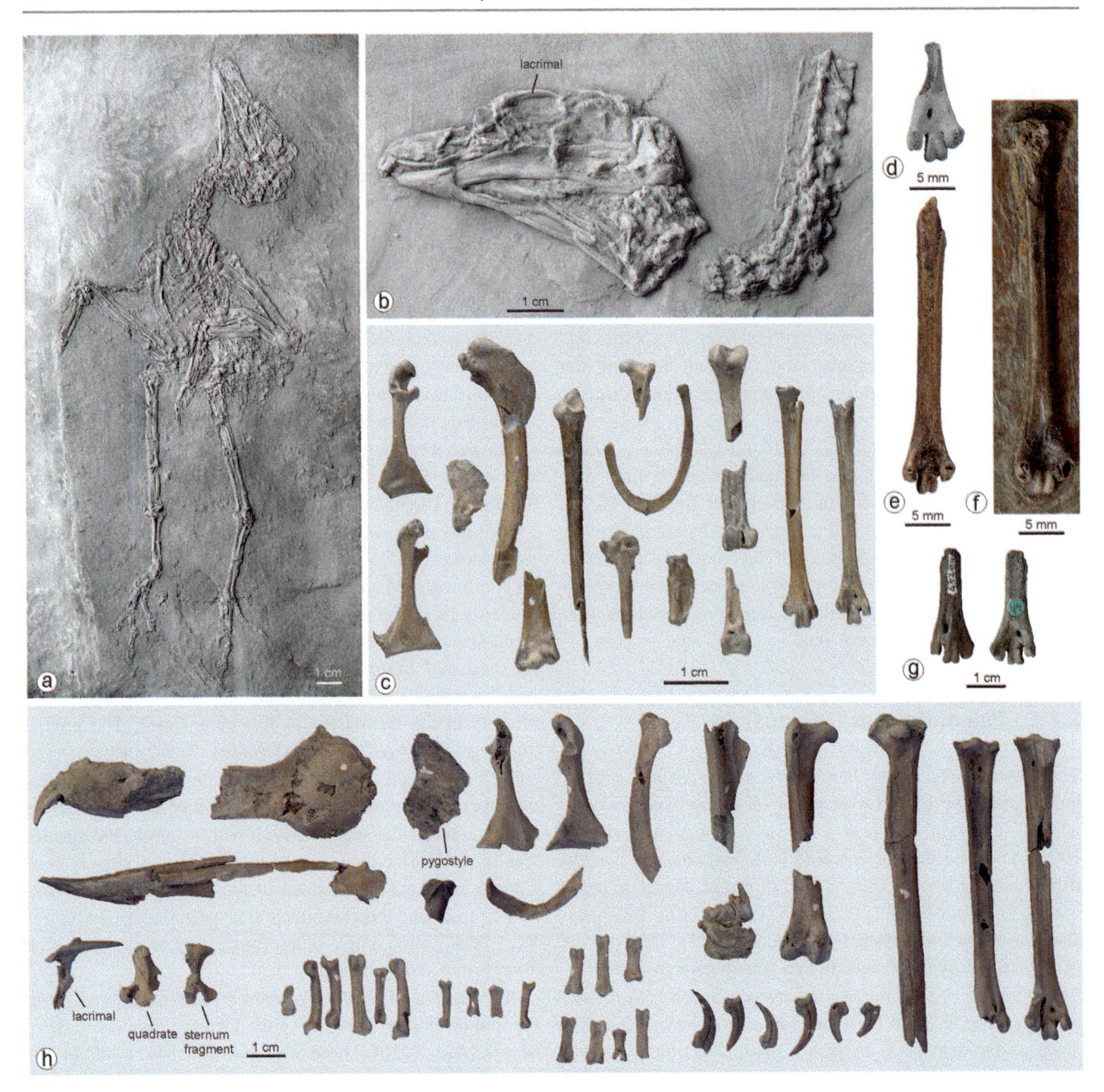

Fig. 8.2 Specimens of the Masillaraptoridae. (**a**) Skeleton of *Masillaraptor parvunguis* from the latest early or earliest middle Eocene of Messel (Senckenberg Research Institute, Frankfurt, SMF-ME 11042), coated with ammonium chloride. (**b**) Skull of *M. parvunguis* from Messel (Royal Belgian Institute of Natural Sciences, Brussels, Belgium, IRSNB Av 83), coated with ammonium chloride. (**c**) Undescribed specimen of *Masillaraptor* from the early Eocene London Clay of Walton-on-the-Naze in the collection of Michael Daniels (WN 93777). (**d**) Distal tarsometatarsus of a masillaraptorid from the early Eocene of Egem in Belgium (Royal Belgian Institute of Natural Sciences, Brussels, Belgium, IRSNB Av 181), coated with ammonium chloride. (**e**) Tarsometatarsus of *Masillaraptor* sp. from the middle Eocene of the Geisel Valley (Geiseltalsammlung, Martin-Luther Universität of Halle-Wittenberg, Germany, GMH NW XIV). (**f**) Tarsometatarsus of *Masillaraptor* sp. from Messel (SMF-ME 1068). (**g**) Distal end of the tarsometatarsus of *Stintonornis mitchelli* from the London Clay of the Isle of Sheppey in dorsal and plantar view (holotype, Natural History Museum, London, NHMUK A 5284). (**h**) Larger undescribed *Masillaraptor*-like species from Walton-on the-Naze (collection of Michael Daniels, Holland-on-Sea, UK, WN 91696). (All photos by Sven Tränkner)

These specimens are here for the first time figured (Fig. 8.2c, h). Both fossils agree well in the overall morphology of the preserved bones and share a deep fossa in the omal extremity of the coracoid. In the larger species (Fig. 8.2h), the skull is well preserved and the beak closely resembles that of *Masillaraptor parvunguis* in its proportions. As in the latter species, the mandible of the London Clay fossil has a slightly upturned tip. Also as in *M. parvunguis*, the lacrimal bears a well-developed supraorbital process. The wing bones are reminiscent of those of the Falconidae, whereas the leg bones exhibit a less "raptor-like" morphology than those of extant Falconiformes. The tarsometatarsus is long and slender, and the hypotarsus is block-like and lacks tendinal sulci or canals. Except for a somewhat smaller trochlea for the second toe, the distal end of the tarsometatarsus resembles that of the putative falconiform *Stintonornis mitchelli* from the London Clay of the Isle of Sheppey (see previous section). The ungual phalanges are curved and sharply pointed, but, unlike in extant Falconiformes they exhibit a laterally open neurovascular sulcus, and the flexor tubercle is less pronounced.

Apart from resembling that of *Masillaraptor*, the beak of the London Clay fossil is also remarkably similar to that of the cariamiform Phorusrhacidae in its proportions. Because molecular sequence data suggest close affinities between the Falconiformes and the Cariamiformes, which are even recovered as sister taxa in some analyses (Braun and Kimball 2021; Kuhl et al. 2021), it is tempting to speculate that masillaraptorids constitute an evolutionary link between both groups (in which case it appears more likely that the fossils are stem group falconiforms with a plesiomorphic phorusrhacid-like beak). However, a well-founded assessment of the evolutionary significance of these birds has to await more data on their osteology and a formal description of the London Clay fossils.

8.4 Cariamiformes (Seriemas and Allies)

The two extant species of the Cariamiformes (Cariamidae [seriemas]) are predominantly carnivorous birds with a terrestrial way of living, which occur in semi-open, dry habitats of South America. By contrast, Paleogene stem group cariamiforms were diversified and widely distributed over Europe and the Americas. These midsize to large and often cursorial species typically have a skull with long supraorbital processes formed by the lacrimal bones, a short carpometacarpus with a wide intermetacarpal space, as well as a block-like hypotarsus that does not enclose bony canals. The entirety of these features is, however, not present in all taxa included in this section, and some are discussed here only because they share a similar overall morphology with unambiguous Cariamiformes.

Even if just those groups are considered whose close relationship to extant Cariamiformes is well established, that is, the Phorusrhacidae, Idiornithidae, and Bathornithidae, the historical biogeography of the Cariamiformes is not easily understood. These birds occurred in the Paleogene of South America, Europe, and North America. The most obvious hypothesis is a dispersal from South America (Phorusrhacidae) to Europe (Idiornithidae) via North America (Bathornithidae), or vice versa (Mourer-Chauviré 1999). As yet, however, no North American Cariamiformes are known before the late Eocene, whereas these birds were already very diversified in the early Eocene of Europe and South America.

The interrelationships of fossil Cariamiformes are not well understood. A recent analysis resulted in a sister group relationship between the Cariamidae and Phorusrhacidae, with the Idiornithidae (*Dynamopterus*) and Bathornithidae (*Bathornis*) branching next (Mayr 2016a). The affinities of other putative Paleogene Cariamiformes, however, remain elusive, and it is anything but certain that all of the birds discussed in the following are correctly assigned to the Cariamiformes.

8.4.1 Phorusrhacidae

The Phorusrhacidae ("terror birds") are one of the most iconic and best-known avian groups from the Cenozoic of South America. Their taxonomy was revised by Alvarenga and Höfling (2003), who recognized five subtaxa, of which only the Psilopterinae, Patagornithinae, and Brontornithinae have a Paleogene fossil record (see also Alvarenga et al. 2011). The taxonomic allocation of the fossils assigned to the Brontornithinae is, however, challenged by the fact that Agnolin (2007) assumed that *Brontornis*, the early Miocene type genus of the taxon, is actually a representative of the Anseriformes and more similar to the Gastornithidae and Dromornithidae than to phorusrhacids (Sects. 4.2 and 4.4). These conclusions were based on the morphology of a quadrate fragment of *Brontornis*, the mandibular process of which exhibits just two condyles as in the Galloanseres instead of three as in most Neoaves. The ungual phalanges of *Brontornis* are furthermore not raptor-like as in typical phorusrhacids. Anseriform affinities of *Brontornis* were accepted by some subsequent authors (e.g., Angst and Buffetaut 2017), but they were disputed by others (e.g., Alvarenga et al. 2011). Analyses by Worthy et al. (2017) supported inclusion of *Brontornis* in the Phorusrhacidae, with these authors also noting that the morphology of the quadrate of this taxon was misinterpreted. Agnolin (2021) countered this re-establishment of phorusrhacid affinities of *Brontornis*, but a definitive placement of this poorly known taxon will only be possible once its skeletal morphology is better understood. If Agnolin's hypothesis is confirmed by future studies, the establishment of a new taxon would be necessary for the Paleogene species listed below under the Brontornithinae, which clearly exhibit a phorusrhacid-like morphology.

The earliest species assigned to the Phorusrhacidae is *Paleopsilopterus itaboraiensis* from the early Eocene of Itaboraí in southeastern Brazil, the fossil material of which includes a proximal tarsometatarsus and distal tibiotarsi (Alvarenga 1985b). *P. itaboraiensis* was classified in the

Psilopterinae, which comprise rather small and gracile species with slender leg bones and a long tarsometatarsus. This assignment is, however, based on overall similarity and needs to be supported by derived characteristics shared by the involved taxa.

A proximal tarsometatarsus and an ungual phalanx of an unnamed species of the Psilopterinae from the late Eocene (Mustersan) of Argentina were reported by Acosta Hospitaleche and Tambussi (2005). Another Paleogene species of the Psilopterinae is *Psilopterus affinis* from the late Oligocene (Deseadan) of Argentina, which is known from an incomplete tarsometatarsus and is among the smallest phorusrhacids. Other taxa from the same formation, which were assigned to the Phorusrhacidae by earlier authors, were considered to be of indeterminable affinities and excluded from the Phorusrhacidae by Alvarenga and Höfling (2003). Regarding these fossils, the authors noted, however, that "*Pseudolarus*" *guaraniticus* (known from a fragmentary proximal humerus) and *Smiliornis penetrans* (the holotype of which is the omal extremity of a coracoid) may actually be synonymous with *P. affinis*. According to Agnolin (2006b), "*P.*" *guaraniticus* indeed belongs to *Psilopterus*, although it represents a larger species than *P. affinis*; Agnolin (2004) also assigned *S. penetrans* to the Cariamiformes. The coracoid of the latter species is of particular interest in that it shares with phorusrhacids the complete reduction of the acrocoracoid process but still exhibits a well-developed and presumably plesiomorphic procoracoid process. Agnolin (2004) furthermore classified the alleged ciconiiform *Ciconiopsis antarctica* into the Psilopterinae. The holotype of this species is an incomplete carpometacarpus, which likewise comes from the Deseadan of Argentina and may well also belong to a species of *Psilopterus*. *Riacama caliginea*, which is known from the shaft and sternal extremity of a coracoid from the Deseadan of Argentina, was classified into the Cariamidae by Agnolin (2004). This species was assigned to the phorusrhacid Psilopterinae by earlier authors, but it was considered to be of indeterminate affinities by Alvarenga and Höfling (2003).

The species of the Patagornithinae are medium-sized phorusrhacids with a long and slender mandibular symphysis and elongated tarsometatarsi (Alvarenga and Höfling 2003). Their sole Paleogene representative is *Andrewsornis abbotti* from the Deseadan of Argentina. Of this species, an incomplete skull, fragments of the mandible, a femur, an incomplete coracoid, as well as pedal phalanges were described.

The Brontornithinae include the largest phorusrhacids, which reached a standing height of more than two meters, given that the taxon *Brontornis* is indeed a phorusrhacid (see above). Brontornithines are more massively built than other phorusrhacids and have a much shorter and stouter tarsometatarsus. The earliest species is *Physornis fortis* from the Deseadan of Argentina, the fossil record of which consists of mandible fragments and a few other incomplete bones. Much better documented is *Paraphysornis brasiliensis*, the description of which was based on a fairly complete skeleton from the late Oligocene/early Miocene of the Taubaté Basin of Brazil (Alvarenga 1982, 1993). This species was very large, with an estimated weight of about 180 kg (Alvarenga and Höfling 2003). Most recently, Buffetaut (2017) identified a fragmentary tibiotarsus of a brontornithine from the late Oligocene (Deseadan) of Bolivia.

An incomplete beak and a distal tarsometatarsus from the late Eocene of Seymour Island (Antarctica), which were tentatively referred to the Phorusrhacidae by earlier authors (Case et al. 1987; Tambussi and Acosta Hospitaleche 2007), are now assigned to the Palaeognathae and Pelagornithidae, respectively (Cenizo 2012). The fragmentary bones of the alleged phorusrhacid *Cunampaia simplex* from the late Eocene (Divisaderan) of Argentina are non-avian and probably belong to the Crocodylomorphae (Agnolin and Pais 2006).

Phorusrhacids show a significant variation in size, with representatives of the Psilopterinae being of similar size to the extant *Cariama cristata*, whereas *Paraphysornis brasiliensis* approached the size of the early Miocene *Brontornis burmeisteri*, a species which had an estimated height of 175 cm "at the level of the back, and the head, when well-raised, could have reached around 280 cm high" (Alvarenga and Höfling 2003: 59). All phorusrhacid species of which the skull is known had very large, mediolaterally compressed, and raptor-like beaks (Fig. 8.3). In contrast to extant Cariamidae, the skull of phorusrhacids exhibits functional basipterygoid processes. Even in the early forms, the wings seem to be greatly reduced, and the coracoid is long and slender, without an acrocoracoid process. As in other large flightless birds (e.g., the Gastornithidae and Dromornithidae), the ribs lack uncinate processes. The phylogenetic interrelationships of the various described taxa are uncertain. There can be little doubt, however, that the small size and long leg bones of the species united in the Psilopterinae are plesiomorphic for the Phorusrhacidae.

It is generally assumed that phorusrhacids were among the top predators in the Cenozoic of South America, and a carnivorous diet of these birds is supported by the sharply hooked tip of the huge beak. The mediolaterally compressed pelvis and beak indicate that the body was very narrow, which led Alvarenga and Höfling (2003) to assume that phorusrhacids hunted in densely vegetated areas. Phorusrhacids greatly differ in the proportions of the limb bones, with the tarsometatarsus being very long and slender in psilopterines, patagornithines, and phorusrhacines but short and stout in brontornithines. Certainly, the latter were therefore graviportal rather than cursorial birds, and Alvarenga and Höfling (2003) hypothesized that brontornithines may have been scavengers.

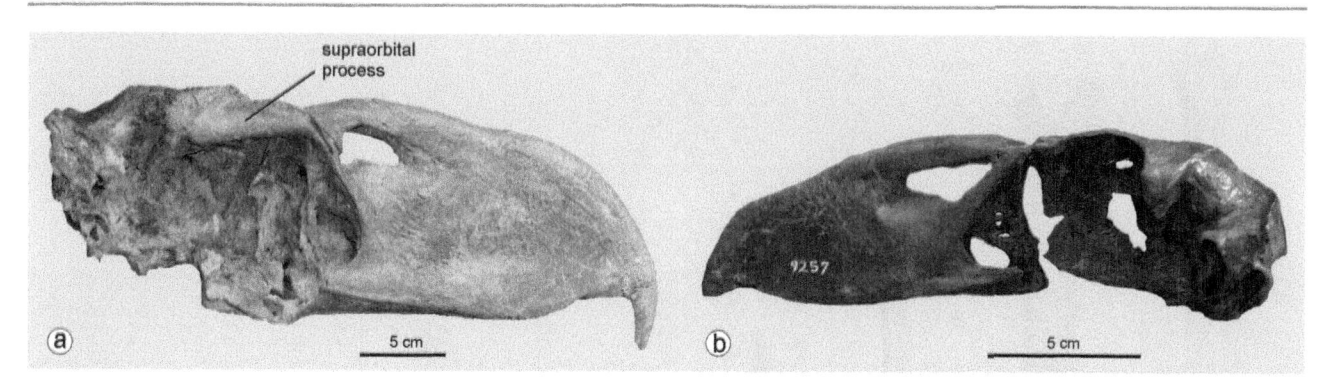

Fig. 8.3 Skulls of phorusrhacids from the early Miocene Santa Cruz Formation in Argentina. (**a**) *Patagornis marshi* (Natural History Museum, London, NHMUK, A 516). (**b**) *Psilopterus lemoinei* (American Museum of Natural History, New York, USA, AMNH 9257). (Photo in (**a**) by Sven Tränkner, in (**b**) by the author)

Whereas Tambussi and Acosta Hospitaleche (2007) assumed flight capabilities for the smaller Psilopterinae, all phorusrhacids were considered flightless by Alvarenga and Höfling (2003). These authors furthermore noted that some phorusrhacids may have been sexually dimorphic in size and morphological features.

A putative North African phorusrhacid, *Lavocatavis africana*, was described from the late early or early middle Eocene of Algeria (Mourer-Chauviré et al. 2011). However, this large-sized species is only known from a femur. Even though this bone closely resembles the femur of phorusrhacids, the femur is unknown for the Eremopezidae (Sect. 3.4), and as pointed out by Mayr (2017), *Lavocatavis* may be more closely related to this flightless North African taxon than to the Phorusrhacidae.

8.4.2 Idiornithidae

Most cariamiform birds from the Paleogene of Europe were assigned to the Idiornithidae. These birds have first been reported from the Quercy fissure fillings, where they occur in middle Eocene to late Oligocene deposits (Mourer-Chauviré 1983, 2006). Idiornithids were also identified in the latest early or earliest middle Eocene of Messel and the in the middle Eocene of the Geisel Valley in Germany (Peters 1995; Mayr 2000a, 2002a, 2020).

As defined by Mourer-Chauviré (1983, 2006), the Idiornithidae include the taxa *Dynamopterus* ("*Idiornis*"), *Elaphrocnemus*, *Oblitavis*, *Occitaniavis*, and *Propelargus*. Only the first two of these are well-represented by numerous postcranial bones, and a clade including all five taxa cannot be established with derived characteristics. Actually, a recent analysis did not support close affinities between *Dynamopterus* and *Elaphrocnemus* (Mayr 2016a), and the latter taxon is discussed in a separate section further below.

Most species of the Idiornithidae have long been assigned to the taxon *Idiornis* (Fig. 8.4). However, Mourer-Chauviré (2013) identified the allegedly cuculiform bird *Dynamopterus velox* as a misclassified representative of the Idiornithidae and noted that the taxon *Dynamopterus* has nomenclatural priority over *Idiornis*. The six species in the Quercy material that were formerly assigned to *Idiornis* (Mourer-Chauviré 1983) are now known as *Dynamopterus velox* ("*Idiornis cursor*"), *D. gaillardi*, *D. gallicus*, *D. gracilis*, *D. minor*, and *D. itardiensis*. The Quercy species of *Dynamopterus* greatly vary in size and most are smaller than extant Cariamidae, with the tarsometatarsus of the smallest species, *D. gracilis*, measuring less than one-third of that of the extant *Cariama cristata*. The largest idiornithid species from the Quercy deposits is *D. itardiensis*.

Idiornithids were also reported from Messel, where they are represented by at least three species. A large-sized species originally described as "*Idiornis*" *tuberculata* is now known as *Dynamopterus tuberculatus* and is represented by a nearly complete but poorly preserved skeleton (Fig. 8.5a; Peters 1995). The holotype of another large-sized species from Messel, *Dynamopterus* ("*Idiornis*") cf. *itardiensis*, is an isolated foot (Fig. 8.5c; Mayr 2000a), and of a smaller, unnamed species likewise only a foot was found (Fig. 8.5b; see Mayr 2016b, who hypothesized that the comparatively abundant isolated feet of larger birds in Messel may represent feeding remains of crocodilians). The large ?*Dynamopterus* ("?*Idiornis*") *anthracinus* from the Geisel Valley is known from the holotype tarsometatarsus and two referred partial legs (Fig. 8.5d; Mayr 2002a, 2020).

Dynamopterus is very similar to extant Cariamidae in skeletal morphology (Fig. 8.4). The small species have long and gracile hindlimbs. As in extant Cariamidae, the hypotarsus is block-like and does not form canals for the flexor tendons of the toes. The coracoid exhibits the characteristic derived morphology of that bone in extant seriemas in that the acrocoracoid and procoracoid processes are connected, or at least nearly so, by an osseous bridge. Other skeletal elements of *Dynamopterus* also closely correspond to those of extant Cariamidae.

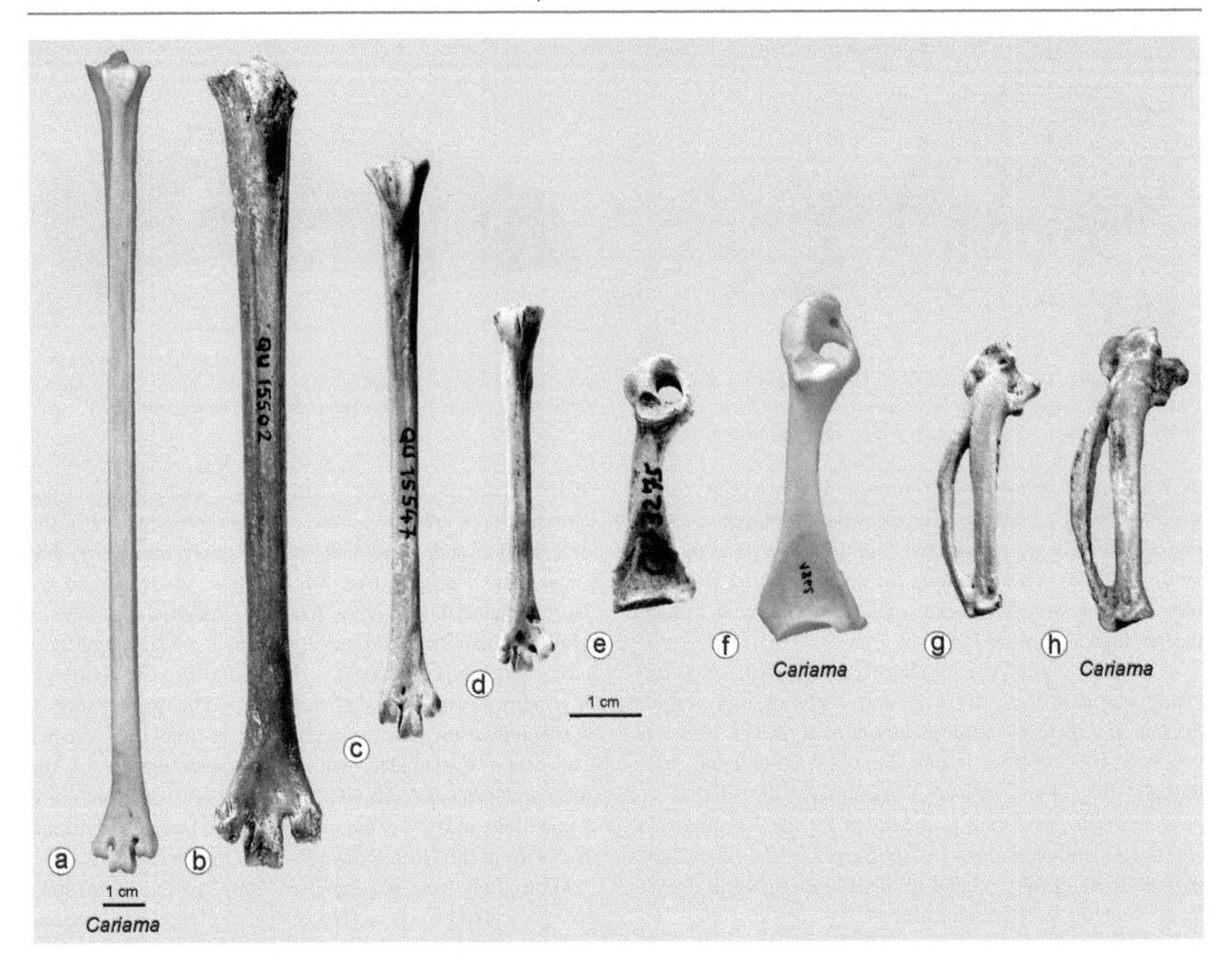

Fig. 8.4 Skeletal elements of some Idiornithidae from the Quercy fissure fillings in France in comparison to the extant Red-legged Seriema, *Cariama cristata* (Cariamidae). (**a**) Tarsometatarsus of *C. cristata*. (**b**) Tarsometatarsus of *Dynamopterus gallicus* (Muséum national d'Histoire naturelle, Paris, MNHN QU 15502 [3002]). (**c**) Tarsometatarsus of *D. minor* (MNHN QU 15547 [3047]). (**d**) Left tarsometatarsus of *D. gaillardi* (MNHN QU 15534 [3034]). (**e**) Coracoid of *D. velox* (MNHN QU 15775 [3275]). (**f**) Coracoid of *C. cristata*. (**g**) Carpometacarpus of *D. velox* (MNHN QU 15751 [3251]). (**h**) Carpometacarpus of *C. cristata*. (Same scale for (**b**)–(**h**); photos of the idiornithid fossils courtesy of Cécile Mourer-Chauviré, those of the extant bones by Sven Tränkner)

The holotype of *Dynamopterus* ("*Idiornis*") *tuberculatus* still constitutes the sole articulated skeleton known of an idiornithid. This species is also the only one in which a cranium is preserved, and the fossil is furthermore remarkable for the occurrence of numerous small bony tubercles on the cervical vertebrae and a few other bones. Similar vertebral tubercles occur in *Perplexicervix microcephalon* from Messel (Sect. 4.5.1) and are also known from an isolated cervical vertebra from the Quercy fissure fillings (Mayr 2007a, 2010). Just because of the unusual nature of these structures, one may ask whether *D. tuberculatus* was correctly assigned to the Idiornithidae or whether it eventually represents a large and flightless relative of *Perplexicervix* (which may or may not show affinities to the anseriform Anhimidae but is unlikely to be closely related to the Cariamiformes). At present, however, there is no compelling evidence for this assumption, so that an independent origin of the tubercles in *D. tuberculatus* and *P. microcephalon* has to be assumed. As evidenced by the *D. tuberculatus* skeleton and a foot of an undetermined species from Messel (Fig. 8.5b), the hallux of *Dynamopterus* is well-developed, whereas it is reduced in extant Cariamidae.

The tarsometatarsi of some of the early and middle Eocene species from Messel and the Geisel Valley (*D. tuberculatus*, *D.* cf. *itardiensis*, and ?*D. anthracinus*) are stouter than the gracile ones of most idiornithids from the Quercy fissure fillings. These species, therefore, were probably less cursorial than the Quercy species of *Dynamopterus*, and the different proportions of the tarsometatarsi may reflect different palaeoenvironments of the fossil sites, which were forested in the German localities but more arid and open in the Quercy area.

8.4 Cariamiformes (Seriemas and Allies)

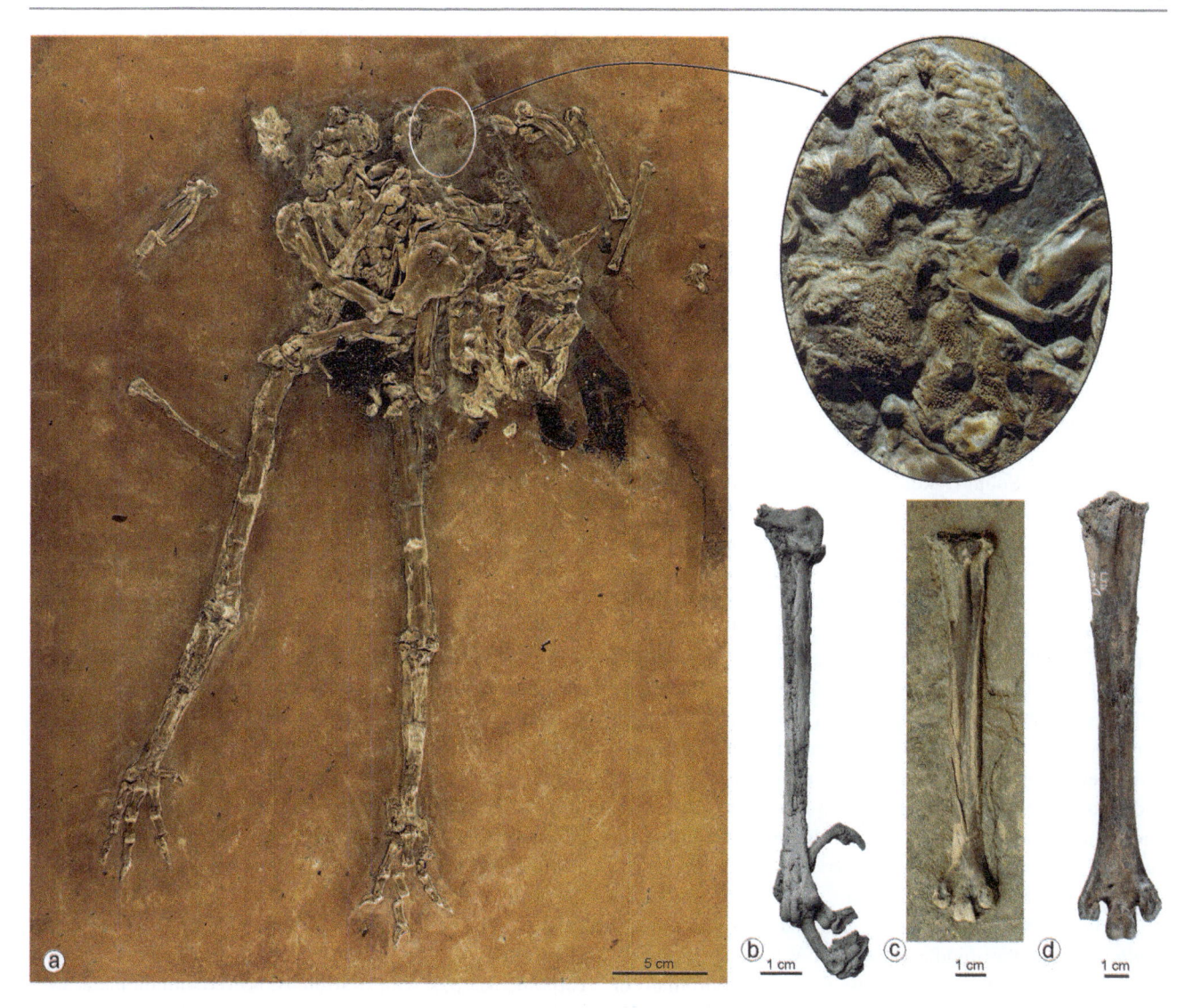

Fig. 8.5 Early and middle Eocene specimens of the Idiornithidae. (**a**) Holotype skeleton of *Dynamopterus tuberculatus* from the latest early or earliest middle Eocene of Messel (Royal Belgian Institute of Natural Sciences, Brussels, Belgium, IRSNB Av 127A); the arrow denotes a detail of the tubercles on the cervical vertebrae. (**b**) Foot of an undetermined idiornithid from Messel (Senckenberg Research Institute, Frankfurt, SMF-ME 1577); coated with ammonium chloride. (**c**) Tarsometatarsus of *D.* cf. *itardiensis* from Messel (SMF-ME 3437a). (**d**) Tarsometatarsus of ?*Dynamopterus anthracinus* from the middle Eocene of the Geisel Valley (holotype, Geiseltalsammlung, Martin-Luther Universität of Halle-Wittenberg, Germany, GMH LXIII-1-1983). (Photo in (**c**) by Anika Vogel, detail of cervical vertebrae by Erwin Haupt, all other photos by Sven Tränkner)

Mourer-Chauviré (1983, 2006) identified three further species of the Idiornithidae among the avian material of the Quercy fissure fillings, namely *Propelargus cayluxensis* (from an unknown stratigraphic horizon), *Occitaniavis elatus* (late Eocene), and *Oblitavis insolitus* (unknown stratigraphic horizon). The fossil material of the large *P. cayluxensis* consists of distal ends of the tarsometatarsus and tibiotarsus. Of the equally large *O. elatus*, a distal tibiotarsus, carpometacarpi, and a femur were found, and the smaller *O. insolitus* is known from humeri and a coracoid, which unlike in other Cariamiformes exhibits a foramen for the supracoracoideus nerve (Mourer-Chauviré 2003: figs. 9, 10).

Gypsornis cuvieri from the late Eocene of the Paris Gypsum was classified into the Rallidae by earlier authors (e.g., Brodkorb 1967; Brunet 1970), but the species was assigned to the Idiornithidae by Cracraft (1973). The fossil material consists only of an incomplete tarsometatarsus and associated pedal phalanges. *G. cuvieri* is distinguished from the Idiornithidae and most other cariamiform birds in the presence of a hypotarsal canal. Its affinities are best considered uncertain until more material is found, but the species more likely is a representative of the Gruoidea and may belong to the roughly coeval and similarly-sized Parvigruidae (Sect. 5.3.4).

8.4.3 Bathornithidae

The taxon *Bathornis* was established by Wetmore (1927) for *B. veredus* from the late Eocene of Colorado. After the initial classification of *Bathornis* into the charadriiform Burhinidae, Wetmore (1933b) established the taxon Bathornithidae, which he considered to be most closely related to the Cariamidae. Cracraft (1968, 1971, 1973) revised the taxonomy of bathornithids and recognized the following nine species from middle Eocene to late Oligocene North American localities: *Eutreptornis uintae* from the middle Eocene (Uinta Formation) of Utah (distal tibiotarsus and proximal tarsometatarsus), *Bathornis cursor* from the late Eocene (Chadronian) of Wyoming (distal tarsometatarsus), *B. veredus* from the late Eocene/early Oligocene (Chadronian and Orellan) of South Dakota, Colorado, and Nebraska (distal tibiotarsi, tarsometatarsi, distal humeri), "*B.*" *celeripes* from the late Eocene/early Oligocene (Chadronian and Orellan) of Wyoming, Nebraska, and South Dakota (tarsometatarsi, tibiotarsi, distal humerus, carpometacarpi), "*B.*" *geographicus* from the early Oligocene (Brule Formation) of South Dakota (tarsometatarsus and tibiotarsus), *B. fricki* from the late Oligocene (Arikareean) of Wyoming (tibiotarsus), *Paracrax antiqua* from the early Oligocene (Orellan) of Colorado (distal humerus and proximal carpometacarpus), as well as *P. gigantea* (distal humerus and incomplete carpometacarpus) and *P. wetmorei* (postcranial bones of a single individual including ulna, humerus, coracoid, scapula, sternum, and pelvis) from the early Oligocene (Whitneyan) of South Dakota.

Olson (1985) furthermore identified "*Neocathartes*" *grallator* from the middle Eocene of Wyoming (Fig. 8.6) and "*Palaeocrex*" *fax* from the late Eocene (Chadronian) of Colorado as representatives of the Bathornithidae. He assumed that both species belong to the taxon *Bathornis*, and that *Bathornis fax* is possibly conspecific with *B. veredus*.

The monophyly of bathornithids has not been satisfactorily established, and even a diagnosis covering all currently recognized taxa does not exist. *Eutreptornis uintae*, for example, is distinguished from other bathornithids in the craniocaudally deeper and mediolaterally narrower distal end of the tibiotarsus. Its classification in the Bathornithidae has already been questioned by Olson (1985), who furthermore detailed that the species assigned to *Bathornis* differ considerably in hypotarsus morphology. The hypotarsus of *Bathornis veredus* is simple and block-like as in *Dynamopterus* and extant Cariamidae, whereas that of "*Bathornis*" *celeripes* and "*B.*" *geographicus* is more complex and exhibits grooves and canals for the flexor tendons of the toes. Because hypotarsus morphology shows little variation within closely related birds, the latter species are therefore unlikely to be within the taxon *Bathornis* (Olson 1985).

Bathornithid affinities of *Paracrax* were questioned by Mayr (2009a), and phylogenetic analyses by Agnolin (2009) and Mayr (2016a) did not support close affinities between this taxon and *Bathornis*. Mayr (2016a) therefore suggested restricting the taxon Bathornithidae to *Bathornis*, which includes *B. grallator*, *B. veredus*, *B. cursor*, and *B. fricki* (the latter two species are only known from a few bones and their placement is uncertain). This classification is followed here and the species of *Paracrax* are discussed in a separate section further below.

Olson (1985) noted that the bones known of *B. grallator* and *B. veredus* are very similar, and Agnolin (2009) synonymized both species, even though the middle Eocene (late Bridgerian/early Uintan) *B. grallator* is at least six million years older than the late Eocene (Chadronian) *B. veredus*. The affinities of *B. cursor* and *B. fricki* cannot be reliably established based on the fossil material assigned to these species.

Although Wetmore (1958) described a humerus of "*Bathornis*" *celeripes* and Cracraft (1968) mentioned humeri of *B. veredus*, the published material of most species of *Bathornis* consists of leg bones. *Bathornis* ("*Neocathartes*") *grallator*, however, is represented by a partial skeleton including the skull, which was restudied by Mayr (2016a). This species was initially described as a flightless species of the Cathartidae (New World Vultures) by Wetmore (1944). The skull of *B. grallator* indeed shows a resemblance to that of cathartids, and as in the latter but in contrast to the Phorusrhacidae and extant Cariamidae, the lacrimal does not form a supraorbital process and is co-ossified with the frontal bone. However, and in addition to other traits, *Bathornis grallator* shares with other Cariamiformes a carpometacarpus with a strongly bowed minor metacarpal and a ball-shaped articular facet for the alular digit (Mayr 2016a). Like other species of *Bathornis*, *B. grallator* has long legs.

Bathornis veredus was about one-third larger than the extant *Cariama cristata* (Wetmore 1927). Regarding *Bathornis veredus*, Wetmore (1933c: 214) noted that "on examination of the two specimens of *veredus* now available it appears to be definitely established that the first toe was missing," whereas *Bathornis grallator* exhibits a well-developed hallux (Wetmore 1944; Mayr 2016a).

Agnolin (2009) proposed a sister group relationship between *Bathornis* and the Phorusrhacidae, which was, however, contested by Mayr (2016a), whose analysis recovered *Bathornis* outside a clade including the Phorusrhacidae and the Cariamidae. A sister group relationship between *Bathornis* and phorusrhacids would suggest that the ancestor of the clade including both taxa was also flightless. This conflicts with the fact that no land connections are known that would have allowed the dispersal of a flightless bird

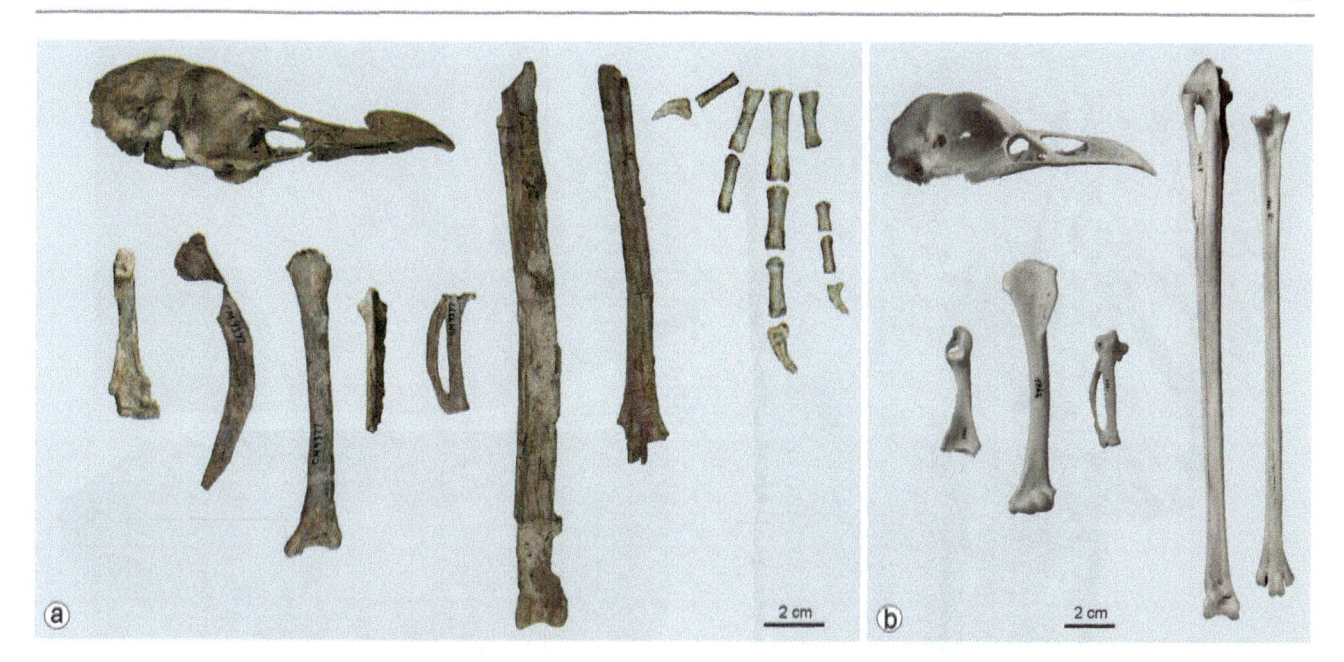

Fig. 8.6 (a) Selected bones (skull, coracoid, scapula, partial humerus, ulna, carpometacarpus, tibiotarsus, and tarsometatarsus as well as pedal phalanges) of the holotype of *Bathornis grallator* (Cariamiformes) from the middle Eocene of Wyoming, USA (Carnegie Museum, Pittsburgh, USA, CM 9377). (b) Selected bones of the extant *Cariama cristata* for comparison. (Photos in (a) by the author, in (b) by Sven Tränkner)

between North and South America during the Paleogene (e.g., Smith et al. 1994).

Because skeletal remains of bathornithids are the most common avian fossils in the Paleogene of the North America Great Plains, it is possible that a peculiar type of eggshell from the late Eocene and early Oligocene of Nebraska and North Dakota, which was assigned to the ootaxon *Metoolithus* (Lawver and Boyd 2018), may belong to these birds. The ornamented shell surface of these eggs, however, markedly differs from the eggs of extant Cariamidae, with an equally ornamented eggshell only being known from the palaeognathous Casuariiformes and non-avian theropods.

The reduced acrocoracoid process of the coracoid shows *B. grallator* to have been flightless. As detailed by Mayr (2016a), the humerus, coracoid, and carpometacarpus exhibit a strong similarity to the corresponding bones of a flightless cariamiform bird from the Quercy fissure fillings, which was described as *Ameghinornis minor* by Mourer-Chauviré (1981). The sole well-dated fossil of this species is 14–18 million years younger than the *B. grallator* holotype, but the early Oligocene *B. veredus* is just some 3–4 million years older than the Quercy fossil (Mayr 2016a). Even though correspondences in the mammalian faunas suggest that dispersal of a flightless bird between North America and Europe would have been possible in the late Eocene or early Oligocene (Mayr 2016a), the incomplete knowledge of the skeletal morphology of *A. minor* as well as the uncertain stratigraphic occurrence of this species (see next section) impede a well-founded assessment of its affinities to *B. grallator*, and the similar morphologies of the wing and pectoral girdle bones may well be the result of convergence in these flightless species.

8.4.4 *Strigogyps* and the Ameghinornithidae

The taxon *Strigogyps* was established by Gaillard (1908) for a distal tibiotarsus from an unknown stratigraphic horizon of the Quercy fissure fillings (Fig. 8.7e). This fossil, which is from a rooster-sized species, was described as *S. dubius* and agrees with owls (Strigiformes) in the absence of a supratendinal bridge. Even though Gaillard (1908) compared the specimen with extant Strigiformes and Accipitriformes, its unusual morphology did not permit an unambiguous identification. More than three decades later, Gaillard (1939) referred a humerus from the Quercy fissure fillings to a new species of *Strigogyps*, *Strigogyps minor*.

Mourer-Chauviré (1981) noted similarities between the humerus of *Strigogyps minor* and that of the Phorusrhacidae and classified the fossil into the new taxon *Ameghinornis*, as *A. minor* (Fig. 8.8a). Further bones assigned to this species by Mourer-Chauviré (1981) include two coracoids and two carpometacarpi. The exact age of most of these fossils is uncertain, but one of the coracoids is from an early Oligocene (MP 30; 30 Ma) locality (Mourer-Chauviré 1983). Mourer-Chauviré (1981) recognized the great similarity between the wing and pectoral girdle bones assigned to *A. minor* and the corresponding elements of the Phorusrhacidae. She considered *A. minor* to be an Old World representative of the

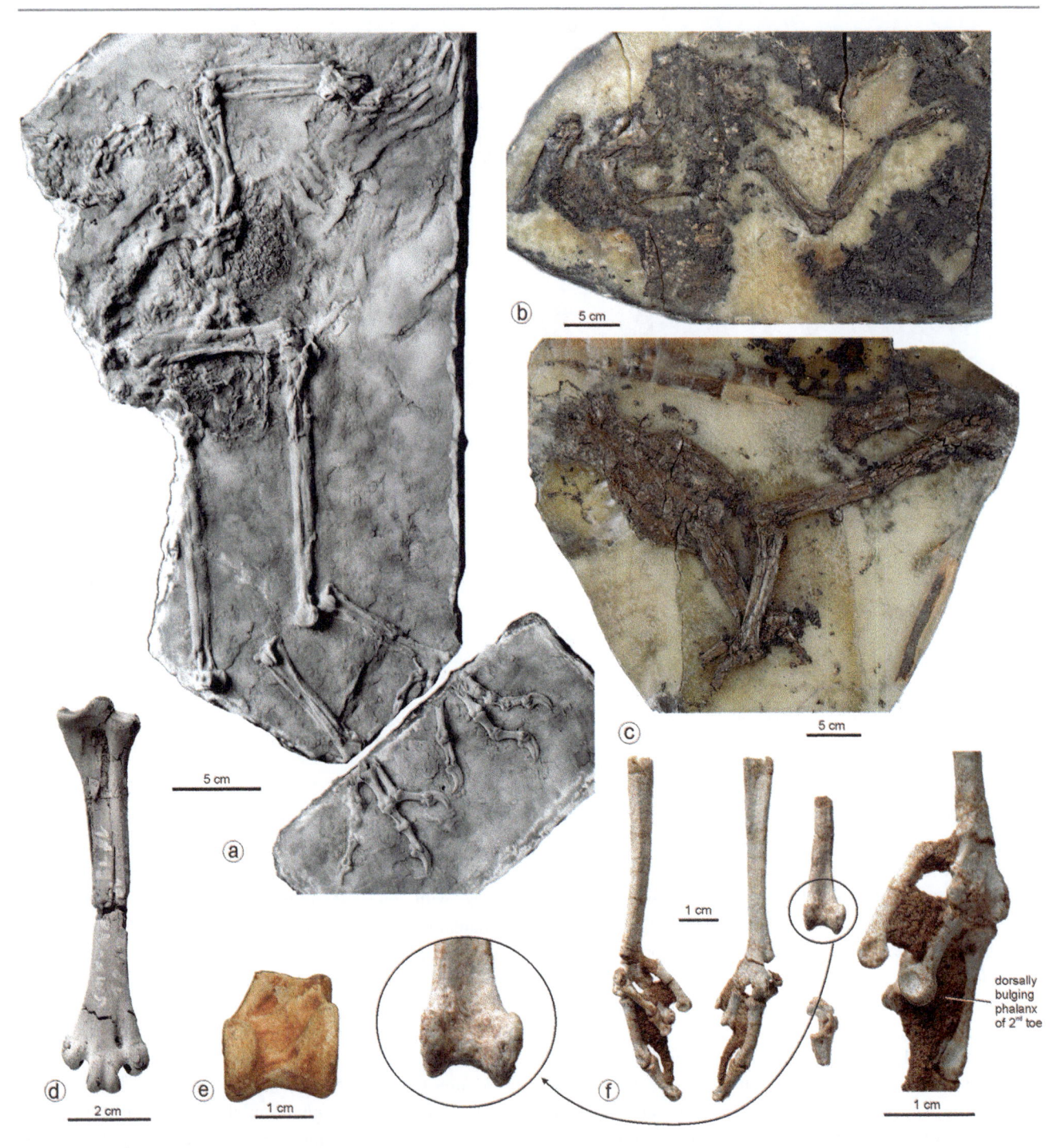

Fig. 8.7 (**a**) Skeleton of *Strigogyps sapea* from the latest early or earliest middle Eocene of Messel (holotype, Senckenberg Research Institute, Frankfurt, SMF-ME 1818); coated with ammonium chloride. (**b**), (**c**) Partial skeleton of *Strigogyps robustus* from the middle Eocene of the Geisel Valley (Geiseltalsammlung, Martin-Luther Universität of Halle-Wittenberg, Germany, **b**: holotype of *Geiseloceros robustus*, GMH 5884, **c**: holotype of *Eocathartes robustus*, GMH 5883). (**d**) Tarsometatarsus of *S. sapea* from Messel (SMF-ME 1819); coated with ammonium chloride. (**e**) Distal tarsometatarsus of *Strigogyps dubius* from an unknown stratigraphic horizon of the Quercy fissure fillings in France (cast of holotype, Musée Guimet d'Histoire naturelle, Lyon, France, MGHNL PQ 1073). (**f**) Holotype of *Qianshanornis rapax* from the mid-Paleocene of China (Institute of Botany, Chinese Academy of Sciences, Beijing, China, IBCAS QS027); the arrow indicates a detail of the distal end of the tibiotarsus. Specimens in (**a**) and (**d**) coated with ammonium chloride. (All photos by Sven Tränkner)

Phorusrhacidae and regarded the phylogenetic affinities of *S. dubius* as uncertain.

Peters (1987) described another putative phorusrhacid from the latest early or earliest middle Eocene of Messel. This species was named *Aenigmavis sapea* and is known

8.4 Cariamiformes (Seriemas and Allies)

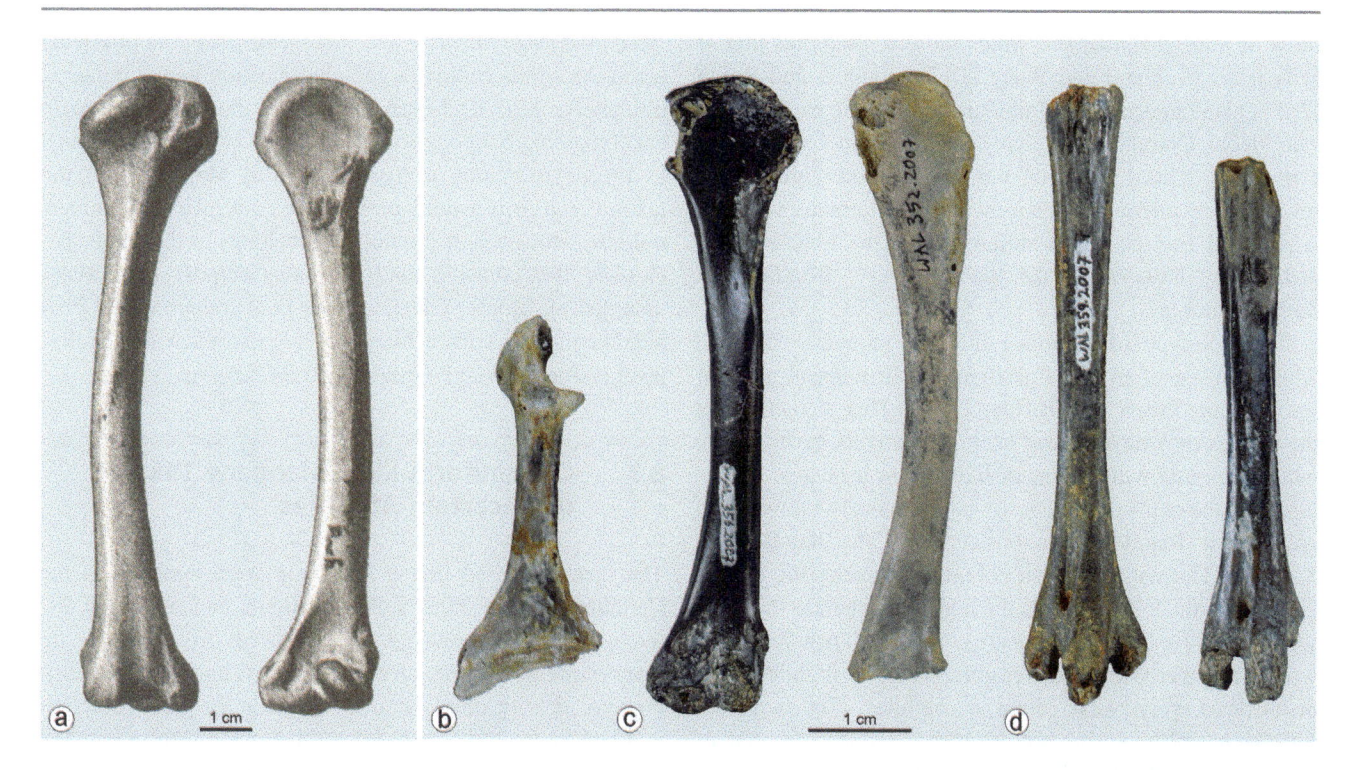

Fig. 8.8 (**a**) Humerus of *Ameghinornis minor* from an unknown stratigraphic horizon of the Quercy fissure fillings in France (from Gaillard 1939: fig. 4). (**b**)–(**c**) Coracoid, humeri, and tarsometatarsi of *Gradiornis walbeckensis* from the late middle Paleocene of Walbeck in Germany (all Institut für Geologische Wissenschaften of Martin-Luther- Universität Halle-Wittenberg, Halle/Saale, Germany; **a**: holotype coracoid, IGWuG WAL349.2007; **b**: humeri, IGWuG WAL356.2007 and WAL352.2007; **c**: tarsometatarsi, IGWuG WAL359.2007 and WAL358.2007). (Photos by Sven Tränkner)

from a postcranial skeleton and fragmentary remains of additional individuals (Fig. 8.7a; Mayr 2005b). It was about that of a male domestic chicken (*Gallus gallus*). The humerus closely resembles that of *Ameghinornis minor*, whereas the distal end of the tibiotarsus lacks an ossified supratendinal bridge and exhibits the characteristic morphology of *S. dubius*. As detailed by Mayr (2005b), the characteristic tibiotarsus morphology clearly shows "*Aenigmavis sapea*" to be a representative of *Strigogyps*, and the species is now classified as *S. sapea*. Because of the similarity in humerus morphology, Mayr (2005b) furthermore considered it likely that *Ameghinornis minor* is a junior synonym of *S. dubius* (see, however, below).

A further representative of *Strigogyps* with an equally confusing taxonomic history occurs in the middle Eocene of the Geisel Valley, in a slightly younger stratigraphic level than Messel (Fig. 8.7b, c; Mayr 2002a, 2020). This species is somewhat larger than *S. sapea* but otherwise very similar in its skeletal morphology (Mayr 2007b, 2020). The most substantial specimen consists of a split partial skeleton on two slabs, which was first studied by Lambrecht (1935). The slab with the wing and shoulder girdle elements was described as "*Geiseloceros robustus*" and compared with the Bucerotidae (hornbills) by Lambrecht (1935). The slab containing the pelvis and leg bones, which was found two meters away from that of the wing bones, formed the holotype of the putative New World vulture "*Eocathartes robustus*." Peter Houde (in Olson 1985) first assumed that "*Eocathartes robustus*" and "*Geiseloceros robustus*" represent a single species, and Olson (1985) noted that neither was probably correctly identified. Mayr (2007b) showed that the fossil closely corresponds to *Strigogyps sapea*, and the species from the Geisel Valley is now classified as *S. robustus*.

In summary, the taxon *Strigogyps* includes three named species, *Strigogyps dubius* (Quercy), *S. sapea* (Messel), and *S. robustus* (Geisel Valley). As yet undescribed remains of a *Strigogyps* species were furthermore found in the middle Eocene maar deposits of Eckfeld in Germany (own observation).

Strigogyps-like birds were also reported from outside Europe. *Qianshanornis rapax* from the mid-Paleocene of eastern China (Mayr et al. 2013) is represented by a partial foot and a few associated bones. As in *Strigogyps*, the distal end of the tibiotarsus lacks an ossified supratendinal bridge (Fig. 8.7f). A notable characteristic of the species is a dorsally bulging distal end of the first phalanx of the second toe, which distinguishes *Q. rapax* from all extant birds and indicates that this toe was dorsally hyperextendible.

Another unnamed species was described from the middle Eocene (48 Ma) of Inner Mongolia in western China on the basis of a distal tibiotarsus, which closely resembles that of *Strigogyps* (Stidham and Wang 2017). Stidham and Smith (2015) also likened a distal tibiotarsus from the early Oligocene Jebel Qatrani Formation of the Fayum in Egypt to *Strigogyps*. This fossil would be of potential significance for the assessment of Paleogene dispersal routes between Africa and Europe. Some differences to the tibiotarsus of *Strigogyps*, however, make the discovery of additional bones desirable.

The holotype of *S. sapea* includes stomach contents, which consist of plant parenchyma in an unusual state of preservation (Mayr and Richter 2011). This suggests that the species may have at least been facultatively herbivorous, which contrasts with the raptor-like pedal phalanges. Unfortunately, the skull of *Strigogyps* is very poorly known, and only small fragments are preserved in the holotype of *S. sapea*. The wing skeleton is highly characteristic in that the ulna is much shorter than the humerus, whereas the hand section is fairly long. The humerus has a small proximal end, which, together with the short ulna, indicates flightlessness, or at least very weak flight capabilities. The wing of *Strigogyps* is, however, not as greatly reduced as that of phorusrhacids. The pelvis is furthermore not as narrow as in phorusrhacids, and the hallux is not as greatly reduced.

Strigogyps is outside a clade including the Idiornithidae, Phorusrhacidae, and Cariamidae, because the hypotarsus does not exhibit the presumably derived morphology found in the latter taxa, in which it is block-like and without any distinct sulci and crests (in *Strigogyps* there are two well-developed crests). The ingested plant matter in the holotype of *S. sapea* (see above) furthermore conflicts with phorusrhacid affinities of *Strigogyps*. Phorusrhacid affinities of *Strigogyps* ("*Aenigmavis*") and *Ameghinornis* were also doubted by Alvarenga and Höfling (2003).

Mayr (2005b) hypothesized that the humerus of "*Strigogyps minor*", which was transferred to the taxon *Ameghinornis* by Mourer-Chauviré (1981), actually belongs to *Strigogyps dubius*, because the bone resembles the humeri of *S. sapea* and *S. robustus*. However, and as noted in the preceding section, the *Ameghinornis minor* humerus is equally similar to that of the North American *Bathornis grallator* (Mayr 2016a). The coracoid and carpometacarpus assigned to "*A. minor*" are clearly distinguished from the corresponding bones of *Strigogyps* and likewise show a close resemblance to *B. grallator*. These latter two bones certainly are not from *Strigogyps* and belong to a flightless species of the Cariamiformes (in size, they would correspond to the large taxon *Propelargus*, which is solely represented by a tibiotarsus and was assigned to the Idiornithidae; see Sect. 8.4.2). Currently, an unambiguous determination of whether the "*A. minor*" humerus belongs to *Strigogyps* (as suggested by Mayr 2005b) or to the other bones referred to "*A. minor*" by Mourer-Chauviré (1981) is not possible.

Because *Ameghinornis* is the type-genus of the family-level taxon Ameghinornithidae, the classification of the humerus ultimately has a bearing on the higher-level taxon encompassing *Strigogyps*. If the *A. minor* humerus belongs to *Strigogyps dubius*, as hypothesized by Mayr (2005b), the taxon *Strigogyps* would be classified into the Ameghinornithidae. However, if the humerus belongs to the same species as the coracoid and carpometacarpus, as initially assumed by Mourer-Chauviré (1981), *A. minor* would be a valid taxon that is distinct from *Strigogyps*, and a new family-level taxon needs to be established for *Strigogyps*.

8.5 Cariamiform-like Paleogene Taxa of Uncertain Affinities

The taxa discussed in the following were assigned to the Cariamiformes by some authors. All are, however, very different from the cariamiform taxa discussed above or are represented by fragmentary material that impedes an unambiguous identification. Their inclusion in the present section is therefore tentative, and a definitive classification requires further fossil material and future revisions.

8.5.1 *Elaphrocnemus* and *Paracrax*

The species of *Elaphrocnemus* (Fig. 8.9a) represent the most abundant medium-sized birds in the nineteenth-century Quercy collections. Mourer-Chauviré (1983) assigned the taxon to the Idiornithidae and recognized three species, *E. phasianus*, *E. crex*, and *E. brodkorbi*, which mainly differ in size. She furthermore noticed that the wing bones assigned to "*Filholornis*" by Milne-Edwards (1892) belong to the same taxon as the leg bones described as *Elaphrocnemus* by that author. New excavations have shown that *E. phasianus* occurs in late Eocene deposits, whereas *E. crex* is only known from Oligocene sites (Mourer-Chauviré 1983). The few fossils of *E. brodkorbi* lack exact stratigraphic data.

The *Elaphrocnemus* species were the size of small to midsize phasianids, and some skeletal elements of these birds are very different from those of *Dynamopterus* and extant seriemas (Cariamidae). Although the humerus, for example, has similar overall proportions to that of unambiguous Cariamiformes, it exhibits a prominent, subtriangular deltopectoral crest. The scapula is peculiar in that it bears an unusually long acromion that served to close the triosseal canal, which in *Dynamopterus* and extant Cariamiformes is achieved by the fusion of the procoracoid and acrocoracoid processes of the coracoid (Fig. 8.9a; Mayr and Mourer-Chauviré 2008). The tarsometatarsus of *Elaphrocnemus* is proportionally shorter than that of *Dynamopterus* and extant

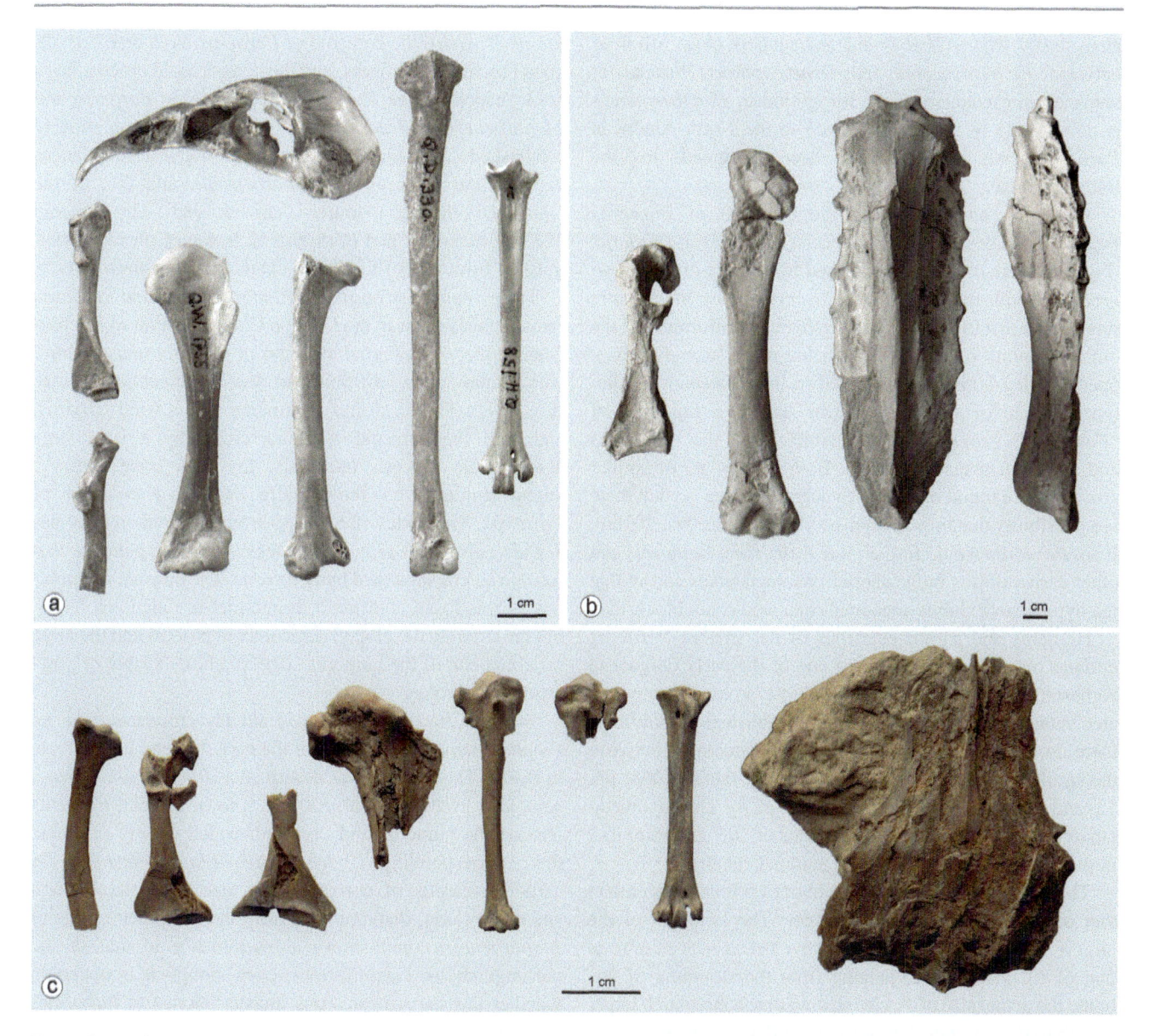

Fig. 8.9 (a) Skeletal elements of *Elaphrocnemus phasianus* from unknown stratigraphic horizons of the Quercy fissure fillings in France (skull in the Muséum national d'Histoire naturelle, Paris, MNHN QU 15688; all other bones in the Naturhistorisches Museum Basel, Switzerland, coracoid: NMB Q.D.242, cranial extremity of scapula: NMB Q.D.293, humerus: NMB Q.W.1755, femur: NMB Q.D.678, tibiotarsus: NMB Q.D.304/330, tarsometatarsus: NMB Q.H.158). (**b**) Selected bones (coracoid, humerus, and sternum in ventral and lateral view) of the holotype of *Paracrax wetmorei* from the early Oligocene of Colorado, USA (American Museum of Natural History, New York, USA, AMNH F.A.M. 42998). (**c**) Undescribed partial skeleton of a bird from the early Eocene of the London Clay of Walton-on-the-Naze, which may be related to one of the above species (collection of Michael Daniels, Holland-on-Sea, UK, WN 95848); the bone on the right is the pelvis in a block of matrix. (All figures to scale; photos in (**b**) by the author, all others by Sven Tränkner)

Cariamidae and exceeds the femur only slightly in length; unlike in *Dynamopterus* and other cariamiforms, the hypotarsus forms a distinct sulcus. Mayr and Mourer-Chauviré (2006) tentatively assigned a three-dimensionally preserved skull and some incomplete crania from unknown localities and stratigraphic horizons of the Quercy fissure fillings to *Elaphrocnemus phasianus*. This fossil exhibits well-developed articular facets for supraorbital processes, but in several features it differs from the skull of the Cariamidae, especially concerning the shape of the more slender beak (Fig. 8.9a).

The species of *Elaphrocnemus* seem to have been less cursorial than those of the Idiornithidae and Cariamidae, and probably they had better flight capabilities. *Elaphrocnemus* lacks characters shared by *Dynamopterus* and extant Cariamidae, such as a well-developed procoracoid process and a hooked acrocoracoid process of the coracoid, as well as a block-like hypotarsus (Mayr 2002b), and an analysis by Mayr (2016a) did not support close affinities between *Elaphrocnemus* and *Dynamopterus*. Instead of being a representative of the Idiornithidae, *Elaphrocnemus* is more likely to be closely related to the North American *Paracrax*, which has long been assigned to the Bathornithidae.

The wing and pectoral girdle elements of *Paracrax wetmorei*, the best-represented species of the taxon *Paracrax* (Fig. 8.9b), are clearly distinguished from those of *Bathornis grallator*, with the humerus of *Paracrax* being much more robust than that of *Bathornis* and other cariamiforms, and the omal extremity of the coracoid being more strongly developed. No leg bones are preserved in the *P. wetmorei* holotype, which is, however, coeval with the holotype of "*Bathornis*" *geographicus*, of which only the tibiotarsus and tarsometatarsus are known. Both species are of similar size and were found in the same geographic area, so that there is a possibility that they are conspecific (Mayr 2009a, 2016a). If so, *Paracrax* would furthermore differ from *Bathornis* and other cariamiforms in the shorter tarsometatarsus and in that the hypotarsus exhibits a tendinal canal and is not block-like.

Humerus and carpometacarpus of the smaller *Paracrax antiqua* may likewise belong to one of the early Oligocene bathornithid species, which are just known from leg elements. At least judging from the published illustrations, there do not seem to be any significant differences between the similarly-sized humeri of *P. antiqua* (Cracraft 1968: fig. 13) and "*B.*" *celeripes* (Wetmore 1958: pl. 4). The tarsometatarsus of "*B.*" *celeripes* agrees with that of "*B.*" *geographicus* in that the hypotarsus encloses a tendinal canal.

The humerus of *P. wetmorei* is about 1.5 times longer than that of *C. cristata* but much stouter. The humerus of the largest species, *P. gigantea*, may have been twice as long as that of *C. cristata*, and judging from the stoutness of this bone, the actual size of *P. gigantea* was probably much larger than that of *C. cristata*. The sternum of *P. wetmorei* exhibits a highly distinctive morphology in that the sternal keel is very low and notably reduced in its midsection. A similar condition otherwise solely occurs in extant Opisthocomiformes (hoatzins), in which the reduction of the sternal keel is due to space requirements of the very large crop of these folivorous birds. The remarkable similarity of the sternum of *Paracrax* and extant Opisthocomiformes suggests that *Paracrax* also had a large crop. In this case, the fossil taxon is likely to have been herbivorous, whereas extant Cariamiformes are carnivorous (as were phorusrhacids).

The humerus of *Paracrax* shows a resemblance to that of *Elaphrocnemus*, and in both taxa, the bone is much stouter than in *Bathornis* and extant Cariamiformes (Mayr 2016a). The coracoid of *Paracrax* also agrees with that of *Elaphrocnemus* in that the articular facet for the scapula is shallow, whereas it is concave in the Bathornithidae, Idiornithidae, and extant Cariamidae. However, unlike in *Elaphrocnemus*, the coracoid of *Paracrax* has a better developed procoracoid process, and the acrocoracoid process has a hook-shaped outline. Cariamiform affinities of *Paracrax* and *Elaphrocnemus* have not yet been unambiguously established, and earlier authors also noted resemblances between the humerus of *Elaphrocnemus* and that of the Opisthocomiformes (Mourer-Chauviré 1983). The coracoid of *Elaphrocnemus* and *Bathornis* is, however, clearly distinguished from that of fossil and extant Opisthocomiformes.

The evolutionary origins of *Elaphrocnemus* and *Paracrax* remain elusive, but in the London Clay collection of Michael Daniels there is a partial skeleton of a highly unusual bird, which may show affinities to the aforementioned taxa (Fig. 8.9c). The fossil has a humerus with a wide proximal end, and the coracoid is characterized by a very large acrocoracoid process. Especially the latter bone shows a vague similarity to the equally distinctive coracoid of *Paracrax*. The undescribed fossil from the London Clay has a wide pelvis and rather short legs, with a hypotarsus that forms a sulcus separated by two crests. The unusual morphology of this fossil is difficult to interpret in functional terms, but the strongly developed acrocoracoid process and the large proximal end of the humerus suggest a bird capable of powerful wing strokes.

Mourer-Chauviré (1999: 87) already commented on an *Elaphrocnemus*-like bird from the early Eocene Itaboraí site in Brazil. This species was described as *Itaboravis elaphrocnemoides* by Mayr et al. (2011). Known skeletal elements include the humerus and coracoid, which closely resemble the corresponding bones of *Elaphrocnemus*. A carpometacarpus of corresponding size from the Itaboraí site is, however, distinguished from the carpometacarpus of *Elaphrocnemus* and shows a resemblance to that of the palaeognathous Tinamiformes, even though it is uncertain whether this carpometacarpus indeed belongs to *Itaboravis* (Mayr et al. 2011).

8.5.2 Eleutherornithidae

A cranial portion of a large pelvis from the middle Eocene of Switzerland was described as *Eleutherornis helveticus* by Schaub (1940), who considered it to be from a palaeognathous bird. *E. helveticus* was about the size of a Lesser Rhea (*Pterocnemia pennata*). No compelling evidence exists, however, that this fossil belongs to a species of the Palaeognathae, and Schaub (1940) noted several features in which it agrees with the pelvis of neognathous birds. Pedal phalanges from the same locality, which were reported some years earlier by Schaub (1929), were considered to be too large to belong to the same species (Schaub 1940).

The size of *E. helveticus* corresponds to that of the coeval "*?Diatryma*" *cotei* from the middle Eocene (MP 14; Mourer-Chauviré 1996) of France, which is known from a distal tarsometatarsus and pedal phalanges of similar size to those of *E. helveticus* (Gaillard 1937). Gastornithid affinities of "*?D.*" *cotei* were already refuted by Andors (1992), with the ungual phalanges of the species being more curved than those of gastornithids and resembling ungual phalanges described by Schaub (1929) from the type locality of *E. helveticus*. In the first edition of the present book (Mayr 2009a), it was hypothesized that *E. helveticus* and "*?D.*" *cotei* may be closely related, which was also indicated by Gaillard (1937). Mlíkovský (2002: 186, 258) considered phorusrhacid affinities of *E. helveticus* and "*?D.*" *cotei*, and this hypothesis was further elaborated by Angst et al. (2013), who synonymized both species and assigned them—under the species name *Eleutherornis cotei*—to the Phorusrhacidae. However, the tarsometatarsus of *E. cotei* distinctly differs from phorusrhacids in the shape of the much wider trochlea for the second toe (Fig. 8.10). The fragmentary nature of the fossil material does not allow a well-founded phylogenetic assignment, but the tarsometatarsus and the curved ungual phalanges also resemble those of *Strigogyps*, and a closer relationship to the latter taxon appears to be more likely from a biogeographic point of view than affinities to the South American Phorusrhacidae.

The fossil material of *Proceriavis martini* from the early Oligocene of England consists of a fragmentary cervical vertebra and a referred pedal phalanx, and the species was tentatively referred to the Eleutherornithidae by Harrison and Walker (1979b), who considered the latter to be palaeognathous birds. Houde and Haubold (1987) confirmed the similarity between these specimens and the corresponding bones of extant Casuariiformes. Nonetheless, the sparse fossil material cannot be considered convincing evidence for the presence of palaeognathous birds in the early Oligocene of Europe. Harrison and Walker (1979b) also described a distal portion of a radius of a very large volant bird, which they regarded as a representative of the Pelagornithidae (Sect. 4.1).

Most recently, Buffetaut and Angst (2021) also tentatively assigned *Macrornis tanaupus* from the late Eocene of England to the Phorusrhacidae. The holotype of this species is a fragmentary tibiotarsus, which was considered non-avian by Harrison and Walker (1976). Judging from the published figures, I am not convinced of an avian identity of this bone and if it is indeed a tibiotarsus, it would differ from other birds (including phorusrhacids) in the presence of a marked ridge on its presumed cranial surface. The fossil may indicate the presence of large flightless birds in the late Eocene of Europe, which is possibly also suggested by the vertebra of *Proceriavis martini* from the early Oligocene of England, but even a tentative assignment to the Phorusrhacidae is not well founded.

8.5.3 Salmilidae

The Salmilidae comprise a single species, *Salmila robusta*, which is represented by three skeletons from the latest early or earliest middle Eocene of Messel (Fig. 8.11; Mayr 2000b, 2002b). In the collection of the Senckenberg Research Institute, there also is an as yet undescribed skeleton of a juvenile bird, which may belong to this species (Peters 1989: fig. 4).

S. robusta reached about two-thirds of the size of the extant Grey-winged Trumpeter (*Psophia crepitans*; Psophiidae) but had much shorter legs. Its phylogenetic affinities have not yet been convincingly established, but the species shares derived characters mainly with the Cariamiformes and the gruiform Psophiidae, the limb bones of which are superficially similar. As in these extant taxa, the carpometacarpus of *S. robusta* is short and with a bowed minor metacarpal, which bears a small tubercle in its proximal section. In contrast to extant Psophiidae and the Cariamiformes, however, the beak appears to have been schizorhinal. The skull lacks supraorbital processes, which are well developed in extant Cariamidae and the Phorusrhacidae. The limb bones have a robust appearance and resemble those of *Elaphrocnemus*; unlike in the latter, however, the deltopectoral crest has a rounded outline, whereas it is more projected in *Elaphrocnemus*. The caudal margin of the sternum exhibits a single pair of deep incisions, and the bone is therefore unlike the narrow sternum of the gruiform Gruoidea to which the Psophiidae belong. The distal end of the tibiotarsus possibly lacks an ossified supratendinal bridge (this is not clearly visible in the fossils), which, if confirmed by future specimens, may indicate affinities to *Strigogyps* (Sect. 8.4.4). The tarsometatarsus is rather short, which indicates that unlike most Cariamiformes, *S. robusta* was not a cursorial species. Also in contrast to most Cariamiformes, the hypotarsus is not block-like but exhibits a distinct sulcus, which separates two hypotarsal crests. As in the Gruoidea but unlike in extant Cariamidae, the first phalanx of the fourth toe bears a large, medially protruding projection. Two specimens with well-preserved feather remains allow the recognition of a long tail, which is similar to that of extant Cariamidae, whereas the tail of the Psophiidae is very short.

An initial phylogenetic analysis resulted in a sister group relationship between *S. robusta* and a clade including *Elaphrocnemus*, *Dynamopterus* as well as the Phorusrhacidae and Cariamidae (Mayr 2002b). However, this analysis included a restricted character and taxon sampling and supported a non-monophyly of the Gruoidea, which can no longer be upheld (Sect. 2.3).

More recently, *S. robusta* was recovered as the sister taxon of the Leptosomiformes in an analysis by Musser and Clarke (2020). Most likely, this placement is the result of inadequate character scoring, and most skeletal elements of *Salmila* and the Leptosomiformes are profoundly different. Overall, the

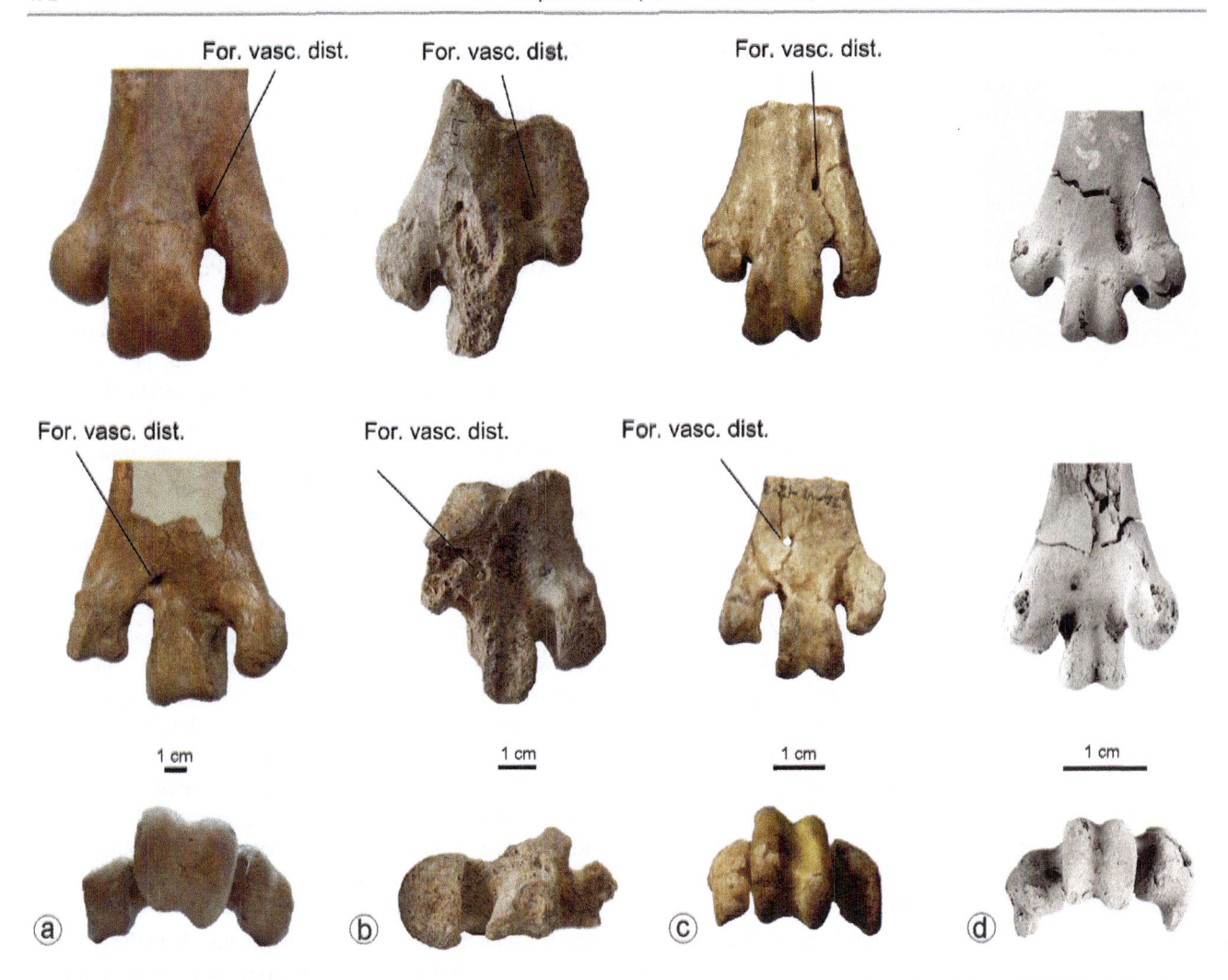

Fig. 8.10 Distal ends of the tarsometatarsus (dorsal, plantar, and distal views) of (**a**) *Gastornis parisiensis* (Gastornithidae), (**b**) *Eleutherornis cotei* (holotype of "?*Diatryma*" *cotei*, Eleutherornithidae), (**c**) *Psilopterus colzecus* (Phorusrhacidae), and (**d**) *Strigogyps sapea* (Senckenberg Research Institute, Frankfurt, SMF-ME 1819). (Photos in (**a**)–(**c**) from Angst et al. (2013: fig. 4), modified and some lettering digitally removed; published under a Creative Commons CC BY 4.0 license. Photos in (**d**) by Sven Tränkner)

skeleton of *Salmila* shows a resemblance to that of *Scandiornis* and *Nahmavis* (Sect. 5.2), and even though there are some differences in the morphology of the feet (Bertelli et al. 2013), these taxa may ultimately turn out to be more closely related than is apparent from their current classifications.

8.5.4 Gradiornis

Another problematic taxon is *Gradiornis walbeckensis*, a species from the late middle Paleocene of Walbeck in Germany (Fig. 8.8b–d), which is represented by a few bones only (humeri, coracoids, and tarsometatarsi; Mayr 2007c). The referred humeri resemble those of *Strigogyps* and *Bathornis*, and the distal ends of the tarsometatarsi are also similar to the Cariamiformes. The peculiar and characteristic morphology of the coracoid (which forms the holotype of the species), however, is not matched by any other avian taxon. The bone has a very straight shaft, lacks a foramen for the supracoracoideus nerve, and exhibits a deeply excavated articular facet for the scapula. *Gradiornis* lacks the derived, block-like hypotarsus of the Cariamiformes, and its phylogenetic affinities are unresolved. A tentatively referred notarium consists of two co-ossified thoracic vertebrae; if it indeed belongs to *Gradiornis*, it would constitute another feature, in which the taxon differs from the Cariamiformes. The long and robust tarsometatarsus indicates that *G. walbeckensis* was a terrestrial bird, and the humerus morphology suggests weak flight capabilities. A *Gradiornis*-like tarsometatarsus was also described by Mayr and Smith (2019b) from the late Paleocene (latest Thanetian) of the locality Rivecourt-Petit Pâtis in France.

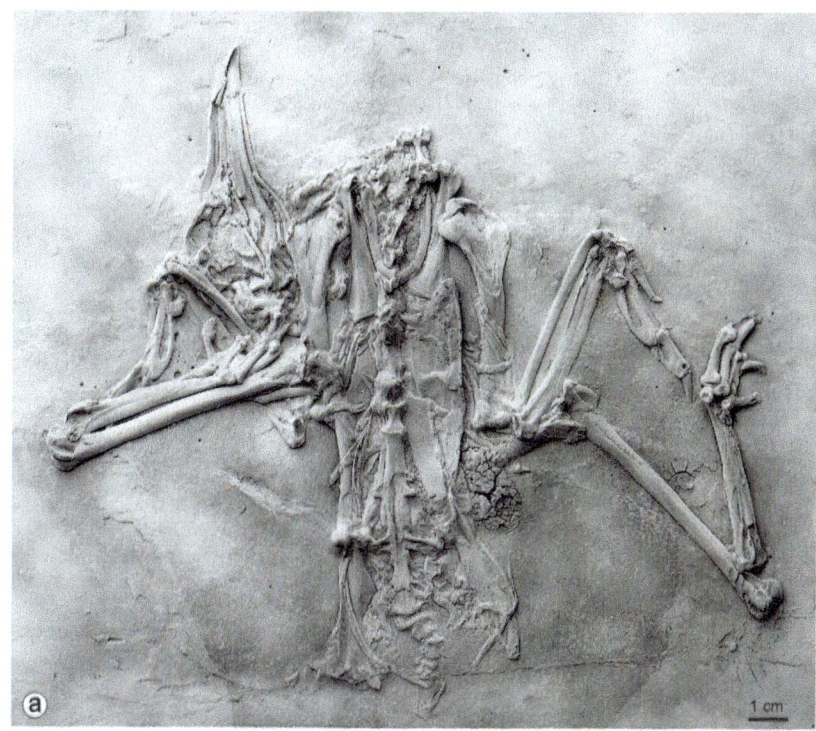

Fig. 8.11 (a), (b) Skeletons of *Salmila robusta* (Salmilidae) from the latest early or earliest middle Eocene of Messel (a: Hessisches Landesmuseum, Darmstadt, Germany, HLMD Be 161, coated with ammonium chloride; b: holotype, Senckenberg Research Institute, Frankfurt, SMF-ME 3014). (Photos by Sven Tränkner)

References

Acosta Hospitaleche C, Tambussi C (2005) Phorusrhacidae Psilopterinae (Aves) en la Formación Sarmiento de la localidad de Gran Hondonada (Eoceno Superior), Patagonia, Argentina. Rev Español Paleontol 20:127–132

Agnolin FL (2004) Revisión systemática de algunas aves deseadenses (Oligoceno Medio) descriptas por Ameghino en 1899. Rev Mus Argent Cienc Nat, n s 6:239–244

Agnolin FL (2006a) Notas sobre el registro de Accipitridae (Aves, Accipitriformes) fósiles Argentinos. Studia Geol Salmanticensia 42:67–80

Agnolin FL (2006b) Posición sistemática de algunas aves fororracoideas (Ralliformes; Cariamae) Argentinas. Rev Mus Argent Cienc Nat, n s 8:27–33

Agnolin FL (2007) *Brontornis burmeisteri* Moreno & Mercerat, un Anseriformes (Aves) gigante del Mioceno Medio de Patagonia, Argentina. Rev Mus Argent Cienc Nat, n s 9:15–25

Agnolin FL (2009) Sistemática y filogenia de las aves fororracoideas (Gruiformes: Cariamae). Fundación de Historia Natural Félix de Azara, Buenos Aires

Agnolin FL (2021) Reappraisal on the phylogenetic relationships of the enigmatic flightless bird (*Brontornis burmeisteri*) Moreno and Mercerat, 1891. Diversity 13(2):90

Agnolin FL, Pais DF (2006) Revisión de *Cunampaia simplex* Rusconi, 1946 (Crocodylomorpha, Mesoeucrocodilia; *non* Aves) del Terciario inferior de Mendoza, Argentina. Rev Mus Argent Cienc Nat, n s 8:35–40

Alvarenga HMF (1982) Uma gigantesca ave fóssil do Cenozóico Brasileiro: *Physornis brasiliensis* sp. n. An Acad Bras Cienc 54:697–712

Alvarenga HMF (1985a) Notas sobre os Cathartidae (Aves) e descrição de um novo gênero do Cenozóico Brasileiro. An Acad Bras Cienc 57:349–357

Alvarenga HMF (1985b) Um novo Psilopteridae (Aves: Gruiformes) dos sedimentos Terciários de Itaboraí, Rio de Janeiro, Brasil. In: Congresso Brasileiro de Paleontologia, 8. Anais. Rio de Janeiro, MME-DNPM, 1983 (Sér Geol 27):17–20

Alvarenga HMF (1993) *Paraphysornis* novo gênero para *Physornis brasiliensis* Alvarenga, 1982 (Aves: Phorusrhacidae). An Acad Bras Cienc 65:403–406

Alvarenga HMF, Höfling E (2003) Systematic revision of the Phorusrhacidae (Aves: Ralliformes). Pap Avulsos Zool 43:55–91

Alvarenga H, Chiappe L, Bertelli S (2011) Phorusrhacids: the terror birds. In: Dyke G, Kaiser G (eds) Living dinosaurs. The evolutionary history of modern birds. John Wiley & Sons, Chichester, pp 187–208

Ameghino F (1899) Sinópsis geológico-paleontológica. Suplemento (adiciones y correciones), La Plata

Andors A (1992a) Reappraisal of the Eocene groundbird *Diatryma* (Aves: Anserimorphae). In: Campbell KE (ed) Papers in avian paleontology honoring Pierce Brodkorb. Nat Hist Mus Los Angeles Cty Sci Ser 36:109–125

Angst D, Buffetaut E (2017) Paleobiology of giant flightless birds. ISTE Press, London

Angst D, Buffetaut E, Lécuyer C, Amiot R (2013) "Terror Birds" (Phorusrhacidae) from the Eocene of Europe imply trans-Tethys dispersal. PLoS ONE 8:e80357

Bertelli S, Lindow BEK, Dyke GJ, Mayr G (2013) Another charadriiform-like bird from the Lower Eocene of Denmark. Paleontol J 47:1282–1301

Boles WE (1993) *Pengana robertbolesi*, a peculiar bird of prey from the Tertiary of Riversleigh, northwestern Queensland, Australia. Alcheringa 17:19–25

Braun EL, Kimball RT (2021) Data types and the phylogeny of Neoaves. Birds 2:1–22

Brodkorb P (1964) Catalogue of fossil birds. Part 2 (Anseriformes through Galliformes). Bull Fla State Mus, Biol Sci 8(3):195–335

Brodkorb P (1967) Catalogue of fossil birds. Part 3 (Ralliformes, Ichthyornithiformes, Charadriiformes). Bull Fla State Mus, Biol Sci 11:99–220

Brunet J (1970) Oiseaux de l'Éocène supérieur du bassin de Paris. Ann Paléontol 56:3–57

Buffetaut E (2017) A brontornithid from the Deseadan (Oligocene) of Bolivia. Contrib Mus Argent Cienc Nat 7:39–47

Buffetaut E, Angst D (2021) *Macrornis tanaupus* Seeley, 1866: an enigmatic giant bird from the upper Eocene of England. Geol Mag 158:1129–1134

Campbell KE Jr, Tonni EP (1980) A new genus of teratorn from the Huayquerian of Argentina (Aves: Teratornithidae). Nat Hist Mus Los Angeles Cty, Contrib Sci 330:59–68

Campbell KE Jr, Tonni EP (1981) Preliminary observations on the paleobiology and evolution of teratorns (Aves: Teratornithidae). J Vertebr Paleontol 1:265–272

Campbell KE Jr, Tonni EP (1983) Size and locomotion in teratorns (Aves: Teratornithidae). Auk 100:390–403

Case JA, Woodburne MO, Chaney DS (1987) A gigantic phororhacoid (?) bird from Antarctica. J Paleontol 61:1280–1284

Cenizo MM (2012) Review of the putative Phorusrhacidae from the Cretaceous and Paleogene of Antarctica: new records of ratites and pelagornithid birds. Pol Polar Res 33:239–258

Cenizo M, Noriega JI, Reguero MA (2016) A stem falconid bird from the Lower Eocene of Antarctica and the early southern radiation of the falcons. J Ornithol 157:885–894

Cracraft J (1968) A review of the Bathornithidae (Aves, Gruiformes), with remarks on the relationships of the suborder Cariamae. Am Mus Novit 2326:1–46

Cracraft J (1971) Systematics and evolution of the Gruiformes (Class Aves) 2. Additional comments on the Bathornithidae, with descriptions of new species. Am Mus Novit 2449:1–14

Cracraft J (1973) Systematics and evolution of the Gruiformes (Class Aves). 3. Phylogeny of the suborder Grues. Bull Am Mus Nat Hist 151:1–127

Cracraft J, Rich PV (1972) The systematics and evolution of the Cathartidae in the Old World Tertiary. Condor 74:272–283

Daxner-Höck G, Erbajeva MA, Göhlich UB, López-Guerrero P, Narantsetseg T, Mennecart B, Oliver A, Vasilyan D, Ziegler R (2019) The Oligocene vertebrate assemblage of Shine Us (Khaliun Basin, south western Mongolia). Ann Naturhist Mus Wien, Ser A 121:195–256

Gaillard C (1908) Les oiseaux des Phosphorites du Quercy. Ann Univ Lyon (Nouv Sér) 23:1–178

Gaillard C (1937) Un oiseau géant dans les dépots éocènes du Mont-d'Or lyonnais. Ann Soc Linn Lyon (Nouv Sér) 80:111–126

Gaillard C (1939) Contribution à l'étude des oiseaux fossiles. Nouv Arch Mus Lyon 15(2):1–100

Grande L (2013) The lost world of Fossil Lake: Snapshots from deep time. University of Chicago Press, Chicago

Griffiths CS (1999) Phylogeny of the Falconidae inferred from molecular and morphological data. Auk 116:116–130

Harrison CJO (1982) A new tiny raptor from the Lower Eocene of England. Ardea 70:77–80

Harrison CJO (1984) Further additions to the Fossil Birds of Sheppey: A new Falconid and three small Rails. Tert Res 5:179–187

Harrison CJO, Walker CA (1976) Birds of the British Upper Eocene. Zool J Linnean Soc 59:323–351

Harrison CJO, Walker CA (1979a) Birds of the British Middle Eocene. In: Harrison CJO, Walker CA (eds) Studies in Tertiary avian paleontology. Tert Res Spec Pap 5:19–27

Harrison CJO, Walker CA (1979b) Birds of the British Lower Oligocene. In: Harrison CJO, Walker CA (eds) Studies in Tertiary avian paleontology. Tert Res Spec Pap 5:29–43

Houde P, Haubold H (1987) *Palaeotis weigelti* restudied: a small Middle Eocene ostrich (Aves: Struthioniformes). Palaeovertebr 17:27–42

Kuhl H, Frankl-Vilches C, Bakker A, Mayr G, Nikolaus G, Boerno ST, Klages S, Timmermann B, Gahr M (2021) An unbiased molecular approach using 3'UTRs resolves the avian family-level tree of life. Mol Biol Evol 38:108–121

Kurochkin EN (1968) [Fossil remains of birds from Mongolia]. Ornitologija 9:323–330 [in Russian]

Kurochkin EN (1976) A survey of the Paleogene birds of Asia. Smithson Contrib Paleobiol 27:5–86

Lambrecht K (1935) Drei neue Vogelformen aus dem Lutétian des Geiseltales. Nova Acta Leopold, N F 3:361–367

Lawver DR, Boyd CA (2018) An avian eggshell from the Brule Formation (Oligocene) of North Dakota. J Vertebr Paleontol 38:e1486848

Mather EK, Lee MS, Camens AB, Worthy TH (in press) An exceptional partial skeleton of a new basal raptor (Aves: Accipitridae) from the late Oligocene Namba formation, South Australia. Hist Biol

Mayr G (2000a) New or previously unrecorded avian taxa from the Middle Eocene of Messel (Hessen, Germany). Mitt Mus Naturkunde Berl, Geowiss Reihe 3:207–219

Mayr G (2000b) A remarkable new 'gruiform' bird from the Middle Eocene of Messel (Hessen, Germany). Paläontol Z 74:187–194

Mayr G (2002a) Avian Remains from the Middle Eocene of the Geiseltal (Sachsen-Anhalt, Germany). In: Zhou Z, Zhang F (eds) Proceedings of the 5th symposium of the Society of Avian Paleontology and Evolution, Beijing, 1–4 June 2000. Science Press, Beijing, pp 77–96

Mayr G (2002b) A new specimen of *Salmila robusta* (Aves: Gruiformes: Salmilidae n. fam.) from the Middle Eocene of Messel. Paläontol Z 76:305–316

Mayr G (2005a) The Paleogene fossil record of birds in Europe. Biol Rev 80:515–542

Mayr G (2005b) "Old World phorusrhacids" (Aves, Phorusrhacidae): a new look at *Strigogyps* ("*Aenigmavis*") *sapea* (Peters 1987). PaleoBios 25:11–16

Mayr G (2006a) An osprey (Aves: Accipitridae: Pandioninae) from the early Oligocene of Germany. Senck leth 86:93–96

Mayr G (2006b) A new raptorial bird from the Middle Eocene of Messel, Germany. Hist Biol 18:95–102

Mayr G (2007a) Bizarre tubercles on the vertebrae of Eocene fossil birds indicate an avian disease without modern counterpart. Naturwiss 94: 681–685

Mayr G (2007b) Synonymy and actual affinities of the putative Middle Eocene "New World vulture" *Eocathartes* Lambrecht, 1935 and "hornbill" *Geiseloceros* Lambrecht, 1935 (Aves, Ameghinornithidae). Paläontol Z 81:457–462

Mayr G (2007c) The birds from the Paleocene fissure filling of Walbeck (Germany). J Vertebr Paleontol 27:394–408

Mayr G (2009a) Paleogene fossil birds, 1st edn. Springer, Heidelberg

Mayr G (2009b) A well-preserved skull of the "falconiform" bird *Masillaraptor* from the middle Eocene of Messel (Germany). Palaeodiv 2:315–320

Mayr G (2010) A new avian species with tubercle-bearing cervical vertebrae from the Middle Eocene of Messel (Germany). Rec Austral Mus 62:21–28

Mayr G (2016a) Osteology and phylogenetic affinities of the middle Eocene North American *Bathornis grallator*—one of the best represented, albeit least known Paleogene cariamiform birds (seriemas and allies). J Paleontol 90:357–374

Mayr G (2016b) Avian feet, crocodilian food and the diversity of larger birds in the early Eocene of Messel. Palaeobiodiv Palaeoenviron 96: 601–609

References

Mayr G (2017) Avian Evolution: The fossil record of birds and its paleobiological significance. Wiley-Blackwell, Chichester

Mayr G (2020) An updated review of the middle Eocene avifauna from the Geiseltal (Germany), with comments on the unusual taphonomy of some bird remains. Geobios 62:45–59

Mayr G, Hurum JH (2020) A tiny, long-legged raptor from the early Oligocene of Poland may be the earliest bird-eating diurnal bird of prey. Sci Nat 107:48

Mayr G, Mourer-Chauviré C (2006) Three-dimensionally preserved cranial remains of *Elaphrocnemus* (Aves, Cariamae) from the Paleogene Quercy fissure fillings in France. Neues Jahrb Geol Paläontol, Mh 2006:15–27

Mayr G, Mourer-Chauviré C (2008) The peculiar scapula of the late Eocene *Elaphrocnemus phasianus* Milne-Edwards, 1892 (Aves, Cariamae). Senck Leth 88:195–198

Mayr G, Perner T (2020) A new species of diurnal birds of prey from the late Eocene of Wyoming (USA)—one of the earliest New World records of the Accipitridae (hawks, eagles, and allies). N Jb Geol Paläont (Abh) 297:205–215

Mayr G, Richter G (2011) Exceptionally preserved plant parenchyma in the digestive tract indicates a herbivorous diet in the Middle Eocene bird *Strigogyps sapea* (Ameghinornithidae). Paläontol Z 85:303–307

Mayr G, Smith R (2002) Avian remains from the lowermost Oligocene of Hoogbutsel (Belgium). Bull Inst Roy Sci Nat Belg 72:139–150

Mayr G, Smith T (2019a) A diverse bird assemblage from the Ypresian of Belgium furthers knowledge of early Eocene avifaunas of the North Sea Basin. N Jb Geol Paläont (Abh) 291:253–281

Mayr G, Smith T (2019b) New Paleocene bird fossils from the North Sea Basin in Belgium and France. Geol Belg 22:35–46

Mayr G, Alvarenga H, Clarke J (2011) An *Elaphrocnemus*-like landbird and other avian remains from the late Paleocene of Brazil. Acta Palaeontol Pol 56:679–684

Mayr G, Yang J, de Bast E, Li C-S, Smith T (2013) A *Strigogyps*-like bird from the middle Paleocene of China with an unusual grasping foot. J Vertebr Paleontol 33:895–901

Miller LH (1909) Teratornis, a new avian genus from Rancho La Brea. Univ California Publ, Bull Dept Geol 5:305–317

Miller LH (1910) The condor-like vultures of Rancho La Brea. Univ California Publ, Bull Dept Geol 6:1–19

Miller AH, Sibley CG (1942) An Oligocene hawk from Colorado. Condor 44:39–40

Milne-Edwards A (1892) Sur les oiseaux fossiles des dépots éocènes de phosphate de chaux du Sud de la France. C R Second Congr Ornithol Internat:60–80

Mindell D, Fuchs J, Johnson J (2018) Phylogeny, taxonomy, and geographic diversity of diurnal raptors: Falconiformes, Accipitriformes, and Cathartiformes. In: Sarasola JH, Grande JM, Negro JJ (eds) Birds of Prey—Biology and conservation in the XXI century. Springer, Cham, pp 3–32

Mlíkovský J (2002) Cenozoic birds of the world. Part 1: Europe. Ninox Press, Praha

Mourer-Chauviré C (1981) Première indication de la présence de Phorusracidés [sic], famille d'oiseaux géants d'Amérique du Sud, dans le Tertiaire européen: *Ameghinornis* nov. gen. (Aves, Ralliformes) des Phosphorites du Quercy, France. Geobios 14:637–647

Mourer-Chauviré C (1983) Les Gruiformes (Aves) des Phosphorites du Quercy (France). 1. Sous-ordre Cariamae (Cariamidae et Phorusrhacidae). Systématique et biostratigraphie. Palaeovertebrata 13(4):83–143

Mourer-Chauviré C (1988) Le gisement du Bretou (Phosphorites du Quercy, Tarn-et-Garonne, France) et sa faune de vertébrés de l'Eocène supérieur. II Oiseaux. Palaeontographica (A) 205:29–50

Mourer-Chauviré C (1991) Les Horusornithidae nov. fam., Accipitriformes (Aves) à articulation intertarsienne hyperflexible de l'Éocene du Quercy. Geobios, mém spéc 13:183–192

Mourer-Chauviré C (1996) Paleogene avian localities of France. In: Mlíkovský J (ed) Tertiary Avian Localities of Europe. Acta Univ Carol Geol 39:567–598

Mourer-Chauviré C (1999) Les relations entre les avifaunes du Tertiaire inférieur d'Europe et d'Amérique du Sud. Bull Soc Géol France 170:85–90

Mourer-Chauviré C (2002) Revision of the Cathartidae (Aves, Ciconiiformes) from the Middle Eocene to the Upper Oligocene Phosphorites du Quercy, France. In: Zhou Z, Zhang F (eds) Proceedings of the 5th symposium of the Society of Avian Paleontology and Evolution, Beijing, 1–4 June 2000. Science Press, Beijing, pp 97–111

Mourer-Chauviré C (2003) Birds (Aves) from the Middle Miocene of Arrisdrift (Namibia). Preliminary study with description of two new genera: *Amanuensis* (Accipitridae, Sagittariidae) and *Namibiavis* (Gruiformes, Idiornithidae). Mem Geol Surv Namibia 19:103–113

Mourer-Chauviré C (2006) The avifauna of the Eocene and Oligocene Phosphorites du Quercy (France): an updated list. Strata, sér 1 13:135–149

Mourer-Chauviré C (2013) *Idiornis* Oberholser, 1899 (Aves, Gruiformes, Cariamae, Idiornithidae): A junior synonym of *Dynamopterus* Milne-Edwards, 1892 (Paleogene, Phosphorites du Quercy, France). N Jb Geol Paläont (Abh) 270:13–22

Mourer-Chauviré C, Cheneval J (1983) Les Sagittariidae fossiles (Aves, Accipitriformes) de l'Oligocène des Phosphorites du Quercy et du Miocène inférieur du Saint-Gérand-le-Puy. Geobios 16:443–459

Mourer-Chauviré C, Tabuce R, Mahboubi MH, Adaci M, Bensalah M (2011) A Phororhacoid bird from the Eocene of Africa. Naturwiss 98:815–823

Musser G, Clarke JA (2020) An exceptionally preserved specimen from the Green River Formation elucidates complex phenotypic evolution in Gruiformes and Charadriiformes. Front Ecol Evol 8:559929

Olson SL (1985) The fossil record of birds. In: Farner DS, King JR, Parkes KC (eds) Avian biology, vol 8. Academic Press, New York, pp 79–238

Olson SL (1999) Early Eocene birds from eastern North America: A faunule from the Nanjemoy Formation of Virginia. In: Weems RE, Grimsley GJ (eds) Early Eocene Vertebrates and Plants from the Fisher/Sullivan Site (Nanjemoy Formation) Stafford County, Virginia. Virginia Div Mineral Resour Publ 152:123–132

Olson SL, Alvarenga HMF (2002) A new genus of small teratorn from the Middle Tertiary of the Taubaté Basin, Brazil (Aves: Teratornithidae). Proc Biol Soc Wash 115:701–705

Peters DS (1987) Ein „Phorusrhacide" aus dem Mittel-Eozän von Messel (Aves: Gruiformes: Cariamae). Doc Lab Géol Lyon 99:71–87

Peters DS (1989) Fossil birds from the oil shale of Messel (Lower Middle Eocene, Lutetian). In: Ouellet H (ed) Acta XIX congressus internationalis ornithologici. University of Ottawa Press, Ottawa, pp 2056–2064

Peters DS (1995) *Idiornis tuberculata* n. spec., ein weiterer ungewöhnlicher Vogel aus der Grube Messel (Aves: Gruiformes: Cariamidae: Idiornithinae). In: Peters DS (ed) Acta palaeornithologica. Cour Forsch-Inst Senckenberg 181:107–119

Rasmussen DT, Olson SL, Simons EL (1987) Fossil birds from the Oligocene Jebel Qatrani Formation, Fayum Province, Egypt. Smithson Contrib Paleobiol 62:1–20

Schaub S (1929) Über eocäne Ratitenreste in der osteologischen Sammlung des Basler Museums. Verh Naturforsch Ges Basel 40:588–598

Schaub S (1940) Ein Ratitenbecken aus dem Bohnerz von Egerkingen. Eclogae Geol Helv 33:274–284

Smith AG, Smith DG, Funnell BM (1994) Atlas of Mesozoic and Cenozoic coastlines. Cambridge University Press, Cambridge

Stidham TA, Smith NA (2015) An ameghinornithid-like bird (Aves, Cariamae, ?Ameghinornithidae) from the early Oligocene of Egypt. Palaeontol Electron 18.1(5A):1–8

Stidham TA, Wang YQ (2017) An ameghinornithid-like bird (Aves: Cariamae: Ameghinornithidae?) from the Middle Eocene of Nei Mongol, China. Vertebr PalAsiat 55:218–226

Tambussi CP, Acosta Hospitaleche C (2007) Antarctic birds (Neornithes) during the Cretaceous-Eocene times. Rev Asoc Geol Argent 62:604–617

Tonni EP (1980) The present state of knowledge of the Cenozoic birds of Argentina. In: Campbell KE (ed) Papers in avian paleontology honoring Hildegarde Howard. Nat Hist Mus Los Angeles Cty Contrib Sci 330:105–114

Wetmore A (1927) Fossil birds from the Oligocene of Colorado. Proc Colo Mus Nat Hist 7:1–13

Wetmore A (1933a) An Oligocene eagle from Wyoming. Smithson Misc Collect 87:1–9

Wetmore A (1933b) Bird remains from the Oligocene deposits of Torrington, Wyoming. Bull Mus Comp Zool 75:297–311

Wetmore A (1933c) A second specimen of the fossil bird *Bathornis veredus*. Auk 50:213–214

Wetmore A (1944) A new terrestrial vulture from the Upper Eocene deposits of Wyoming. Ann Carnegie Mus 30:57–69

Wetmore A (1958) Miscellaneous notes on fossil birds. Smithson Misc Collect 135(8):1–11

Wetmore A, Case EC (1934) A new fossil hawk from the Oligocene beds of South Dakota. Contrib Mus Paleontol, Univ Michigan 4:129–132

Worthy TH, Degrange FJ, Handley WD, Lee MS (2017) The evolution of giant flightless birds and novel phylogenetic relationships for extinct fowl (Aves, Galloanseres). R Soc Open Sci 4(10):170975

Zelenkov NV, Kurochkin EN (2015) Class aves. In Kurochkin EN, Lopatin AV, Zelenkov NV (edes) [Fossil vertebrates of Russia and neighbouring countries: fossil reptiles and birds. Part 2]. GEOS, Moscow, pp 86–290 [in Russian]

9. Psittacopasseres: Psittaciformes (Parrots) and Passeriformes (Passerines)

Earlier analyses of morphological data recovered the Piciformes or the Coraciiformes as the closest relatives of the Passeriformes (e.g., Mayr and Clarke 2003; Livezey and Zusi 2007), whereas all current molecular data sets congruently support a sister group relationship between the Psittaciformes and the Passeriformes (Hackett et al. 2008; Suh et al. 2011; Yuri et al. 2013; Jarvis et al. 2014; Prum et al. 2015; Kuhl et al. 2021). The clade including these two latter taxa was termed Psittacopasserae (Suh et al. 2011), which is here emended to the grammatically correct name Psittacopasseres.

Parrots and passerines differ in many aspects of their external and internal morphology. Extant Psittaciformes are short-legged, zygodactyl birds with a reversed fourth toe, and as in other zygodactyl birds, the tarsometatarsal trochlea for the fourth toe bears a large accessory trochlea. The long-legged Passeriformes, by contrast, have anisodactyl feet, in which the fourth toe directs forward as it does in most other birds. As already noted in the first edition of this book (Mayr 2009), the hypothesis of a sister group relationship between the Passeriformes and the Psittaciformes is of particular interest, because the Passeriformes have an extinct sister taxon with zygodactyl feet, the aptly named Zygodactylidae, which are among the more common small arboreal birds in Paleogene sites of the Northern Hemisphere.

The distal end of the tarsometatarsus of the Zygodactylidae closely resembles the distal tarsometatarsus of psittaciform birds, which suggests that the stem species of the clade including the Passeriformes and Psittaciformes had parrot-like, zygodactyl feet and that passerines secondarily regained anisodactyl feet in their evolutionary history. Based on these new insights in the evolution of parrots and passerines, some fossil taxa, which were initially identified as stem group representatives of the Psittaciformes (Mayr 2009), are now considered to be zygodactyl stem group passerines (Fig. 9.1).

Botelho et al. (2014) identified a possible developmental mechanism, which may explain a secondary loss of the tarsometatarsal structures associated with a zygodactyl foot. According to these authors, a zygodactyl foot is formed through a reduction of the extensor muscles of the fourth toe. The resultant asymmetric muscular forces lead to a reversion of this toe, and during the ontogenetic development of the embryo, the osseous structures associated with a zygodactyl foot are formed by this changed orientation of the fourth toe. The Passeriformes are characterized by a reduction of many of the foot muscles, which, according to Botelho et al. (2014), restores muscular balance and prevents the fourth toe from turning back in early ontogeny. As a consequence, the skeletal correlates of a zygodactyl foot are not formed. However, plausible as this hypothesis may be, the extensor muscle of the fourth toe appears to have been well-developed in the Zygodactylidae (Olson 1985), which places a caveat on the above-outlined developmental model.

The functional reasons for a secondary loss of a zygodactyl foot in the stem lineage of the Passeriformes remain elusive. Traditionally, this foot morphology was considered a perching adaptation. Because passerines are the most diversified group of perching birds, a secondary loss of zygodactyl feet would be difficult to explain under this hypothesis. However, zygodactyl feet often occur in birds that nest in tree cavities and, in doing so, need to cling to vertical surfaces (i.e., tree trunks) to enter their nesting sites. Parrots are also cavity nesters, and if zygodactyl feet are an adaptation to these breeding habits, their loss in the stem lines of passerines may have been due to a transition from cavity-nesting to open nests (Mayr 2015).

9.1 Halcyornithidae and Messelornithidae: Owl/Parrot Mosaics of Uncertain Affinities

These two taxa exhibit zygodactyl (Halcyornithidae) or semi-zygodactyl (Messelasturidae) feet and were assigned to the stem group of the Psittaciformes in the first edition of this book (Mayr 2009). However, as noted above and further detailed below, passerines had stem group representatives with zygodactyl feet and a tarsometatarsus resembling that of the Psittaciformes. The characteristic tarsometatarsus morphology shared by the Psittaciformes and zygodactyl stem group representatives of the Passeriformes is therefore likely to be plesiomorphic for the Psittacopasseres, especially concerning the separation of the accessory trochlea for the fourth toe from the main portion of this trochlea by a distinct furrow. This furrow is absent in the Halcyornithidae and Messelasturidae, which suggests a position outside the Psittacopasseres for these birds.

An earlier analysis supported a sister group relationship between the Messelasturidae and strigiform birds (Mayr 2005a), but the results of this analysis were only weakly supported. With regard to the overall proportions of the limb bones, the skeleton of halcyornithids likewise resembles that of the Strigiformes, and similarities to owls were highlighted in an earlier description of a specimen from Messel (Hoch 1988). Distinctive owl-like traits of halcyornithids and messelasturids include the presence of long supraorbital processes of the lacrimal bone and the fact that the coracoid exhibits a foramen for the supracoracoideus nerve. Messelasturids furthermore have raptor-like ungual phalanges.

Supraorbital processes and a coracoideal foramen for the supracoracoideus nerve are also present in the Falconiformes. If the latter are the sister taxon of the Psittacopasseres, as suggested by most current molecular analyses (Sect. 2.3), these features may be plesiomorphic for the Psittacopasseres. However, and as also noted in Sect. 2.3, a clade including the Falconidae and Psittacopasseres is not recovered in all molecular analyses (Kuhl et al. 2021; Braun and Kimball 2021), and a more robust placement of the extant taxa is needed for a definitive assessment of the unusual character mosaic shown by halcyornithids and messelasturids.

Unambiguous representatives of the Halcyornithidae and Messelasturidae are only known from Eocene sites, but a skeleton of an as-yet unnamed species from the Paleocene of Menat in France was considered to be possibly related to these birds (Mayr et al. 2020).

9.1.1 Halcyornithidae

The small zygodactyl birds assigned to the Halcyornithidae were found in various early and middle Eocene sites in Europe and North America and have a characteristic skeletal morphology (Fig. 9.2). Based on a skeleton from the North American Green River Formation, the first species was described as "*Primobucco*" *olsoni* by Feduccia and Martin (1976), who erroneously assigned it to the Primobucconidae. The latter were then considered to be piciform birds but actually are stem group representatives of the coraciiform Coracii (Sect. 10.3.4.1). "*P.*" *olsoni* is clearly distinguished from coraciiform birds in numerous features and has now been assigned to the genus *Cyrilavis*, which includes a further species from the Green River Formation, *C. colburnorum* (Ksepka et al. 2011).

More than two decades ago, close relatives of *Cyrilavis* ("*Primobucco*") *olsoni* were also reported from the early Eocene of Europe. In Messel, these birds are represented by complete skeletons, which belong to *Pseudasturides* ("*Pseudastur*") *macrocephalus*, *Serudaptus pohli*, and at least two other unnamed species (Fig. 9.2a–c; Mayr 1998a, 2000a). A halcyornithid species from the London Clay of Walton-on-the-Naze was described as *Pulchrapollia gracilis* by Dyke and Cooper (2000), and further specimens of this species and other halcornithids from Walton-on-the-Naze are in the collection of Michael Daniels (Fig. 9.2d). Halcyornithid fossils were also reported from the London Clay of the Isle of Sheppey and include a partial skeleton that was assigned to *Pseudasturides* cf. *macrocephalus* (Dyke 2001; Mayr 2007).

A halcyornithid tarsometatarsus from the early Eocene Nanjemoy Formation in Virginia (USA) was tentatively referred to *Pulchrapollia* by Mayr (2016), and bones of unidentified halcyornithids are known from the early Eocene of Egem in Belgium (Mayr and Smith 2019) and the coeval Danish Fur Formation (Waterhouse et al. 2008). Halcyornithid tarsometatarsi were also found in the Geisel Valley, and a tentatively identified partial mandible from middle Eocene deposits of this locality may be the latest record of the group (Mayr 2002a, 2020a).

Mayr (1998a) introduced the name "Pseudasturidae" for these birds, which is now regarded as a junior synonym of the Halcyornithidae (Mayr 2007). The latter taxon was established by Harrison and Walker (1972) for the first fossil bird to be scientifically named, *Halcyornis toliapicus*, a species based on a cranium from the London Clay of the Isle of Sheppey.

Halcyornithids exhibit a characteristic skeletal morphology, with a short, parrot-like tarsometatarsus, which bears a well-developed accessory trochlea for the reversed fourth toe. As noted above and unlike in psittaciform birds, this accessory trochlea is not separated by a furrow from the main portion of the trochlea. The skull exhibits long supraorbital processes and well-developed temporal fossae. The humerus is long and slender. The coracoid is similar to that of extant Strigiformes in overall morphology; in contrast to taxa of the Psittacopasseres, it exhibits a foramen for the supracoracoideus nerve.

Initially, the phylogenetic affinities of *Pseudasturides* and its allies were considered uncertain (Mayr 1998a), but subsequently it was suggested that these birds are stem group representatives of the Psittaciformes (Mayr 2002b). As

9.1 Halcyornithidae and Messelornithidae: Owl/Parrot Mosaics of Uncertain Affinities

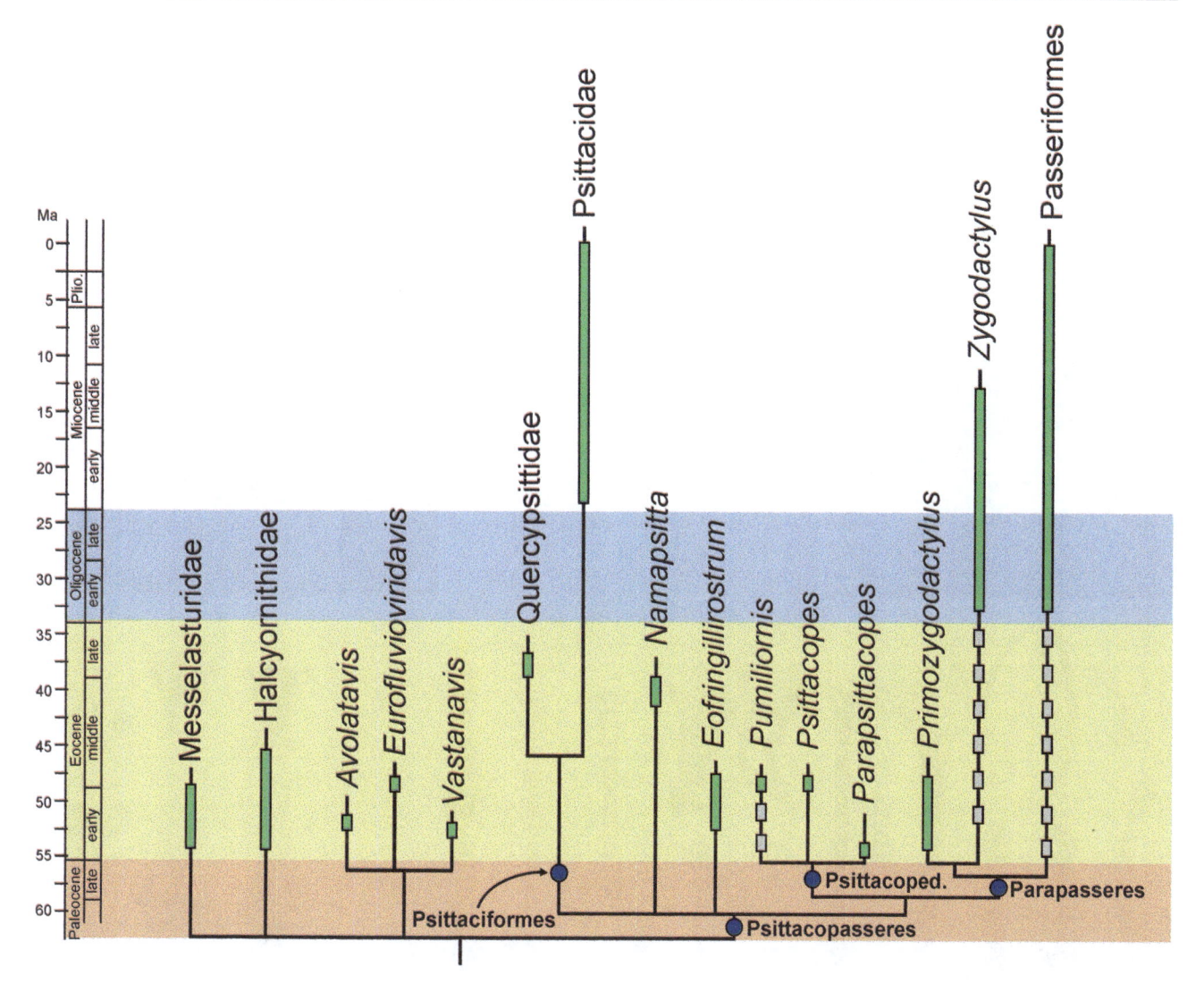

Fig. 9.1 Phylogenetic interrelationships and stratigraphic occurrences of selected taxa of the Psittacopasseres. The dotted gray lines denote stratigraphic ranges based on fossils of uncertain affinities. The affinities of taxa shown in a polytomy are controversial, divergence dates are hypothetical; see text for further details

detailed above, this classification had to be revised after molecular analyses provided support for a clade including the Psittaciformes and the Passeriformes, and in analyses constrained to a molecular scaffold, the Halcyornithidae resulted as the sister taxon of the Psittacopasseres (Mayr 2015; Ksepka et al. 2019).

9.1.2 Messelasturidae

The Messelasturidae include three named species from early Eocene localities: *Tynskya eocaena* from the Green River Formation, *T. waltonensis* from the London Clay of Walton-on-the-Naze, and *Messelastur gratulator* from Messel (Fig. 9.3). *T. eocaena* is represented by a single skeleton on a slab (Mayr 2000b, 2021), whereas the holotype of *T. waltonensis* consists of various isolated bones of a single individual (Fig. 9.3f; Mayr 2021); of *M. gratulator* two partial skeletons and two skulls were identified (Peters 1994; Mayr 2005a, 2011a). Multiple skeletons of undescribed messelasturids were furthermore collected by Michael Daniels in the London Clay of Walton-on-the-Naze, where they are among the more common avian fossils (Fig. 9.3c–e). These birds are listed as owls in Feduccia (1999: tab. 4.1). Remains of two messelasturid species were also found in the Nanjemoy Formation in Virginia, USA (Mayr et al. 2022).

M. gratulator was tentatively assigned to the Accipitridae in the original description (Peters 1994). The phylogenetic affinities of *T. eocaena* were initially considered uncertain (Mayr 2000b), but comparisons were made with raptorial birds and the Halcyornithidae ("Pseudasturidae"). A close relationship between *Tynskya* and *Messelastur* was assumed

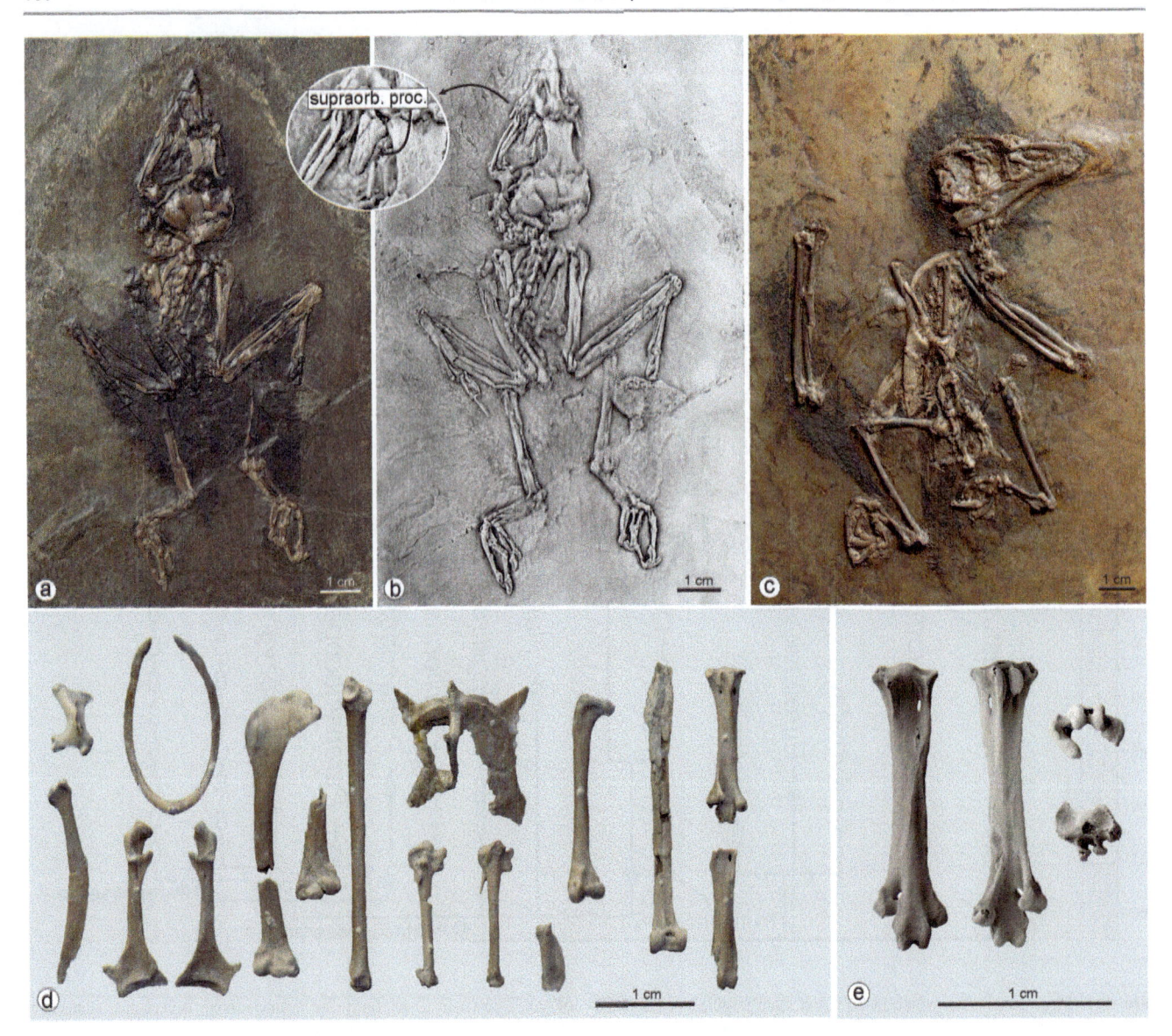

Fig. 9.2 Fossils of the Halcyornithidae. (**a**), (**b**) *Pseudasturides macrocephalus* from the latest early or earliest middle Eocene of Messel, Germany (Staatliches Museum für Naturkunde Karlsruhe, Germany, SMNK-PAL 2373a); in (**b**) the fossil was coated with ammonium chloride, the arrow indicates an enlarged detail showing the supraorbital process. (**c**) Holotype of *Serudaptus pohli* from Messel (Wyoming Dinosaur Center, Thermopolis, USA, WDC-C-MG 201a). (**d**) Partial skeleton of a halcyornithid from the early Eocene London Clay of Walton-on-the-Naze, UK (collection of Michael Daniels, Holland-on-Sea, UK, WN 85508). (**e**) Tarsometatarsus of an unnamed halcyornithid from the early Eocene Nanjemoy Formation in Virginia, USA, in dorsal, plantar, distal, and proximal view (Senckenberg Research Institute Frankfurt, SMF Av 628). (Photo in (**b**) by Anika Vogel, all others by Sven Tränkner)

by Mayr (2005a), who established the taxon Messelasturidae for a clade including these birds.

M. gratulator and the slightly smaller species of *Tynskya* were about the size of the extant Scops Owl, *Otus scops*. As in halcyornithids, the skull of *Messelastur* exhibits long, caudally directed supraorbital processes, the presence of which cannot be determined in the fossils of the two *Tynskya* species owing to the poor preservation of the skull. In both, messelasturids and halcyornithids, there are furthermore no pneumatic openings in the pneumotricipital fossa of the humerus, the tarsometatarsal trochlea for the second toe is very small, and the coracoid exhibits a foramen for the supracoracoideus nerve.

Messelasturids are distinguished from halcyornithids by very deep mandibular rami, owl-like ungual phalanges, and a distinctive morphology of the tarsometatarsus, which does not form a large accessory trochlea for the fourth toe (Mayr 2000b, 2011a, 2021). Unlike in halcyornithids, the feet of messelasturids were therefore probably not fully zygodactyl. The beak of messelasturids is also proportionally shorter and more raptor-like than that of halcyornithids. The deep mandibular rami and the raptor-like feet indicate a specialized

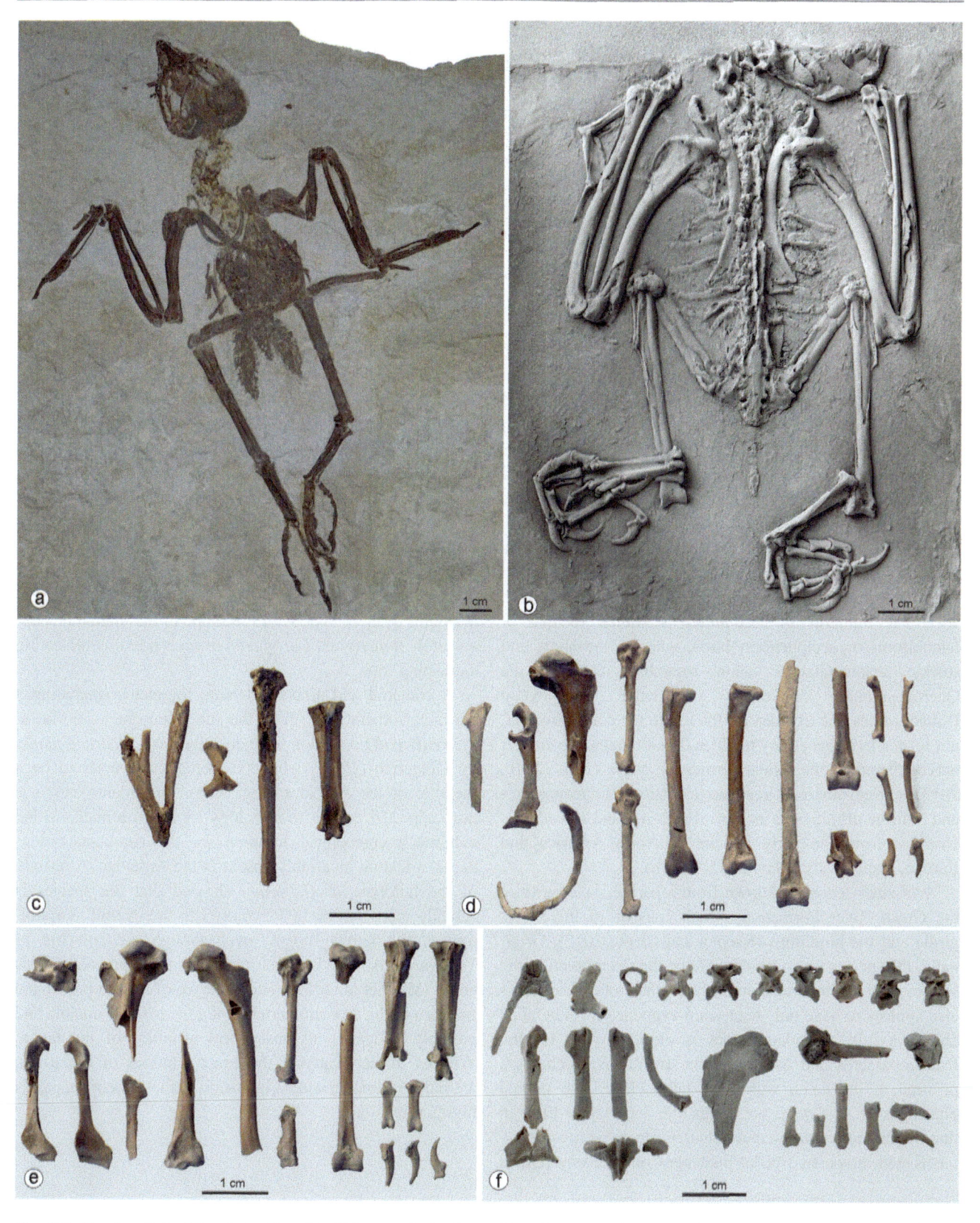

Fig. 9.3 Fossils of the Messelasturidae. (**a**) Holotype of *Tynskya eocaena* from the early Eocene Green River Formation, Wyoming, USA (Bayerische Staatssammlung für Paläontologie und Geologie, Munich, Germany, BSPG 1997 I 6). (**b**) Postcranial skeleton of *Messelastur gratulator* from the latest early or earliest middle Eocene of Messel, Germany (Senckenberg Research Institute Frankfurt, SMF-ME 11348); coated with ammonium chloride. (**c**)–(**e**) Three individuals of undescribed messelasturids from the early Eocene

feeding ecology of messelasturids. and these birds may have exploited a feeding niche no longer available to extant birds (Mayr 2021). An analysis constrained to a molecular scaffold resulted in a sister group relationship between the Messelasturidae and a clade including the Halcyornithidae and Psittacopasseres, but this placement was weakly supported (Mayr 2015).

9.2 *Vastanavis, Avolatavis, Eurofluvioviridavis*: Further Eocene parrot-like Birds of Controversial Affinities

India has a very scarce Paleogene avian fossil record, but numerous bones of a distinctive, parrot-like taxon were found in early Eocene deposits of the Vastan Lignite Mine in Gujarat Province (Fig. 9.4d; Mayr et al. 2007, 2010, 2013). *Vastanavis* includes two species, *V. eocaena* and *V. cambayensis*, which were semi-zygodactyl arboreal birds the size of an average parrot or falcon (Mayr et al. 2010, 2013). The coracoid has a deeply excavated scapular cotyla and bears a resemblance to the coracoid of the psittaciform Quercypsittidae from the late Eocene of France (see Sect. 9.4). The humerus of *Vastanavis* is similar to that of falconiform or accipitriform birds, whereas the short and stocky tarsometatarsus again resembles that of the Quercypsittidae. Unlike in the latter and extant Psittaciformes, the trochlea for the fourth toe does, however, not form a large accessory trochlea. Even though *Vastanavis* was assigned to the Psittaciformes by Mayr et al. (2010, 2013), current analyses recover it outside Psittacopasseres and indicate affinities to the taxa *Avolatavis* and *Eurofluvioviridavis* from the early Eocene of North America and Europe, respectively (Mayr 2015, 2020b).

Avolatavis tenens is known from a partial skeleton from the Green River Formation, which consists of the pelvic girdle and the hindlimbs (Ksepka and Clarke 2012). Originally, the species was considered to be the sister taxon of the Quercypsittidae, but unlike in the latter, the foot of *Avolatavis* also appears to have only been semi-zygodactyl. As detailed above, a fully zygodactyl foot is plesiomorphic for the Psittacopasseres and its absence in *Avolatavis* precludes a position of the taxon within the latter clade. The ungual phalanges of *Avolatavis* are distinctive and "raptor-like" in that the lateral neurovascular sulcus is closed. A tarsometatarsus with associated pedal phalanges of a closely related species from the London Clay of Walton-on-the-Naze was reported by Mayr and Daniels (1998) and was likewise compared with the Quercypsittidae by these authors (Fig. 9.4c).

The description of *Eurofluvioviridavis robustipes* was based on a skeleton from Messel (Fig. 9.4a). In the original description (Mayr 2005b), the species was likened to the Fluvioviridavidae (Sect. 6.8.5), but it was also noted that its short and stout tarsometatarsus resembles that of the Quercypsittidae in the shape of the large trochlea for the second toe and in the presence of a wing-like flange on the trochlea for the fourth toe. Again, however, and as evidenced by the position of the feet of the holotype skeleton, *Eurofluvioviridavis* was at best semi-zygodactyl. The beak of *E. robustipes* is fairly long and broad and bears a close resemblance to that of *Fluvioviridavis*. The feet of both taxa are, however, very different, and with regard to the shape of the stout tarsometatarsus and the raptor-like ungual phalanges, *Eurofluvioviridavis* closely resembles *Avolatavis* and *Vastanavis* (in which, however, only some of the ungual phalanges have a closed neurovascular sulcus).

Even though some current analyses support a clade including *Vastanavis*, *Avolatavis*, and *Eurofluvioviridavis* (Mayr 2015, 2020b), the affinities of these three taxa are far from being well resolved. The same is true for the ecological attributes of these taxa, with the shape of the ungual phalanges of *Avolatavis* and *Eurofluvioviridavis* possibly indicating a grasping foot.

A coracoid and associated bone fragments (including a partial tarsometatarsus) from the late Paleocene (~56 Ma) of the Willwood Formation were described as *Calcardea junnei* by Gingerich (1987), who considered the species to be a member of the Ardeidae (Fig. 9.5a). This identification is not supported by the morphology of the specimen, which exhibits a coracoideal foramen for the supracoracoideus nerve, a feature invariably absent in the Ardeidae. A restudy of the holotype of *C. junnei* showed that the species is actually more similar to *Vastanavis* in the overall morphology of the known bones, even though the coracoid of *Vastanavis* likewise lacks a foramen for the supracoracoideus nerve (Mayr et al. 2019). The presence of a large pneumatic foramen in the thoracic vertebra of *Calcardea* is notable and possibly indicative of anseriform affinities of the taxon. Without further material, the exact affinities of *Calcardea* are indeterminable, and its inclusion in the present chapter is highly provisional.

Fig. 9.3 (Continued) London Clay of Walton-on-the-Naze (collection of Michael Daniels, Holland-on-Sea, UK, **c**: WN 86528, **d**: WN 81302, **e**: WN 91686). (**f**) Selected bones (partial mandible, vertebrae, and fragments of major pectoral girdle and wing bones as well as a proximal femur and pedal phalanges) of the holotype of *Tynskya waltonensis* from the early Eocene London Clay of Walton-on-the-Naze (SMF Av 652); coated with ammonium chloride. (All photos by Sven Tränkner)

Fig. 9.4 (a) *Eurofluvioviridavis robustipes* from the latest early or earliest middle Eocene of Messel (Staatliches Museum für Naturkunde Karlsruhe, Germany, SMNK-PAL 3835). (b) Foot of a *Vastanavis*-like bird from Messel, Germany (holotype, Senckenberg Research Institute, Frankfurt, SMF-ME 11286); coated with ammonium chloride. (c) Leg bones of an *Avolatavis*- or *Eurofluvioviridavis*-like bird from the early Eocene London Clay of Walton-on-the-Naze (collection of Michael Daniels, Holland-on-Sea, UK, WN 87563). (d) Bones of different individuals of *Vastanavis* sp. from the early Eocene Cambay Shale Formation of the Vastan Lignite Mine in India (H.N.B. Garhwal University, Department of Geology, Uttarakhand, India; coracoid: GU/RSR/VAS 1802; humerus: GU/RSR/VAS 1803; radius: GU/RSR/VAS 1688; ulna: GU/RSR/VAS 1805; carpometacarpi: GU/RSR/VAS 1706 and 1806; tibiotarsus: GU/RSR/VAS 1808; tarsometatarsus: GU/RSR/VAS 1809; phalanges: GU/RSR/VAS 1812 and 1813); the specimens were coated with ammonium chloride. (e) Coracoid and tarsometatarsi of *Quercypsitta sudrei* from the late Eocene of the Quercy fissure fillings in France (Université Claude Bernard, Lyon, France, UCBL FSL 367080, 367081, and 367082). (f) Humerus, coracoid, and tarsometatarsus of the extant *Nestor notabilis* (Psittacidae). (All photos by Sven Tränkner)

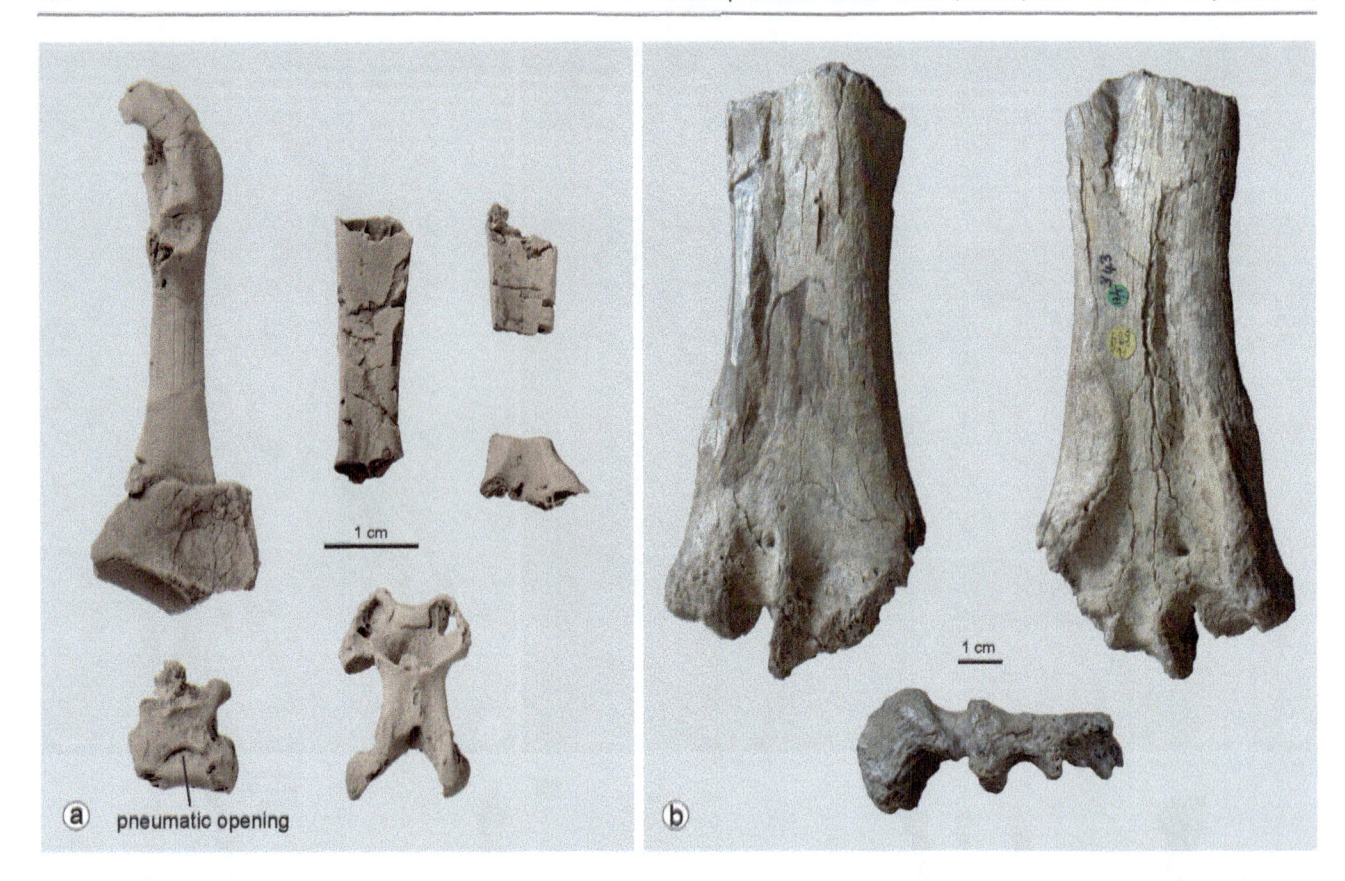

Fig. 9.5 (a) Selected bones (coracoid, two vertebrae, fragments of both tarsometatarsi) of *Calcardea junnei* from the late Paleocene Willwood Formation, Wyoming, USA (holotype, University of Michigan, Museum of Paleontology, Ann Arbor, UM 76882); coated with ammonium chloride. (b) Right tarsometatarsus of *Cladornis pachypus* from the late Oligocene of Patagonia in dorsal, plantar, and distal view (holotype, Natural History Museum, London, NHMUK A 589). (Photos by Sven Tränkner)

9.3 Cladornithidae

Often, merely tentative assignments of Paleogene birds to extant clades are possible, and the classifications of a fair number of the taxa discussed in the present book are provisional. The phylogenetic affinities of few species are, however, as enigmatic as those of *Cladornis pachypus* (Fig. 9.5b). The holotype and only known specimen of this large bird is an incomplete tarsometatarsus from the late Oligocene (Deseadan) of the Argentinean part of Patagonia. Ameghino (1895) identified *C. pachypus* as a penguin-like aquatic bird, but Simpson (1946) already rebutted sphenisciform affinities of *Cladornis*. Olson (1985: 193) remarked that the bone "appears to be from some sort of very large and extremely weird land bird." The shaft of the tarsometatarsus of *C. pachypus* is very flat, and the bone appears to have been quite short (even though the proximal end is not preserved, the proportions of the preserved part indicate that probably not more than about one-fourth of the bone is missing). Judging from the large articular facet for the first metatarsal, *C. pachypus* had a well-developed hallux; the distal vascular foramen is very small. The rims of the trochlea for the third toe are separated by a marked furrow, and the trochlea for the fourth toe bears a plantarly directed flange, which led Olson (1985: 193) to assume that *C. pachypus* may have been "tending towards being zygodactyl". The plantar flange on the trochlea for the fourth toe and the presumably well-developed hallux indicate a grasping foot, and because of its large size, *C. pachypus* may have been a raptorial bird. Overall, the tarsometatarsus shows a remote similarity to that of the much smaller taxa *Vastanavis*, *Avolatavis*, and *Eurofluvioviridavis*, but the bone also bears a resemblance to the tarsometatarsus of the strigiform taxon *Berruornis* (Sect. 10.1.). Further fossil material is needed to conclusively resolve the phylogenetic affinities of this unusual species. Its placement in the present chapter is merely in the absence of a better-supported alternative, and it is highly likely that the taxon will ultimately end in a very different higher-level clade.

9.4 Psittaciformes (Parrots)

Various putative Paleogene psittaciform taxa were listed in the first edition of this book (Mayr 2009), but many of these are now placed outside Psittacopasseres (see the preceding sections). Psittaciform affinities are, however, maintained for the Quercypsittidae (Fig. 9.4e), which are only known from the late Eocene (MP 17) locality La Bouffie of the Quercy fissure fillings (Mourer-Chauviré 1992). Two species, *Quercypsitta ivani* and *Q. sudrei*, are distinguished, and the described skeletal elements include an incomplete carpometacarpus as well as coracoids, distal tibiotarsi, tarsometatarsi, and pedal phalanges. A distal tarsometatarsus from the middle Eocene (MP 11–13; Mlíkovský 2002) of England, which was referred to the non-psittaciform *Palaeopsittacus georgei* (Sect. 6.8.5) by Harrison (1982), resembles that of *Quercypsitta* (Mayr and Daniels 1998).

The short and stout tarsometatarsus of quercypsittids is reminiscent of the corresponding bone of some extant parrots in its proportions, but the accessory trochlea for the reversed fourth toe is less developed than that of crown group parrots. Unlike in extant parrots, the coracoid furthermore exhibits a plesiomorphic, cup-like articular facet for the scapula (Mourer-Chauviré 1992), which indicates a position of quercypsittids outside crown group Psittaciformes.

Even though fossil parrots are known since a long time from the Neogene of the Northern Hemisphere (Mayr 2017a), no representatives of crown group Psittaciformes have been found in early Paleogene fossil deposits. The earliest modern-type psittaciform bird is *Mogontiacopsitta miocaena* from the late Oligocene or early Miocene of the Mainz Basin in Germany (Mayr 2010).

9.5 Passeriformes (Passerines)

The recognition of the aptly named Zygodactylidae as stem group representatives of the Passeriformes has shown that the latter had zygodactyl stem group representatives. Apart from their foot morphology, the species of the Zygodactylidae are very similar to crown group passerines in their skeletal morphology. In the past years, however, various other stem group Passeriformes with zygodactyl feet were identified, which are more distantly related to the Zygodactylidae and crown group passerines. Even more unexpectedly, these newly identified Eocene taxa show that zygodactyl stem group passerines underwent a notable radiation, which parallels that of the crown group regarding a high diversity of different feeding adaptations.

9.5.1 Psittacopedidae and Allies

Psittacopes lepidus is a small bird, which is known from two skeletons from Messel (Fig. 9.6a; Mayr and Daniels 1998). This species has zygodactyl feet and a short and deep beak with large nostrils. Mostly because of these latter traits, it was considered to be a stem group representative of the Psittaciformes in the initial description. It was then already noted that very similar birds occur in the London Clay of Walton-on-the-Naze. These fossils belong to at least three species, which are represented by three-dimensionally preserved skulls and major limb elements (Mayr and Daniels 1998). Most of these fossils are in the collection of Michael Daniels, but one has recently been acquired by the Senckenberg Research Institute and was described as *Parapsittacopes bergdahli* (Mayr 2020b). *P. bergdahli* is very similar to the sparrow-sized *P. lepidus*, and as in the latter, the short beak of *Parapsittacopes* exhibits very large nostrils; the mandible is weak and has low rami (Fig. 9.7a). The tarsometatarsus exhibits an accessory trochlea for the reversed fourth toe, which, as in the Zygodactylidae and extant Psittaciformes, is separated from the main trochlea by a distinct furrow. As detailed above, this derived tarsometatarsus morphology is now considered to be plesiomorphic for the clade including the Psittaciformes and Passeriformes, and the taxa of the Psittacopedidae share some derived traits with passerines, which are absent in parrots. Among these is a furcula with a well-developed apophysis, a dorsal tubecle on the distal end of the humerus, a hook-like process on the distal end of the radius, as well as pneumatic openings in the caudal surface of the otic process of the quadrate (Mayr 2020b). The beak of *P. bergdahli* resembles that of extant waxwings (Passeriformes, Bombycilllidae) in its shape, and the species may therefore have been an opportunistic feeder, which caught insects by sallying flights from perches but may have also fed on fruits or berries.

Another taxon that is now considered to be closely related to *Psittacopes* is *Pumiliornis tessellatus*, of which three skeletons were found in Messel (Fig. 9.6b; Mayr 1999, 2008a; Mayr and Wilde 2014). This tiny, wren-sized species has a long beak, a caudally positioned sternal keel, and lacks an ossified supratendinal bridge on the distal end of the tibiotarsus. The feet are zygodactyl, with the first phalanx of the fourth toe being unusually wide and flat. As in *Psittacopes* and *Parapsittacopes*, the bodies of the thoracic vertebrae bear marked lateral fossae (pleurocoels). The phylogenetic affinities of *Pumiliornis* were long unresolved, and the taxon was compared with a number of only distantly related extant and fossil taxa (Mayr 2009). After the discovery of a well-preserved new skeleton, however, the close affinities between *Psittacopes* and *Pumiliornis* became evident, and in current analyses, *Pumiliornis* is recovered in a clade with *Psittacopes*, the Zygodactylidae, and crown group

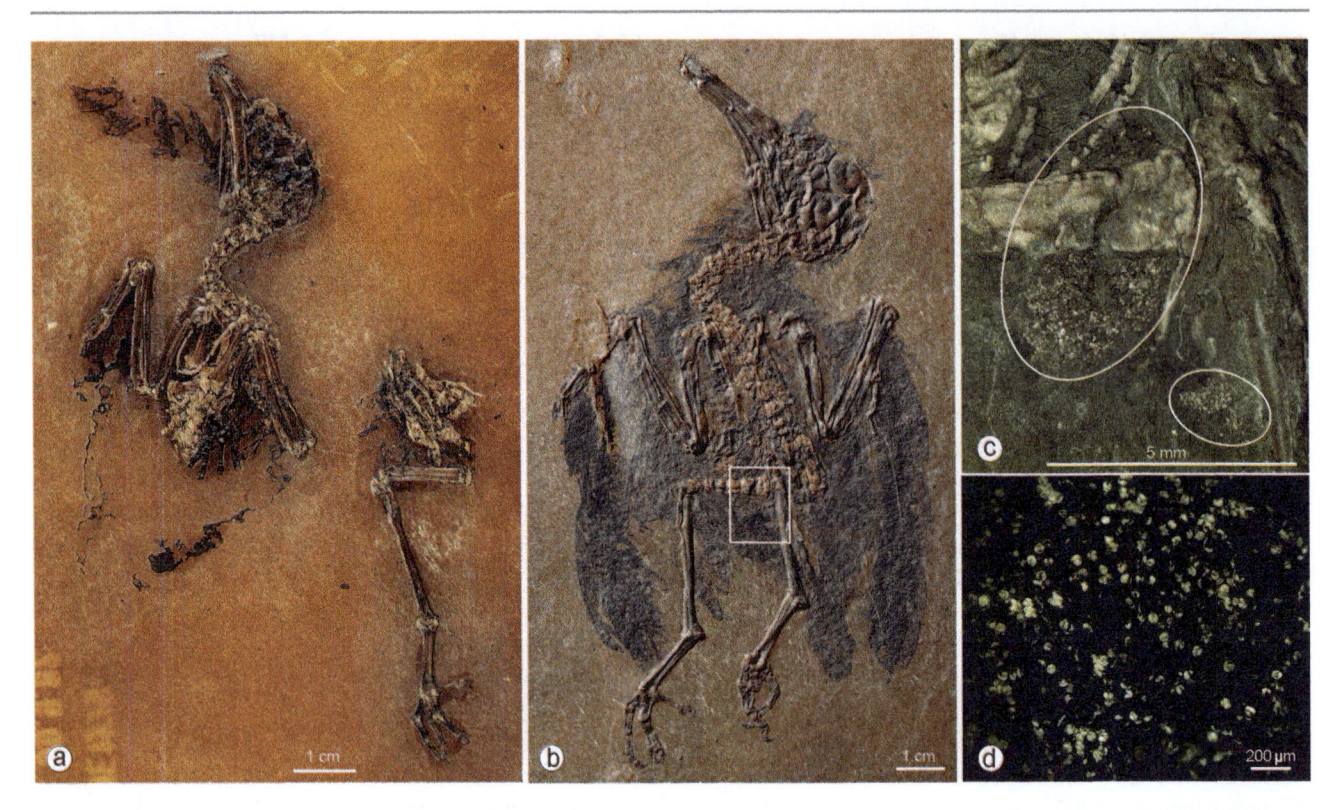

Fig. 9.6 Zygodactyl stem group representatives of the Passeriformes. (**a**) Holotype of *Psittacopes lepidus* from the latest early or earliest middle Eocene of Messel, Germany (Senckenberg Research Institute Frankfurt, SMF-ME 1279). (**b**) Specimen of *Pumiliornis tessellatus* from Messel (SMF-ME 11414A), with (**c**) and (**d**) details of the preserved stomach contents containing numerous pollen grains; the frame in (**b**) denotes the position of the detail shown in (**c**), the circles indicate the areas with pollen grains. (Photos in (**a**) and (**b**) by Sven Tränkner, microscopic images in (**c**) and (**d**) by Volker Wilde)

passerines (Mayr 2015, 2020b; Ksepka et al. 2019). One of the *Pumiliornis* fossils is preserved with stomach contents consisting of numerous pollen grains, which belong to at least two different plant species (Fig. 9.6c, d; Mayr and Wilde 2014). Together with the long and slender beak, these stomach contents suggest nectarivorous feeding habits of *Pumiliornis*. The distal ends of a tibiotarsus and a tarsometatarsus from the early Eocene Nanjemoy Formation (Virginia, USA) were tentatively assigned to *Pumiliornis* by Mayr et al. (2022).

A further species with a convoluted taxonomic history is *Morsoravis sedilis* from the early Eocene of the Danish Fur Formation (Fig. 9.8). The exceptionally well-preserved holotype of this small bird was first studied by Bertelli et al. (2010), who referred the species to the Charadriiformes. In part, this conclusion was based on the presence of lateral fossae on the body of the thoracic vertebrae, which are, however, also present in the above-discussed taxa assigned to the Psittacopedidae. The foot of *Morsoravis* was at least semi-zygodactyl, with a wide and flat first phalanx of the fourth toe (Mayr 2011b). The holotype of *M. sedilis* lacks wing and pectoral elements, but these bones were preserved in another specimen from the Fur Formation, which now appears to be lost (Mayr 2011b). *M. sedilis* agrees with *P. tessellatus* in the presence of pleurocoels on the thoracic vertebrae, a fairly long tibiotarsus, the morphology and proportions of the tarsometatarsus, and the robust feet. Phylogenetic analyses by Mayr (2015, 2020b) and Ksepka et al. (2019) supported a clade including *Psittacopes*, *Parapsittacopes*, *Pumiliornis*, *Morsoravis*, the Zygodactylidae and crown group Passeriformes, but the exact interrelationships of the former four taxa are poorly resolved. An undescribed *Morsoravis*-like bird was also found in the Green River Formation (Grande 2013: fig. 143B). A distal tarsometatarsus from the Nanjemoy Formation in Virginia (USA) was tentatively assigned to *Morsoravis* (Mayr 2016; Mayr et al. 2022).

Eofringillirostrum boudreauxi from the Green River Formation and *E. parvulum* from Messel (Fig. 9.9; Ksepka et al. 2019) are two further species that appear to be zygodactyl stem group passerines. Of the latter species, just a crushed partial skeleton is known, but the holotype of *E. boudreauxi* is exquisitely preserved. Both species have a deep, conical, and finch-like beak, which indicates a predominantly granivorous diet. *Eofringillirostrum* furthermore differs from *Psittacopes* and *Parapsittacopes* in that the first phalanx of the second toe is abbreviated, which may indicate that the foot had a grasping function and was used to manipulate food

9.5 Passeriformes (Passerines)

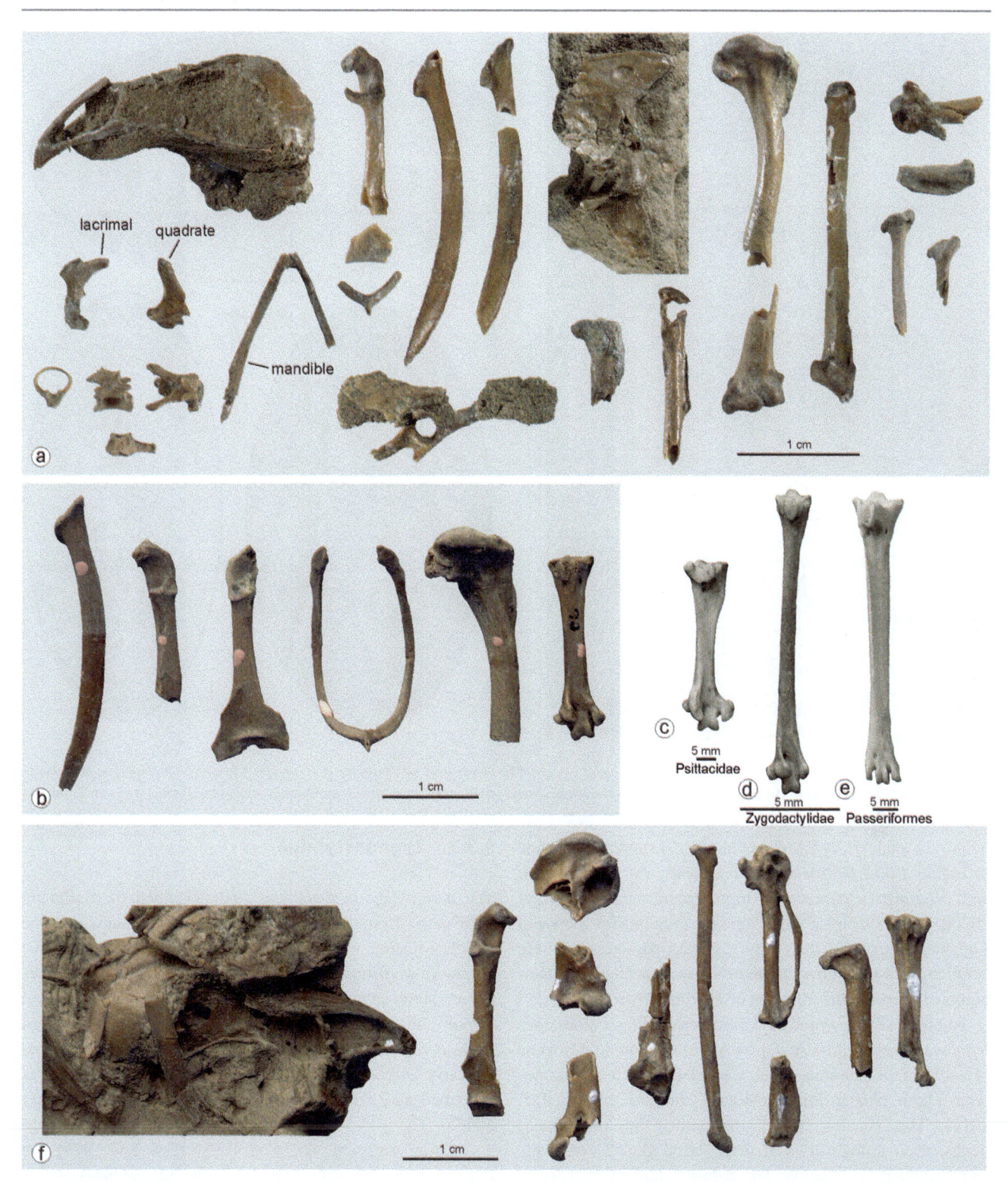

Fig. 9.7 Further zygodactyl stem group representatives of the Passeriformes. (**a**) Holotype of *Parapsittacopes bergdahli* from the early Eocene London Clay of Walton-on-the-Naze, UK (Senckenberg Research Institute, Frankfurt, SMF Av 653, formerly collection of Paul Bergdahl, Kirby-le-Soken, UK). (**b**) Partial skeleton of a closely related, undescribed species from Walton-on-the-Naze (collection of Michael Daniels, Holland-on-Sea, UK, WN 85506). Tarsometatarsi of (**c**) the extant *Nestor notabilis* (Psittacidae), (**d**) an undescribed species of the Zygodactylidae from the London Clay (WN 92747), and (**e**) the passeriform *Corvus frugilegus* (Corvidae). (**f**) Another undescribed *Parapsittacopes*-like species from the London Clay of Walton-on-the-Naze (WN 91711), the skull is in the block of matrix on the left. (All photos by Sven Tränkner)

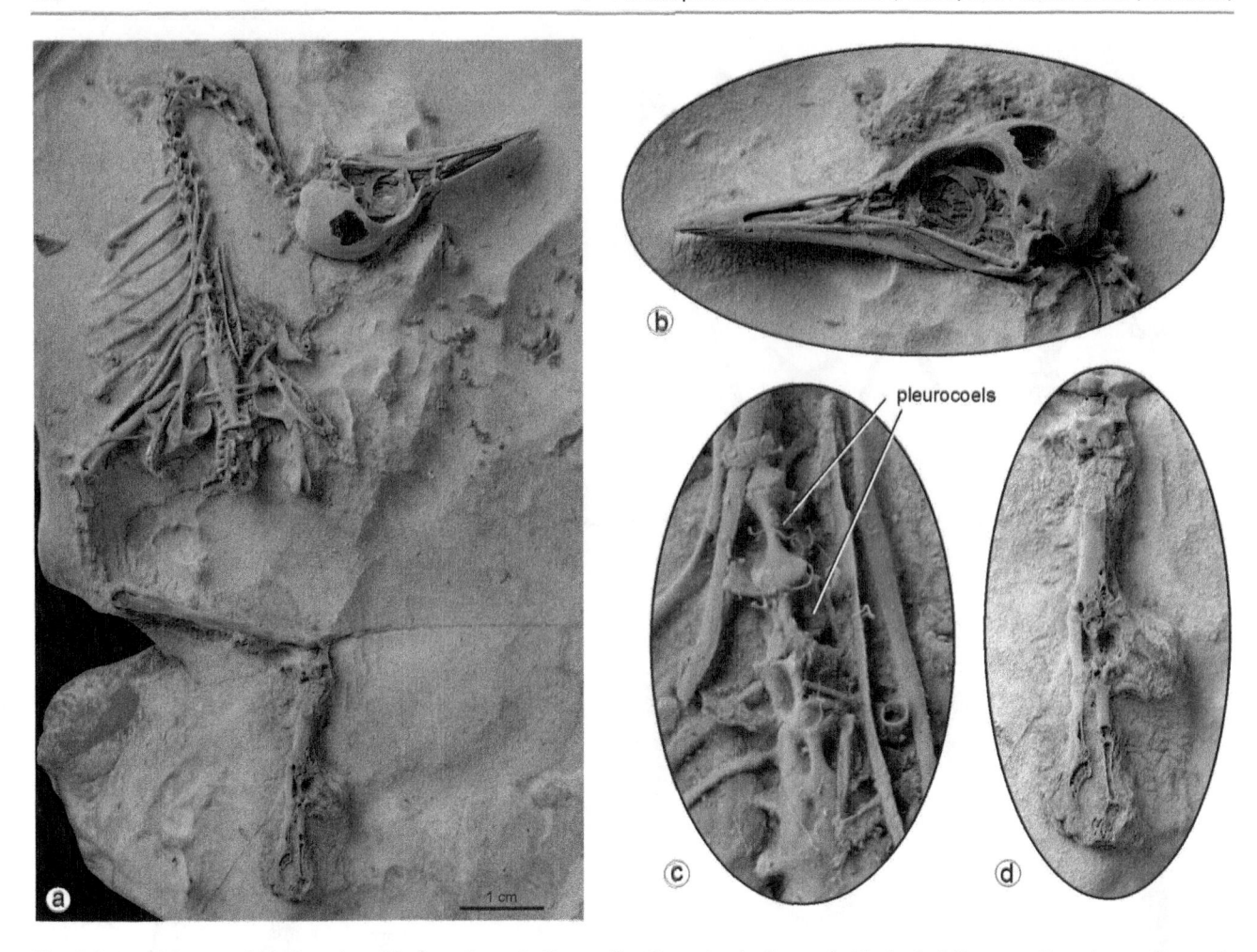

Fig. 9.8 (a) Holotype of *Morsoravis sedilis* from the early Eocene Fur Formation in Denmark (Geological Museum, Copenhagen, Denmark, MGUH 28930), with details of (b) the skull, (c) the thoracic vertebrae, and (d) the foot; coated with ammonium chloride. (Photos by Sven Tränkner)

items. An analysis by Ksepka et al. (2019) supported a clade including *Eofringillirostrum*, *Psittacopes*, *Parapsittacopes*, and *Morsoravis*, whereas a more recent analysis (Mayr 2020b) resulted in a sister group relationship between *Eofringillirostrum* and the Psittacopasseres. These conflicting results show that more data are needed to unambiguously resolve the affinities of *Eofringillirostrum*.

Another small, zygodactyl bird resembling *Psittacopes lepidus* is *Namapsitta praeruptorum* from the middle Eocene (47–49 Ma) of Namibia, which was assigned to a new monotypic taxon Namapsittidae (Mourer-Chauviré et al. 2015, 2017). This species is well represented by most major bones, even though none of the referred skeletal elements were found in association. The coracoid has a flat articular surface for the scapula and a long procoracoid process. The tarsometatarsus is long and slender. The humerus, however, is stouter than in *P. lepidus* and differs in that the dorsal tubercle (tuberculum dorsale) on its proximal end is much better developed.

9.5.2 Zygodactylidae

The taxon *Zygodactylus* was first described by Ballmann (1969a, b) from the early Miocene of Germany and the middle Miocene of France, and for a long time, it was just known from distal tibiotarsi and tarsometatarsi. These bones did not allow an unambiguous classification of the taxon, and it took almost four decades until an articulated skeleton of *Zygodactylus* could be identified. This specimen comes from the early Oligocene of the Luberon area in France and was described as *Zygodactylus luberonensis* (Fig. 9.10c; Mayr 2008b). The osteological data obtained from this fossil support earlier assumptions (Mayr 1998b; Feduccia 1999) that *Zygodactylus* is closely related to early Eocene birds, which were initially classified into the taxon "Primoscenidae." The latter taxon, which was established by Harrison and Walker (1977), was therefore synonymized with the Zygodactylidae (Mayr 2008b).

Early Eocene zygodactylids are known from many complete skeletons and isolated, three-dimensionally preserved

9.5 Passeriformes (Passerines)

Fig. 9.9 Fossils of the early Eocene *Eofringillirostrum*. (**a**) Holotype of *E. boudreauxi* from the Green River Formation (Field Museum, Chicago, USA, FMNH PA 793). (**b**), (**c**) Holotype (slab and counter slab) of *E. parvulum* from Messel (Royal Belgian Institute of Natural Sciences, Brussels, Belgium, IRSNB Av 128a+b); coated with ammonium chloride. (Photo in (**a**) courtesy of Lance Grande, (**b**) and (**c**) by Sven Tränkner)

bones (Figs. 9.10a, b). The first species was described by Harrison and Walker (1977) as *Primoscens minutus*, the holotype of which is an incomplete carpometacarpus from the London Clay. Additional and much more complete London Clay specimens were collected by Michael Daniels from Walton-on-the-Naze (Fig. 9.10d, f; Mayr 1998b; Daniels in Feduccia 1999: table 4.1). Complete skeletons were furthermore identified in Messel and the Green River Formation. In Messel, zygodactylids are among the most abundant small birds (Fig. 9.10a), and the six currently recognized species (*Primozygodactylus danielsi*, *P. ballmanni*, *P. major*, *P. eunjooae*, *P. quintus*, and *P. longibrachium*) mainly differ in size and limb or toe proportions (Mayr 1998b, 2017b; Mayr and Zelenkov 2009). Two species from the Green River Formation were described as *Eozygodactylus americanus* and *Zygodactylus grandei* (Fig. 9.10b; Weidig 2010; Smith et al. 2018). The assignment of the latter to the taxon *Zygodactylus* was based

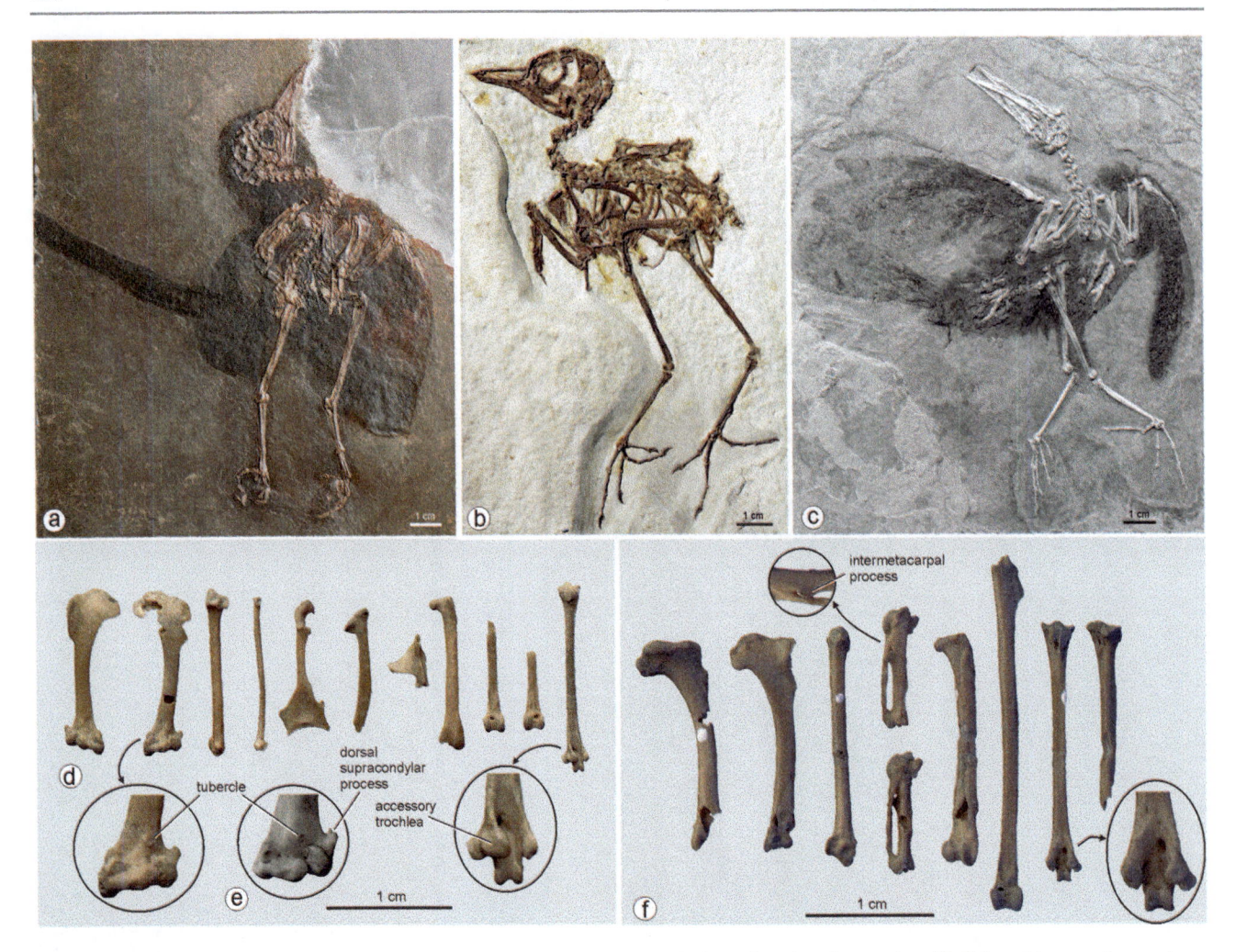

Fig. 9.10 Fossils of the Zygodactylidae. (**a**) Holotype of *Primozygodactylus quintus* from the latest early or earliest middle Eocene of Messel (Senckenberg Research Institute Frankfurt, SMF-ME 11091A). (**b**) Holotype of *Zygodactylus grandei* from the early Eocene Green River Formation (from Smith et al. 2018: fig. 2; published under a Creative Commons CC BY 4.0 license). (**c**) Holotype of *Zygodactylus luberonensis* from the early Oligocene of the Luberon in France (holotype, SMF Av 519), ultraviolet-induced fluorescence photograph; note that the skull of this specimen was inserted in the slab and can only be tentatively assigned to the species. (**d**) Selected bones of an undetermined species of the Zygodactylidae from the early Eocene London Clay of Walton-on-the-Naze (collection of Michael Daniels, Holland-on-Sea, UK, WN 92747). (**e**) Distal end of the humerus of *Corvus bennetti* (Corvidae, Passeriformes). (**f**) Selected bones of *Primozygodactylus* sp. (collection of Michael Daniels, Holland-on-Sea, UK, WN 88583A). The arrows in (**d**) and (**f**) indicate osteological details of the carpometacarpus and the distal ends of the humerus and tarsometatarsus. (Photo in (**a**) by Anika Vogel, in (**c**)–(**f**) by Sven Tränkner)

on a lateral convexity on the distal end of the tarsometatarsus, but with regard to the size and shape of the much smaller accessory trochlea for the fourth toe, the North American species is actually quite different from the species of *Zygodactylus*. With an age of about 52 Ma, the early Eocene *Z. grandei* would furthermore be some 20 million years younger than the earliest European record of *Zygodactylus*, the early Oligocene *Z. luberonensis*, and almost 40 million years older than the youngest *Zygodactylus* species, the middle Miocene (about 14 Ma) *Z. grivensis*. Such an extreme longevity does not seem likely for an avian genus.

Weidig (2010: fig. 11) figured an undescribed small zygodactylid from the Green River Formation, which was also shown in the first edition of the present book (Mayr 2009: fig. 16.10c). This fossil differs from other zygodactylids in the proportions of its limb bones, with the ulna being longer than the tarsometatarsus. Usually, the ulna of zygodactylids is much shorter than the tarsometatarsus, so that the affinities of this North American fossil may need to be revised. An unusually short tarsometatarsus of an unnamed zygodactylid was, however, also reported from the early Eocene Nanjemoy Formation (Virginia, USA) by Mayr et al. (2022).

The latest North American record of the Zygodactylidae is *Zygodactylus ochlurus* from the early Oligocene of Montana (Hieronymus et al. 2019). The holotype and only known specimen of this species is a dissociated and rather poorly preserved skeleton, which allows the recognition of just a few osteological details. *Z. ochlurus* was distinctly smaller than other zygodactylids, from which it also differs in some osteological details, with the sternum, for example, being distinguished from that of *Z. luberonensis* in its proportions. Whether *Z. ochlurus* is indeed a member of the Zygodactylidae, let alone a member of the taxon *Zygodactylus*, is uncertain. Judging from its very small size, the species may as well be a representative of the long-legged Sylphornithidae, which occur in coeval European fossil localities (Sect. 10.3.5.1).

Zygodactylids were small birds with a long and slender tarsometatarsus and long toes. As indicated by the taxon name, they had a permanently reversed fourth toe. The tarsometatarsal trochlea for this toe exhibits a large accessory trochlea, which in *Zygodactylus* is distally elongated as in extant Psittaciformes and the piciform Pici. The carpometacarpus is short and exhibits a very large intermetacarpal process, similar to that of extant Passeriformes and Piciformes. The beak of *Primozygodactylus* has a thrush-like shape and is neither very long nor very short. In some Messel specimens of *Primozygodactylus* long tail feathers are preserved, and it can be discerned that the two central rectrices are greatly elongated (Mayr and Zelenkov 2009).

In contrast to the distal end of the tarsometatarsus, the carpometacarpus of zygodactylids is very similar to that of passeriform birds. The high degree of derived similarity between the carpometacarpi of zygodactylids and passerines is exemplified by the fact that the carpometacarpus of zygodactylids has twice been assigned to the Passeriformes, by Harrison and Walker (1977) in the case of *Primoscens minutus*, and by Ballmann (1969a) concerning *Zygodactylus ignotus* (Mayr 2008b). As in crown group passerines but unlike in the Psittacopedidae, the carpometacarpus of zygodactylids exhibits a large intermetacarpal process. The distal end of the humerus of one of the London Clay zygodactylids in the Daniels collection likewise exhibits a derived morphology otherwise only found in passeriform birds, in that the dorsal supracondylar process is very large and in that there is a marked tubercle in the center of the cranial surface of the distal humerus, above the dorsal condyle (Fig. 9.10d; Mayr 1998b: fig. 25A).

As in passeriform birds, the hypotarsus of zygodactylids exhibits a closed canal for the flexor hallucis longus muscle, and the tarsometatarsus is greatly elongated and bears a plantar crest. A sister group relationship between the Zygodactylidae and the Passeriformes was supported by multiple phylogenetic analyses (Mayr 2008b, 2015, 2020b; Smith et al. 2018; Ksepka et al. 2019), and the clade including both taxa was termed Parapasseres (Mayr 2015).

The proportionally longer toes and larger accessory trochlea for the fourth toe indicate that *Zygodactylus* had a different way of living than early Eocene zygodactylids. The greater span width of the toes may indicate a more terrestrial way of living (Mayr and Zelenkov 2009). The holotype of *Primozygodactylus major* is preserved with stomach contents consisting of grape seeds (Vitaceae) (Mayr 1998b), and ingested seeds are also known from several other zygodactylids from Messel.

9.5.3 Passeriformes

Crown group Passeriformes include more than half of all extant avian species. Molecular analyses provided a phylogenetic framework for the major clades, and according to these studies the Acanthisittidae (New Zealand wrens) are the sister taxon of all other extant passerines, the Eupasseres, which comprise the Suboscines and Oscines (Ericson et al. 2003; Barker et al. 2004; Prum et al. 2015; Kuhl et al. 2021). The Suboscines (tyrant flycatchers, antbirds, manakins, and allies) are mainly restricted to the New World, their sole Old World representatives being pittas (Pittidae) as well as broadbills and kin (Eurylaimidae, Calyptomenidae, Philepittidae, and Sapayoidae). Phylogenies derived from molecular data indicate that the globally distributed Oscines, the most species-rich passeriform group, originated on the Australian continental plate (Barker et al. 2002; Ericson et al. 2002).

In the past years, the Paleogene fossil record of passerines has been substantially improved, but pre-Oligocene fossils still are extremely scarce. In fact, the sole record are two fragmentary bones, a proximal carpometacarpus, and a distal tibiotarsus, from the early Eocene Tingamarra Local Fauna of Australia, which were assigned to the Passeriformes by Boles (1995, 1997). As in extant passerines, the carpometacarpus exhibits a large intermetacarpal process. It differs, however, from extant passerines in the proportionally larger ventral portion of the carpal trochlea, and a depression between the major metacarpal and the pisiform process is shallower in the fossil (Mayr 2013). Additional material is needed for unambiguous identification of these fossils, but from a biogeographic point of view, it is well possible that they belong to stem group representatives of the Passeriformes (see below and Sect. 11.2.1).

Most other Paleogene fossils of passerines are from the Oligocene of Europe, from where a number of different species are meanwhile known. The earliest specimens come from early Oligocene (Rupelian) sites in Germany, Poland, and France. From the early Oligocene of Wiesloch-Frauenweiler in Germany, Mayr and Manegold (2004, 2006a) described a skeleton of a small passerine as *Wieslochia* weissi (Fig. 9.11c–g). This species lacks derived characteristics of crown group Oscines, such as a bifurcated acromion of the scapula, and it shares with extant Suboscines the presence of a well-developed tubercle on the ulna for the

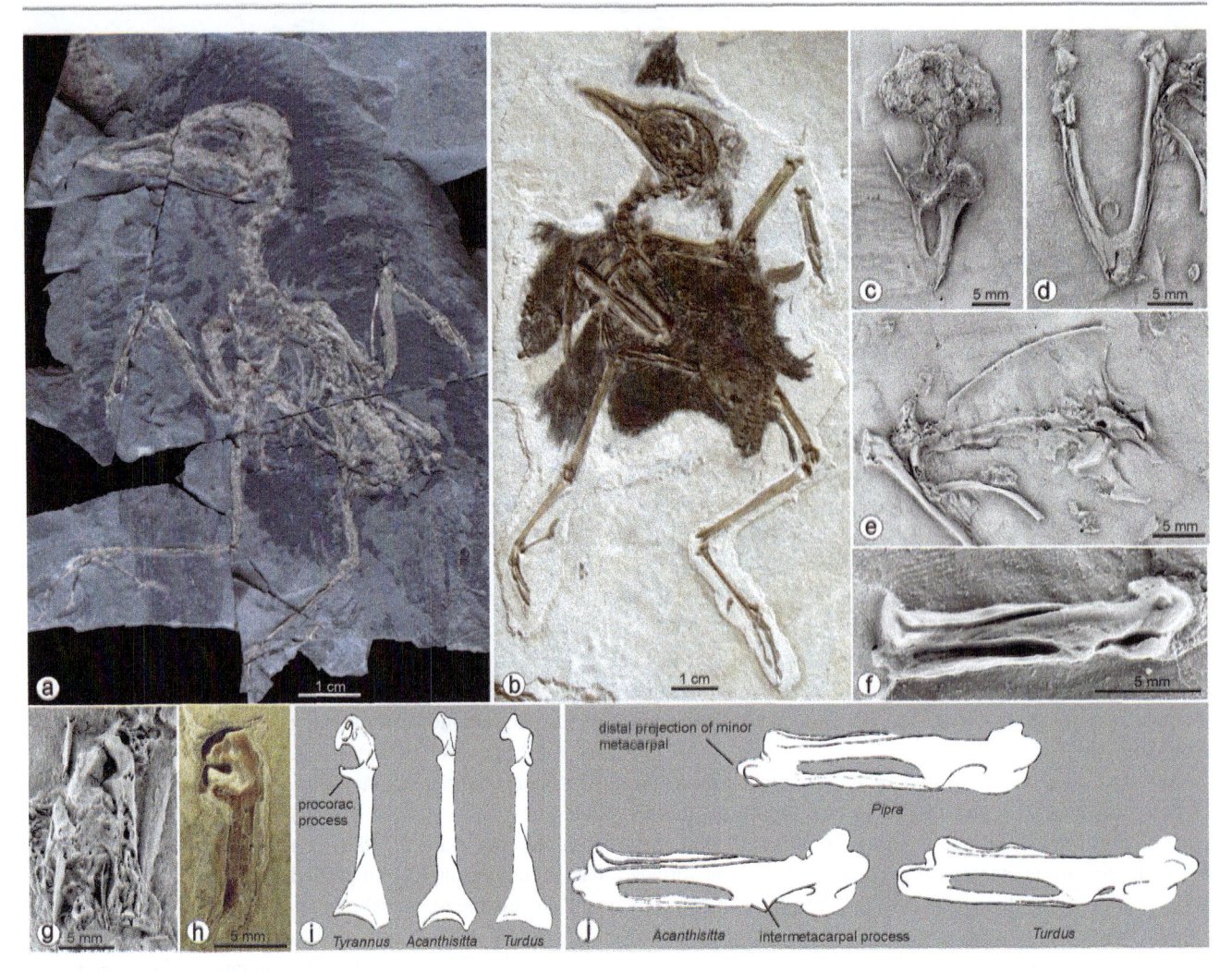

Fig. 9.11 Fossils of early Oligocene Passeriformes. (**a**) *Crosnoornis nargizia* from the Carpathian Basin in Poland (holotype, Institute of Systematics and Evolution of Animals, PAS, Kraków, Poland, MSFK RR 01/2013); from Bochenski et al. (2021: fig. 1), published under a Creative Commons CC BY 4.0 license. (**b**) Fossil of an unnamed suboscine passeriform from the Luberon area in France, which was described by Riamon et al. (2020) (collection of Nicholas Tourment, Marseille, NT-LBR-014). (**c**)–(**g**) Selected skeletal elements of *Wieslochia weissi* from Wiesloch-Frauenweiler in Germany (Staatliches Museum für Naturkunde Karlsruhe, Germany, SMNK-PAL 3980; **c**: skull, **d**: mandible, **e**: sternum, **f**: carpometacarpus, **g**: coracoid); specimen coated with ammonium chloride. (**h**) Coracoid of the suboscine fossil from the Luberon (NT-LBR-014, from Riamon et al. 2020: fig. 4a; published under a Creative Commons CC BY 4.0 license). (**i**) Coracoids of extant passeriform taxa belonging to the Suboscines (*Tyrannus*), Acanthisittidae (*Acanthisitta*), and Oscines (*Turdus*). (**j**) Carpometacarpi of the Suboscines (*Pipra*), Acanthisittidae (*Acanthisitta*), and Oscines (*Turdus*). (Photos in (**b**) by the author, in (**c**)–(**f**) by Sven Tränkner)

ventral collateral ligament (tuberculum ligamenti collateralis ventralis). *Wieslochia* also most closely resembles extant Suboscines in overall skeletal morphology, but Mayr and Manegold (2004, 2006a) noted some features that may support its position outside crown group Eupasseres: the minor metacarpal of the carpometacarpus does not protrude as far distally as the articular facet for the minor wing digit, the coracoid lacks a hooked acrocoracoid process, and the hypotarsus exhibits furrows instead of canals for some the flexor tendons of the toes.

The fossil record of passerines is particularly rich in Rupelian strata of the Carpathian Basin in Poland. From the various localities in this area, several skeletons of small-sized passeriform species have been collected. The holotype of *Jamna szybiaki* is an articulated skeleton lacking the feet (Bochenski et al. 2011). This species is characterized by the fact that the spina externa of the sternum (the spine formed by its cranial end) is not bifurcated; among extant passerines, this condition is only found in the suboscine Eurylaimidae and some Cotingidae. Even though an identification of passeriform fossils is usually straightforward, the holotype of *J. szybiaki* lacks the feet and does not allow the recognition of important skeletal details. The species resulted as the sister taxon of the Zygodactylidae in a phylogenetic analysis by

Smith et al. (2018), but passeriform affinities are supported by the presence of only two notches in the caudal margin of the sternum. The fossil material of another passerine from the Rupelian of Poland, *Resoviaornis jamrozi*, likewise consists of a partial skeleton lacking the feet (Bochenski et al. 2013), and of a third species, *Winnicavis gorskii*, just some wing and pectoral girdle bones are known (Bochenski et al. 2018). The most recent addition to the passeriform record from the Rupelian of Poland is *Crosnoornis nargizia* (Bochenski et al. 2021), which is based on a complete skeleton and was assigned to the Suboscines (Fig. 9.11a). Further passerine remains from the localities in the Outer Carpathians consist of isolated feet (Bochenski et al. 2014a, b).

A wing fragment of a very small passeriform bird was reported from the early Oligocene of the Luberon area in France (Mayr and Manegold 2006b). The well-preserved carpometacarpus of this specimen most closely resembles that of extant Suboscines. The species it belonged to was of similar size to the Madagascan sunbird-asities (*Neodrepanis* spp.), the smallest extant Old World representatives of the Suboscines. Also from the Luberon area, a complete skeleton of a passerine was described by Riamon et al. (2020). This fossil (Fig. 9.11b), which is in a private collection, has not been named and was assigned to the suboscine subclade Tyrannida, which includes all New World Suboscines. This assignment is mainly based on the shape of the coracoid, which exhibits a long procoracoid process. Riamon et al. (2020) argued that the fossil shows no close resemblance to *Wieslochia*, but the specimen from the Luberon is actually not that dissimilar to the latter taxon. Among others, the unnamed passerine from the Oligocene of France and *Wieslochia weissi* share a relatively short and stout humerus as well as a coracoid without a hooked acrocoracoid process. *Wieslochia*, *Crosnoornis*, and the unnamed passerine from the Luberon area furthermore have short legs, with the tibiotarsus being not much longer than the ulna, whereas the tibiotarsus is the longest limb bone in *Resoviaornis* and most extant passerines. The beak shapes of *Wieslochia*, *Crosnoornis*, and the Luberon passerine are also very similar and it is likely that these birds are closely related.

Late Oligocene passerine fossils are known from sites in France (Mourer-Chauviré et al. 1989) and Germany (Manegold 2008). Some of the French specimens were assigned to the Oscines (Mourer-Chauviré et al. 1989), but Mourer-Chauviré (2006) also noted the presence of a suboscine passerine in the late Oligocene (MP 26) of France. Manegold (2008) described a diverse passerine assemblage from the late Oligocene of Germany and likewise identified representatives of both Oscines and Suboscines. It is noteworthy that all of these specimens belong to small species.

The fact that the earliest European Passeriformes already closely resemble their extant counterparts supports the hypothesis that passerines dispersed into the Northern Hemisphere from one of the southern continents (e.g., Olson 1989; Ericson et al. 2003). All early Oligocene passerines known so far furthermore do not belong to the Oscines, the only passeriform taxon which occurs in Europe today. The earliest European fossil record of oscine passerines is from the late Oligocene (Mourer-Chauviré et al. 1989; Manegold 2008), and non-oscine passerines seem to have colonized Europe before the arrival of the Oscines from the Australian continental plate. The fossil record, therefore, suggests that there were at least two major dispersal events of passerines into Europe, with suboscine-like taxa having arrived toward the early Oligocene, some 32–34 Ma, whereas oscine passerines appeared in Europe several million years later.

Outside Europe, Paleogene fossils of passerines are very scant, and apart from the above-mentioned specimens from the early Eocene of Australia, the sole fossils are bones of the logrunner *Orthonyx kaldowinyeri* (Orthonychidae) from late Oligocene deposits of the Riversleigh fossil site in Australia (Nguyen et al. 2014). This species is the oldest passerine that can be assigned to an extant oscine subclade and was assigned to a genus-level taxon, which is still existent. As yet, Paleogene passerines are altogether unknown from Africa, Asia, and the Americas.

References

Ameghino F (1895) Sur les oiseaux fossiles de la Patagonie et la faune mammalogique des couches a *Pyrotherium*. Bol Inst Geogr Argent 15 [for 1894]:501–602

Ballmann P (1969a) Die Vögel aus der altburdigalen Spaltenfüllung von Wintershof (West) bei Eichstätt in Bayern. Zitteliana 1:5–60

Ballmann P (1969b) Les oiseaux miocènes de La Grive-Saint-Alban (Isère). Geobios 2:157–204

Barker FK, Barrowclough GF, Groth JG (2002) A phylogenetic hypothesis for passerine birds: taxonomic and biogeographic implications of an analysis of nuclear DNA sequence. Proc R Soc Lond Ser B 269:295–308

Barker FK, Cibois A, Schikler P, Feinstein J, Cracraft J (2004) Phylogeny and diversification of the largest avian radiation. Proc Natl Acad Sci U S A 101:11040–11045

Bertelli S, Lindow BEK, Dyke GJ, Chiappe LM (2010) A well-preserved 'charadriiform-like' fossil bird from the Lower Eocene Fur Formation of Denmark. Palaeontology 53:507–531

Bochenski ZM, Tomek T, Bujoczek M, Wertz K (2011) A new passerine bird from the early Oligocene of Poland. J Ornithol 152:1045–1053

Bochenski ZM, Tomek T, Wertz K, Swidnicka E (2013) The third nearly complete passerine bird from the early Oligocene of Europe. J Ornithol 154:923–931

Bochenski ZM, Tomek T, Swidnicka E (2014a) The first complete leg of a passerine bird from the early Oligocene of Poland. Acta Palaeontol Pol 59:281–285

Bochenski ZM, Tomek T, Swidnicka E (2014b) A complete passerine foot from the late Oligocene of Poland. Palaeontol Electron 17.1 (6A):1–7

Bochenski ZM, Tomek T, Wertz K, Happ J, Bujoczek M, Swidnicka E (2018) Articulated avian remains from the early Oligocene of Poland adds to our understanding of passerine evolution. Palaeontol Electron 21.2(32A):1–12

Bochenski ZM, Tomek T, Bujoczek M, Salwa G (2021) A new passeriform (Aves: Passeriformes) from the early Oligocene of Poland sheds light on the beginnings of Suboscines. J Ornithol 162:593–604

Boles WE (1995) The world's oldest songbird. Nature 374:21–22

Boles WE (1997) Fossil songbirds (Passeriformes) from the early Eocene of Australia. Emu 97:43–50

Botelho JF, Smith-Paredes D, Nuñez-Leon D, Soto-Acuña S, Vargas AO (2014) The developmental origin of zygodactyl feet and its possible loss in the evolution of Passeriformes. Proc R Soc Lond Ser B 281:20140765

Braun EL, Kimball RT (2021) Data types and the phylogeny of Neoaves. Birds 2:1–22

Dyke GJ (2001) Fossil pseudasturid birds (Aves, Pseudasturidae) from the London Clay. Bull Nat Hist Mus Lond (Geol Ser) 57:1–4

Dyke GJ, Cooper JH (2000) A new psittaciform bird from the London Clay (Lower Eocene) of England. Palaeontology 43:271–285

Ericson PGP, Christidis L, Cooper A, Irestedt M, Jackson J, Johansson US, Norman JA (2002) A Gondwanan origin of passerine birds supported by DNA sequences of the endemic New Zealand wrens. Proc R Soc Lond Ser B 269:235–241

Ericson PGP, Irestedt M, Johansson US (2003) Evolution, biogeography, and patterns of diversification in passerine birds. J Avian Biol 34:3–15

Feduccia A (1999) The origin and evolution of birds, 2nd edn. Yale University Press, New Haven

Feduccia A, Martin LD (1976) The Eocene zygodactyl birds of North America (Aves: Piciformes). Smithson Contrib Paleobiol 27:101–110

Gingerich PD (1987) Early Eocene bats (Mammalia, Chiroptera) and other vertebrates in freshwater limestones of the Willwood Formation, Clark's Fork Basin, Wyoming. Contrib Mus Paleontol Univ Michigan 27:275–320

Grande L (2013) The lost world of Fossil lake: Snapshots from deep time. University of Chicago Press, Chicago

Hackett SJ, Kimball RT, Reddy S, Bowie RCK, Braun EL, Braun MJ, Chojnowski JL, Cox WA, Han K-L, Harshman J, Huddleston CJ, Marks BD, Miglia KJ, Moore WS, Sheldon FH, Steadman DW, Witt CC, Yuri T (2008) A phylogenomic study of birds reveals their evolutionary history. Science 320:1763–1767

Harrison CJO (1982) The earliest parrot: a new species from the British Eocene. Ibis 124:203–210

Harrison CJO, Walker CA (1972) The affinities of *Halcyornis* from the Lower Eocene. Bull Brit Mus (Nat Hist) Geol 21:153–170

Harrison CJO, Walker CA (1977) Birds of the British Lower Eocene. Tert Res Spec Pap 3:1–52

Hieronymus TL, Waugh DA, Clarke JA (2019) A new zygodactylid species indicates the persistence of stem passerines into the early Oligocene in North America. BMC Evol Biol 19:3

Hoch E (1988) On the ecological role of an Eocene bird from Messel, West Germany. Cour Forsch-Inst Senckenberg 107:249–261

Jarvis ED, Mirarab S, Aberer AJ, Li B, Houde P, Li C, Ho SYW, Faircloth BC, Nabholz B, Howard JT, Suh A, Weber CC, da Fonseca RR, Li J, Zhang F, Li H, Zhou L, Narula N, Liu L, Ganapathy G, Boussau B, Bayzid MS, Zavidovych V, Subramanian S, Gabaldón T, Capella-Gutiérrez S, Huerta-Cepas J, Rekepalli B, Munch K, Schierup M. et al. (75 further co-authors) (2014) Whole-genome analyses resolve early branches in the tree of life of modern birds. Science 346:1320–1331

Ksepka DT, Clarke JA (2012) A new stem parrot from the Green River Formation and the complex evolution of the grasping foot in Pan-Psittaciformes. J Vertebr Paleontol 32:395–406

Ksepka DT, Clarke JA, Grande L (2011) Stem parrots (Aves, Halcyornithidae) from the Green River Formation and a combined phylogeny of Pan-Psittaciformes. J Paleontol 85:835–852

Ksepka DT, Grande L, Mayr G (2019) Oldest finch-beaked birds reveal parallel ecological radiations in the earliest evolution of passerines. Curr Biol 29:657–663

Kuhl H, Frankl-Vilches C, Bakker A, Mayr G, Nikolaus G, Boerno ST, Klages S, Timmermann B, Gahr M (2021) An unbiased molecular approach using 3'UTRs resolves the avian family-level tree of life. Mol Biol Evol 38:108–121

Livezey BC, Zusi RL (2007) Higher-order phylogeny of modern birds (Theropoda, Aves: Neornithes) based on comparative anatomy: II.—Analysis and discussion. Zool J Linnean Soc 149:1–94

Manegold A (2008) Passerine diversity in the late Oligocene of Germany: earliest evidence for the sympatric coexistence of Suboscines and Oscines. Ibis 150:377–387

Mayr G (1998a) A new family of Eocene zygodactyl birds. Senck leth 78:199–209

Mayr G (1998b) „Coraciiforme" und „piciforme" Kleinvögel aus dem Mittel-Eozän der Grube Messel (Hessen, Deutschland). Cour Forsch-Inst Senckenberg 205:1–101

Mayr G (1999) *Pumiliornis tessellatus* n. gen. n. sp., a new enigmatic bird from the Middle Eocene of Grube Messel (Hessen, Germany). Cour Forsch-Inst Senckenberg 216:75–83

Mayr G (2000a) New or previously unrecorded avian taxa from the Middle Eocene of Messel (Hessen, Germany). Mitt Mus Naturkunde Berl, Geowiss Reihe 3:207–219

Mayr G (2000b) A new raptor-like bird from the Lower Eocene of North America and Europe. Senck leth 80:59–65

Mayr G (2002a) Avian Remains from the Middle Eocene of the Geiseltal (Sachsen-Anhalt, Germany). In: Zhou Z, Zhang F (eds) Proceedings of the 5th symposium of the Society of Avian Paleontology and Evolution, Beijing, 1–4 June 2000. Science Press, Beijing, pp 77–96

Mayr G (2002b) On the osteology and phylogenetic affinities of the Pseudasturidae—Lower Eocene stem-group representatives of parrots (Aves, Psittaciformes). Zool J Linnean Soc 136:715–729

Mayr G (2005a) The postcranial osteology and phylogenetic position of the Middle Eocene *Messelastur gratulator* Peters, 1994—a morphological link between owls (Strigiformes) and falconiform birds? J Vertebr Paleontol 25:635–645

Mayr G (2005b) A *Fluvioviridavis*-like bird from the Middle Eocene of Messel, Germany. Can J Earth Sci 42:2021–2037

Mayr G (2007) New specimens of Eocene stem-group psittaciform birds may shed light on the affinities of the first named fossil bird, *Halcyornis toliapicus* Koenig, 1825. Neues Jahrb Geol Palaontol Abh 244:207–213

Mayr G (2008a) *Pumiliornis tessellatus* Mayr, 1999 revisited—new data on the osteology and possible phylogenetic affinities of an enigmatic Middle Eocene bird. Paläontol Z 82:247–253

Mayr G (2008b) Phylogenetic affinities of the enigmatic avian taxon *Zygodactylus* based on new material from the early Oligocene of France. J Syst Palaeontol 6:333–344

Mayr G (2009) Paleogene fossil birds, 1st edn. Springer, Heidelberg

Mayr G (2010) Mousebirds (Coliiformes), parrots (Psittaciformes), and other small birds from the late Oligocene/early Miocene of the Mainz Basin, Germany. N Jb Geol Paläont (Abh) 258:129–144

Mayr G (2011a) Well-preserved new skeleton of the Middle Eocene *Messelastur* substantiates sister group relationship between Messelasturidae and Halcyornithidae (Aves, ?Pan-Psittaciformes). J Syst Palaeontol 9:159–171

Mayr G (2011b) On the osteology and phylogenetic affinities of *Morsoravis sedilis* (Aves) from the early Eocene Fur Formation of Denmark. Bull Geol Soc Denmark 59:23–35

Mayr G (2013) The age of the crown group of passerine birds and its evolutionary significance—molecular calibrations versus the fossil record. Syst Biodivers 11:7–13

Mayr G (2015) A reassessment of Eocene parrotlike fossils indicates a previously undetected radiation of zygodactyl stem group representatives of passerines (Passeriformes). Zool Scr 44:587–602

References

Mayr G (2016) The world's smallest owl, the earliest unambiguous charadriiform bird, and other avian remains from the early Eocene Nanjemoy Formation of Virginia (USA). Paläontol Z 90:747–763

Mayr G (2017a) Avian Evolution: The fossil record of birds and its paleobiological significance. Wiley-Blackwell, Chichester

Mayr G (2017b) New species of *Primozygodactylus* from Messel and the ecomorphology and evolutionary significance of early Eocene zygodactylid birds (Aves, Zygodactylidae). Hist Biol 29:875–884

Mayr G (2020a) An updated review of the middle Eocene avifauna from the Geiseltal (Germany), with comments on the unusual taphonomy of some bird remains. Geobios 62:45–59

Mayr G (2020b) A remarkably complete skeleton from the London Clay provides insights into the morphology and diversity of early Eocene zygodactyl near-passerine birds. J Syst Palaeontol 18:1891–1906

Mayr G (2021) A partial skeleton of a new species of *Tynskya* Mayr, 2000 (Aves, Messelasturidae) from the London Clay highlights the osteological distinctness of a poorly known early Eocene "owl/parrot mosaic". Pal Z 95:337–357

Mayr G, Clarke J (2003) The deep divergences of neornithine birds: a phylogenetic analysis of morphological characters. Cladistics 19:527–553

Mayr G, Daniels M (1998) Eocene parrots from Messel (Hessen, Germany) and the London Clay of Walton-on-the-Naze (Essex, England). Senck leth 78:157–177

Mayr G, Manegold A (2004) The oldest European fossil songbird from the early Oligocene of Germany. Naturwiss 91:173–177

Mayr G, Manegold A (2006a) New specimens of the earliest European passeriform bird. Acta Palaeontol Pol 51:315–323

Mayr G, Manegold A (2006b) A small suboscine-like passeriform bird from the early Oligocene of France. Condor 108:717–720

Mayr G, Smith T (2019) A diverse bird assemblage from the Ypresian of Belgium furthers knowledge of early Eocene avifaunas of the North Sea Basin. N Jb Geol Paläont (Abh) 291:253–281

Mayr G, Wilde V (2014) Eocene fossil is earliest evidence of flower-visiting by birds. Biol Lett 10:20140223

Mayr G, Zelenkov N (2009) New specimens of zygodactylid birds from the middle Eocene of Messel, with description of a new species of *Primozygodactylus* Mayr, 1998. Acta Palaeontol Pol 54:15–20

Mayr G, Rana RS, Sahni A, Smith T (2007) Oldest fossil avian remains from the Indian subcontinental plate. Curr Sci 92:1266–1269

Mayr G, Rana RS, Rose KD, Sahni A, Kumar K, Singh L, Smith T (2010) *Quercypsitta*-like birds from the early Eocene of India (Aves, ?Psittaciformes). J Vertebr Paleontol 30:467–478

Mayr G, Rana RS, Rose KD, Sahni A, Kumar K, Smith T (2013) New specimens of the early Eocene bird *Vastanavis* and the interrelationships of stem group Psittaciformes. Paleontol J 4:1308–1314

Mayr G, Gingerich PD, Smith T (2019) *Calcardea junnei* Gingerich, 1987 from the late Paleocene of North America is not a heron, but resembles the early Eocene Indian taxon *Vastanavis* Mayr et al., 2007 J Vertebr Paleontol 93:359–367

Mayr G, Hervet S, Buffetaut E (2020) On the diverse and widely ignored Paleocene avifauna of Menat (Puy-de-Dôme, France): new taxonomic records and unusual soft tissue preservation. Geol Mag 156:572–584

Mayr G, De Pietri V, Scofield RP (2022) New bird remains from the early Eocene Nanjemoy Formation of Virginia (USA), including the first records of the Messelasturidae, Psittacopedidae, and Zygodactylidae from the Fisher/Sullivan site. Hist Biol 34:322–334

Mlíkovský J (2002) Cenozoic birds of the world: Part 1: Europe. Ninox Press, Praha

Mourer-Chauviré C (1992) Une nouvelle famille de perroquets (Aves, Psittaciformes) dans l'Éocène supérieur des Phosphorites du Quercy, France. Geobios, mém spéc 14:169–177

Mourer-Chauviré C (2006) The avifauna of the Eocene and Oligocene Phosphorites du Quercy (France): an updated list. Strata, sér 1 13:135–149

Mourer-Chauviré C, Hugueney M, Jonet P (1989) Découverte de Passeriformes dans l'Oligocène supérieur de France. C R Acad Sci Sér II 309:843–849

Mourer-Chauviré C, Pickford M, Senut B (2015) Stem group galliform and stem group psittaciform birds (Aves, Galliformes, Paraortygidae, and Psittaciformes, family incertae sedis) from the Middle Eocene of Namibia. J Ornithol 156:275–286

Mourer-Chauviré C, Pickford M, Senut B (2017) New data on stem group Galliformes, Charadriiformes, and Psittaciformes from the middle Eocene of Namibia. Contrib MACN 7:99–131

Nguyen JMT, Boles WE, Worthy TH, Hand SJ, Archer M (2014) New specimens of the logrunner *Orthonyx kaldowinyeri* (Passeriformes: Orthonychidae) from the Oligo-Miocene of Australia. Alcheringa 38:245–255

Olson SL (1985) The fossil record of birds. In: Farner DS, King JR, Parkes KC (eds) Avian Biology, vol 8. Academic Press, New York, pp 79–238

Olson SL (1989 ["1988"]) Aspects of global avifaunal dynamics during the Cenozoic. In: Ouellet H (ed) Acta XIX Congressus Internationalis Ornithologici. University of Ottawa Press, Ottawa, pp 2023–2029

Peters DS (1994) *Messelastur gratulator* n. gen. n. spec., ein Greifvogel aus der Grube Messel (Aves: Accipitridae). Cour Forsch-Inst Senckenberg 170:3–9

Prum RO, Berv JS, Dornburg A, Field DJ, Townsend JP, Lemmon EM, Lemmon AR (2015) A comprehensive phylogeny of birds (Aves) using targeted next-generation DNA sequencing. Nature 526:569–573

Riamon S, Tourment N, Louchart A (2020) The earliest Tyrannida (Aves, Passeriformes), from the Oligocene of France. Sci Rep 10:9776

Simpson GG (1946) Fossil penguins. Bull Am Mus Nat Hist 87:1–100

Smith NA, DeBee AM, Clarke JA (2018) Systematics and phylogeny of the Zygodactylidae (Aves, Neognathae) with description of a new species from the early Eocene of Wyoming, USA. PeerJ 6:e4950

Suh A, Paus M, Kiefmann M, Churakov G, Franke F, Brosius J, Kriegs J, Schmitz J (2011) Mesozoic retroposons reveal parrots as the closest living relatives of passerine birds. Nature Comm 2:443

Waterhouse DW, Lindow BEK, Zelenkov N, Dyke GJ (2008) Two new parrots (Psittaciformes) from the Lower Eocene Fur Formation of Denmark. Palaeontology 51:575–582

Weidig I (2010) New birds from the lower Eocene Green River Formation, North America. Rec Austral Mus 62:29–44

Yuri T, Kimball RT, Harshman J, Bowie RC, Braun MJ, Chojnowski JL, Han K-L, Hackett SJ, Huddleston CJ, Moore WS, Reddy S, Sheldon FH, Steadman DW, Witt CC, Braun EL (2013) Parsimony and model-based analyses of indels in avian nuclear genes reveal congruent and incongruent phylogenetic signals. Biol 2:419–444

10 Strigiformes (Owls), Coliiformes (Mousebirds), and Cavitaves (Trogons, Rollers, Woodpeckers, and Allies)

The taxa included in the present chapter form a clade in most current analyses of molecular data (under some settings, the Psittacopasseres are likewise part of this clade; e.g., Braun and Kimball 2021: fig. 4D). These birds are part of the Anomalogonatae, or "higher land bird assemblage," of earlier authors and exhibit disparate feeding adaptations and habitat preferences, even though most species are adapted to an arboreal way of living.

The representatives of this clade have a comprehensive fossil record in early Eocene fossil localities of the Northern Hemisphere and appear to have been among the first Neornithes, which underwent a radiation in the angiosperm-dominated forested palaeohabitats that evolved after the K/Pg mass extinction. Some of the fossils described in the present chapter furthermore offer insights into the historical biogeography of clades with a restricted extant distribution.

10.1 Strigiformes (Owls)

Crown group Strigiformes are divided into the Tytonidae (barn owls) and Strigidae (true owls), which both have a worldwide distribution. All owls possess semi-zygodactyl feet and most extant species are crepuscular or nocturnal. The skeletal morphology of owls is distinctive and allows an unambiguous identification of the major postcranial bones.

The fossil record of strigiform birds goes back to the Paleocene and is quite extensive in Eocene and Oligocene deposits of the Northern Hemisphere. The Paleogene fossil record of owls from the Southern Hemisphere, by contrast, is very scant. Although many fossil strigiform taxa have been reported from European and North American sites, their interrelationships are still poorly understood. The two known Paleocene taxa are markedly different, and whereas one (the North American *Ogygoptynx*) is quite similar to crown group Strigiformes, the other (the European *Berruornis*) is not. Apart from their proportionally shorter legs, most late Eocene and Oligocene owls most closely resemble extant Tytonidae in overall morphology, but these similarities are likely to be plesiomorphic.

10.1.1 *Berruornis* and Sophiornithidae

The taxon *Berruornis* was established by Mourer-Chauviré (1994) for fossils from the late Paleocene of the Mont Berru area in France. The species from this locality, *Berruornis orbisantiqui*, is known from several tarsometatarsi and an incomplete distal tibiotarsus (Mourer-Chauviré 1994). A second species of *Berruornis*, *B. halbedeli*, occurs in the late middle Paleocene of Walbeck in Germany and is represented by a tarsometatarsus and a tentatively referred partial upper beak (Mayr 2002a, 2007). According to Mourer-Chauviré (1994), a large phalanx of a strigiform-like bird reported by Nessov (1992) from the late Paleocene of Kazakhstan may also belong to *Berruornis*.

The species of *Berruornis* were very large, about the size of the extant Snowy Owl, *Bubo scandiacus*. Their assignment to the Strigiformes can be established with the absence of a supratendinal bridge on the distal tibiotarsus, the morphology of the hypotarsus, which exhibits two widely separated crests, and the large tarsometatarsal trochlea for the second toe (see Mourer-Chauviré 1994). The tarsometatarsus is stout and the medial hypotarsal crest is pierced by a small foramen. The wing-like flange formed by the tarsometatarsal trochlea for the fourth toe is much less developed than in extant owls. If the assignment of the Walbeck beak to *Berruornis* is correct, the taxon also has a proportionally longer and narrower beak than crown group Strigiformes. The ventral surface of the rostral end of the premaxillary symphysis is furthermore distinctive in that it bears a pair of grooves along its sides, which among extant birds is found in the Falconidae and, albeit in a much less pronounced form, in the Cathartidae.

Berruornis was assigned to the Sophiornithidae by Mourer-Chauviré (1994). This latter taxon was originally introduced for *Sophiornis quercynus*, an equally large species, which is known from a tarsometatarsus, a distal humerus, and pedal phalanges from an unknown locality and stratigraphic horizon of the Quercy fissure fillings (Mourer-Chauviré 1987). Classification of *Berruornis* into the Sophiornithidae is, however, largely based on overall similarity, and as detailed by Mourer-Chauviré (1994), the tarsometatarsi of *Berruornis* and *Sophiornis* differ in several aspects. Unlike in *Sophiornis*, the tarsometatarsal trochleae for the second and fourth toes are hardly plantarly deflected in *Berruornis*, which may indicate a sister group relationship between *Berruornis* and all other strigiform birds.

10.1.2 Ogygoptyngidae

The single species classified into this taxon is *Ogygoptynx wetmorei*, which is the earliest strigiform bird as yet reported (Fig. 10.1a). *O. wetmorei* is represented by a tarsometatarsus from the late Paleocene (Tiffanian) of Colorado (Rich and Bohaska 1976, 1981). The bone is long and slender as in crown group Tytonidae and more closely resembles the tarsometatarsus of extant owls than do the tarsometatarsi of *Berruornis* and the Protostrigidae. In particular, the enlarged trochlea for the second toe distally exceeds the trochlea for the third toe in length (unlike in the Protostrigidae; see next section) and, in distal view, the trochleae are arranged on a markedly curved line (unlike in *Berruornis*). A presumably autapomorphic feature of *Ogygoptynx* concerns the shape of the proximal tasometatarsus end, which is shaped like a parallelogram in proximal view, whereas it is rectangular in other Strigiformes. Despite being the earliest known strigiform bird, the distally prominent trochlea for the second toe and the arrangement of the trochleae indicate that *O. wetmorei* is more closely related to crown group Strigiformes than are *Berruornis* and the Protostrigidae (Mourer-Chauviré 1987). Although *Ogygoptynx* is likely to be outside crown group Strigiformes, the fossil material does not allow an unambiguous further assessment of its affinities.

10.1.3 Protostrigidae, *Primoptynx*, and *Eostrix*

The Protostrigidae occur in the middle and late Eocene of North America and in the early Oligocene of Europe. The taxon was originally established for *Minerva* ("*Protostrix*"), which includes four species from the middle and late Eocene of North America, that is, *Minerva antiqua*, *M. leptosteus* (Fig. 10.1b), and *M. saurodosis* from the middle Eocene (Bridgerian) of Wyoming, as well as *M. californiensis* from the late Eocene of California. Whereas *M. antiqua* and *M. leptosteus* are represented by leg bones (Wetmore 1933; Rich 1982; Mourer-Chauviré 1983), the similarly-sized *M. saurodosis* and *M. californiensis* are only known from humeri (Wetmore 1921; Howard 1965). Therefore, it is well possible that *M. saurodosis* is a junior synonym of either *M. antiqua* or *M. leptosteus*, which have a similar temporal and geographical distribution.

In Europe, the Protostrigidae are represented by *Oligostrix rupelensis*, the holotype of which is a distal tibiotarsus from the early Oligocene of Germany (Fischer 1983). *O. rupelensis* was a small species, which measured just half the size of the much larger species of *Minerva*. The description of another species of *Oligostrix*, *O. bergeri* from the late Oligocene of Switzerland, was based on a tarsometatarsus (De Pietri et al. 2013). This species constitutes the latest record of the Protostrigidae, and its tarsometatarsus is much narrower than that of the species of the North American taxon *Minerva*.

Protostrigid owls are characterized by a greatly widened medial condyle of the tibiotarsus and, possibly functionally correlated therewith, a strongly developed first and second toe (Mourer-Chauviré 1983). The morphology of the ungual phalanx of the hallux of *Minerva* is peculiar in that the articular facet is markedly concave and the extensor process strongly proximally protruding (Mourer-Chauviré 1983: fig. 3). This morphology even led Wetmore (1933) to erroneously assume that the ungual phalanges of *Minerva antiqua* are from an edentate mammal (see Mourer-Chauviré 1983). The tarsometatarsal trochlea for the second toe of protostrigids is furthermore proportionally shorter and mediolaterally wider than in most other owls. Its shape resembles that of the corresponding trochlea of *Berruornis* and may constitute the plesiomorphic condition for the Strigiformes.

The above taxa occur in middle Eocene to late Oligocene localities, but a similar foot morphology to that of protostrigid owls has also been reported for *Primoptynx poliotauros* from the early Eocene (Wasatchian) of the Willwood Formation in Wyoming, USA (Mayr et al. 2020a). This large species was about the size of the extant Spectacled Owl (*Pulsatrix perspicillata*) and is one of the best-represented Paleogene owls. The holotype consists of a fairly complete postcranial skeleton but lacks most of the skull (Fig. 10.1c, d). A fragment of the mandible indicates that the beak was narrower than in most extant owls, and the wings were proportionally shorter. The most characteristic feature of *Primoptynx*, however, is an accipitrid-like foot morphology with enlarged ungual phalanges of the first and second toes. *Primoptynx* therefore probably used its feet to kill prey items in a hawk-like manner, which pin down prey with their body weight and then puncture it with the large ungual phalanges of the first and second toe. In extant owls, by contrast, all ungual phalanges have a similar size and prey

10.1 Strigiformes (Owls)

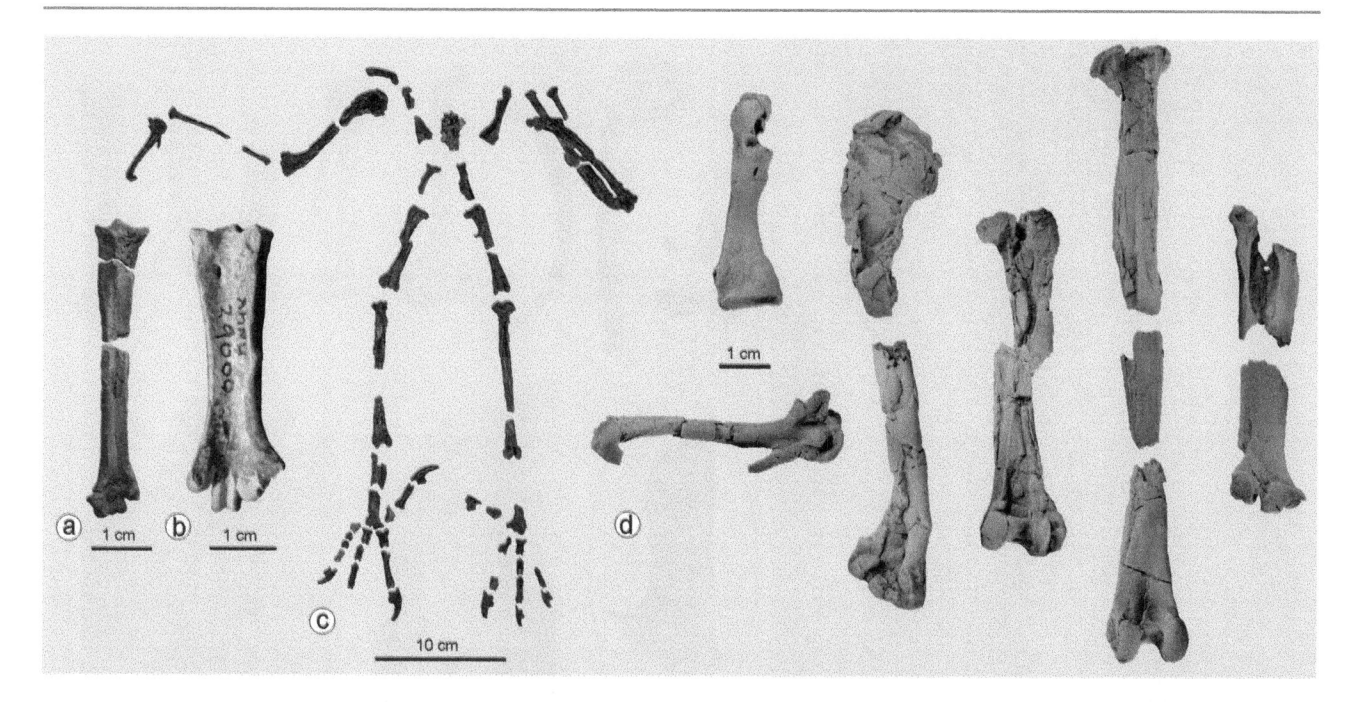

Fig. 10.1 Paleocene and Eocene owls (Strigiformes) from North America. (**a**) Tarsometatarsus of *Ogygptynx wetmorei* from the late Paleocene of Colorado (holotype, American Museum of Natural History, New York, USA, AMNH 2653). (**b**) Tarsometatarsus of *Minerva* cf. *leptosteus* from the late early to early middle Eocene Bridger Formation in Wyoming (AMNH 29000). (**c**) Holotype skeleton of *Primoptynx poliotauros* from the early Eocene Willwood Formation of the Bighorn Basin, Wyoming (University of Michigan, Museum of Paleontology, Ann Arbor, Michigan, USA, UMMP 96195). (**d**) Selected bones (coracoid, partial carpometacarpus, humerus, femur, tibiotarsus, and tarsometatarsus) of the *P. poliotauros* holotype; coated with ammonium chloride. (Photos in (**c**) and (**d**) by Sven Tränkner, others by the author)

is killed with the beak. *Primoptynx* and protostrigid owls therefore may have been specialized in feeding on prey that required an accipitrid-like killing strategy, such as larger-sized or more defensive mammals, and these peculiar owls may have succumbed to competition with accipitrid diurnal birds of prey, which diversified towards the late Eocene and early Oligocene. The early Eocene *Primoptynx* lacks the widened medial condyle of the tibiotarsus, which characterizes the geologically younger taxa *Minerva* and *Oligostrix*, even though *Primoptynx* may well be the sister taxon of the Protostrigidae (Mayr et al. 2020a).

Earlier authors also assigned the taxon *Eostrix* to the Protostrigidae, which includes various small-sized species. The holotypes of *Eostrix mimica* and *E. martinellii* from the early Eocene of Wyoming (Willwood and Wind River formations) consist of leg bones (distal tarsometatarsus and tibiotarsus for *E. mimica* and distal tarsometatarsus for *E. martinellii*; Wetmore 1938; Martin and Black 1972). *E. gulottai* from the early Eocene Nanjemoy Formation (Virginia, USA) is likewise just known from a distal tarsometatarsus (Fig. 10.2c, d; Mayr 2016). This species is the smallest fossil owl described so far and is also in the size range of the smallest extant owl. Outside North America, two species of *Eostrix* were described: *E. vincenti* from the early Eocene of England, which is represented by a proximal tarsometatarsus and pedal phalanx only (Harrison 1980), and *E. tsaganica* from the early Eocene of Mongolia, the description of which was based on the proximal and distal ends of the tarsometatarsus and a distal tibiotarsus (Kurochkin and Dyke 2011).

Eostrix does not exhibit the derived foot morphology found in the above taxa of the Protostrigidae, that is, the medial condyle of the tibiotarsus is not widened and the tarsometatarsal trochlea for the second toe is not greatly enlarged (the morphology of the ungual phalanges of *Eostrix* is unknown). Overall, the tarsometatarsus is more similar to that of extant Strigiformes, even though the trochlea for the second toe is still proportionally shorter. Of the known taxa of Paleogene owls, *Eostrix* is therefore the most likely candidate for a taxon on the lineage leading to crown group Strigiformes. Its assignment to the Protostrigidae is likely to be incorrect and needs to be revised, once further data on the skeletal morphology of *Eostrix* become available.

A partial skeleton of an undescribed *Eostrix*-like owl from the London Clay of Walton-on-the-Naze in the collection of Michael Daniels includes an isolated lacrimal bone with a well-developed supraorbital process (Fig. 10.2b; Mayr 2017). The presence of supraorbital processes was also noted for an unpublished owl from the middle Eocene North American Bridger Formation (Fowler et al. 2018). In extant Strigiformes, supraorbital processes are present in the embryo but are subsequently reduced, being vestigial (Fig. 10.2f) or altogether absent in the adult birds, in which the lacrimals are furthermore co-ossified with the frontal

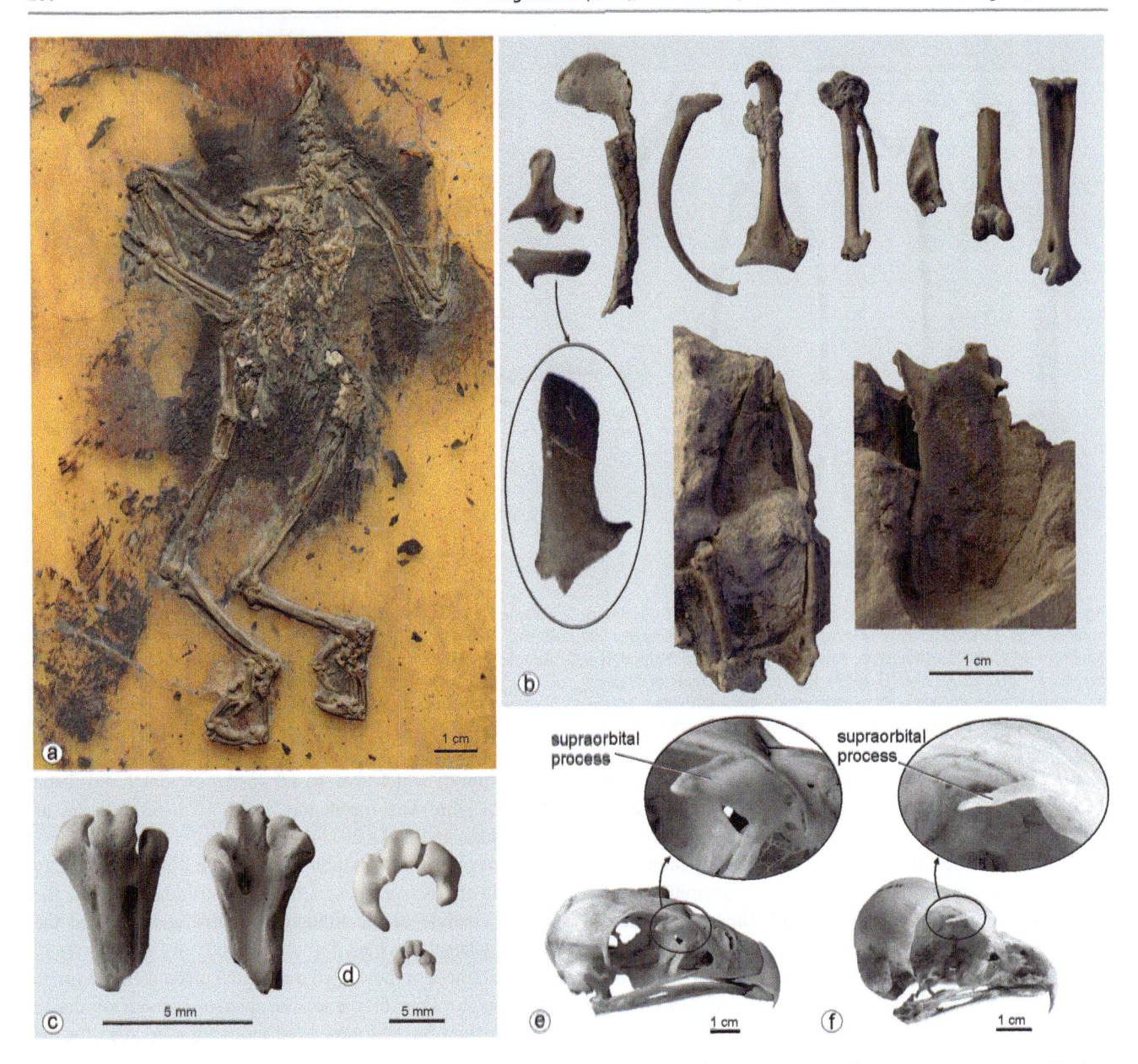

Fig. 10.2 (a) Holotype of the small owl *Palaeoglaux artophoron* (Strigiformes) from the latest early or earliest middle Eocene of Messel (Senckenberg Research Institute, Frankfurt, SMF-ME 1144a). (b) Selected bones of an undescribed, *Eostrix*-like strigiform species from the early Eocene London Clay of Walton-on-the-Naze (collection of Michael Daniels, Holland-on-Sea, UK, WN 92717); the fossil includes various isolated bones as well as the mandible (bottom, center) and sternum (bottom, right), the arrow indicates an enlarged detail of the supraorbital process. (c) Holotype distal tarsometatarsus of *Eostrix gulottai* from the early Eocene Nanjemoy Formation in Virginia, USA, in dorsal and plantar view (SMF Av 627). (d) Distal ends of the tarsometatarsi of *E. gulottai* (bottom) and the extant Barn Owl (*Tyto alba*; Tytonidae) shown at the same scale. (e) Skull of the extant Crested Caracara, *Caracara plancus* (Falconidae). (f) Skull of the extant Hawk-Owl, *Surnia ulula* (Strigidae), which is a diurnal species of extant owls with unusually large supraorbital processes. The arrows in (e) and (f) indicate enlarged details of the supraorbital processes. (Photo in (b) by Anika Vogel, all others by Sven Tränkner)

bones. The occurrence of large supraorbital processes in Paleogene stem group representatives of the Strigiformes shows that these processes were reduced in the stem lineage of owls, which is likely to have been due to an enlargement of the eyeballs in relation to nocturnal activity. Their presence in Eocene Strigiformes, therefore, suggests that these owls had smaller eyes than their extant relatives and may have had a diurnal foraging activity. Differences in the feeding ecology of early Eocene owls are also indicated by the narrow mandible of the London Clay owl, and the nocturnal way of living or crown group Strigiformes may have evolved in concert with the radiation of muroid rodents, which constitute the main prey of most extant owl species.

10.1.4 Necrobyinae, Palaeoglaucidae, and Selenornithinae

These three taxa were named by Mourer-Chauviré (1987) for the inclusion of various modern-type Strigiformes from the Quercy fissure fillings, which were considered stem group representatives of the Tytonidae (Mourer-Chauviré 1987: fig. 8). According to Mourer-Chauviré (1987, 2006), the Necrobyinae include the taxa *Necrobyas*, *Nocturnavis*, *Palaeobyas*, and *Palaeotyto*. All major limb bones of *Necrobyas* were described by Mourer-Chauviré (1987), and the following species were recognized in the taxon by Mourer-Chauviré (2006): *N. rossignoli* (late Eocene), *N. harpax* (early Oligocene), *N. edwardsi* (late Oligocene), and *N. medius* (from deposits of unknown age). Mourer-Chauviré (2006) furthermore listed *N. minimus* (early Oligocene) in the Quercy avifauna, but the latter species was synonymized with *Prosybris antiqua* by Mlíkovský (1998), and this taxonomic act was accepted by Mourer-Chauviré (1999). *P. antiqua* is otherwise known from the early Miocene of France, but a tentative record also comes from the early Oligocene of Belgium (Mayr and Smith 2002).

The species of *Necrobyas* agree with the Tytonidae and differ from the Strigidae in that the coracoid has a wide procoracoid process and an acrocoracoid process without pneumatic foramina; the proximal end of the humerus bears a marked transverse groove, and the proximal end of the tarsometatarsus lacks an ossified supratendinal bridge (arcus extensorius) (Mourer-Chauviré 1987). All of these features are, however, likely to be plesiomorphic for the Strigiformes, and in contrast to crown group Tytonidae, the tarsometatarsus of *Necrobyas* is rather short and stout. *N. edwardsi* was about the size of a Southern Boobook (*Ninox boobook*), the other species were slightly smaller.

Two large owls from unknown localities of the Quercy fissure fillings were described as *Palaeotyto cadurcensis* and *Palaeobyas cracrafti* by Mourer-Chauviré (1987). The former is known from a coracoid and the latter from a tarsometatarsus; Mourer-Chauviré (1987: 111) already considered it possible that both bones actually belong to a single species. The coracoid of *Palaeotyto* is characterized by a straight acrocoracoid process, which shows little medial deflection, and a long foramen for the supracoracoideus nerve. The tarsometatarsus of *Palaeobyas* is very robust, and the distal end is just slightly curved across the trochleae; the proportions of the bone superficially resemble those of the tarsometatarsus of the Paleocene *Berruornis*. The taxonomic status of *P. cadurcensis* and *P. cracrafti*, and their classification into the Necrobyinae, need to be verified with further material. A smaller species of the Necrobyinae from the late Eocene of the Quercy fissure fillings, *Nocturnavis incerta*, is only known from humeri.

The taxon Palaeoglaucidae was established for *Palaeoglaux perrierensis*, the fossil material of which consists of several limb and pectoral girdle bones from late Eocene (MP 17) deposits of the Quercy fissure fillings (Mourer-Chauviré 1987). *P. perrierensis* differs from the species of *Necrobyas* in a more globular ventral condyle of the humerus and in the presence of pneumatic foramina in the acrocoracoid process of the coracoid. The latter feature is shared with crown group Strigidae. A *Palaeoglaux*-like fragmentary distal tarsometatarsus was reported from the late Eocene of Switzerland by De Pietri et al. (2013); this unnamed species was smaller than *P. perrierensis*.

Peters (1992) assigned a small, Little Owl (*Athene noctua*)-sized, strigiform species from the latest early or earliest middle Eocene of Messel to *Palaeoglaux* and described it as *P. artophoron* (Fig. 10.2a). The two postcranial skeletons of *P. artophoron* identified so far (Peters 1992; Mayr 2009a) constitute the only articulated skeletons of Paleogene owls, although Olson (1985: 131) mentioned "several complete and perfectly preserved skeletons of a small species of owl" from the "middle" Oligocene of Wyoming. *P. artophoron* is distinguished from all crown group Strigiformes by the plesiomorphic absence of an osseous arch on the radius (Peters 1992: fig. 4), the presence of which is a characteristic derived feature of extant owls. The first phalanx of the second toe is furthermore not as strongly abbreviated as in strigid owls. The Messel owl was assigned to the taxon *Palaeoglaux* owing to the presence of deeply excavated fossa in the supracoracoideus sulcus of the coracoid. However, this trait also occurs in *Primoptynx* and may be plesiomorphic for the Strigiformes. Close comparisons between the Messel owl and the Quercy specimens of *P. perrierensis* are impeded by the fact that most bones of the *P. artophoron* specimens are crushed. Whether *P. artophoron* is indeed a representative of *Palaeoglaux* therefore needs to be verified by future specimens, and it appears possible that the species actually belongs to the taxon *Eostrix*. Peters (1992) noted "ribbon-like" feathers in the holotype of *P. artophoron*, which were assumed to be an autapomorphic feature of the species, possibly reflecting an ornamental plumage. However, a similar type of feather preservation is found in other birds from Messel and may either be a taphonomic artifact or due to the fossilization of growing feathers.

The holotype and only known specimen of *Selenornis henrici* is a distal tibiotarsus from an unknown stratigraphic horizon of the Quercy fissure fillings. The species was assigned to the monotypic taxon Selenornithinae by Mourer-Chauviré (1987) and is of similar size to *Necrobyas rossignoli*. Another species of the taxon *Selenornis*, *S. steendorpensis* from the early Oligocene of Belgium, is known from a distal tibiotarsus and a tarsometatarsus

(Fig. 10.3; Mayr 2009b). A distal tibiotarsus of a large owl from the early Oligocene Jebel Qatrani Formation of Egypt (Smith et al. 2020) was also likened to the Selenornithidae and is the earliest fossil record of the Strigiformes in Africa.

Most species of crown group Strigiformes belong to the Strigidae. Strigid owls are characterized by an osseous arch on the proximal end of the tarsometatarsus (Fig. 10.3g), which guides the tendon of an extensor muscle of the toes and first occurs in *Heterostrix tatsinensis* from the early Oligocene of Mongolia. The holotype of this species is a tarsometatarsus, which exhibits such a distinctive morphology that *H. tatsinensis* was assigned to a new taxon Heterostrigidae (Kurochkin and Dyke 2011). In particular, *Heterostrix* is characterized by a derived shape of the tarsometatarsal trochlea for the second toe, which forms a well-delimited accessory trochlea and indicates an increased mobility of this toe. As noted by Mayr (2017), *Aurorornis taurica* from the late Eocene of the Crimea Peninsula, which was likewise based on a tarsometatarsus and was initially assigned to the Protostrigidae (Panteleyev 2011), is another representative of the Heterostrigidae. In *Aurorornis*, the bony tarsometatarsal arch is only partially ossified, but otherwise, the taxon closely resembles *Heterostrix*.

Pellets, which possibly stem from owls, were found in the latest early or earliest middle Eocene of Messel (Mayr and Schaal 2016) and in the late Oligocene of Enspel in Germany (Smith and Wuttke 2015).

10.2 Coliiformes (Mousebirds)

The Coliiformes include six predominantly frugivorous extant species, which occur in Africa south of the Sahara. These sparrow-sized birds have a short, finch-like beak and facultatively pamprodactyl and zygodactyl feet, that is, the first and fourth toes can be moved forward and backward. Not least due to this distinctive foot morphology, mousebirds are very agile and acrobatic birds, which often feed upside down. They have an unusually long tail with very stiff central feathers, which serves as a propping device for birds dangling from their perches.

Most coliiform birds are characterized by an ulna that is subequal to the humerus in length and has a large dorsal cotyla. Functionally related to the facultatively pamprodactyl foot, the fossa for the first metatarsal is located on the medial (rather than plantar) surface of the tarsometatarsus shaft, and the hindlimbs exhibit various other derived features that rarely occur in other birds (Houde and Olson 1992; Mayr and Peters 1998).

The fairly extensive Paleogene fossil record of coliiform birds shows the extant taxa to be relics of what was once a much more diversified group. All Paleogene specimens are from fossil sites in Europe and North America. Archaic stem group taxa predominated in the early and middle Eocene, whereas only modern-type Coliiformes are known from the late Eocene on. Two major clades can be distinguished (Fig. 10.4): the Sandcoleidae, which exhibit a plesiomorphic skeletal morphology and have unambiguously identified records exclusively from Eocene sites, and the Coliidae, which also first occur in the Eocene and include the extant species.

The earliest fossil assigned to the Coliiformes is *Tsidiiyazhi abini* from the mid-Paleocene (62.5 Ma) Nacimiento Formation of New Mexico (USA), the holotype of which includes various fragmentary bones of a single individual (Ksepka et al. 2017). *T. abini* was a small bird with zygodactyl feet. The species was considered to be a representative of the Sandcoleidae in the original description, from which it differs, however, in the wider tarsometatarsal trochlea for the second toe.

10.2.1 Sandcoleidae

These morphologically distinctive stem group representatives of the Coliiformes occur in the early and middle Eocene of North America and Europe. North American sandcoleids include the species *Sandcoleus copiosus* (early Eocene of the Willwood Formation) and *Anneavis anneae* (early Eocene of the Willwood and Green River formations), as well as *Uintornis lucaris*, *U. marionae*, *Botauroides parvus*, and *Eobucco brodkorbi* from the middle Eocene of the Bridger Formation. The fossil record of the latter four species consists only of incomplete tarsometatarsi, whereas the skeletal morphology of *S. copiosus* and *A. anneae* is well known (Houde and Olson 1992).

In Europe, sandcoleids are represented by *Eoglaucidium pallas*, a species that was first described as an owl and occurs in Messel and the Geisel Valley (Fig. 10.5; Fischer 1987; Mayr and Peters 1998; Mayr 2002b, 2018, 2020). *E. pallas* and a further unnamed sandcoleid species (Mayr 2000a) are known from complete articulated skeletons, and the former is among the most abundant medium-sized birds in Messel. Sandcoleids also occur in the early Eocene of France (Mayr and Mourer-Chauviré 2004) and in the London Clay (Daniels in Feduccia 1999: table 4.1).

The species of *Sandcoleus*, *Anneavis*, and *Eoglaucidium* share a similar skeletal morphology, with robust feet and a thrush-like bill. The caudal portion of the sternum has a distinctive shape in that the intermediate bars (trabeculae intermediae) originate from the lateral ones (trabeculae laterales), so that the medial incisions are deeper than the lateral ones. The deltopectoral crest of the humerus is much longer than that of extant mousebirds, and unlike in the latter, the coracoid exhibits a foramen for the supracoracoideus nerve. The tarsometatarsus has characteristic proportions,

10.2 Coliiformes (Mousebirds)

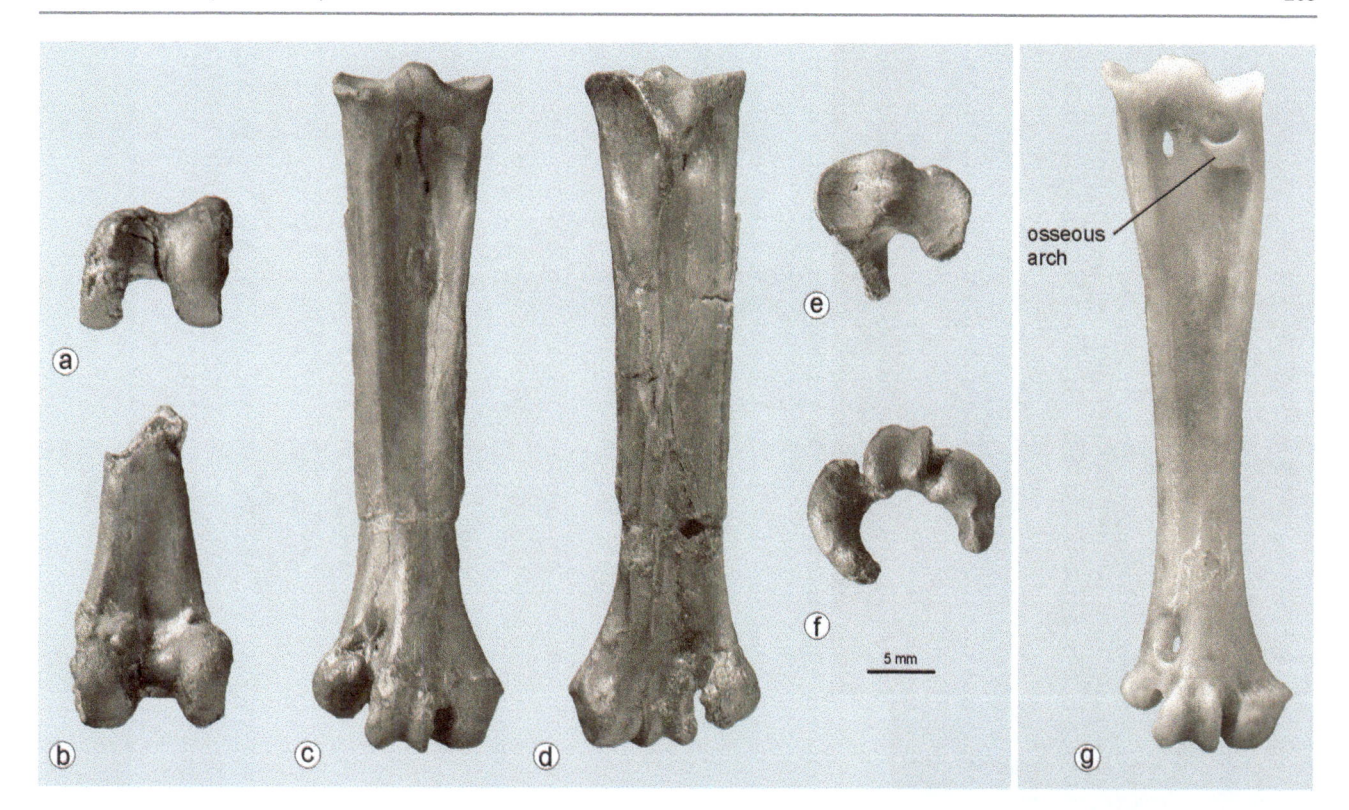

Fig. 10.3 (a), (b) Distal end of the tibiotarsus and (c)–(f) tarsometatarsus (dorsal, plantar, proximal, and distal view) of *Selenornis steendorpensis* from the early Oligocene of Belgium (holotype, Royal Belgian Institute of Natural Sciences, Brussels, Belgium, IRSNB Av 84 a+b). (g) Tarsometatarsus of the extant Ural Owl, *Strix uralensis*. (Photos by Sven Tränkner)

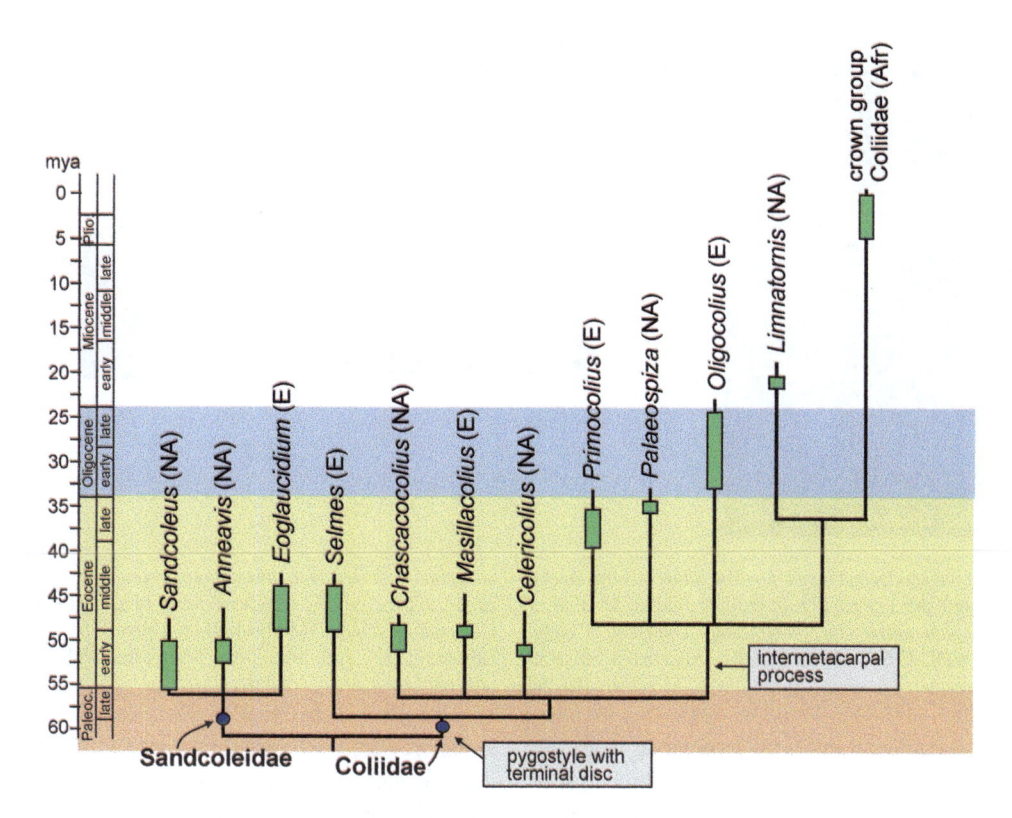

Fig. 10.4 Interrelationships and known stratigraphic occurrences of coliiform birds. The divergence dates are hypothetical

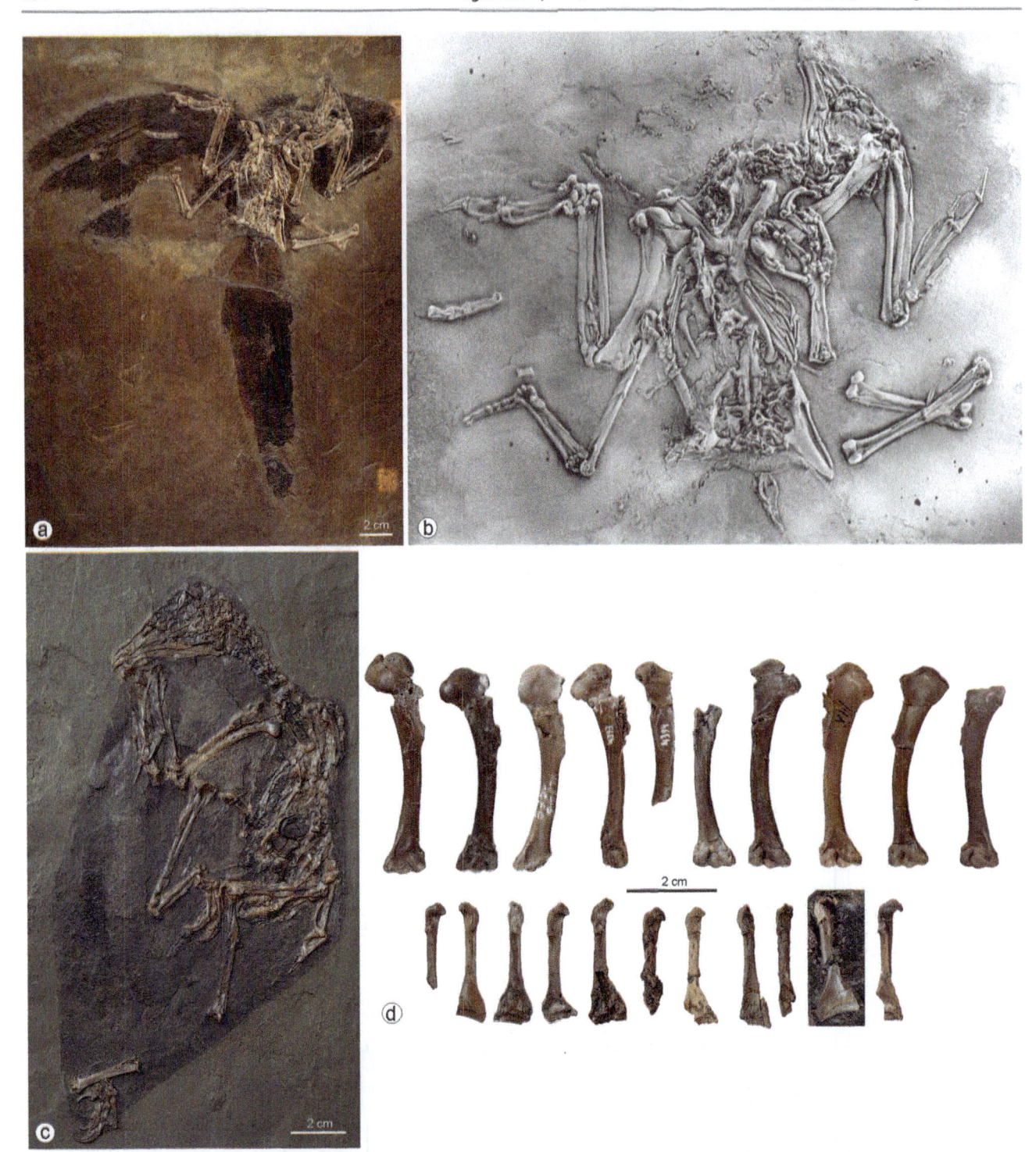

Fig. 10.5 Fossils of the sandcoleid taxon *Eoglaucidium* (Coliiformes). (**a**)–(**c**) Skeletons from the latest early or earliest middle Eocene of Messel (**a**, **b**: *Eoglaucidium* sp., Wyoming Dinosaur Center, Thermopolis, USA, WDC-C-MG 149; **c**: *E. pallas*, Staatliches Museum für Naturkunde Karlsruhe, Germany, SMNK-PAL 553a); in (**b**) the specimen is coated with ammonium chloride. (**d**) Humeri and coracoids of *E. pallas* from the middle Eocene of the Geisel Valley (all in the Geiseltalsammlung, Martin-Luther Universität of Halle-Wittenberg, Germany). The specimen in (**b**) was coated with ammonium chloride. (All photos by Sven Tränkner)

and in contrast to extant mousebirds the hypotarsus bears two furrows for the deep flexor tendons of the toes, which may be closed to form one or two canals (in extant Coliiformes there is just a single canal). The first phalanges of all anterior toes are abbreviated, whereas in extant Coliiformes only the first three phalanges of the fourth toe are shortened. The ungual phalanges are very long and "raptor-like."

Some specimens of *A. anneae* and *E. pallas* are preserved with feather remains, which allow the recognition of a long, staggered tail similar to that of extant mousebirds (Fig. 10.5a). As in the latter, there seem to have been ten rectrices. These were, however, not as stiff as in the extant species, so that sandcoleids probably did not use their tail as a propping device and assumed the usual, upright posture when perching and foraging (Mayr 2018).

In several specimens of *Eoglaucidium* from Messel, large seeds are furthermore preserved as stomach contents (Mayr and Peters 1998; Mayr 2018). This indicates that a frugivorous feeding ecology evolved early in the evolution of coliiform birds. Seeds are also known from the stomach contents of Eocene representatives of the Coliidae. However, whereas stem group Coliidae exhibit feeding specializations in the morphology of the beak (see next section), such are unknown for the Sandcoleidae. It has therefore been hypothesized that the "raptor-like" grasping feet of sandcoleids constitute an adaptation for the manipulation of food items (e.g., larger fruits or infructescences), as they do in extant parrots (Mayr 2018). In the feeding ecology of stem group representatives of the Coliidae, by contrast, the beak may have been more important. Why sandcoleids became extinct towards the end of the Eocene remains elusive, but ecological competition with ecologically similar avian groups may have played a role.

10.2.2 Coliidae

Two hallmark features distinguish most species of the Coliidae from those of the Sandcoleidae: the legs are proportionally longer, with the tarsometatarsus being longer and narrower, and the pygostyle terminates in a large, disk-like expansion, which serves as the attachment site for well-developed depressor muscles of the tail. Both features are likely to be functionally linked to the unique perching posture of mousebirds, which use their tail as a propping device when dangling from perches.

The putative sandcoleid *Chascacocolius oscitans* was shown to be more closely related to crown group Coliiformes than to the Sandcoleidae (Mayr and Peters 1998, Mayr 2001a, Mayr and Mourer-Chauviré 2004). This species occurs in the early Eocene of the Willwood Formation, and except for the skull and tarsometatarsus all major skeletal elements were found (Houde and Olson 1992). An incomplete wing from the Fur Formation was assigned to *Chascacocolius* cf. *oscitans* by Dyke et al. (2004), and an isolated skull from Messel was described as a new species *Chascacocolius cacicirostris* (Mayr 2005a).

Chascacocolius is characterized by a mandible with very long retroarticular processes (Fig. 10.6c). These processes serve as the attachment site of a muscle, which depresses the mandible and, by increasing the lever arm of this muscle, retroarticular processes enable gaping, that is, opening of the beak within a substrate. The upper beak of *C. oscitans* is unknown, but that of *C. cacicirostris* is very similar to the beak of some New World blackbirds (Passeriformes, Icteridae), which can forcefully open their bill within fruits (Mayr 2005a).

In its postcranial skeleton, *Chascacocolius* resembles another early Eocene coliiform species, *Selmes absurdipes*, which is known from two skeletons from Messel, a humerus and a poorly preserved skeleton from the Geisel Valley (Fig. 10.7; Mayr 2001a, 2020), as well as a tentatively referred tarsometatarsus from an unknown locality of the Quercy fissure fillings (Mayr and Mourer-Chauviré 2004). In the holotype of *S. absurdipes*, a dense package of as yet undetermined seeds is preserved as stomach contents. This species has a long and slender tarsometatarsus, which exhibits a distinctive morphology with a short and wide trochlea for the third toe. Because the first phalanges of all three anterior toes are greatly abbreviated, *S. absurdipes* was classified into the Sandcoleidae by Peters (1999). However, the species also exhibits the characteristic derived morphology of the pygostyle of extant Coliidae (Mayr 2001a), and as in extant Coliidae the hypotarsus includes only a single canal (Mayr and Mourer-Chauviré 2004). Phylogenetic analyses by Mayr and Mourer-Chauviré (2004) and Ksepka and Clarke (2009) support its identification as an early diverging stem group representative of the Coliidae.

Of another Eocene stem group representative of the Coliidae, *Masillacolius brevidactylus*, three skeletons were found in the Messel site (Fig. 10.6a, b; Mayr and Peters 1998; Mayr 2015). This species had the proportionally longest tarsometatarsus of all Coliiformes and unusually short tarsometatarsal trochleae. Judging from the position of the toes in the articulated Messel skeletons, the feet seem to have been fully pamprodactyl, that is, the hallux permanently directed forwards. Mayr and Peters (1998) assumed that *M. brevidactylus* was specialized for clinging to vertical surfaces, which also concords with its robust and stout ungual phalanges. The caudal end of the mandible of *M. brevidactylus* bears long retroarticular processes, and in one of the specimens from Messel a large seed is preserved as stomach content (Mayr 2015).

The description of *Eocolius walkeri* from the London Clay was based on several isolated bones (Dyke and Waterhouse 2001). The species lacks the enlarged dorsal condyle of the ulna, which is an apomorphy of the Coliiformes (Ksepka and Clarke 2009; contra Dyke and Waterhouse 2001). It is therefore unlikely that it is a representative of the Coliiformes, from which it also differs in the shorter flexor process of the humerus.

The holotype of *Celericolius acriala* is a skeleton from the early Eocene (52 Ma) of the Green River Formation (Ksepka

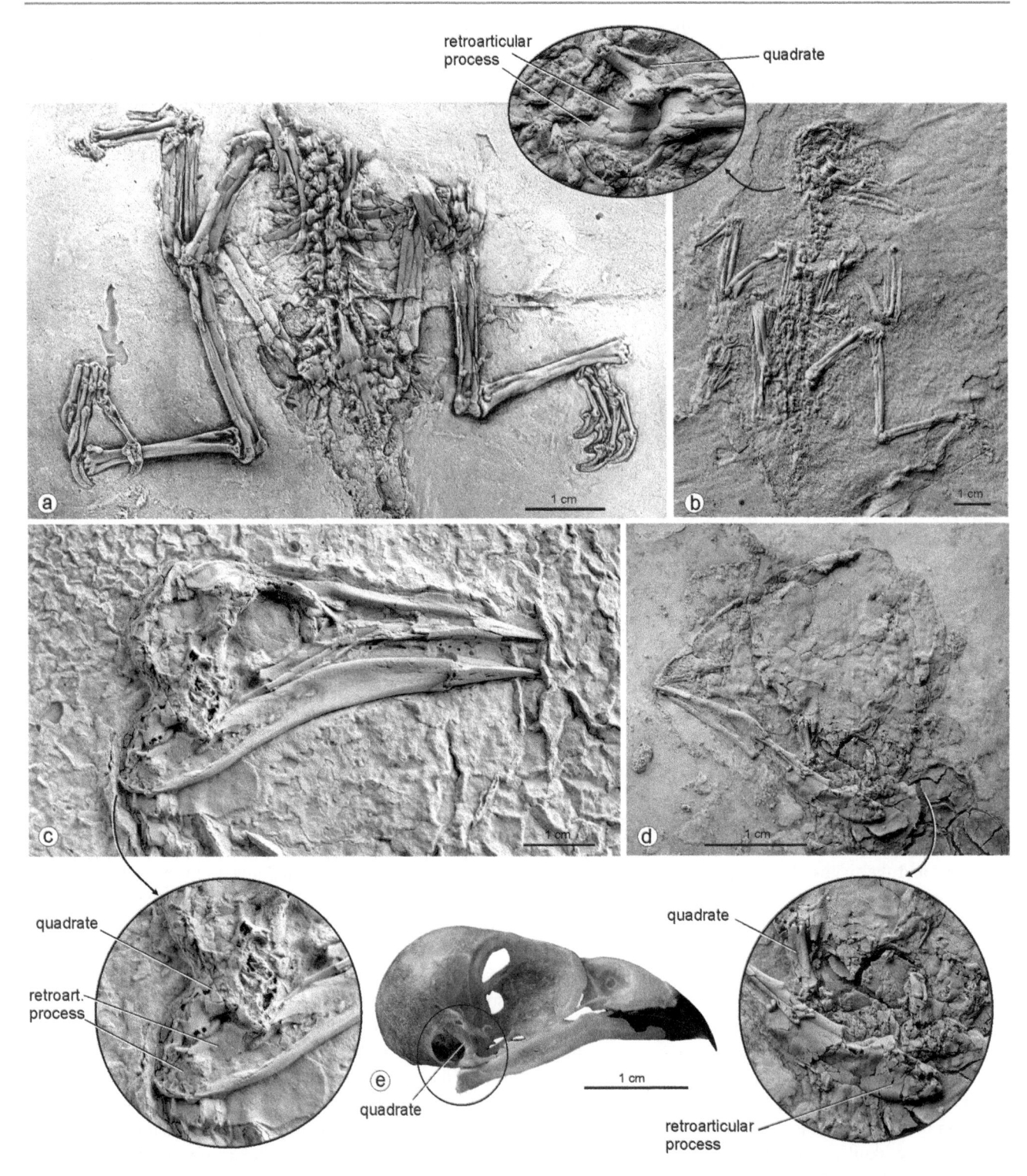

Fig. 10.6 Fossils of Paleogene coliiform birds with elongated retroarticular processes. (**a**), (**b**) Skeletons of *Masillacolius brevidactylus* (Coliidae) from the latest early or earliest middle Eocene of Messel (**a**: holotype, Hessisches Landesmuseum, Darmstadt, Germany, HLMD Me 10472; **b**: Senckenberg Research Institute, Frankfurt, SMF-ME 11322). (**c**) Skull of *Chascacocolius cacicirostris* (Coliidae) from the latest early or earliest middle Eocene of Messel (holotype, SMF-ME 3790). (**d**) Skull of *Oligocolius psittacocephalon* from the late Oligocene of Enspel in Germany (holotype, Generaldirektion Kulturelles Erbe Rheinland-Pfalz, Mainz, Germany, PW 2012/5052-LS). (**e**) Skull of the extant *Urocolius macrourus* (Coliidae). The arrows in (**b**)–(**d**) indicate details of the elongated retroarticular processes of the mandible; the skull area shown in the details of the fossil species is encircled in (**e**). All fossil specimens were coated with ammonium chloride. (Photos by Sven Tränkner)

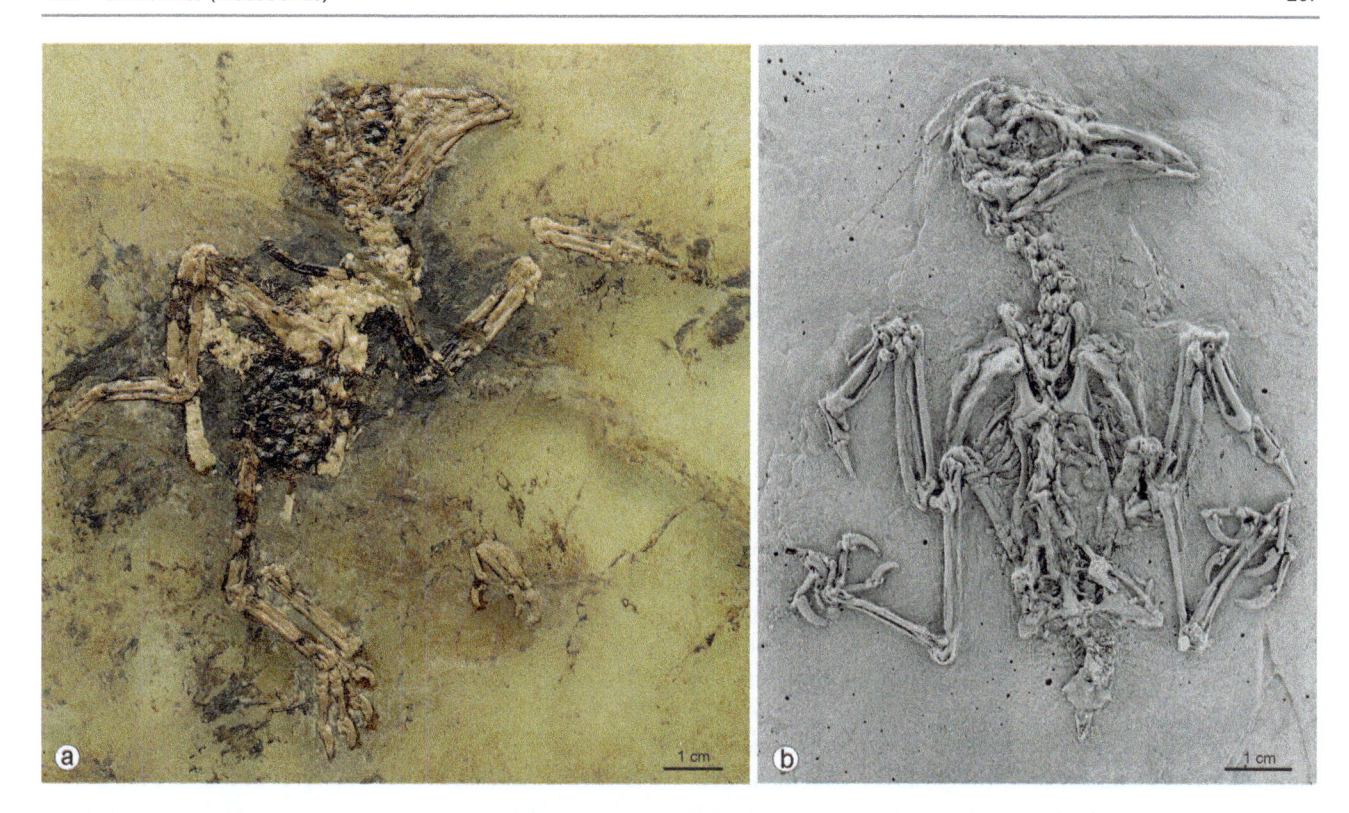

Fig. 10.7 (**a**), (**b**) Skeletons of *Selmes absurdipes* (Coliidae) from the latest early or earliest middle Eocene of Messel (**a**: holotype, Senckenberg Research Institute, Frankfurt, SMF-ME 2375; **b**: Wyoming Dinosaur Center, Thermopolis, USA, WDC-C-MG 147); the fossil in (**b**) was coated with ammonium chloride. (Photo in (**a**) by Anika Vogel, in (**b**) by Sven Tränkner)

and Clarke 2010a). This species has much longer wings than extant mousebirds, which indicates that it was adapted to flight in more open areas. The tail feathers of *Celericolius* likewise appear to have been very long.

Essentially modern-type stem group representatives of the Coliidae first occur in the middle Eocene. Among these are two species of *Primocolius*, *Primocolius sigei* and *P. minor*, which were described by Mourer-Chauviré (1988) from the middle and late Eocene of the Quercy fissure fillings. The species of *Primocolius* were the first Paleogene coliiform birds recognized as such, but both are known from a few bones only (humerus, tarsometatarsus, proximal carpometacarpus). A poorly preserved skeleton from the late Eocene deposits of the Paris Gypsum probably also belongs to *Primocolius* (Mayr 1998, 2000b).

Equally advanced stem group representatives of the Coliiformes were also present in the late Eocene of North America. It was first noted by Mayr (2001a) that *Palaeospiza bella* from the late Eocene of the Florissant shales of Colorado, which was originally described as a passeriform bird, belongs to the Coliiformes and closely resembles *Primocolius sigei* in size, limb proportions, and morphological features (such as a carpometacarpus with an enlarged intermetacarpal process). An assignment of this species to the Coliiformes was confirmed by Ksepka and Clarke (2009), who re-described the holotype. Whether *Primocolius* and *Palaeospiza* are distinct on the genus level has yet to be shown, and the single skeleton known of the latter taxon does not allow the recognition of subtle morphological details.

A fragmentary tarsometatarsus of a seemingly modern-type mousebird was reported from the earliest Oligocene of Belgium (Mayr and Smith 2001). More substantial is the fossil record of a small coliiform bird from the early Oligocene of Wiesloch-Frauenweiler in Germany, which was described as *Oligocolius brevitarsus* by Mayr (2000b). This species is known from a disarticulated postcranial skeleton and has a proportionally more strongly developed wing and a shorter tarsometatarsus than its extant relatives. Another species of *Oligocolius*, *O. psittacocephalon*, occurs in the late Oligocene Enspel size in Germany (Fig. 10.6d; Mayr 2013). Of this species the skull is known, and the mandible bears long retroarticular processes. The holotype is notable in that it is preserved with very large seeds as crop and stomach contents. In contrast to other Coliiformes except for *Celericolius*, the ulna of *Oliogocolius* distinctly exceeds the humerus in length, which may indicate that the taxon was adapted to a more sustained flight than extant mousebirds.

Primocolius, *Palaeospiza*, *Oligocolius*, and extant mousebirds form a clade, which is characterized by a well-developed intermetacarpal process on the carpometacarpus (Mourer-Chauviré 1988; Mayr 2000b, 2001a). This process is only incipiently developed in *Selmes*, *Celericolius*, and *Masillacolius* and serves as the insertion site of a muscle, which flexes the hand section of the wing. A large intermetacarpal process occurs in birds, in which the hand section is exposed to high aerodynamic loads, but the functional reasons for its development in the Coliidae are unknown.

The occurrence of long retroarticular processes in three taxa of stem group Coliidae (*Chascacocolius*, *Masillacolius*, and *Oligocolius*) is notable (Fig. 10.6). However, for a well-founded reconstruction of the evolution of retroarticular processes in the Coliidae, an improved knowledge of the mandible of some critical taxa, such as *Celericolius* and *Primocolius*, is required. Because *Oligocolius* and crown group Coliidae form a clade to the exclusion of *Chascacocolius* and *Masillacolius*, this unusual feeding specialization may be plesiomorphic for a coliiform subclade, which also includes the extant species. In this case, long retroarticular processes must have been secondarily reduced in crown group Coliiformes, which still calls for an explanation.

Stem group representative of the Coliidae occurred in Europe until the late Miocene (Mayr 2011), and their final disappearance is likely to have been due to the emergence of the cold Northern Hemispheric winters, which did not allow the persistence of frugivorous birds with poor migration capabilities. The earliest African record of the Coliiformes is from the early Miocene of Namibia (Mourer-Chauviré 2008).

10.3 Cavitaves: Birds that Nest in Burrows and Tree Cavities

Molecular analyses congruently recovered a clade including the Leptosomiformes, Trogoniformes, Bucerotes, Coraciiformes, and Piciformes. This clade was termed Cavitaves, in reference to the fact that all of these birds nest in tree cavities or self-excavated burrows (Yuri et al. 2013). It is notable that many extant taxa of the Cavitaves either have a restricted geographical distribution (e.g., the Leptosomiformes, as well as the coraciiform Brachypteraciidae, Todidae, and Momotidae, and the piciform Galbulae) or exhibit highly derived feeding specializations (e.g., the Bucerotidae as well as the piciform Ramphastidae and Picidae). This may indicate that the evolution of these birds was driven by ecological competition, and the radiation of passerines into arboreal habitats may have had a major impact on their diversity.

The clade including non-leptosomiform Cavitaves was termed Eucavitaves (Fig. 10.8; Yuri et al. 2013). One of the few known morphological apomorphies, which characterize members of this clade, is a derived morphology of the radial carpal bone of the wing (Mayr 2014).

Many fossil representatives of the Cavitaves can be assigned to one of the extant higher-level clades, and the evolutionary origin of the clade remains obscure from a fossil perspective. *Foshanornis songi* from the early Eocene Buxin Formation of China is known from a largely complete skeleton, which is preserved in a dissociated state, so that the morphologies of most major bones are well visible (Zhao et al. 2015). The species has a roller-like skeletal morphology, but it differs from rollers and other taxa of the Eucavitaves in that the radial carpal bone does not exhibit the derived morphology characterizing the species of this clade. *Lapillavis incubarens* from Messel was likened to *Foshanornis* based on the overall morphologies of the bones of the partial skeleton which constitutes the holotype. Even though the affinities of *L. incubarens* are elusive, the holotype is notable, because it exhibits medullary bone, which indicates that the fossil stems from a breeding female (Mayr 2016).

10.3.1 Leptosomiformes (Courols)

The single extant species of the Leptosomiformes, the Courol or "Cuckoo-roller" (*Leptosomus discolor*), is a carnivorous, forest-dwelling bird of Madagascar and the Comoro islands. Fossil stem group representatives from the early and middle Eocene of Europe and North America show that Pan-Leptosomiformes had a wide distribution over the Northern Hemisphere in the early Paleogene. These fossils were classified into the taxon *Plesiocathartes*, which was assigned to the Cathartidae by earlier authors (Gaillard 1908; Cracraft and Rich 1972). Five species of *Plesiocathartes* have been named: *Plesiocathartes europaeus* from an unknown stratigraphic horizon of the Quercy fissure fillings, *P. geiselensis* from the Geisel Valley (Fig. 10.9d), *P. kelleri* from Messel (Fig. 10.9a, b), as well as *P. wyomingensis* (Fig. 10.9c) and *P. major* from the Green River Formation (Mayr 2002b, 2002c, 2020; Mourer-Chauviré 2002; Weidig 2006). An as-yet unnamed species of *Plesiocathartes* occurs in the London Clay (Fig. 10.9f; Mayr 2002c), and there are also multiple specimens of stem group Leptosomiformes from Walton-on-the-Naze in the collection of Michael Daniels (Fig. 10.9g). *Plesiocathartes*-like tarsometatarsi were furthermore reported from the latest Thanetian of the locality Rivecourt-Petit Pâtis in France (Mayr and Smith 2019a), the early Eocene of Egem in Belgium (Fig. 10.9e; Mayr and Smith 2019b), and the early Eocene of the Vastan Lignite Mine in India (Mayr et al. 2010).

The species from Messel and the Green River Formation are represented by complete skeletons and are remarkably similar to the Courol in their skeletal morphology (Mayr 2002b, c, 2008; Weidig 2006). As in extant Leptosomiformes, the coracoid exhibits a foramen for the

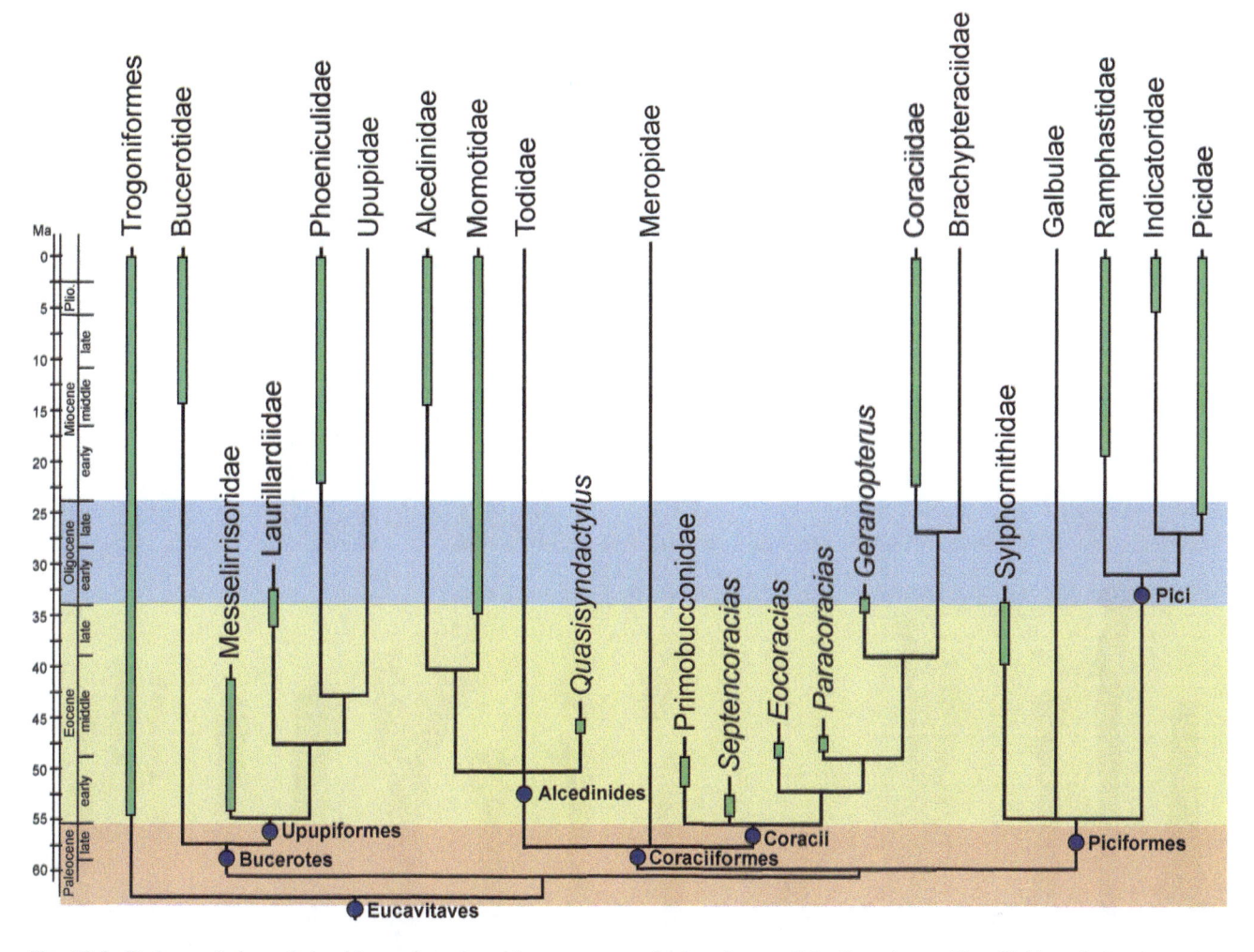

Fig. 10.8 Phylogenetic interrelationships and stratigraphic occurrences of selected taxa of the Eucavitaves. The affinities of taxa shown in a polytomy are controversial, divergence dates are hypothetical; see text for further details

supracoracoideus nerve and a concave articular facet for the scapula. These presumably plesiomorphic features are absent in coraciiform birds. *Plesiocathartes* furthermore shares with the extant Courol a derived morphology of the tibiotarsus (the condyles of which are very low and widely separated) and tarsometatarsus (hypotarsus with two furrows, which are closed to canals in *Leptosomus*, and trochlea for the second toe reaching farther distally than that for the fourth toe; Mayr 2002b, c). As in extant Leptosomiformes, the postorbital processes of the skull are strongly elongated.

These Eocene stem group Leptosomiformes differ from *Leptosomus* in the morphology of the furcula, which has a smaller omal extremity, is more widely U-shaped, and has shafts that are not as strap-like as in the extant Courol. They are furthermore distinguished from *Leptosomus* by their proportionally longer legs, details of the pelvis morphology (the pubis is not reduced in its midsection and the obturator foramen caudally open), and by the absence of an accessory tarsometatarsal trochlea for the fourth toe. The latter characteristic indicates that the feet were anisodactyl and not semi-zygodactyl as in the extant Courol.

All stratigraphically well-constrained specimens of fossil stem group representatives of the Leptosomiformes come from Eocene deposits, and these birds may have already disappeared from the Northern Hemisphere towards the Oligocene.

10.3.2 Trogoniformes (Trogons)

The Trogoniformes are insectivorous or frugivorous arboreal birds, which today occur in the tropical and subtropical regions of continental Africa, Asia, and the New World. Trogons are the only avian group, in which the second toe permanently directs backward. This heterodactyl foot is accompanied by a plantarly reversed tarsometatarsal trochlea for the second toe.

Skeletal remains of several individuals of early Eocene trogons were collected by Michael Daniels in the London

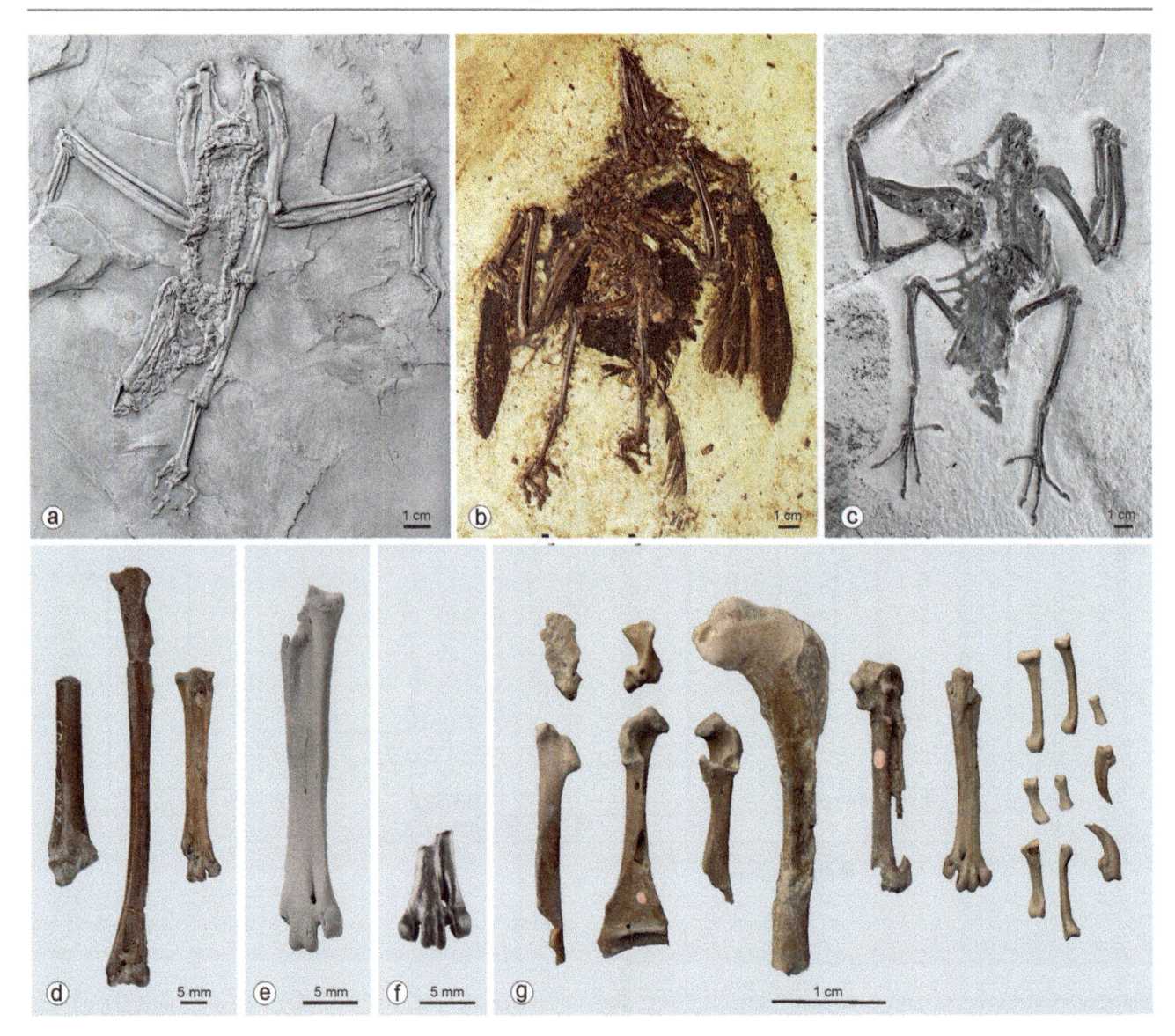

Fig. 10.9 Fossil of the Leptosomiformes (courols). (**a**), (**b**) Skeletons of *Plesiocathartes kelleri* from the latest early or earliest middle Eocene of Messel (**a**: holotype, Senckenberg Research Institute, Frankfurt, SMF-ME 3639, **b**: Hessisches Landesmuseum Darmstadt, HLMD Be 162). (**c**) Holotype of *P. wyomingensis* from the early Eocene Green River Formation in Wyoming USA (Wyoming Dinosaur Center, WDC-2001-CGR-021). (**d**) Holotype of *P. geiselensis* from the middle Eocene of the Geisel Valley (Geiseltalsammlung, Martin-Luther Universität of Halle-Wittenberg, Germany, GMH XXXV-559). (**e**) Tarsometatarsus of *Plesiocathartes* sp. from the early Eocene of Egem in Belgium (Royal Belgian Institute of Natural Sciences, Brussels, Belgium, IRSNB Av 185). (**f**) Distal tarsometatarsus of an unnamed species of *Plesiocathartes* from the early Eocene London Clay of the Isle of Sheppey (Natural History Museum, London, NHMUK A 6178). (**g**) Undescribed stem group leptosomiform from the early Eocene London Clay of Walton-on-the-Naze (collection of Michael Daniels, Holland-on-Sea, UK, WN 88585). Specimens in (**a**) and (**e**) were coated with ammonium chloride. (Photo in (**b**) by Wolfgang Fuhrmannek, in (**c**) by Ilka Weidig, all others by Sven Tränkner)

Clay of Walton-on-the-Naze (Fig. 10.10d, e). These as yet undescribed specimens have already been mentioned earlier (Mayr 1999), and a completely preserved tarsometatarsus, which is here figured for the first time, shows a plantarly deflected trochlea for the second toe (Fig. 10.10e). The oldest formally described trogoniform fossil is a cranium from the Fur Formation in Denmark, which was described as *Septentrogon madseni* by Kristoffersen (2002). The cranium of *S. madseni* shares with crown group Trogoniformes the presence of basipterygoid processes and a derived morphology of the zygomatic process, but the fossil is distinguished from extant trogons in a proportionally narrower nasofrontal hinge.

Masillatrogon pumilio from Messel (Fig. 10.10a, b) is likewise distinguished from crown group Trogoniformes in the shape of its narrower beak (Mayr 2005b, 2009c). The species was initially assigned to the taxon *Primotrogon* (see below) from which it, however, differs in some

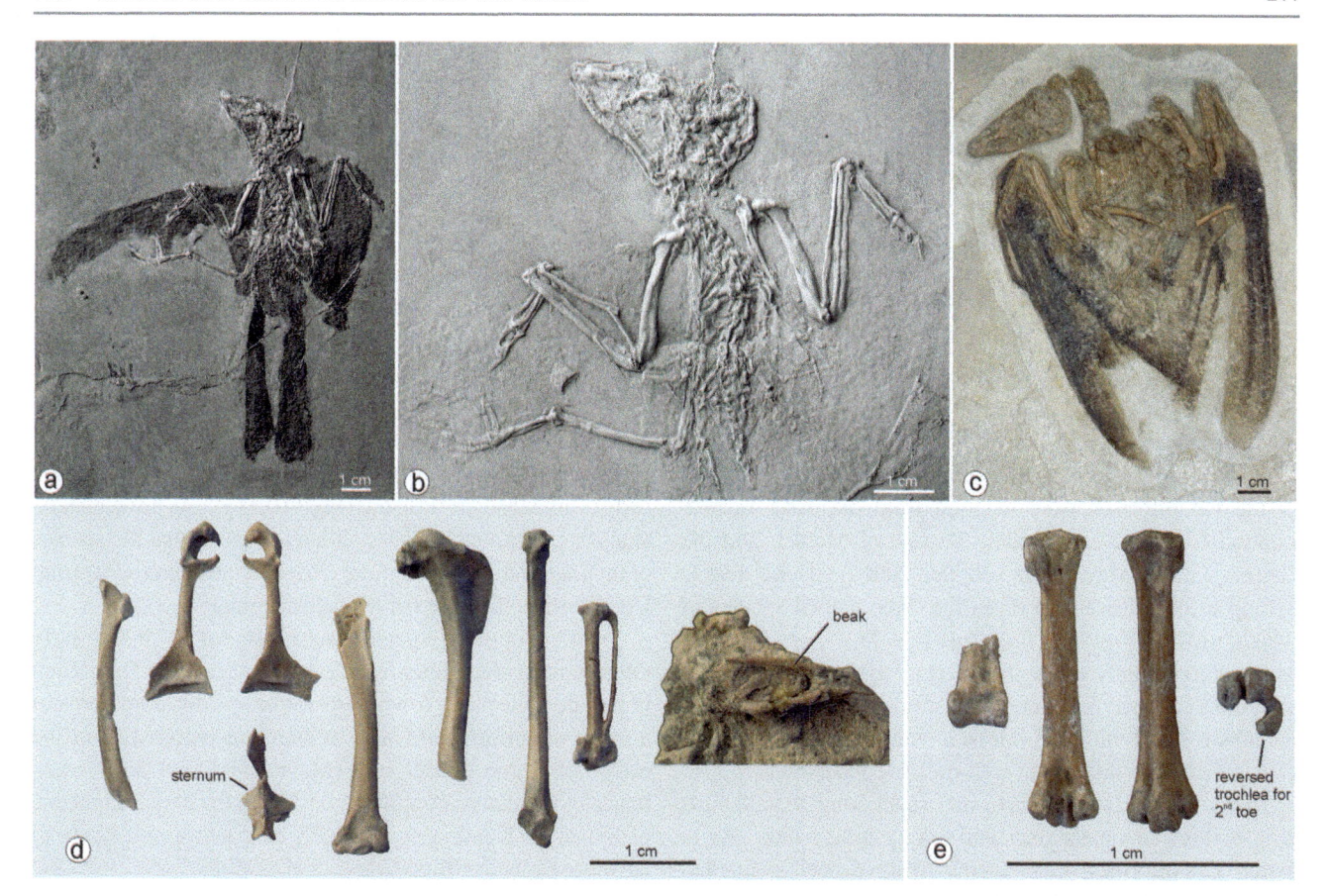

Fig. 10.10 Fossils of the Trogoniformes (trogons). (**a**), (**b**) Skeleton of *Masillatrogon pumilio* from the latest early or earliest middle Eocene of Messel (Royal Belgian Institute of Natural Sciences, Brussels, Belgium, IRSNB Av 81), in (**b**) the specimen was coated with ammonium chloride. (**c**) Holotype of *Primotrogon wintersteini* from the early Oligocene of the Luberon area in France (Bayerische Staatssammlung für Paläontologie und Historische Geologie, München, Germany, BSP 1997 I 38). (**d**) Partial skeleton of an undescribed *Masillatrogon*-like trogon from the early Eocene London Clay of Walton-on-the-Naze (collection of Michael Daniels, Holland-on-Sea, UK, WN 94814). (**e**) Distal tibiotarsus and tarsometatarsus of an undescribed trogon from the early Eocene London Clay of Walton-on-the-Naze, with the tarsometatarsus being shown in plantar, dorsal, and distal view (collection of Michael Daniels, Holland-on-Sea, UK, WN 89608). (Photo in (**c**) by the author, all others by Sven Tränkner)

plesiomorphic features including a proportionally longer hallux. The dorsal tubercle (tuberculum dorsale) of the humerus of *M. pumilio* is smaller than in extant Trogoniformes (Mayr 2009c). This tubercle serves for the attachment of the supracoracoideus muscle, and its less pronounced development in *M. pumilio* suggests that this fossil taxon differed in its flight technique and foraging behavior from modern trogons, which are able to perform short-term hovering flight to feed on fruits or to pick up insects from leaves and trunks. The heterodactyl position of the toes is clearly visible in the two known skeletons of the species, and one of the specimens exhibits well-preserved feather remains, which show the tail of *M. pumilio* to have been very long as in extant trogons.

Skeletons of early Oligocene trogons were furthermore described from Switzerland (Olson 1976) and the Luberon area in France (Mayr 1999, 2001b). The Luberon trogon, *Primotrogon wintersteini* (Fig. 10.10c), is represented by two specimens, which allow the recognition of a heterodactyl foot. It is shown to be outside crown group Trogoniformes by its plesiomorphic skull morphology (the cranium and beak are narrower than in crown group Trogoniformes and the orbitae are proportionally smaller) and by the absence of derived characters of the coracoid (Mayr 1999, 2005b). Like *Masillatrogon pumilio*, *Primotrogon wintersteini* is furthermore smaller than its extant relatives. An isolated wing of a trogon is also known from the early Oligocene of Wiesloch-Frauenweiler (Mayr 2005b), and a distal humerus of an unnamed trogon was reported from the early Oligocene of Belgium (Mayr and Smith 2013).

With *Masillatrogon* and *Primotrogon* being successive sister taxa of crown group Trogoniformes (Mayr 2009c), the fossil record suggests an Old World origin of the total group of trogons (Pan-Trogoniformes). Apparently, trogons underwent few morphological changes during the past 50 million years and were therefore well adapted to their ecological niches early on, remaining relatively unaffected by subsequent competition with other avian groups. As insectivorous or frugivorous birds with poor capabilities for long-distance migration, these birds were, however, impacted by climatic cooling in the Neogene, and the emergence of the cold Northern Hemispheric winters in the late Miocene confined their range to the tropical and subtropical zones.

10.3.3 Bucerotes (Hornbills, Hoopoes, and Woodhoopoes)

Extant Bucerotes comprise the Bucerotiformes (hornbills), which occur in Africa and Asia and have no Paleogene fossil record, as well as the Upupiformes, which include the African and Eurasian Upupidae (hoopoes) and the African Phoeniculidae (woodhoopoes). Stem group representatives of the Upupiformes were reported from the early Eocene London Clay, the early/middle Eocene of Messel, and the Geisel Valley (Mayr 1998, 2000c, 2006a). These fossils belong to the Messelirrisoridae, the three named species of which (*Messelirrisor halcyrostris*, *M. parvus*, and *M. grandis*) mainly differ in size (Fig. 10.11a–f). Together with the Zygodactylidae, messelirrisorids are among the most abundant small birds in Messel, and their fossil record consists of a fair number of well-preserved skeletons.

Messelirrisorids are small to very small, long-beaked birds with a short tarsometatarsus and a very long hallux. As in extant Upupiformes, there seems to have been a marked sexual dimorphism in beak length, which appears to have been much longer in male individuals. Unlike extant Upupiformes, the carpometacarpus bears an intermetacarpal process. Crown group Upupiformes either forage on the ground (Upupidae) or are adapted to trunk climbing (Phoeniculidae), whereas messelirrisorids appear to have been perching birds without specialized locomotory habits (Mayr 1998). In one specimen of *Messelirrisor* from Messel the tail feathers are distinctly barred, which reflects the original color pattern through preservation of the melanosomes (Fig. 10.11a; Vinther et al. 2008). In other specimens, fossilized residues of the uropygial gland waxes are preserved (Mayr 2006a; O'Reilly et al. 2017).

A sister group relationship between the Messelirrisoridae and crown group Upupiformes is supported by a number of derived characters including large, blade-like retroarticular processes on the mandible (Mayr 1998, 2006a). Messelirrisorids lack several derived characters shared by extant Bucerotes, the more inclusive clade containing the Upupiformes and Bucerotiformes (Mayr 2006a), but their original identification as stem group representatives of the Upupiformes was supported by a phylogenetic analysis based on a comprehensive data set (Mayr 2006a).

An as yet undescribed skeleton from the Green River Formation (Feduccia 1999: 335; Grande 2013: fig. 133) also seems to be a member of the Bucerotes, which is indicated by the very long hallux and the distinctive morphologies of the carpometacarpus and distal tarsometatarsus; this specimen would constitute the first New World record of the taxon. A fragmentary quadrate from the early Eocene Tingamarra Local Fauna of Australia was likened to the Bucerotes by Elzanowski and Boles (2015), but an unambiguous identification of this fossil requires the discovery of more material.

Mayr (1998) noted that *Laurillardia longirostris* and *L. munieri* from the late Eocene of the Paris Gypsum are stem group representatives of the Upupiformes. These species were before regarded as passeriform birds and were assigned to the taxon Laurillardiidae by Harrison (1979). Skeletons of two species of *Laurillardia*, *L. smoleni* and *L.* cf. *munieri*, were described by Mayr et al. (2020b) from the early Oligocene of Poland (Fig. 10.11g). A phylogenetic analysis performed by these authors suggests that the species of the Laurillardiidae are more closely related to crown group Upupiformes than are the Messelirrisoridae.

Small upupiform species were also found in the early Oligocene of the Luberon area in France; these fossils are in private collections (own observation). A distal humerus of a small upupiform bird was furthermore reported from the early Oligocene of Belgium (Mayr and Smith 2013). Leg bones of a putative upupiform bird from the early Oligocene of Slovakia (Kundrát et al. 2015), by contrast, are likely to have been misidentified (Mayr et al. 2020b).

10.3.4 Coraciiformes (Rollers, Bee-eaters, Kingfishers, and Allies)

Like the species of the Upupiformes and Trogoniformes, many coraciiform birds have syndactyl feet, in which the basal phalanges of the anterior toes are tightly joined over much of their length by connective tissue. This distinctive trait may be an adaptation to the breeding biology of these birds, many of which dig burrows or tunnels for their nesting sites.

10.3.4.1 Coracii (Rollers and Ground Rollers)

Extant rollers have an Old World distribution and include the Coraciidae (true rollers) and the more terrestrial, long-legged Madagascan Brachypteraciidae (ground rollers). Among others, these birds are characterized by very long postorbital processes of the skull, which constitute an adaptation to the feeding biology of these carnivorous birds.

Several Paleogene stem group representatives of the Coracii were described, which are successive sister taxa of the crown group. From a biogeographic point of view, it is remarkable that stem group representatives of the Coracii are also known from the early Eocene of North America, where rollers do not occur today. A putative record of a stem group representative of the Coracii from the early Eocene (52 Ma)

10.3 Cavitaves: Birds that Nest in Burrows and Tree Cavities

Fig. 10.11 Fossils of Paleogene Upupiformes (hoopoes and allies). (a)–(c) Specimens of *Messelirrisor grandis* (Messelirrisoridae) from the latest early or earliest middle Eocene of Messel (a: Hessisches Landesmuseum, Darmstadt, Germany, HLMD Be 178, b: Senckenberg Research Institute, Frankfurt, SMF-ME 10956A, c: Staatliches Museum für Naturkunde Karlsruhe, Germany, SMNK-PAL 3802). (d) Humerus of a messelirrisorid from the middle Eocene of the Geisel Valley (Geiseltalsammlung, Martin-Luther Universität of Halle-Wittenberg, Germany, GMH L-9-1969). (e) Distal humerus, carpometacarpus, and partial coracoid of an undescribed messelirrisorid from the early Eocene London Clay of Walton-on-the-Naze (collection of Michael Daniels, Holland-on-Sea, UK, WN 90655). (f) Humerus, coracoid, and furcula of *M. halcyrostris* from Messel (SMF-ME 2245). (g) Skeleton of *Laurillardia* cf. *munieri* (Laurillardiidae) from the early Oligocene of the Carpathian Basin in Poland (SMNK-PAL 9200a). Note the marked barring of the tail feathers of the specimen in (a). (Photos by Sven Tränkner)

of the Argentinean part of Patagonia (Degrange et al. 2021), however, was misidentified. The description of this species, *Ueekenkcoracias tambussiae*, was based on a rather poorly preserved leg, which distinctly differs from that of rollers in its proportions (the tarsometatarsus is stouter and the pedal phalanges are more robust, with the ungual phalanx of the second toe being unusually large), and the fossil actually more closely resembles the leg of *Palaeopsittacus* (Mayr in press).

One of the geologically oldest species assigned to the Coracii is *Septencoracias morsensis*, the holotype of which is a well preserved partial skeleton from early Eocene (54 Ma) deposits of the Fur Formation in Denmark (Fig. 10.12a; Bourdon et al. 2016). A partial skeleton from the London Clay of Walton-on-the-Naze was described as *Septencoracias* cf. *morsensis* by Mayr (in press), and there are further undescribed skeletons of this species in the collection of Michael Daniels (Fig. 10.12b, e); a tarsometatarsus

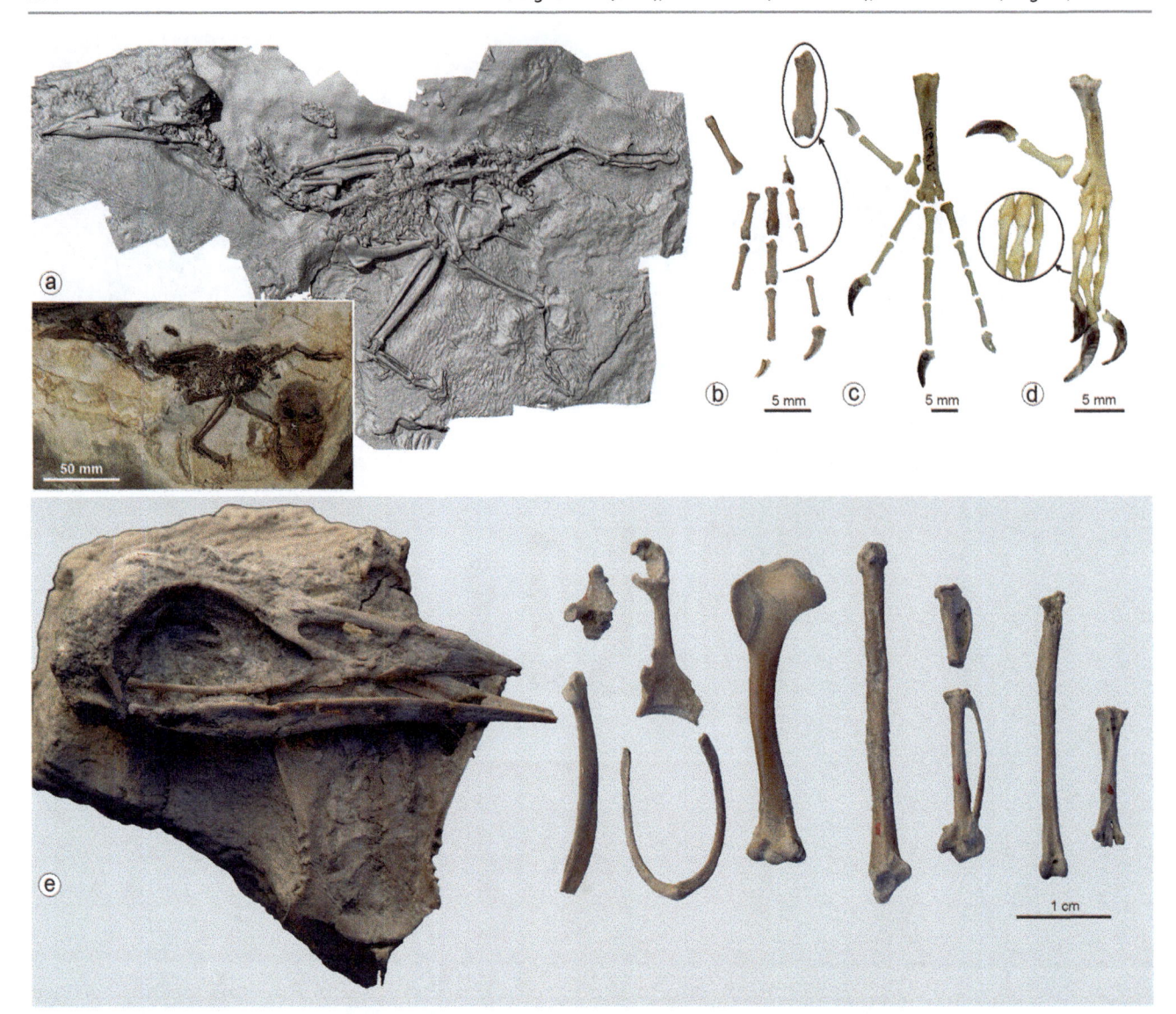

Fig. 10.12 Fossils of the early Eocene stem group roller *Septencoracias* (Coracii). (**a**) Holotype of *Septencoracias morsensis* from the early Eocene Fur Formation in Denmark (from Bourdon et al. 2016: fig. 1; published under a Creative Commons CC BY 4.0 license); the larger image is a surface scan. (**b**) Foot of *Septencoracias* cf. *morsensis* from the early Eocene London Clay of Walton-on-the-Naze (Senckenberg Research Institute, Frankfurt, SMF Av 655); the arrow denotes an enlarged view of the second phalanx of the third toe, which has an unusually widened distal end. (**c**) Foot of the extant *Coracias garrulus* (Coraciidae). (**d**) Foot of the extant *Merops apiaster* (Meropidae); the arrow denotes an enlarged view of the central phalanges of the anterior toes, which have widened distal ends. (**e**) Undescribed specimen of *Septencoracias* from Walton-on-the-Naze (collection of Michael Daniels, Holland-on-Sea, UK, WN 83456). (Photos in (**b**)–(**e**) by Sven Tränkner)

from the early Eocene of Egem (Belgium) was furthermore referred to *Septencoracias* by Mayr and Smith (2019b). *Septencoracias* was tentatively assigned to the Primobucconidae by Bourdon et al. (2016). Intriguingly, the taxon also shows some derived traits of the Meropidae, such as a quadrate with a shallow (rather than cup-shaped) articular surface for the jugal bar and widened distal ends of the basal phalanges of the anterior toes (Mayr in press). These similarities are of particular interest, because current molecular analyses support a sister group relationship between the Coracii and the Meropidae (Prum et al. 2015; Kuhl et al. 2021). However, the obvious conclusion that *Septencoracias* is an early stem group representative of the Meropidae, which retained a plesiomorphic Coracii-like morphology, is not supported by current phylogenetic analyses (Bourdon et al. 2016; Mayr in press). Most likely, the above features, therefore, evolved convergently in *Septencoracias* and the Meropidae, but they indicate that the fossil taxon differed from extant rollers in its foraging behavior and way of living.

The Primobucconidae (Fig. 10.13a) occur in early Eocene (MP 8/9-11) sites of North America and Europe. The taxon was initially established by Feduccia and Martin (1976) for

Primobucco mcgrewi. This species was described by Brodkorb (1970) on the basis of a wing from the Green River Formation, and in the original description, it was considered to be the earliest representative of the Neotropical piciform puffbirds (Bucconidae). Houde and Olson (1989) identified primobucconids as stem group representatives of the Coracii and mentioned a complete skeleton of *P. mcgrewi* from the Green River Formation. The latter fossil was studied in greater detail by Mayr et al. (2004), who described two very similar species from Messel as *P. perneri* and *P. frugilegus*. Further specimens of *P. mcgrewi* from the Green River Formation were reported by Ksepka and Clarke (2010b), and a tarsometatarsus of an unidentified primobucconid species was also found in the early Eocene of France (Mayr et al. 2004). The species of *Primobucco* are short-legged birds and much smaller than extant rollers, from which they are further distinguished in the less robust beak and a proportionally shorter postorbital process. Specimens from the early Eocene of England, which were assigned to the Primobucconidae by Olson and Feduccia (1979) and Harrison (1982), have been misidentified (Mayr et al. 2004).

Other Paleogene rollers were more similar to crown group Coracii in their skeletal morphology. One of these, *Eocoracias brachyptera*, is represented by several skeletons from Messel (Fig. 10.13d; Mayr and Mourer-Chauviré 2000). As in extant rollers, the postorbital processes of *E. brachyptera* are strongly elongated, but a number of plesiomorphic characteristics clearly distinguish the fossil species from crown group rollers (Mayr and Mourer-Chauviré 2000). The tarsometatarsus is very short and resembles that of the extant taxon *Eurystomus* (Coraciidae) in its proportions. Some of the specimens exhibit excellent feather preservation, and with regard to the short wings and the long and graduated tail, the feathering of *E. brachyptera* resembles that of the Brachypteraciidae. The wings of the Coraciidae, by contrast, are proportionally longer, and unlike in *E. brachyptera* the tail of true rollers is more or less deeply forked. Outgroup comparisons with *E. brachyptera* therefore suggest that the feathering of the Brachypteraciidae, which is well-suited for agile maneuvering in forested environments, is plesiomorphic for crown group rollers. The long wings and forked tail of the Coraciidae, by contrast, are derived traits that enable true rollers a more rapid flight in their open habitats (Mayr and Mourer-Chauviré 2000).

Paracoracias occidentalis is a stem group roller from the North American Green River Formation (Clarke et al. 2009), and the holotype skeleton and only known specimen of the species presumably comes from the middle Eocene Laney Shale Member (Grande 2013: 243). *P. occidentalis* was recovered as more closely related to crown group Coracii than is *E. brachyptera*, but the sole character supporting this placement is rather subtle and concerns the shape of the nostrils.

Mayr and Walsh (2018) described a three-dimensionally preserved partial skull and associated postcranial bones of an unnamed stem group roller from the early Eocene (51.5–53.5 Ma) London Clay of the Isle of Sheppey and identified plesiomorphic differences to extant rollers that indicate a disparate feeding behavior (Fig. 10.13b, c). This species was larger than *Septencoracias morsensis* and of similar size to *Paracoracias occidentalis*.

Another taxon of stem group rollers, the Geranopteridae, occurs in the late Eocene of the Quercy fissure fillings and includes *Geranopterus alatus*, the smaller *G. milneedwardsi*, and an unnamed species (Mayr and Mourer-Chauviré 2000; Mourer-Chauviré and Sigé 2006). Most major postcranial bones of *G. alatus* are known, and Mayr and Mourer-Chauviré (2000) also referred a partial skull to this species. The latter specimen shows the postorbital process of *Geranopterus* to exhibit a small cranial projection, which is absent in other stem group rollers but also occurs in extant Coracii. The carpometacarpus of *Geranopterus* furthermore shares with the extant species a small intermetacarpal process, which is likewise not found in other stem group rollers. *Cryptornis antiquus*, which has already been described in the mid-nineteenth century and is represented a largely complete but very poorly preserved skeleton from the late Eocene of the Paris Gypsum in France, appears to be closely related to, or even conspecific with, *G. alatus* (Mayr and Mourer-Chauviré 2000).

Extant rollers are almost exclusively carnivorous and feed on invertebrates and small vertebrates. It is therefore notable that in the two known specimens of *Primobucco frugilegus* (Primobucconidae) and in one specimen of *Eocoracias brachyptera* (Eocoraciidae) seeds are preserved as stomach contents (Mayr and Mourer-Chauviré 2000; Mayr et al. 2004). These fossils indicate that stem group representatives of the Coracii had a more versatile feeding ecology than the extant species (Mayr et al. 2004). Eocene rollers furthermore had less developed temporal fossae for the adductor muscles of the mandible and also differed in the morphology of the cranial cervical vertebrae. This led Mayr and Walsh (2018) to hypothesize that there were also differences in the functional morphology of the neck musculature, which were related to behavioral differences between stem and crown group Coracii.

10.3.4.2 Meropidae (Bee-eaters) and Alcedinides (Kingfishers, Todies, and Motmots)

The taxa united in this section were long considered to be closely related and share a derived morphology of the ear ossicle, the columella (Feduccia 1977). However, even though analyses of molecular sequence data supported a clade including the Alcedinidae (kingfishers), Todidae (todies), and Momotidae (motmots), the Meropidae (bee-eaters) were not recovered as part of this clade and

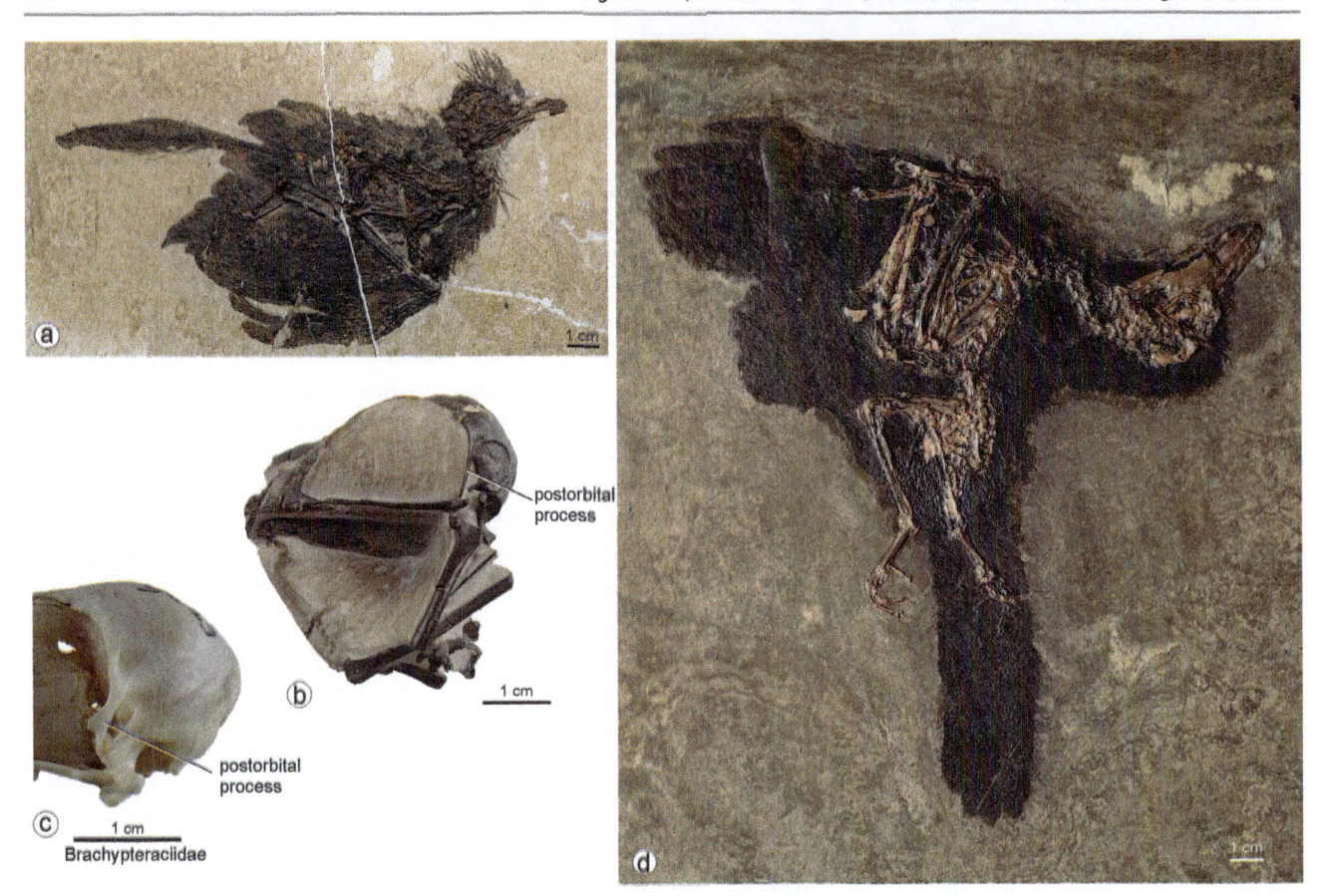

Fig. 10.13 Early Eocene stem group representatives of rollers (Coracii). (**a**) An undescribed primobucconid from the latest early or earliest middle Eocene of Messel (Senckenberg Research Institute, Frankfurt, SMF-ME 11349A). (**b**) Skull of an unnamed roller from the London Clay of the Isle of Sheppey, UK (National Museums of Scotland, Edinburgh, UK, NMS G.2014.54.1). (**c**) Caudal portion of the skull of the extant *Atelornis pittoides* (Brachypteraciidae). (**d**) Skeleton of *Eocoracias brachyptera* (Eocoraciidae) from Messel (Staatliches Museum für Naturkunde Karlsruhe, Germany, SMNK-PAL 2663a). (Photo in (**a**) by Anika Vogel, all others by Sven Tränkner)

resulted as the sister taxon of either the Coracii (Prum et al. 2015; Kuhl et al. 2021) or the clade (Coracii + Alcedinides) (Hackett et al. 2008). The Todidae and Momotidae were traditionally assumed to be sister groups and share a derived morphology of the caudal end of the mandible (Olson 1976), but molecular analyses supported a clade (Todidae + (Momotidae + Alcedinidae)) (Ericson et al. 2006; Hackett et al. 2008; Prum et al. 2015; Kuhl et al. 2021).

Most extant species of the Alcedinidae and all Meropidae occur in the Old World, whereas the extant distribution of the Todidae and Momotidae is confined to the New World. The fossil record shows, however, that at least the distribution of the Todidae is a relict one.

The geologically oldest species for which affinities to the Alcedinides were proposed is *Quasisyndactylus longibrachis* from Messel, which is a small bird and comparatively abundant in the locality (Fig. 10.14a; Mayr 1998). *Q. longibrachis* has a dorsoventrally flattened, very long, and remarkably tody-like beak. In contrast to crown group Todidae, however, the tarsometatarsus is rather short and the carpometacarpus lacks a well-developed intermetacarpal process. The proximal end of the first phalanx of the hind toe bears a lateral projection, which is characteristic of the Meropidae and Alcedinides. In all skeletons, the three anterior toes are preserved in tight attachment, which indicates the former presence of a syndactyl foot. Because of the presumably plesiomorphic morphology of the furcula, it was hypothesized that *Q. longibrachis* is the sister taxon of a clade formed by the Meropidae and Alcedinides (Mayr 1998, 2004a). Because the latter clade is not supported by current molecular analyses, this classification can no longer be upheld, and the affinities of *Quasisyndactylus* need to be revisited.

A partial humerus and a tarsometatarsus of an early Eocene representative of the Alcedinides were also reported from Egem in Belgium (Mayr and Smith 2019b). All other fossils, however, are from late Eocene and early Oligocene sites.

Various fossils have been reported from the late Eocene and early Oligocene Europe and North America, which show derived characteristics of the Todidae, the extant distribution of which is restricted to the Greater Antilles. All were

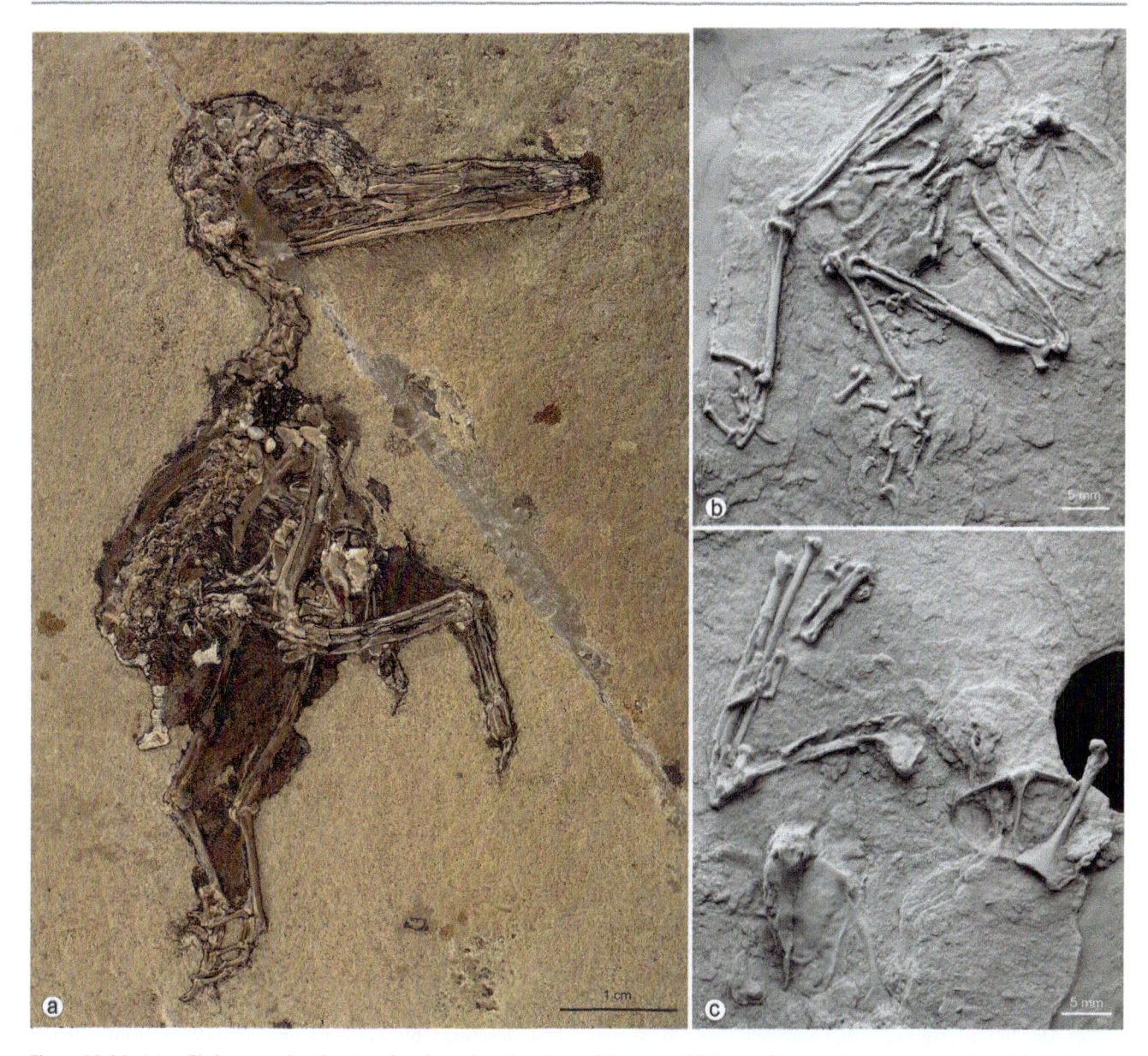

10.3 Cavitaves: Birds that Nest in Burrows and Tree Cavities

Fig. 10.14 (a) Skeleton of *Quasisyndactylus longibrachis* (Alcedinides) from the latest early or earliest middle Eocene of Messel (Senckenberg Research Institute, Frankfurt, SMF-ME 3543a). (b), (c) Partial skeleton of *Palaeotodus itaboraiensis* (Todidae) from the early Oligocene of Wiesloch-Frauenweiler in Germany (b: pelvis and legs, c: wings and pectoral girdle; SMF Av 505); coated with ammonium chloride. (All photos by Sven Tränkner)

assigned to the taxon *Palaeotodus*, which was originally introduced by Olson (1976) for a species from the early Oligocene (Brule Formation) of Wyoming. The description of this North American species, *Palaeotodus emryi*, was based on a skull and the proximal portion of a humerus (Olson 1976), but there are undescribed postcranial remains of a second individual, which also comes from the Brule Formation (Mayr and Knopf 2007). Because the humerus of *P. emryi* is proportionally larger than that of extant Todidae, Olson (1976) hypothesized that Paleogene todies had better-developed wings and dispersal capabilities than their extant relatives.

Two species of *Palaeotodus*, *P. escampsiensis* and *P. itardiensis*, were described by Mourer-Chauviré (1985) from the late Eocene (MP 19: *P. escampsiensis*) and early Oligocene (MP 23: *P. itardiensis*) of France. The description of *P. escampsiensis* was based on an incomplete humerus; of *P. itardiensis* a proximal ulna, distal tibiotarsus, and proximal tarsometatarsus were found. *P. itardiensis* is of similar size to *P. emryi* and distinctly larger than extant todies, whereas *P. escampsiensis* is as small as the species of crown group Todidae. *Palaeotodus itardiensis* also occurs in the early Oligocene of Wiesloch-Frauenweiler (Germany), where postcranial skeletons of two individuals have been found

(Fig. 10.14b, c; Mayr and Knopf 2007; Mayr and Micklich 2010). These fossils show that *Palaeotodus* agrees with extant Momotidae and Todidae in its long hindlimbs (the very long and slender tarsometatarsus measures almost the length of the humerus) and in the reduction of the procoracoid process of the coracoid. As in crown group Todidae, the carpometacarpus furthermore bears a large intermetacarpal process. The distal end of the tarsometatarsus, however, differs in some plesiomorphic features from that of extant todies. When exactly todies disappeared from the Old World is unknown, as is the date of their first occurrence in the New World.

The presence of stem group Todidae in the late Eocene indicates that the Meropidae and the stem species of the clade (Momotidae + Alcedinidae) must have also diverged by that time. Indeed, *Protornis glarniensis* from the early Oligocene of Switzerland was assigned to the Momotidae by Olson (1976). The species is known from an incomplete skeleton and was before classified into the Alcedinidae by Peyer (1957). A second specimen referred to *P. glarniensis* by Peyer (1957) was identified as a trogon by Olson (1976). Among others, the caudal ends of the mandible of *P. glarniensis* exhibit deep notches, which are a derived characteristic of extant Momotidae and Todidae. Except for the proportionally shorter symphysis, the wide and flat mandible of *Protornis* closely resembles that of the extant momotid taxa *Electron* and *Hylomanes*. However, owing to the poor preservation of the only known skeleton, more data on the skeletal morphology of *Protornis* are needed for a well-founded classification.

The Meropidae and Alcedinidae have no published Paleogene fossil record (Mourer-Chauviré 1982, 2006 tentatively identified fossils of both taxa in the Quercy fissure fillings, but the fragmentary remains have not yet been described). Molecular data indicate an Indomalayan origin of crown group Alcedinidae, some 27 Ma (Andersen et al. 2018), and it is well possible that Paleogene stem group representatives of the taxon were more similar to the Momotidae in their skeletal morphology.

10.3.5 Piciformes (Jacamars, Puffbirds, Woodpeckers, and Allies)

The zygodactyl Piciformes comprise the Neotropic Galbulae (jacamars and puffbirds) and the Pici (barbets, toucans, honeyguides, woodpeckers and allies), which have a nearly worldwide distribution but do not occur in the Australian region and on Madagascar. In addition to a unique arrangement of the digital flexor tendons, piciform birds exhibit a derived morphology of the proximal phalanx of the major digit of the wing (Mayr et al. 2003).

The Paleogene fossil record of piciform birds is still quite scanty. Remarkably, all specimens known so far are from very small species, some of which are even smaller than the smallest extant Piciformes. A very poorly known species with possible piciform affinities is "*Neanis*" *kistneri*, a small zygodactyl bird from the Green River Formation. "*N.*" *kistneri* was considered a piciform bird in the original description (Feduccia 1973) and it was regarded as closely related to the Galbulae by Houde and Olson (1989). The latter authors also first recognized that it is not a congener of *N. schucherti*, the type species of the taxon *Neanis* (see Sect. 10.3.5.2). The holotype of "*N.*" *kistneri* was meanwhile transferred on a resin slab, so that additional skeletal details are visible, and a second specimen from the Green River Formation was described by Weidig (2010). The feet of "*N.*" *kistneri* exhibit a zygodactyl arrangement, and the trochlea for the fourth toe bears a well-developed accessory trochlea. The tarsometatarsus is shorter than that of the Zygodactylidae, and the procoracoid process of the coracoid is long and slender. However, despite some overall similarity to extant Galbulae, piciform affinities of "*N.*" *kistneri* have not yet been well based, and better-preserved specimens are needed to firmly establish the phylogenetic affinities of this species.

The "coraciiform-like" overall morphology of *Neanis* is notable, and without the zygodactyl feet, the fossils would certainly not have been considered to be related to piciform birds. A more "coraciiform-like" skeletal morphology is, however, also found in the Galbulae and is plesiomorphic for the Piciformes. Its pronounced occurrence in *Neanis* may suggest that the taxon is temporally close to the divergence of piciform and coraciiform birds.

10.3.5.1 Sylphornithidae

The Sylphornithidae are tiny, long-legged birds, which were first described by Mourer-Chauviré (1988) from the late Eocene (MP 16) locality Bretou of the Quercy fissure fillings. Of the species from this French site, *Sylphornis bretouensis*, major portions of all limb bones are known (Mourer-Chauviré 1988; Mayr 1998). As noted by Mayr (1998), the coeval and similarly-sized *Palaegithalus cuvieri* from the late Eocene of the Paris Gypsum, which was originally described as a passeriform bird (see Brunet 1970), may also be a member of the Sylphornithidae. The very poor preservation of the only known skeleton of *P. cuvieri* does, however, not allow a definitive assignment, and alternatively, the fossil could be from a representative of the Zygodactylidae (Sect. 9.5.2). A further species, *Oligosylphe mourerchauvireae*, was described by Mayr and Smith (2002) from the early Oligocene of Belgium on the basis of a distal tarsometatarsus and tibiotarsus (Fig. 10.15c). Another early Oligocene sylphornithid, *Jacamatia luberonensis* from the Luberon in France, is represented by an isolated wing (Duhamel et al.

2020); whether this species is distinct from the coeval *Oligosylphe mourerchauvireae* and *Palaegithalus cuvieri* has yet to be shown.

Sylphornithids are well characterized by an extraordinarily long and slender tarsometatarsus, the distal end of which is very wide and has short trochleae. The trochlea for the fourth toe exhibits a plantarly directed wing-like flange, which indicates that the foot was at least facultatively zygodactyl. The carpometacarpus of *Sylphornis* bears a large intermetacarpal process similar to that of extant Piciformes and Passeriformes (Mayr 1998; Duhamel et al. 2020). Sylphornithids differ from the equally long-legged Zygodactylidae not only in the absence of an accessory trochlea for the fourth toe but also in other aspects of the morphology of the tarsometatarsal trochleae. Furthermore unlike in zygodactylids, the hypotarsus does not enclose a canal for the tendon of the flexor hallucis longus muscle. Mourer-Chauviré (1988) already recognized similarities in the hypotarsus morphology of *Sylphornis* and the piciform Galbulae, and an analysis by Mayr (2004b) also supported piciform affinities of the Sylphornithidae, which resulted as the sister taxon of either the Galbulae or the clade (Pici + Galbulae). A more recent analysis by Duhamel et al. (2020) likewise suggested affinities to the Galbulae. It is, however, well possible that future findings may lead to a revision of the phylogenetic affinities of these enigmatic birds.

10.3.5.2 Gracilitarsidae

The taxon Gracilitarsidae was introduced for *Gracilitarsus mirabilis*, a tiny, long-legged bird known from three skeletons from Messel (Fig. 10.15a, b; Mayr 1998, 2001c, 2005c). The limb proportions of this species have no counterpart among extant birds. The stout and short humerus and the very long hand section of the wing indicate that it may have been an aerial specialist, similar to the Apodiformes and the passeriform Hirundinidae (swallows). Whereas, however, the latter have short legs, the tarsometatarsus of *Gracilitarsus* is very long and slender. Owing to the fact that the first phalanges of the third and fourth toes are shortened, the three anterior toes, which bear large ungual phalanges, are furthermore of nearly equal length. The alular phalanx of the wing exhibits an unusually well-developed ungual phalanx.

As in the Sylphornithidae (Sect. 10.3.5.1), the minor metacarpal distinctly exceeds the major metacarpal in distal extent. *Gracilitarsus* is, however, distinguished from the likewise tiny and long-legged *Sylphornis* in the more elongated carpometacarpus, the narrower distal end of the tarsometatarsus, and in that the trochlea for the third toe bears a deep furrow, which is also found in some extant trunk-climbing birds.

According to the shape of its beak, which is of similar proportions to that of some short-billed passeriform Nectariniidae (sunbirds), *G. mirabilis* was probably either insectivorous or nectarivorous. Owing to the lack of an extant analog, the peculiar foot morphology is difficult to interpret, but *Gracilitarsus* may have used its feet for clinging to vertical surfaces, possibly owing to a specialized feeding technique (Mayr 2001c). In light of the fact that a similar foot structure, that is, a greatly elongated tarsometatarsus with rather short anterior toes of subequal length, is not found in extant birds, it is notable that the coliiform taxon *Masillacolius*, which also occurs in Messel (Sect. 10.2.2), exhibits a very similar foot morphology.

A phylogenetic analysis supported a clade including the Gracilitarsidae, Sylphornithidae, and crown group Piciformes (Mayr 2005c), even though this clade is not strongly supported. The Gracilitarsidae and extant Piciformes exhibit a distinct medial notch on the sternal end of the coracoid, which occurs in piciform and coraciiform birds but is also found in other more distantly related taxa.

Mayr (2001c, 2005c) compared *Gracilitarsus* with two other Paleogene birds, *Eutreptodactylus itaboraiensis* from the early Eocene of Itaboraí in Brazil and *Neanis schucherti* from the early Eocene Green River Formation. The holotype of *E. itaboraiensis* is a morphologically distinctive tarsometatarsus. The bone has a strongly medially projecting trochlea for the second toe and a plantarly-directed, wing-like flange on the trochlea for the fourth toe (Baird and Vickers-Rich 1997). The latter feature indicates the presence of semi- or facultatively zygodactyl feet. Whether or not this feature was present in *Gracilitarsus* cannot be assessed in the known specimens of the taxon. However, the tarsometatarsus of *E. itaboraiensis* resembles that of *G. mirabilis* in that it exhibits a marked furrow on the dorsal surface of the distal end, between the trochleae for the third and fourth toes, and the tarsometatarsi of both taxa are also similar in their overall shape. *E. itaboraiensis* was considered to be a representative of the Cuculiformes in the original description, but this hypothesis is not supported by the morphology of the specimen. A very small partial tibiotarsus from the Itaboraí site, which may possibly belong to *E. itaboraiensis*, was reported by Mayr et al. (2011).

N. schucherti was described on the basis of wing and pectoral girdle elements from the Green River Formation. The wing shows similar proportions to that of *G. mirabilis*, and the coracoid exhibits a notch in the medial margin of the sternal extremity. *N. schucherti* was originally described as a passeriform bird, but it was considered to be an aerial specialist by Houde and Olson (1989). Although the bones known of this species are very similar to those of *Gracilitarsus*, the discovery of hindlimb elements is needed to further assess the hypothesis of a closer relationship.

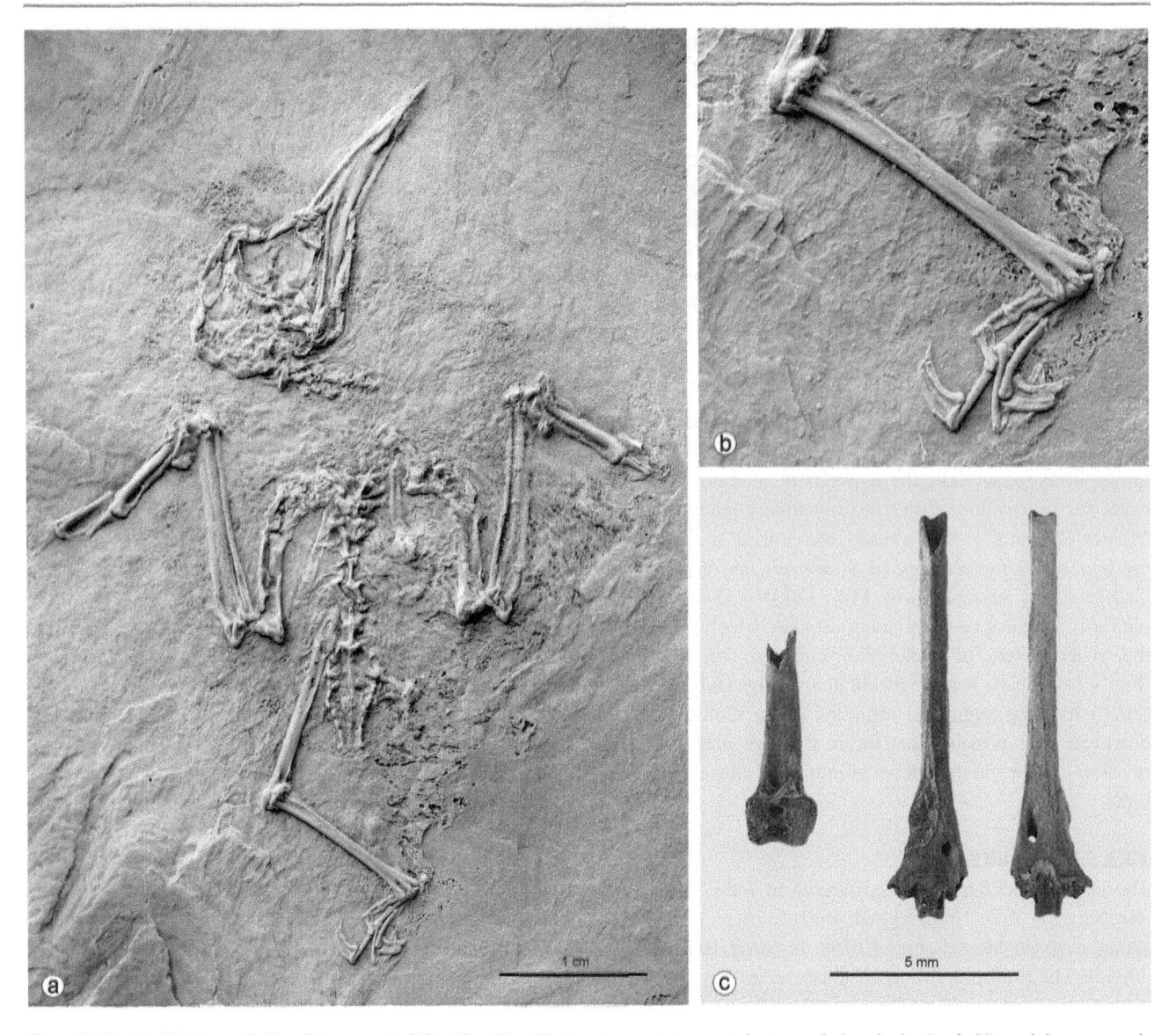

Fig. 10.15 (a) Skeleton of *Gracilitarsus mirabilis* (Gracilitarsidae) from the latest early or earliest middle Eocene of Messel (Senckenberg Research Institute, Frankfurt, SMF-ME 3547), with (**b**) a detail of the foot; fossil coated with ammonium chloride. (**c**) Tibiotarsus and tarsometatarsus (plantar and dorsal view) of *Oligosylphe mourerchauvireae* from the early Oligocene of Belgium (Royal Belgian Institute of Natural Sciences, Brussels, Belgium, IRSNB Av 76 and 77). (Photos by Sven Tränkner)

10.3.5.3 Pici

All Paleogene fossils of the Pici are from Europe and most belong to species with a modern-type morphology. A notable exception, however, is *Picavus litencicensis*, the holotype of which is a postcranial skeleton from the early Oligocene of the Czech Republic (Fig. 10.16a, b; Mayr and Gregorová 2012). This very small species was assigned to the new taxon Picavidae and shares some derived characteristics of the tarsometatarsus with the Pici, including a distinctive morphology of the trochlea for the fourth toe and a narrow trochlea for the second toe. As in the Pici, Galbulae, and Zygodactylidae but unlike in other zygodactyl birds, the carpometacarpus seems to exhibit an intermetacarpal process. The tarsometatarsus, however, lacks the distally elongated accessory trochlea, which characterizes the crown group representatives of the Pici.

The earliest unambiguously identified fossils of modern-type Pici are from the early Oligocene of Belgium and Germany and belong to species, which are likewise much smaller than any extant European species of the Pici. The only named species, *Rupelramphastoides knopfi* from the early Oligocene Wiesloch-Frauenweiler locality in southern Germany, is represented by two partial skeletons, which constitute the most substantial remains of Paleogene Pici (Fig. 10.16c–g; Mayr 2005d, 2006b). *R. knopfi* is the smallest known species of the Pici, being smaller even than extant piculets (Picumninae). It is clearly identified as a representative of the Pici by a large accessory tarsometatarsal trochlea

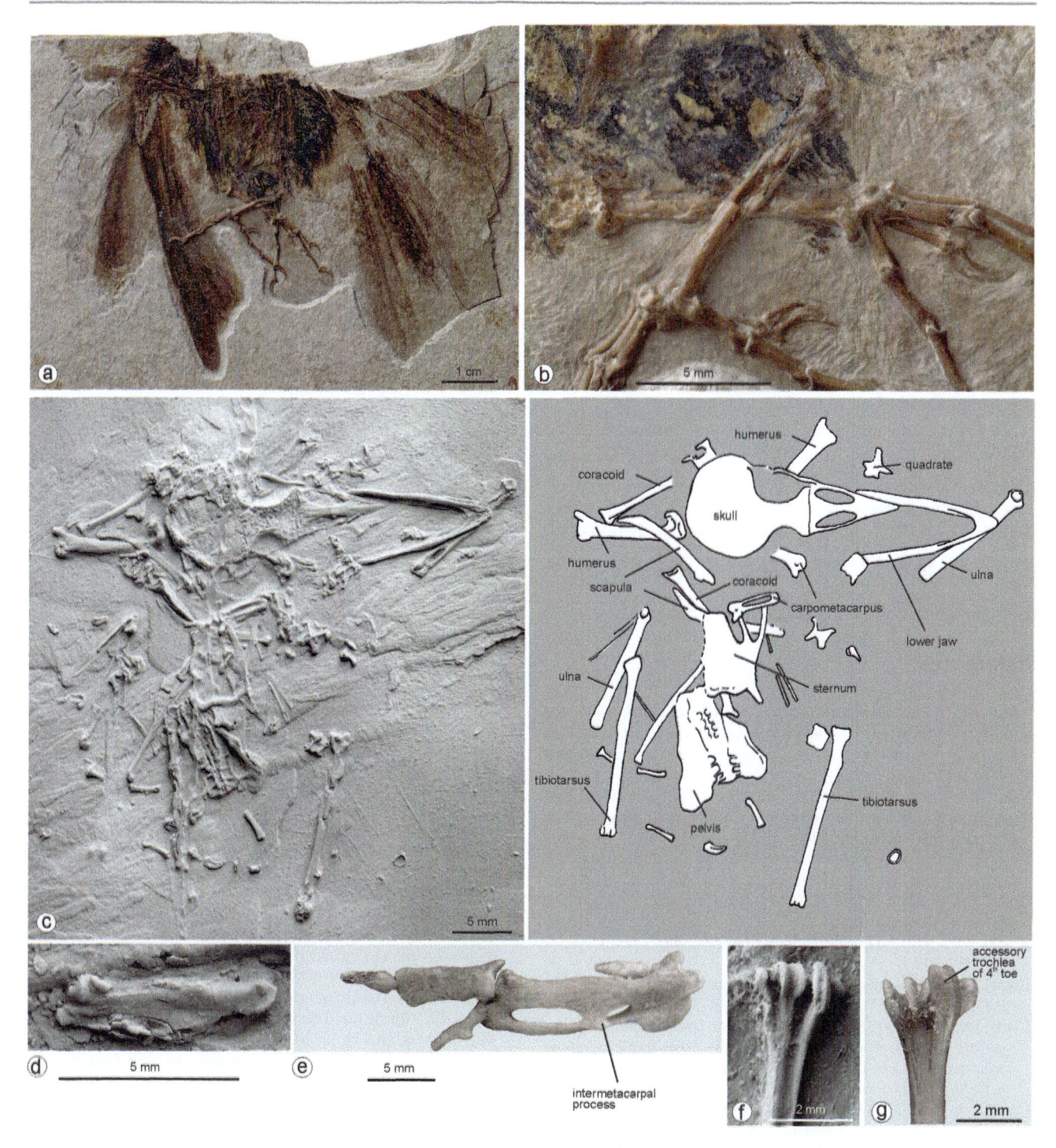

10.3 Cavitaves: Birds that Nest in Burrows and Tree Cavities

Fig. 10.16 Paleogene Piciformes. (**a**) Partial Skeleton of the holotype of *Picavus litenticensis* from the early Oligocene of the Czech Republic (Moravian Museum, Brno, Czech Republic, MM Ge 29982a), with (**b**) a detail of the feet. (**c**) Partial skeleton of *Rupelramphastoides knopfi* from the early Oligocene of Wiesloch-Frauenweiler in Germany (Staatliches Museum für Naturkunde Karlsruhe, Germany, SMNK-PAL 4334 a) with an interpretive drawing of the skeleton; specimen coated with ammonium chloride. (**d**) Carpometacarpus of *R. knopfi* (holotype, Senckenberg Research Institute, Frankfurt, SMF Av 500a). (**e**) Hand section of the wing of the extant *Psilopogon pyrolophus* (Megalaimidae). (**f**), (**g**) Distal tarsometatarsus of (**f**) *R. knopfi* (holotype) and (**g**) the extant *Dendropicos fuscescens* (Picidae). (All photos by Sven Tränkner)

for the reversed fourth toe, a very narrow trochlea for the second toe, and by the presence of marked tubercles on the shaft of the ulna for the attachment of the secondary feathers.

In its proportions, the skull resembles that of extant honeyguides (Indicatoridae). The tarsometatarsus of *Rupelramphastoides* is long as it is in many extant barbets and toucans (Megalaimidae, Lybiidae, Ramphastidae, and allies), but the bill morphology clearly shows the fossil taxon to be outside the crown clade including these birds, which have more strongly ossified nostrils. The sternum of the fossil taxon exhibits a bifurcated spina externa, which is a derived characteristic of the Picidae, whereas the plesiomorphic absence of an ossified extensor bridge (arcus extensorius) on the proximal end of the tarsometatarsus supports a position outside a clade including the Indicatoridae and Picidae (Mayr 2005d, 2006b). As suggested by its long tarsometatarsus, *R. knopfi* may be a stem group representative of the clade including barbets and toucans, but the leg proportions of the fossil may well be plesiomorphic for the Pici, and a phylogenetic placement is not straightforward.

A tarsometatarsus of an unnamed representative of the Pici was also reported from the late Oligocene of Germany (Mayr 2001d). In its size and proportions, this bone resembles the tarsometatarsus of the extant African Cardinal Woodpecker (*Dendropicos fuscescens*), but an unambiguous assignment to the Picidae is not possible on the basis of the available fossil material.

References

Andersen MJ, McCullough JM, Mauck WM III, Smith BT, Moyle RG (2018) A phylogeny of kingfishers reveals an Indomalayan origin and elevated rates of diversification on oceanic islands. J Biogeogr 45:269–281

Baird RF, Vickers-Rich P (1997) *Eutreptodactylus itaboraiensis* gen. et. sp. nov., an early cuckoo (Aves: Cuculiformes) from the Late Paleocene of Brazil. Alcheringa 21:123–127

Bourdon E, Kristoffersen AV, Bonde N (2016) A roller-like bird (Coracii) from the Early Eocene of Denmark. Sci Rep 6:34050

Braun EL, Kimball RT (2021) Data types and the phylogeny of Neoaves. Birds 2:1–22

Brodkorb P (1970) An Eocene puffbird from Wyoming. Contrib Geol 9: 13–15

Brunet J (1970) Oiseaux de l'Éocène supérieur du bassin de Paris. Ann Paléontol 56:3–57

Clarke JA, Ksepka DT, Smith NA, Norell MA (2009) Combined phylogenetic analysis of a new North American fossil species confirms widespread Eocene distribution rollers (Aves, Coracii). Zool J Linnean Soc 157:586–611

Cracraft J, Rich PV (1972) The systematics and evolution of the Cathartidae in the Old World Tertiary. Condor 74:272–283

De Pietri VL, Mourer-Chauviré C, Menkveld-Gfeller U, Meyer CA, Costeur L (2013) An assessment of the Cenozoic avifauna of Switzerland, with a description of two fossil owls (Aves, Strigiformes). Swiss J Geosci 106:187–197

Degrange FJ, Pol D, Puerta P, Wilf P (2021) Unexpected larger distribution of Paleogene stem-rollers (Aves, Coracii): new evidence from the Eocene of Patagonia, Argentina. Sci Rep 11:1363

Duhamel A, Balme C, Legal S, Riamon S, Louchart A (2020) An early Oligocene stem Galbulae (jacamars and puffbirds) from southern France, and the position of the Paleogene family Sylphornithidae. Auk 137:ukaa023

Dyke GJ, Waterhouse DM (2001) A mousebird (Aves: Coliiformes) from the Eocene of England. J Ornithol 142:7–15

Dyke GJ, Waterhouse DM, Kristoffersen AV (2004) Three new fossil landbirds from the early Paleogene of Denmark. Bull Geol Soc Denmark 51:47–56

Elzanowski A, Boles WE (2015) A coraciiform-like bird quadrate from the Early Eocene Tingamarra local fauna of Queensland, Australia. Emu 115:110–116

Ericson PGP, Anderson CL, Britton T, Elzanowski A, Johansson US, Källersjö M, Ohlson JI, Parsons TJ, Zuccon D, Mayr G (2006) Diversification of Neoaves: integration of molecular sequence data and fossils. Biol Lett 2:543–547

Feduccia A (1973) A new Eocene zygodactyl bird. J Paleontol 47:501–503

Feduccia A (1977) A model for the evolution of perching birds. Syst Zool 26:19–31

Feduccia A (1999) The origin and evolution of birds, 2nd edn. Yale University Press, New Haven

Feduccia A, Martin LD (1976) The Eocene zygodactyl birds of North America (Aves: Piciformes). Smithson Contrib Paleobiol 27:101–110

Fischer K (1983) *Oligostrix rupelensis* n. gen., n. sp., eine neue Ureule (Protostrigidae, Strigiformes, Aves) aus dem marinen Mitteloligozän des Weißelsterbeckens bei Leipzig (DDR). Z Geol Wiss 11:483–487

Fischer K (1987) Eulenreste (*Eoglaucidium pallas* nov. gen., nov. sp., Strigiformes, Aves) aus der mitteleozänen Braunkohle des Geiseltals bei Halle (DDR). Mitt Zool Mus Berlin 63, Suppl: Ann Ornithol 11: 137–142

Fowler DW, Freedman Fowler EA, Alexander JM (2018) The finest fossil owl. J Vertebr Paleontol, Prog Abs 2018:129

Gaillard C (1908) Les oiseaux des Phosphorites du Quercy. Ann Univ Lyon (Nouv Sér) 23:1–178

Grande L (2013) The lost world of Fossil lake. Snapshots from deep time. University of Chicago Press, Chicago

Hackett SJ, Kimball RT, Reddy S, Bowie RCK, Braun EL, Braun MJ, Chojnowski JL, Cox WA, Han K-L, Harshman J, Huddleston CJ, Marks BD, Miglia KJ, Moore WS, Sheldon FH, Steadman DW, Witt CC, Yuri T (2008) A phylogenomic study of birds reveals their evolutionary history. Science 320:1763–1767

Harrison CJO (1979) The Upper Eocene birds of the Paris basin: a brief re-appraisal. Tert Res 2:105–109

Harrison CJO (1980) A small owl from the Lower Eocene of Britain. Tert Res 3:83–87

Harrison CJO (1982) Cuculiform, piciform and passeriform birds in the Lower Eocene of England. Tert Res 4:71–81

Houde P, Olson SL (1989 [1988]) Small arboreal nonpasserine birds from the Early Tertiary of western North America. In: Ouellet H (ed) Acta XIX Congressus Internationalis Ornithologici. University of Ottawa Press, Ottawa, pp 2030–2036

Houde P, Olson SL (1992) A radiation of coly-like birds from the early Eocene of North America (Aves: Sandcoleiformes new order). In: Campbell KE (ed) Papers in avian paleontology honoring Pierce Brodkorb. Nat Hist Mus Los Angeles Cty Sci Ser 36:137–160

Howard H (1965) First record of avian fossils from the Eocene of California. J Paleontol 39:350–354

Kristoffersen AV (2002) An early Paleogene trogon (Aves: Trogoniformes) from the Fur Formation, Denmark. J Vertebr Paleontol 22:661–666

Ksepka DT, Clarke JA (2009) Affinities of *Palaeospiza bella* and the phylogeny and biogeography of mousebirds (Coliiformes). Auk 126: 245–259

Ksepka DT, Clarke JA (2010a) New fossil mousebird (Aves: Coliiformes) with feather preservation provides insight into the ecological diversity of an Eocene North American avifauna. Zool J Linnean Soc 160:685–706

References

Ksepka DT, Clarke JA (2010b) *Primobucco mcgrewi* (Aves: Coracii) from the Eocene Green River Formation: new anatomical data from the earliest definitive record of stem rollers. J Vertebr Paleontol 30:215–225

Ksepka DT, Stidham TA, Williamson TE (2017) Early Paleocene landbird supports rapid phylogenetic and morphological diversification of crown birds after the K-Pg mass extinction. Proc Natl Acad Sci U S A 114:8047–8052

Kuhl H, Frankl-Vilches C, Bakker A, Mayr G, Nikolaus G, Boerno ST, Klages S, Timmermann B, Gahr M (2021) An unbiased molecular approach using 3'UTRs resolves the avian family-level tree of life. Mol Biol Evol 38:108–121

Kundrát M, Soták J, Ahlberg PE (2015) A putative upupiform bird from the Early Oligocene of the Central Western Carpathians and a review of fossil birds unearthed in Slovakia. Acta Zool 96:45–59

Kurochkin EN, Dyke GJ (2011) The first fossil owls (Aves: Strigiformes) from the Paleogene of Asia and a review of the fossil record of Strigiformes. Paleontol J 45:445–458

Martin LD, Black CC (1972) A new owl from the Eocene of Wyoming. Auk 89:887–888

Mayr G (1998) „Coraciiforme" und „piciforme" Kleinvögel aus dem Mittel-Eozän der Grube Messel (Hessen, Deutschland). Cour Forsch-Inst Senckenberg 205:1–101

Mayr G (1999) A new trogon from the Middle Oligocene of Céreste, France. Auk 116:427–434

Mayr G (2000a) New or previously unrecorded avian taxa from the Middle Eocene of Messel (Hessen, Germany). Mitt Mus Naturkunde Berl, Geowiss Reihe 3:207–219

Mayr G (2000b) A new mousebird (Coliiformes: Coliidae) from the Oligocene of Germany. J Ornithol 141:85–92

Mayr G (2000c) Tiny hoopoe-like birds from the Middle Eocene of Messel (Germany). Auk 117:968–974

Mayr G (2001a) New specimens of the Middle Eocene fossil mousebird *Selmes absurdipes* Peters 1999. Ibis 143:427–434

Mayr G (2001b) A second skeleton of the early Oligocene trogon *Primotrogon wintersteini* Mayr 1999 (Aves: Trogoniformes: Trogonidae) in an unusual state of preservation. Senck leth 81:335–338

Mayr G (2001c) A new specimen of the tiny Middle Eocene bird *Gracilitarsus mirabilis* (new family: Gracilitarsidae). Condor 103:78–84

Mayr G (2001d) The earliest fossil record of a modern-type piciform bird from the late Oligocene of Germany. J Ornithol 142:2–6

Mayr G (2002a) An owl from the Paleocene of Walbeck, Germany. Mitt Mus Naturkunde Berl, Geowiss Reihe 5:283–288

Mayr G (2002b) Avian Remains from the Middle Eocene of the Geiseltal (Sachsen-Anhalt, Germany). In: Zhou Z, Zhang F (eds) Proceedings of the 5th symposium of the Society of Avian Paleontology and Evolution, Beijing, 1–4 June 2000. Science Press, Beijing, pp 77–96

Mayr G (2002c) A new species of *Plesiocathartes* (Aves: ? Leptosomidae) from the Middle Eocene of Messel, Germany. PaleoBios 22:10–20

Mayr G (2004a) New specimens of *Hassiavis laticauda* (Aves: Cypselomorphae) and *Quasisyndactylus longibrachis* (Aves: Alcediniformes) from the Middle Eocene of Messel, Germany. Cour Forsch-Inst Senckenberg 252:23–28

Mayr G (2004b) The phylogenetic relationships of the early Tertiary Primoscenidae and Sylphornithidae and the sister taxon of crown group piciform birds. J Ornithol 145:188–198

Mayr G (2005a) A new Eocene *Chascacocolius*-like mousebird (Aves: Coliiformes) with a remarkable gaping adaptation. Org Divers Evol 5:167–171

Mayr G (2005b) New trogons from the early Tertiary of Germany. Ibis 147:512–518

Mayr G (2005c) Phylogenetic affinities and composition of the early Eocene Gracilitarsidae (Aves, ?Piciformes). Neues Jahrb Geol Paläontol, Mh 2005:1–16

Mayr G (2005d) A tiny barbet-like bird from the Lower Oligocene of Germany: the smallest species and earliest substantial fossil record of the Pici (woodpeckers and allies). Auk 122:1055–1063

Mayr G (2006a) New specimens of the Eocene Messelirrisoridae (Aves: Bucerotes), with comments on the preservation of uropygial gland waxes in fossil birds from Messel and the phylogenetic affinities of Bucerotes. Paläontol Z 80:405–420

Mayr G (2006b) First fossil skull of a Paleogene representative of the Pici (woodpeckers and allies) and its evolutionary implications. Ibis 148:824–827

Mayr G (2007) The birds from the Paleocene fissure filling of Walbeck (Germany). J Vertebr Paleontol 27:394–408

Mayr G (2008) The Madagascan "cuckoo-roller" (Aves: Leptosomidae) is not a roller—notes on the phylogenetic affinities and evolutionary history of a "living fossil". Acta Ornithol 43:226–230

Mayr G (2009a) Paleogene fossil birds, 1st edn. Springer, Heidelberg

Mayr G (2009b) A small loon and a new species of large owl from the Rupelian of Belgium (Aves: Gaviiformes, Strigiformes). Paläontol Z 83:247–254

Mayr G (2009c) A well-preserved second trogon skeleton (Aves, Trogonidae) from the middle Eocene of Messel, Germany. Palaeobiodiv Palaeoenv 89:1–6

Mayr G (2011) Two-phase extinction of "Southern Hemispheric" birds in the Cenozoic of Europe and the origin of the Neotropic avifauna. Palaeobiodiv Palaeoenv 91:325–333

Mayr G (2013) Late Oligocene mousebird converges on parrots in skull morphology. Ibis 155:384–396

Mayr G (2014) Comparative morphology of the radial carpal bone of neornithine birds and the phylogenetic significance of character variation. Zoomorphol 133:425–434

Mayr G (2015) A new specimen of the Early Eocene *Masillacolius brevidactylus* and its implications for the evolution of feeding specializations in mousebirds (Coliiformes). C R Palevol 14:363–370

Mayr G (2016) The world's smallest owl, the earliest unambiguous charadriiform bird, and other avian remains from the early Eocene Nanjemoy Formation of Virginia (USA). Paläontol Z 90:747–763

Mayr G (2017) Avian Evolution: The fossil record of birds and its paleobiological significance. Wiley-Blackwell, Chichester

Mayr G (2018) New data on the anatomy and paleobiology of sandcoleid mousebirds (Aves, Coliiformes) from the early Eocene of Messel. Palaeobiodiv Palaeoenv 98:639–651

Mayr G (2020) An updated review of the middle Eocene avifauna from the Geiseltal (Germany), with comments on the unusual taphonomy of some bird remains. Geobios 62:45–59

Mayr G (in press) A partial skeleton of *Septencoracias* from the early Eocene London Clay reveals derived features of bee-eaters (Meropidae) in a putative stem group roller (Aves, Coracii). Palaeobiodiv Palaeoenv

Mayr G, Gregorová R (2012) A tiny stem group representative of Pici (Aves, Piciformes) from the early Oligocene of the Czech Republic. Paläontol Z 86:333–343

Mayr G, Knopf C (2007) A tody (Alcediniformes, Todidae) from the early Oligocene of Germany. Auk 124:1294–1304

Mayr G, Micklich N (2010) New specimens of the avian taxa *Eurotrochilus* (Trochilidae) and *Palaeotodus* (Todidae) from the early Oligocene of Germany. Paläontol Z 84:387–395

Mayr G, Mourer-Chauviré C (2000) Rollers (Aves: Coraciiformes s.s.) from the Middle Eocene of Messel (Germany) and the Upper Eocene of the Quercy (France). J Vertebr Paleontol 20:533–546

Mayr G, Mourer-Chauviré C (2004) Unusual tarsometatarsus of a mousebird from the Paleogene of France and the relationships of *Selmes* Peters, 1999. J Vertebr Paleontol 24:366–372

Mayr G, Peters DS (1998) The mousebirds (Aves: Coliiformes) from the Middle Eocene of Grube Messel (Hessen, Germany). Senck leth 78: 179–197

Mayr G, Schaal SKF (2016) Gastric pellets with bird remains from the early Eocene of Messel. Palaios 31:447–451

Mayr G, Smith R (2001) Ducks, rails, and limicoline waders (Aves: Anseriformes, Gruiformes, Charadriiformes) from the lowermost Oligocene of Belgium. Geobios 34:547–561

Mayr G, Smith R (2002) Avian remains from the lowermost Oligocene of Hoogbutsel (Belgium). Bull Inst Roy Sci Nat Belg 72:139–150

Mayr G, Smith T (2013) Galliformes, Upupiformes, Trogoniformes, and other avian remains (?Phaethontiformes and ?Threskiornithidae) from the Rupelian stratotype in Belgium, with comments on the identity of "*Anas*" *benedeni* Sharpe, 1899. In: Göhlich UB, Kroh A (eds) Paleornithological Research 2013—Proceedings of the 8th international meeting of the Society of Avian Paleontology and Evolution. Natural History Museum Vienna, Vienna, pp 23–35

Mayr G, Smith T (2019a) New Paleocene bird fossils from the North Sea Basin in Belgium and France. Geol Belg 22:35–46

Mayr G, Smith T (2019b) A diverse bird assemblage from the Ypresian of Belgium furthers knowledge of early Eocene avifaunas of the North Sea Basin. N Jb Geol Paläontol, Abh 291:253–281

Mayr G, Walsh S (2018) Exceptionally well-preserved early Eocene fossil reveals cranial and vertebral features of a stem group roller (Aves, Coraciiformes). Paläontol Z 92:715–726

Mayr G, Manegold A, Johansson U (2003) Monophyletic groups within "higher land birds"—comparison of morphological and molecular data. J Zool Syst Evol Res 41:233–248

Mayr G, Mourer-Chauviré C, Weidig I (2004) Osteology and systematic position of the Eocene Primobucconidae (Aves, Coraciiformes sensu stricto), with first records from Europe. J Syst Palaeontol 2:1–12

Mayr G, Rana RS, Rose KD, Sahni A, Kumar K, Singh L, Smith T (2010) *Quercypsitta*-like birds from the early Eocene of India (Aves, ?Psittaciformes). J Vertebr Paleontol 30:467–478

Mayr G, Alvarenga H, Clarke J (2011) An *Elaphrocnemus*-like landbird and other avian remains from the late Paleocene of Brazil. Acta Palaeontol Pol 56:679–684

Mayr G, Gingerich PD, Smith T (2020a) Skeleton of a new owl from the early Eocene of North America (Aves, Strigiformes) with an accipitrid-like foot morphology. J Vertebr Paleontol 40:e1769116

Mayr G, Bochenski ZM, Tomek T, Wertz K, Bienkowska-Wasiluk M, Manegold A (2020b) Skeletons from the early Oligocene of Poland fill a significant temporal gap in the fossil record of upupiform birds (hoopoes and allies). Hist Biol 32:1163–1175

Mlíkovský J (1998) A new barn owl (Aves: Strigidae) from the early Miocene of Germany, with comments on the fossil history of the Tytonidae. J Ornithol 139:247–261

Mourer-Chauviré C (1982) Les oiseaux fossiles des Phosphorites du Quercy (Eocène supérieur à Oligocène supérieur): implications paléobiogéographiques. Geobios, mém spéc 6:413–426

Mourer-Chauviré C (1983) *Minerva antiqua* (Aves, Strigiformes), an owl mistaken for an edentate mammal. Am Mus Novit 2773:1–11

Mourer-Chauviré C (1985) Les Todidae (Aves, Coraciiformes) des Phosphorites du Quercy (France). Proc Kon Ned Akad Wet, Ser B 88:407–414

Mourer-Chauviré C (1987) Les Strigiformes (Aves) des Phosphorites du Quercy (France): Systématique, biostratigraphie et paléobiogéographie. Doc Lab Géol Lyon 99:89–135

Mourer-Chauviré C (1988) Le gisement du Bretou (Phosphorites du Quercy, Tarn-et-Garonne, France) et sa faune de vertébrés de l'Eocène supérieur. II Oiseaux Palaeontographica (A) 205:29–50

Mourer-Chauviré C (1994) A large owl from the Palaeocene of France. Palaeontol 37:339–348

Mourer-Chauviré C (1999) Comments on "Tertiary barn owls of Europe". J Ornithol 140:363–364

Mourer-Chauviré C (2002) Revision of the Cathartidae (Aves, Ciconiiformes) from the Middle Eocene to the Upper Oligocene Phosphorites du Quercy, France. In: Zhou Z, Zhang F (eds) Proceedings of the 5th symposium of the Society of Avian Paleontology and Evolution, Beijing, 1–4 June 2000. Science Press, Beijing, pp 97–111

Mourer-Chauviré C (2006) The avifauna of the Eocene and Oligocene Phosphorites du Quercy (France): an updated list. Strata, sér 1 13: 135–149

Mourer-Chauviré C (2008) Birds (Aves) from the Early Miocene of the Northern Sperrgebiet, Namibia. Mem Geol Surv Namibia 20:147–167

Mourer-Chauviré C, Sigé B (2006) Une nouvelle espèce de *Jungornis* (Aves, Apodiformes) et de nouvelles formes de Coraciiformes s.s. dans l'Éocène supérieur du Quercy. Strata, ser 1 13:151–159

Nessov LA (1992b) Mesozoic and Paleogene birds of the USSR and their paleoenvironments. In: Campbell KE (ed) Papers in avian paleontology honoring Pierce Brodkorb. Nat Hist Mus Los Angeles Cty Sci Ser 36:465–478

Olson SL (1976) Oligocene fossils bearing on the origins of the Todidae and the Momotidae (Aves: Coraciiformes). Smithson Contrib Paleobiol 27:111–119

Olson SL (1985) The fossil record of birds. In: Farner DS, King JR, Parkes KC (eds) Avian biology, vol 8. Academic Press, New York, pp 79–238

Olson SL, Feduccia A (1979) An Old-World occurrence of the Eocene avian family Primobucconidae. Proc Biol Soc Wash 92:494–497

O'Reilly S, Summons R, Mayr G, Vinther J (2017) Preservation of uropygial gland lipids in a 48-million-year-old bird. Proc R Soc Lond Ser B 284:20171050

Panteleyev AV (2011) [First bird remains from the Paleogene of Crimea]. In: Batashev MS, Makarov NP, Martinovich NV (eds) [Dedicated to Arkadiy Yakovlevich Tugarinov, a selection of scientific articles]. Krasnoyarsk Regional Museum, Krasnoyarsk, pp 83–91 [in Russian]

Peters DS (1992) A new species of owl (Aves: Strigiformes) from the Middle Eocene Messel oil shale. In: Campbell KE (ed) Papers in Avian paleontology honoring Pierce Brodkorb. Nat Hist Mus Los Angeles Cty, Sci Ser 36:161–169

Peters DS (1999) *Selmes absurdipes*, new genus, new species, a sandcoleiform bird from the oil shale of Messel (Germany, Middle Eocene). In: Olson SL (ed) Avian Paleontology at the Close of the 20th Century: Proceedings of the 4th international meeting of the Society of Avian Paleontology and Evolution, Washington, D.C., 4–7 June 1996. Smithson Contrib Paleobiol 89:217–222

Peyer B (1957) Protornis glaronensis H. v. Meyer. Neubeschreibung des Typusexemplars und eines weiteren Fundes. Schweiz Paläontol Abh 73:1–47

Prum RO, Berv JS, Dornburg A, Field DJ, Townsend JP, Lemmon EM, Lemmon AR (2015) A comprehensive phylogeny of birds (Aves) using targeted next-generation DNA sequencing. Nature 526:569–573

Rich PV (1982) Tarsometatarsus of *Protostrix* from the mid-Eocene of Wyoming. Auk 99:576–579

Rich PV, Bohaska DJ (1976) The world's oldest owl: a new strigiform from the Paleocene of southwestern Colorado. Smithson Contrib Paleobiol 27:87–93

Rich PV, Bohaska DJ (1981) The Ogygoptyngidae, a new family of owls from the Paleocene of North America. Alcheringa 5:95–102

Smith KT, Wuttke M (2015) Avian pellets from the late Oligocene of Enspel, Germany—ecological interactions in deep time. Palaeobiodiv Palaeoenv 95:103–113

Smith NA, Stidham TA, Mitchell JS (2020) The first fossil owl (Aves, Strigiformes) from the Paleogene of Africa. Diversity 12(4):163

References

Vinther J, Briggs DEG, Prum RO, Saranathan V (2008) The colour of fossil feathers. Biol Lett 4:522–525

Weidig I (2006) The first New World occurrence of the Eocene bird *Plesiocathartes* (Aves: ?Leptosomidae). Paläontol Z 80:230–237

Weidig I (2010) New birds from the lower Eocene Green River Formation, North America. Rec Austral Mus 62:29–44

Wetmore A (1921) A fossil owl from the Bridger Eocene. Proc Acad Natl Sci Phila 73:455–458

Wetmore A (1933) The status of *Minerva antiqua*, *Aquila ferox* and *Aquila lydekkeri* as fossil birds. Am Mus Novit 680:1–4

Wetmore A (1938) Another fossil owl from the Eocene of Wyoming. Proc US Nat Mus 85:27–29

Yuri T, Kimball RT, Harshman J, Bowie RC, Braun MJ, Chojnowski JL, Han K-L, Hackett SJ, Huddleston CJ, Moore WS, Reddy S, Sheldon FH, Steadman DW, Witt CC, Braun EL (2013) Parsimony and model-based analyses of indels in avian nuclear genes reveal congruent and incongruent phylogenetic signals. Biol 2:419–444

Zhao T, Mayr G, Wang M, Wang W (2015) A trogon-like arboreal bird from the early Eocene of China. Alcheringa 39:287–294

Paleogene Avifaunas: A Synopsis of General Biogeographic and Paleoecological Aspects

The Paleogene period witnessed major tectonic and climatic events that are likely to have shaped various aspects of avian evolution. Owing to the formation of the Antarctic Circumpolar Current after the separation of South America and Australia from Antarctica, there was an onset of global cooling toward the early Oligocene (e.g., Berggren and Prothero 1992). Changes in continental geography due to plate tectonics and sea-level fluctuations furthermore affected faunal interchanges between major geographical areas. Many publications deal with the effect of these events on mammalian evolution (e.g., Janis 1993; Cox 2000; Rose 2006, and references therein), but their impact on Paleogene avifaunas is still insufficiently understood. Even fewer published data exist on ecological interactions of Paleogene bird communities, such as competition with mammals or between different avian taxa. In the following sections, some general aspects of the evolution of Paleogene avifaunas are outlined, which pertain to these issues.

11.1 Continental avifaunas of the Northern Hemisphere

The pre-Oligocene avifaunas of Europe and North America include many archaic taxa, which are clearly distinguished from extant bird groups. No crown group representatives of extant family-level taxa are known from Eocene or earlier fossil sites, even though some early Eocene taxa were already very similar to their extant relatives, such as *Scaniacypselus* and crown group swifts (Apodidae), *Plesiocathartes* and the extant Courol (Leptosomiformes), and stem group Coracii (*Eocoracias*, *Paracoracias*, *Geranopterus*) and extant rollers.

11.1.1 Biogeography

Europe and North America were connected by northern latitude land corridors via Greenland in the late Paleocene and early Eocene, the existence of which was terminated by the opening of the North Atlantic (e.g., Smith et al. 1994). Until the early Oligocene, Europe was furthermore separated from Asia by the Turgai Strait, an epicontinental seaway, which divided the Eurasian landmass. Overland dispersal between Europe and Asia was therefore impeded in the early Paleogene, whereas intermittent land connections between Asia and North America via Beringia existed during the whole Cenozoic (e.g., Rose 2006).

Given the occurrence of similar mammalian taxa in the Paleocene of North America and Europe (Rose 2006), it would be surprising if there were not also significant concordances in the avifaunas of that time. Indeed, the palaeognathous lithornithids are known from the Paleocene of both, Europe and North America (see Sect. 3.1), but otherwise the known non-marine Paleocene avifaunas of Europe and North America share few closely related taxa. Although strigiform birds occur on both continents, the North American *Ogygoptynx* and the European *Berruornis* represent very different lineages (Sects. 10.1.1 and 10.1.2). The Gastornithidae seem to have been restricted to Europe in the Paleocene, whereas Paleocene Presbyornithidae have so far only been reported from North America and Asia. Admittedly, it has to be kept in mind that the Paleocene fossil record of birds is still much less known than that of coeval mammals. Therefore, general conclusions cannot be drawn yet, and the biogeographic picture may change with future discoveries.

There was a maximum interchange of mammalian faunas between Europe and North America in the early Eocene (e.g., Janis 1993; Rose 2006), and previous authors already recognized that the early Eocene avifaunas of Europe and North America were likewise very similar (e.g., Houde and Olson 1989; Blondel and Mourer-Chauviré 1998). The occurrence of the flightless Gastornithidae in the early Eocene of North America and Europe actually provides direct evidence for the existence of a land connection by that time or shortly before (Andors 1992). Other avian groups, of which closely related taxa occur in the early Eocene of both Europe and North America, are the Lithornithidae, Gallinuloididae, Messelornithidae, Juncitarsidae, Eocypselidae, stem group Leptosomiformes (*Plesiocathartes*), Sandcoleidae and other stem group Coliiformes (*Chascacocolius*), stem group Coracii, as well as the Halcyornithidae, Messelasturidae, Zygodactylidae, and the zygodactyl stem group passeriform taxon *Eofringillirostrum*. The species of most of these taxa were forest-dwelling birds, which is in concordance with the presence of extensive paratropical forests in the early Eocene of at least the western part of the Northern Hemisphere (Blondel and Mourer-Chauviré 1998). With regard to possible high latitude dispersal corridors across Greenland, it is noteworthy that fossils of the Gastornithidae were discovered in early Eocene deposits of Ellesmere Island in the Canadian Arctic (Stidham and Eberle 2016).

Several avian taxa were only found in the early and middle Eocene of either North America or Europe, but because some of the most productive fossil localities on these two continents represent different palaeoenvironments (Sect. 2.5), this may be an artifact of the incomplete fossil record. Taxa, which occur in the early Eocene of North America but are unknown from coeval European sites include the frigatebird *Limnofregata* and the Presbyornithidae. Early Eocene species that were only reported from European deposits belong to the Remiornithidae, Idiornithidae, Archaeotrogonidae, stem group Nyctibiiformes and Trochilidae, as well as the apodiform Aegialornithidae. In the first edition of this book, the Palaeotididae were also listed as a taxon that is only known from the early and middle Eocene of Europe, but as detailed in Sect. 3.2.2, it now seems likely that the North American Geranoididae are close relatives of these palaeognathous birds.

Opening of the North Atlantic after the early Eocene restricted an overland dispersal between Europe and North America to land connections via Beringia. Still, the early Oligocene avifaunas of Europe and North America were remarkably similar, and taxa reported from the latest Eocene and early Oligocene of both North America and Europe include the enigmatic *Eocuculus*, as well as stem group representatives of the Todidae (*Palaeotodus*) and Coliidae (*Palaeospiza* in North America versus *Primocolius* and *Oligocolius* in Europe). Again, the correspondences seem to be greatest for forest-dwelling taxa, whereas the terrestrial avifaunas of Europe and North America already appear to have been more disparate in the early Oligocene. Whether the North American Bathornithidae are closely related to flightless Cariamiformes from the late Eocene and early Oligocene of Europe remains to be determined (Sect. 8.4.4). The cariamiform Idiornithidae, which are fairly abundant in some early Eocene European localities, however, have not been reported from North American fossil sites.

Because Europe and Asia were separated by the Turgai Strait in the early Eocene, its is likely that the flightless Gastornithidae dispersed into Asia from western North America rather than Europe, where the taxon seems to have originated (Sect. 4.2). An exact stratigraphic correlation of the Asian and North American records of these birds would, however, be required to confirm this hypothesis.

The closure of the Turgai Strait at the beginning of the Oligocene facilitated an exchange between European and Asian faunal elements. Dispersal of new taxa from Asia was probably one of the major factors which led to a marked faunal turnover in Europe, known as the "Grande Coupure" (e.g., Prothero 1994; Hooker et al. 2004). Little has so far been published on the impact of this event on European avifaunas at the Eocene/Oligocene boundary, and all observations were made for the Quercy taxa (Mourer-Chauviré 1980, 1988). As yet, only a few avian groups with a comprehensive early Paleogene fossil record are known that disappear at the Eocene/Oligocene boundary, that is, the Quercymegapodiidae, Quercypsittidae, the putative cariamiform taxon *Elaphrocnemus*, and the apodiform Aegialornithidae. By contrast, there are a fair number of taxa that have their earliest European fossil record in the latest Eocene/earliest Oligocene, including the Anatidae, Phasianoidea, Phalacrocoracidae, Palaelodidae, Gaviiformes (*Colymboides*), Rallidae, Parvigruidae, Charadriiformes, Pici, and Passeriformes. Many of these latter groups are aquatic or semi-aquatic, so that the Turgai Strait may not have formed an insurmountable barrier. Hence, their occurrence in Europe cannot be convincingly ascribed to the disappearance of this epicontinental seaway. The absence of some of these birds in earlier deposits may be an artifact of the fossil record, since most of our knowledge about late Eocene European avifaunas is based on fossils from the Quercy deposits, which mainly include terrestrial or arboreal forms. Dispersal from Asia is, however, likely for the Passeriformes, which appear to have originated in the Australian region (Oliveros et al. 2019). Extant Phasianoidea and Pici have poor capabilities for crossing the open sea, and the occurrence of these taxa in the early Oligocene of Europe may have also been due to dispersal from Asia after the closure of the Turgai Strait.

11.1.2 Climatic Cooling and avifaunal Turnovers

The Paleocene-Eocene Thermal Maximum, which is dated at 55.8 Ma (Woodburne et al. 2009), was followed by a temperature increase toward the early Eocene Climatic Optimum, some 52–50 Ma (Figueirido et al. 2012). The equable climate of the early Eocene Climatic Optimum lasted for about two million years and marks the warmest period of the Cenozoic, after which global temperatures began to decrease, with a major temperature drop characterizing the Eocene/Oligocene boundary (Zachos et al. 2001; Figueirido et al. 2012). For mammals, it was shown that these climatic events impacted faunal dynamics in the Eocene (Woodburne et al. 2009; Figueirido et al. 2012), with extinction events having been reported at the end of the early Eocene—the most severe of all pre-human Cenozoic mass extinctions ("Bridgerian crash")—and in the earliest Oligocene (Prothero 1994; Woodburne et al. 2009; Figueirido et al. 2012).

An assessment of faunal dynamics during the early Paleogene is more difficult for birds owing to the sparse fossil record from the Paleocene period and the patchy stratigraphic coverage of Eocene avifaunas. With regard to the Eocene extinctions, Lindow and Dyke (2006: 483) assumed that climatic changes did play a major role and that "deteriorations in climate beginning in the middle Eocene appear to be responsible for the demise of previously widespread avian lineages like Lithornithiformes and Gastornithidae." The factual basis for such general conclusions is, however, rather slim, and the composition of the bird assemblage from Walton-on-the-Naze, which dates back some 53 Ma, is very similar to that of Messel, which has an age of 48 Ma.

Unquestionably, there are many avian taxa that seem to have been widespread in the early and middle Eocene of Europe and North America and do not have an unambiguous later fossil record. In addition to the Gastornithidae and Lithornithidae, these include the Presbyornithidae, Gallinuloididae, Prophaethontidae, Eocypselidae, Sandcoleidae, Halcyornithidae, Messelasturidae, and Primobucconidae. Other taxa, such as the Leptosomiformes, Nyctibiiformes, and Podargiformes disappear from the Northern Hemisphere during the middle or late Eocene. At present, however, our knowledge of the exact stratigraphic occurrences of these taxa is simply too poor to correlate their extinctions with other biotic and abiotic events. As detailed in the preceding section and at least with regard to European avifaunas, the early Oligocene extinction events likewise appear to be only indirectly related to climatic changes but can be attributed to faunal exchanges owing to the disappearance of major geographic barriers.

In the Northern Hemisphere, climatic cooling had a profound impact on the vegetation of North America, where paratropical forests were replaced by open and more arid woodlands toward the late Eocene and Oligocene. These environmental changes affected the diversity and composition of mammalian faunas and led to a decimation of arboreal species toward the Oligocene (Webb 1977; Janis 1993; Cox 2000). Likewise, they are probably responsible for the disappearance of many lineages of early Paleogene forest-dwelling birds from North America. The Zygodactylidae, for example, have no fossil record in North America after the early Oligocene but occurred in Europe until the middle Miocene. Coliiform birds are known from the early Eocene of Europe and North America, and during this epoch, very similar species lived on both continents. Their latest record in North America is the late Eocene *Palaeospiza bella* (Sect. 10.2.2), whereas coliiform birds also persisted into the Neogene in Europe (e.g., Ballmann 1969). The same is true for stem group representatives of the Coracii, which are only known from the early and middle Eocene of North America but evolved into present-day rollers in the Old World. The Protostrigidae may also have survived longer in Europe, where they were found in late Oligocene deposits, whereas all North American records are of Eocene age.

In Europe, the Tethys Sea had a moderating effect, so that "European floras never reached the extremes of cooling or drying seen in North America" (Prothero 1994: 154), and as yet, no convincing evidence has been presented that climatic cooling had a major impact on the demise of avian higher-level taxa in the Eocene of Europe (contra Lindow and Dyke 2006). Mayr (2011) hypothesized that there was a two-phase extinction of Southern Hemispheric endemics in the Cenozoic of the Northern Hemisphere (Fig. 11.1). Various frugivorous or insectivorous taxa with poor migration capabilities disappeared toward the late Miocene, when the emergence of cold Northern Hemispheric winters seasonally restricted the food resources for these birds. Climatic cooling explains today's absence of the Trogoniformes, Coliiformes, and Psittaciformes in natural higher latitude habitats of the Northern Hemisphere. However, the Oligocene climate was still rather equable, and in Europe, these three taxa persisted into the Neogene. As outlined above, other taxa became extinct much earlier, and their disappearance from the Northern Hemisphere cannot be explained by climatic factors. This is, for example, true for stem group representatives of the Trochilidae, Nyctibiiformes, and Leptosomiformes, which were widely distributed in the early Paleogene and could have survived in tropical regions of the Old World. For these groups, which have a restricted extant distribution in the Southern Hemisphere, an extinction owing to biotic factors appears to be more likely than one caused by climatic cooling alone (see Sect. 11.3)

With regard to mammalian evolution, Alroy et al. (2000: 259) concluded that "over the scale of the whole Cenozoic, intrinsic, biotic factors like logistic diversity dynamics and within-lineage evolutionary trends seem to be far more important" than climatic changes. The same is also likely to be true in the case of birds, in particular those of the Eocene epoch.

Climatic cooling may, however, have had an indirect impact on the evolution of lacustrine avifaunas of temperate regions through the disappearance of crocodilians. In the

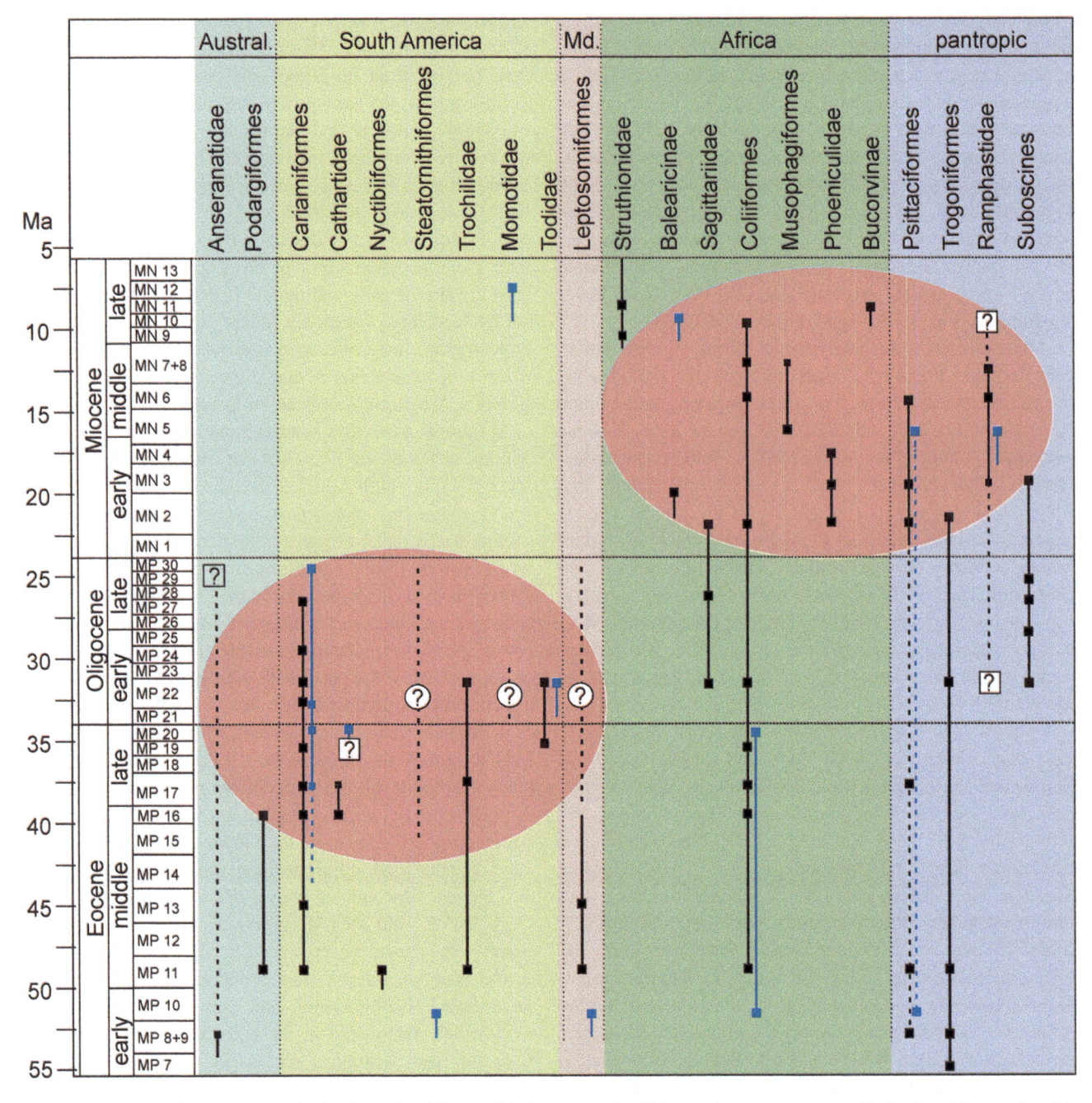

Fig. 11.1 Stratigraphic occurrences in the Cenozoic of Europe (black lines) and North America (blue lines) of avian taxa with extant Southern Hemispheric relatives. Note the earlier extinction dates of taxa that today have distribution in South America, the Australian region (Austral), and Madagascar (Md) compared to those, which today occur in Africa or have a pantropic distribution. The stratigraphic positions of actual fossils are indicated by squares; question marks denote uncertain identifications (squares) or the lack of exact stratigraphic data (encircled); after Mayr (2011)

early Paleogene, crocodilians were very abundant and diversified in many fossil sites of lacustrine origin, with the small Messel lake, for example, having yielded abundant remains of no less than seven sympatric crocodilian species (Schaal et al. 2018). At the same time, there is a conspicuous absence of aquatic birds in many early Paleogene sites of lacustrine origin (a notable exception is a fossil of a putative gaviiform bird from the Okanagan Highlands, which lived in a high altitude palaeolake in a microthermal climate that did not permit the existence of crocodilians). Crocodilians disappeared from northern latitudes in the course of climatic cooling toward the mid-Cenozoic, and at least in Central Europe, the diversification of aquatic avifaunas seems to have been correlated with the demise of crocodilians during the Miocene.

Today, truly aquatic birds, such as ducks, coots, and grebes, are conspicuously less abundant in tropical lakes than they are in higher latitude lacustrine environments. Crocodilians evidently coexisted with aquatic birds in some late Oligocene localities (Sect. 7.2) and do so today in various tropical lakes. It is, however, unlikely that the early evolution of lacustrine ducks, grebes, and coots would have commenced in tropical environments with numerous crocodilians, and these birds either evolved in high latitudes or altitudes with at least seasonally cold climates or in geographic regions without lacustrine crocodilians. It may therefore be no coincidence that certain groups of aquatic birds are today diversified and abundant in high altitude lacustrine environments (e.g., grebes and coots in the Andean region) or in hypersaline lakes (e.g., flamingos).

11.2 Continental avifaunas of the Southern Hemisphere

Despite a rather sparse Paleogene fossil record, the southern continents played an important role in discussions on the historical biogeography of extant avian groups. A number of extant higher-level taxa with species in the Northern Hemisphere furthermore seem to have had their origin in the Southern Hemisphere. Such was assumed by Olson (1989) for the Anseriformes, Psittaciformes, Podicipediformes, Columbiformes, and Passeriformes. However, for such biogeographic conclusions, it is important to distinguish between stem group and crown group representatives of a clade. Whereas crown group Passeriformes, for example, are indeed likely to have originated in the Southern Hemisphere (e.g., Claramunt and Cracraft 2015), stem group passerines, such as the Zygodactylidae and Psittacopedidae, were widespread in the early Paleogene of the Northern Hemisphere. The same is true for anseriform birds and, possibly, the Psittaciformes. Only of the Columbiformes and Podicipediformes, no stem group representatives have as yet been found in early Paleogene Northern Hemispheric sites. A Southern Hemispheric origin is also likely for crown group Falconiformes and the Pelecanidae (see taxonomic sections above).

11.2.1 Biogeography

The break-up of the Southern Hemispheric supercontinent Gondwana was largely completed at the beginning of the Paleogene (e.g., Smith et al. 1994). Africa separated from South America about 100 Ma, and Madagascar split from continental Africa in the middle to early late Jurassic, about 155–160 Ma. Through the Lord Howe Rise, New Zealand was possibly connected with Australia until 75 Ma (Haddrath and Baker 2001). A land connection between South America and Antarctica may have persisted until the late Eocene, about 36 Ma, when the Drake Passage opened, and overland dispersal between Antarctica and Australia was possible at least until 64 Ma (Woodburne and Case 1996).

Cracraft (1973) was among the first to set the extant distribution of birds in relation to the Mesozoic and Paleogene geography of the landmasses. In particular, he assumed that the early divergences of several extant clades with Southern Hemispheric representatives were due to the break-up of the southern continents in the Late Cretaceous. Cracraft (2001) further expounded this hypothesis of a vicariant origin of extant avian higher-level taxa, concluding that "trans-Antarctic distribution patterns" of several neornithine taxa indicate their origin in the Cretaceous of Gondwana. Similar hypotheses were also set up by molecular systematists for the initial divergences within some neoavian taxa, e.g., crown group Psittaciformes (Miyaki et al. 1998) and Passeriformes (Ericson et al. 2002, 2003).

The flightless palaeognathous birds have long been the classic avian example of a diversification owing to the break-up of Gondwana. Under the traditional assumption that these birds constitute a monophyletic group with a flightless stem species, the dispersal of the ancestors of the extant lineages would have strongly depended on continental geography (van Tuinen et al. 1998; Haddrath and Baker 2001). However, paraphyly of the flightless Palaeognathae with respect to the volant Tinamiformes is now supported by multiple independent molecular data (Hackett et al. 2008; Harshman et al. 2008; Prum et al. 2015; Kuhl et al. 2021), and flightlessness evolved at least three times independently, in the stem lineages of the Struthioniformes, Rheiformes, and Apterygiformes/Casuariiformes. As a volant bird, the stem species of these latter taxa would not have depended on land corridors for its dispersal.

The distributions of extant Megapodiidae (mainly Australian continental plate) and Cracidae (South and Central America) have also been taken as evidence for a vicariant diversification of crown group Galliformes owing to the break-up of Gondwana in the Late Cretaceous (Cracraft 1973, 2001). This hypothesis conflicts, however, with the global absence of crown group Galliformes in pre-Oligocene fossil deposits (Mayr and Weidig 2004), and the fact that all Paleogene galliform birds from South America and Africa are stem group representatives of the clade (Sect. 4.3.2). Stem group Galliformes were furthermore reported from the Northern Hemisphere and are successively more closely related to crown group Galliformes. The fossil record, therefore, suggests a Northern Hemispheric origin of crown group Galliformes, the stem species of which may have lived in the late Eocene/early Oligocene of Asia. From there, the Phasianoidea could have dispersed into Europe after the closure of the Turgai Strait in the early Oligocene, and a species in the stem lineage of the Megapodiidae could have reached Australia after northward drift brought this continent close to Asia in the Oligocene. The stem species of the Cracidae most likely dispersed into South America via North America, which was connected with Asia through Beringia during most of the Paleogene.

The evidence for diversification of neoavian lineages owing to the break-up of Gondwana is very weak. The traditional "Gruiformes," which were cited as an example of a neoavian taxon with a "trans-Antarctic distribution pattern" by Cracraft (2001), constitute a polyphyletic assemblage (Sect. 2.3). Likewise, the "Caprimulgiformes," another taxon listed by Cracraft (2001), are not monophyletic, and, except for the Aegotheliformes, all extant taxa of the Strisores with a Southern Hemispheric distribution (Steatornithiformes, Podargiformes, and Nyctibiiformes) have stem group representatives in the Northern Hemisphere. Although crown group Passeriformes probably originated in the Southern Hemisphere, an early diversification of these birds due to the break-up of Gondwana is also not well supported (contra Cracraft 2001; Ericson et al. 2002, 2003). There is strong evidence for an origin of the passeriform Oscines on the Australian continental plate (Barker et al. 2002; Ericson et al. 2002, 2003), but the area of origin of the Suboscines is less certain, and the distribution of the Acanthisittidae is restricted to New Zealand. As noted in the first edition of this book (Mayr 2009), the fossil data would be in agreement with a Paleogene origin of crown group Passeriformes on the Australian continental plate, which is also suggested by recent analyses of calibrated molecular data (Oliveros et al. 2019). The Suboscines could have dispersed into South America via Antarctica only before the onset of glaciation of this continent, and the fossil record (Sect. 9.5.3) rather indicates dispersal over the Northern Hemisphere and a late, i.e., Neogene, arrival in South America and Africa.

Owing to the then limited fossil record, Cracraft's earlier studies were mainly based on the distribution of extant taxa. That biogeographic inferences based on the distribution of the modern representatives can be profoundly misleading is, for example, shown by marsupial mammals, which are today restricted to Australia and the Americas but originated in Asia (Rose 2006). To address such shortcomings, Claramunt and Cracraft (2015) included 130 fossil taxa in a new biogeographic analysis. Based on these data, the authors concluded that the stem species of Neornithes lived in the Late Cretaceous of South America and that the early diversification of crown group birds was restricted to South America and Antarctica for 30 million years before neornithine lineages dispersed across the globe toward the Cretaceous/Paleogene boundary, some 66 million years ago. Under this biogeographic scenario, archaic stem group representatives of early diverging crown group lineages are expected to have been restricted to South America and Antarctica, whereas the Northern Hemisphere would have witnessed a sudden occurrence of more advanced taxa. As evident from the present book and as detailed by Mayr (2017a), this hypothesis does not conform to the fossil record. Paleogene avifaunas of the Northern Hemisphere not only include various stem group Galliformes and Anseriformes, which are successive sister taxa of their extant relatives, but they also feature stem group representatives of taxa, the extant species of which are today restricted to South America, with the Trochilidae, Opisthocomiformes, and Cariamiformes being among the most notable examples. Even in cases where multiple stem group taxa were described that are successive sister taxa of a crown clade, Claramunt and Cracraft (2015) included only a single representative, and this selective consideration of fossil taxa is likely to have biased their analysis (Mayr 2017a). Multiple range extensions of "South American" taxa into the Northern Hemisphere are well possible, but no plausible explanations were put forth to explain a subsequent restriction of their distribution to South America again. This biogeographic pattern would be so much the more unusual, because the Northern Hemispheric disappearance of many of these groups preceded the onset of the pronounced climatic cooling toward the mid-Cenozoic (Mayr 2011).

Not many of the characteristic groups which occur in the early Paleogene of the Northern Hemisphere have been reported from the Southern Hemisphere. In particular, our knowledge of early Paleogene avifaunas of Australia is too incomplete for well-founded conclusions on their composition and biogeographic affinities.

With a few exceptions, such as the anseriform Presbyornithidae, the known Paleogene non-marine avifauna of South America is clearly distinguished from that of other continents. Taxa such as the Lithornithidae, Sandcoleidae, and Halcyornithidae have an abundant fossil record in the Eocene of Europe and North America but were not yet identified in Southern Hemispheric deposits. As detailed by Mayr (in press), the record of a putative stem group roller from the early Eocene of Argentina (Degrange et al. 2021) is erroneous. Campbell and Tonni (1981) noted that Cenozoic avifaunas of South America are characterized by the dominance of large carnivorous birds, especially phorusrhacids and teratorns, which may be correlated with a low diversity of large mammalian carnivores (van Valkenburgh 1999). It is, however, remarkable, that no fossils of the Strigiformes were as yet reported from the Paleogene of South America, whereas owls are among the more abundant avian taxa in the Paleogene of the Northern Hemisphere.

Paleogene avifaunas of Africa remain poorly known but appear to have featured distinctive taxa unknown from coeval European localities. Such is, for example, true for the putative stem group cuculiform *Chambicuculus* from the late early or early middle Eocene of Namibia. The late Eocene/early

Oligocene avifauna of the Fayum in Egypt includes several higher-level taxa, which seem to represent African endemics (e.g., Eremopezidae, Xenerodiopidae, and Musophagidae) or are unknown from the early Oligocene of Europe (e.g., Ciconiidae and Jacanidae). If the identifications of the often fragmentary remains from the Fayum are confirmed by future findings, some taxa are, however, shared with coeval avifaunas of Europe (e.g., Palaelodidae, Pandionidae). More substantial Paleogene avifaunas of Africa south of the Sahara have only recently been described. These include at least some taxa that are similar to fossils found in Northern Hemispheric sites of similar age, such as stem group Galliformes and parrot-like birds (Mourer-Chauviré et al. 2015, 2017).

India was also part of the southern supercontinent Gondwana and reached its current position toward the early Eocene. Paleogene avifaunas of the Indian subcontinent are very scarce, but the taxon *Vastanavis* shows a resemblance to fossils from the early Eocene of Europe and North America (Mayr et al. 2010; Mayr 2015).

11.2.2 Extant Southern Hemispheric "endemics" in the Paleogene of the Northern Hemisphere

It has long been recognized that the distribution of many extant avian taxa is very different from that of their Paleogene stem group representatives (Mourer-Chauviré 1982; Olson 1989). Mourer-Chauviré (1982) already pointed out that the closest extant relatives of several avian taxa from the Quercy fissure fillings in France are restricted to the Southern Hemisphere. In particular, there is a notably high number of birds in these localities and other Paleogene Northern Hemispheric fossil sites, the closest extant relatives of which only occur in Central or South America (Mourer-Chauviré 1999; James 2005; Mayr 2005). Such is true for stem group representatives of the Cariamiformes (middle Eocene to early Miocene of Europe and late Eocene to late Oligocene of North America), Steatornithiformes (early Eocene of North America), Nyctibiiformes (Eocene of Europe), Trochilidae (middle Eocene to early Oligocene of Europe), Todidae (late Eocene/early Oligocene of Europe and North America), and, possibly (see comments in the taxonomic sections), the Momotidae and Cathartidae.

By contrast, the extant distribution of just two avian taxa with unambiguous Northern Hemispheric Paleogene stem group representatives is restricted to continental Africa south of the Sahara, that is, the Coliidae (early Paleogene of North America and Paleogene and Neogene of Europe) and Sagittariidae (Oligocene and Neogene of Europe). Stem group representatives of the Madagascan Leptosomiformes occurred in the early Eocene of Europe and North America.

With the possible exception of the Anseranatidae (Sect. 4.5.3), no avian taxa were reported from the Paleogene of the Northern Hemisphere, the crown group representatives of which are restricted to Australia. Earlier identifications of stem group representatives of the Megapodiidae in the Paleogene of Europe (Mourer-Chauviré 1982) were preliminary, and these fossils are now assigned to stem group Galliformes (Quercymegapodiidae).

Many of the above taxa probably had a wider distribution in the early Paleogene than is evident from their current fossil record. Because of the high degree of similarity of the early Eocene avifaunas of Europe and North America (see Sect. 11.1.1), stem group representatives of the Nyctibiiformes and Trochilidae, for example, may have also occurred in the Paleogene of North America. These taxa could also have been present in the Paleogene of South America and other southern continents. Rather than there having been profound shifts in the ranges of the stem- and crown group representatives of so many taxa, it is more likely that the crown group representatives of these birds have relict distributions (pro Olson 1989; contra Cracraft 2001).

The extant species of the above taxa with a South or Central American distribution have very different ecological requirements. Therefore, and because the modern descendants of these groups do not occur in the Old World tropics, the disappearance of their stem group representatives from the Northern Hemisphere is not easily explained by climatic changes alone (Sect. 11.1.2). Instead, biotic factors may have played a major role in the extinction of these groups. South America was isolated from other continents during most of the Cenozoic (e.g., Smith et al. 1994). Until the formation of the Panamanian Isthmus at the end of the Pliocene, this "splendid isolation" impeded the invasion of potential competitors and other species, which could have impacted prevailing ecosystems. Thereby, certain avian taxa may have persisted in South America and other, equally isolated geographic regions, such as Madagascar, whereas these birds succumbed to more competitive species in other parts of their range.

Because the above avian groups have very different ecological preferences, ecological interactions, which led to their extinction in the Northern Hemisphere, must have been complex and multifactorial. Examples of such interactions were detailed by Cristoffer and Peres (2003), who hypothesized that the higher density of large mammalian herbivores in Old World tropical forests is likely to be responsible for, e.g., a lower abundance of snails and butterflies than in the New World tropics. In a similar way, the much lower taxonomic diversity of placental mammals may have affected South and Central American bird communities by the absence of key predators and potential competitors. Clearly, however, much more research has to be conducted to identify potential factors, which impacted the distribution of, for example, nectarivorous stem group Trochilidae in a similar way as that of insectivorous and probably nocturnal stem group representatives of the Nyctibiiformes.

11.3 Ecological Interactions

This section summarizes some considerations on possible ecological interactions between birds and other organisms, which may have had an impact on the evolution of Paleogene avifaunas, focusing on some aspects of bird/mammal and bird/bird interactions. Coevolution with certain groups of insects and plants may have also played an important role in the early diversification of neornithine birds, but in most cases, the scarce data at hand do not even allow ad hoc hypotheses (see, however, Sect. 6.8).

11.3.1 Mammalian Evolution and Terrestrial avifaunas

In the first edition of this book (Mayr 2009), it was detailed that one of the notable aspects of Paleogene avifaunas concerns the occurrence of multiple lineages of flightless birds in continental avifaunas. This was attributed to a reduced abundance of larger terrestrial predators after the K/Pg extinction event and throughout the early Paleogene.

Mammals are among the main predators of birds, and it is well-known that an environment with little or no predation pressure frequently led to the evolution of flightlessness in birds (e.g., Feduccia 1999; Mayr 2017b). Today, flightless birds therefore mainly occur on remote, predator-free oceanic islands.

The Paleogene diversification of carnivorous mammals is likely to have had a significant impact on the evolution of birds, especially those with limited or no flight capabilities, and it was hypothesized that the dispersal of mammalian carnivores into continental Europe, specifically taxa of the clade Carnivora (cats, dogs, mustelids, and allies), terminated the existence of flightless birds (Mayr 2011). The evolutionary history of carnivorans was reviewed by van Valkenburgh (1999), Rose (2006), Flynn et al. (2010), Solé et al. (2016), and Hassanin et al. (2021), on whose surveys the following notes are based. The earliest larger carnivorous mammals in the Paleogene of the Northern Hemisphere were the coyote-sized ungulate Mesonychia, which first occur in the mid-Paleocene of North America and Europe and are considered to be representatives of the Artiodactyla and, hence, are more closely related to even-toed ungulates and whales than to carnivorans (e.g., Rose 2006). More specialized carnivores were the taxa of the Creodonta, which existed throughout the Eocene but had their greatest diversity in the early and middle Eocene. Creodonts are often united with the Carnivora in a clade Ferae, the monophyly of which is, however, disputed (e.g., Solé et al. 2016). The first true stem group Carnivora are from the mid-Paleocene of North America, from where they dispersed into Europe and Asia toward the early Eocene. These early taxa belong to the Miacidae and Viverravidae and were only fox-sized animals. The earliest crown group carnivorans, that is, representatives of the clades Caniformia (dogs, bears, mustelids, and allies) and Feliformia (cats, hyenas, civets, and allies), occurred in the late Eocene and early Oligocene. The Caniformia evolved in North America and did not disperse into the Old World before the Neogene, whereas the Feliformia originated in Asia and first appeared in Europe in the early Oligocene, just after the Grande Coupure.

After the extinction of non-avian dinosaurs and marine reptiles in the Late Cretaceous, several million years passed before the first specialized mammalian carnivores appeared (van Valkenburgh 1999). The absence of larger terrestrial predators in the early Paleocene permitted the evolution of flightlessness in various avian lineages, and it is probably no coincidence that the first occurrences of penguins, flightless Palaeognathae, gastornithids, and phorusrhacids are from the mid or late Paleocene (it is unlikely that the ages of the oldest known fossils correspond exactly to the dates of flight loss in these taxa, and assuming a ghost lineage of a few million years, the origin of flightlessness in these birds is likely to have fallen into the early Paleocene). Of the major lineages of flightless birds in Cenozoic avifaunas, at least penguins and phorusrhacids have an origin in the Southern Hemisphere, which correlates with the fact that no placental carnivores occurred in the Paleocene and Eocene of this part of the globe (Rose 2006: 342). Even though larger carnivorous marsupials (Sparassodonta) already existed in the early Paleogene of South America, only opossum-sized carnivorous mammals are known from the early Eocene of Antarctica (Goin et al. 1999). These dispersed into the continent from South America, and there may have been a predator-free period in the earliest Paleogene of the Antarctic region, which facilitated the evolution of flightless birds. New Zealand in particular was devoid of larger terrestrial predators throughout most of the Cenozoic, which, together with the Late Cretaceous extinction of marine reptiles, appears to have been an important factor in the early evolution of penguins, the earliest fossils of which come from the New Zealand region.

The origin of flightlessness in early Paleogene birds of the Northern Hemisphere is also likely to have been due to reduced predation pressure. Europe was largely isolated from other continents in the Late Cretaceous and Paleocene (e.g., Smith et al. 1994), and it is a plausible scenario that a species in the evolutionary lineage of the graviportal Gastornithidae lost its flight capability on this continent in the Late Cretaceous/earliest Paleocene. Gastornithids and the flightless Remiornithidae occur in the Paleocene of Europe but have not been reported from deposits of that epoch in North America. This indicates a lower predation pressure in the earliest Paleogene of Europe compared to the situation in North America. There were still a large number of flightless birds in the early and middle Eocene of Europe, including the Gastornithidae, Palaeotididae, the taxon *Strigogyps*, and possibly some larger species of the Idiornithidae. Of these taxa, only the Gastornithidae, which were probably less susceptible to predation owing to their large size, were reported from North America, where their fossil record is restricted to the early Eocene. Provided that the Geranoididae are indeed close relatives of the palaeognathous Palaeotididae, they add a further taxon of flightless birds to the early Eocene North American avifauna.

As detailed by Mayr (2011), the latest fossil record of a Paleogene flightless bird in Europe is a coracoid from the early Oligocene (MP 23) of France, which was assigned to *Ameghinornis minor* by Mourer-Chauviré (1983). By contrast, no unambiguous evidence exists for the presence of flightless birds in the Oligocene of North America. Therefore, it is likely that the early diversity of at least non-cursorial flightless birds in the continental Northern Hemispheric avifauna was greatly reduced (Europe), if not terminated (North America), by the occurrence of larger carnivorans toward the late Eocene/early Oligocene. Owing to the geographic isolation of some of the southern continents throughout most of the Cenozoic, flightless birds persisted longer in the Southern Hemisphere than in Europe and North America. In the extant (pre-human) avifauna, however, only cursorial flightless palaeognathous birds, which are able to outrun predators, coexist with large carnivorous mammals (Rheiformes in South America, Struthioniformes in Africa and Asia), whereas most other flightless birds are (were) restricted to areas free of larger mammalian predators (Australia, New Zealand, Madagascar, and various smaller oceanic islands).

Buffetaut and Angst (2014) challenged the above scenario on the evolution of flightlessness in Paleogene avifaunas. According to these authors, the occurrence of flightless birds in Europe a few million years after the Eocene/Oligocene boundary conflicts with the hypothesis that immigration of carnivorans after the Grande Coupure terminated the existence of flightless birds in Europe. Instead, Buffetaut and Angst (2014) hypothesized that a lack of competition with large herbivorous mammals may have played a greater role in the evolution of flightlessness in Paleogene avifaunas. This is, however, not a likely assumption, because there is no such correlation between the abundance of large mammalian herbivores and flightless birds in extant continental avifaunas. A herbivorous diet is furthermore reasonably well-established only for gastornithids, whereas the dietary preferences of palaeotidids are unknown. The early Oligocene *Ameghinornis minor* coracoid is from a flightless species of the Cariamiformes, with these birds likely having been carnivorous rather than herbivorous.

As already noted, just a few species lost their flight capabilities in a continental environment. Except for those of the Palaeognathae and the presumably galloanserine Gastornithidae, most belong to the Cariamiformes, flightless representatives of which appear to have been widespread and occurred in South America (Phorusrhacidae), North America (Bathornithidae), Europe (some Idiornithidae), and, possibly, Africa and Asia (if *Strigogyps* is a cariamiform bird and was flightless). Apparently, therefore, cariamiforms appear to have been particularly prone to the evolution of flightlessness.

Extant seriemas include two species, which have a predominantly terrestrial way of living and forage for larger insects and small vertebrates, and most extinct Cariamiformes probably had similar feeding habits. Foraging on or near the ground is one of the preconditions for the loss of flight capabilities in land birds (Mayr 2017b). In suitable environments, cariamiform birds, therefore, appear to have been predestined for the loss of flight capabilities by their way of living and their ecological preferences.

Mammals probably have an even greater impact on avian populations through predation on eggs and chicks, and today some of the greatest threats to endemic avifaunas are due to nest predation by invasive mammalian species. In particular, introduced rats and mice are formidable predators of eggs and nestlings, which account for high chick mortality in seabird colonies and for the extinction of numerous ground-breeding terrestrial birds on remote oceanic islands (e.g., Fukami et al. 2006). Primates and tree squirrels likewise are important nest predators, and there is a negative correlation between the abundance of tree squirrels and parrots in extant tropical and subtropical latitudes, with parrots being species-poor in Africa and Asia, where squirrels are abundant and diversified (Corlett and Primack 2006). The impact of predators is particularly severe if these expand their ranges and disperse into new areas, where birds had no chance to evolve adequate nest protection and escape strategies. Such can not only be seen in large-scale extinctions on remote islands but may also have been the case in the geologic past, when newly emerged land corridors enabled intercontinental mammalian dispersal.

Finally, it should be noted that mammals probably not only had an impact on the evolution of birds as predators but also as prey. Among the main extant avian predators of mammals are owls and diurnal birds of prey. The former include predominantly nocturnal species and have a significantly older fossil record than diurnal birds of prey, which dates back into the Paleocene. If most small mammals of that epoch were nocturnal, mammalian predation by owls may have preceded that by diurnal birds of prey. Alternatively, owls may have become nocturnal at a later stage of their evolution owing to an increasing feeding competition with diurnal birds of prey, the evolution of which was probably also intimately connected with that of mammalian prey.

11.3.2 The Impact of Passerines on the Diversity of Paleogene Avian Insectivores

Today, most small insectivorous or omnivorous birds belong to the Passeriformes, which also constitute the majority of perching birds. As noted in Sect. 9.5.3, there are bones of putative passerines from the early Eocene of Australia. Outside the latter continent, passerines first occur in the early Oligocene of Europe, where they did not become the predominant group of small arboreal birds before the Oligocene. Passerines have no Paleogene fossil record in the New World and Africa, and in the latter continent, they have not even been reported from early Miocene deposits (Mourer-Chauviré 2003).

Both, the phylogenetic relationships of the extant taxa and the fossil record, are in concordance with an origin of Pan-Passeriformes in the Australian region (Sect. 9.5.3 and Oliveros et al. 2019). Dispersal of passerines into Europe seems to have been from Asia and may have been related to the northward drift of the Australian continental plate in the mid-Paleogene and the closure of the Turgai Strait in the early Oligocene.

Before the Oligocene, small non-passerines most likely filled ecological niches, which today are occupied by passerines, and the arrival of the latter probably had a major impact on these Paleogene non-passerine birds (Harrison 1979; Mayr 2005). If Paleogene Passeriformes had similar feeding preferences to their extant relatives, it is to be expected that feeding competition with passeriform birds during periods of restricted food availability had its greatest effect on small insectivorous taxa. Indeed, there seems to have been a greater diversity of insectivorous taxa of the Strisores in the early Paleogene. The crown group members of some taxa with putatively insectivorous or omnivorous Paleogene stem group representatives furthermore exhibit feeding specializations or foraging strategies, which are not, or only rarely, found in passeriform birds (Mayr 2005).

For example, crown group Upupiformes either forage on the ground (Upupidae) or are specialized for trunk climbing (Phoeniculidae), whereas the Messelirrisoridae, as evidenced by their foot structure, seems to have been perching birds (Mayr 1998). Within crown group Pici, the Picumninae, and Picinae are specialized for trunk-climbing and woodpecking, the Indicatoridae mainly feed on beeswax, whereas toucans and many barbets mostly eat fruits. By contrast, and according to its bill morphology, the early Oligocene *Rupelramphastoides* appears to have been a more opportunistic, insectivorous or omnivorous, feeder (Mayr 2006). Judging from their bill shapes, stem group representatives of the Coliiformes exhibited a great diversity of feeding specializations in the Paleogene (Houde and Olson 1992; Mayr and Peters 1998; Mayr 2001), whereas the six extant species of mousebirds are mainly herbivorous. The beak of *Chascacocolius* (Sect. 10.2.2), which closely resembles that of some New World blackbirds (Icteridae), is a particularly striking example of a passerine-like feeding adaptation in Paleogene non-Passeriformes.

It is notable that one of the most species-rich groups of arboreal birds, that is, oscine Passeriformes, evolved in the Australian region (e.g., Claramunt and Cracraft 2015). On the other hand, two widely distributed groups of arboreal birds, the Trogoniformes and the Piciformes, do not occur east of Wallace's Line, which separates Asia from Australia. The absence of these two avian groups in Australia has puzzled biogeographers for a long time. In view of our improved understanding of the early evolution of oscine passerines, it seems possible that the latter occupied ecological niches for small arboreal birds early on, which impeded the dispersal of piciform and trogoniform birds into the Australian region.

11.3.3 Marine avifaunas

The evolution of many taxa of aquatic birds appears to have commenced in marine habitats. This is not only true for avian groups that are today still restricted to the marine realm, such as the Sphenisciformes, Phaethontiformes, and Procellariiformes but also for those that are today more diversified in lacustrine environments, such as the Anseriformes.

Major aspects of the evolution of marine avifaunas were summarized by Warheit (1992, 2002), although these reviews mainly focused on the Neogene fossil record. Among the most outstanding characteristics of Paleogene marine avifaunas is the existence of giant soaring and diving birds without extant counterparts. Whereas the former (Pelagornithidae) seem to have achieved a worldwide distribution already by the early Eocene, the latter were restricted to the Southern Hemisphere (Sphenisciformes) and the North Pacific (Plotopteridae).

11.3.3.1 Volant Seabirds

Today, the Procellariiformes and Phaethontiformes, as well as the suliform Fregatidae and some Charadriiformes (Stercorariidae, Laridae, and Alcidae) are the predominant taxa among the volant seabirds, and many species of these taxa regularly feed offshore. In the early Paleogene, by contrast, the most abundant and widespread marine birds seem to have been stem group Phaethontiformes and the Pelagornithidae, which are found together at several fossil sites, such as the early Eocene of the Ouled Abdoun Basin in Morocco and the London Clay. Pelagornithids undoubtedly were highly pelagic and regularly fed offshore, but the habitat preferences of stem group Phaethontiformes are less clear (Sect. 7.1), and these birds may also have occurred in littoral regions on a regular basis.

Small to medium-sized pelagornithids coexisted with very large forms in the early Paleogene, whereas all Neogene species of the Pelagornithidae had a large size. Concerning a trend toward gigantism, pelagornithids paralleled the Mesozoic pterosaurs, of which also only the giant forms survived into the Late Cretaceous. In the case of pterosaurs, it was hypothesized that competition with birds led to the extinction of the smaller species (e.g., Paul 2002). In analogy, ecological competition with other marine birds may have played a role in the extinction of smaller pelagornithids. Just because of their very large size, competitive pressure may have been lower for the giant species, and intraspecific or interspecific competition within pelagornithids may have driven a size increase in these birds. However, the extreme specializations of the flight and feeding apparatus made giant pelagornithids prone to environmental changes and ultimately proved to be an evolutionary dead end. More data have to be gathered on the stratigraphic distribution of these birds to identify factors that contributed to their extinction, but increased predation at their breeding sites or competition with other seabird taxa for suitable nesting sites may have played a role.

11.3.3.2 Large Diving Birds

In penguins, flightlessness evolved around the K/Pg boundary, and even some of the earliest stem group representatives of the Sphenisciformes had a large size compared to extant flightless diving birds (Sect. 7.4). Species of the Plotopteridae likewise achieved a very large size, with some of the earliest known (late Eocene) plotopterids already having been very large-sized. The radiation of penguins may have been facilitated by the absence of terrestrial predators in the earliest Cenozoic of the Antarctic/New Zealand region, whereas the radiation of plotopterids was possibly related to the evolution of kelp ecosystems in the North Pacific (Sect. 7.6). In both cases, highly productive marine ecosystems provided abundant food resources for marine diving birds.

Giant penguins became extinct toward the late Oligocene (e.g., Clarke et al. 2007). Large plotopterids also disappeared toward the late Oligocene and only some small species appear to have survived into the early Neogene. Similarly sized diving birds do not exist in late Neogene and extant avifaunas, and various factors have been suggested to explain their disappearance. Goedert (1988), in the case of plotopterids, and Vickers-Rich (1991: 748), regarding penguins, assumed that the demise of these birds may have had climatic reasons. Simpson (1971) considered it possible that the giant penguins succumbed to competition with pinnipeds and cetaceans, which was also assumed by Olson and Hasegawa (1979) in the case of plotopterids.

In the eastern North Pacific, plotopterids coexisted with odontocete cetaceans from at least the late Eocene on (Goedert and Cornish 2002), which makes it unlikely that competition with whales played a major role in their extinction. However, Warheit and Lindberg (1988) detailed that interference competition at breeding sites with land-based gregarious pinnipeds, such as elephant seals, has a significant impact on large flightless seabirds. They hypothesized that in New Zealand and Australian penguins this led to shifts to smaller body sizes, which temporally coincide with the first local appearance of gregarious pinnipeds (Warheit and Lindberg 1988). As further set forth by these authors, many extant penguins either nest in hilly or rocky terrains, which do not allow large aggregations of pinnipeds, or they are fossorial, that is, they nest in earth tunnels and thereby avoid competition with pinnipeds.

The evolutionary history of plotopterids is much less understood than that of penguins, but their extinction coincided temporally with that of giant penguins. It was hypothesized that plotopterids may have been outcompeted by pinnipeds, either through predation or by competition for breeding sites on offshore islands (Goedert and Cornish 2002). According to Goedert and Cornish (2002), the local extinction of plotopterids at the Oregon and Washington coastlines may have been driven by the demonstrable disappearance of offshore volcanic islands as potential breeding sites toward the late Oligocene. Sea level rising first led to the submersion of these islands along the coasts of Washington and Oregon, whereas they persisted longer in California, from where the last North American plotopterid species, *Plotopterum joaquinensis*, stems (Goedert and Cornish 2002). Whether a loss of suitable offshore breeding sites also occurred in the western North Pacific remains to be studied, and more data are needed for an understanding of ecological drivers that caused the extinction of plotoperids.

More recently, Ando and Fordyce (2014) addressed potential drivers for the evolution of flightless marine diving birds. These authors did not find a correlation between an increase of pinniped diversity and a decrease of plotopterids, whereas their data indicated a correlation between a late Oligocene/early Miocene diversity decrease of flightless seabirds and a coeval increase of the numbers of odontocete whales. Ando and Fordyce (2014) hypothesized that feeding competition may have played a role, but they noted that various other factors may have also contributed to the demise of large diving seabirds.

References

Alroy J, Koch PL, Zachos JC (2000) Global climate change and North American mammalian evolution. Paleobiol 26(Suppl):259–288

Ando T, Fordyce RE (2014) Evolutionary drivers for flightless, wing-propelled divers in the Northern and Southern Hemispheres. Palaeogeogr Palaeoclimatol Palaeoecol 400:50–61

Andors A (1992) Reappraisal of the Eocene groundbird *Diatryma* (Aves: Anserimorphae). In: Campbell KE (ed) Papers in avian paleontology honoring Pierce Brodkorb. Nat Hist Mus Los Angeles Cty Sci Ser 36:109–125

Ballmann P (1969) Les oiseaux miocènes de La Grive-Saint-Alban (Isère). Geobios 2:157–204

Barker FK, Barrowclough GF, Groth JG (2002) A phylogenetic hypothesis for passerine birds: taxonomic and biogeographic implications of an analysis of nuclear DNA sequence. Proc R Soc Lond Ser B 269:295–308

Berggren WA, Prothero DR (1992) Eocene-Oligocene climatic and biotic evolution: an overview. In: Prothero DR, Berggren WA (eds) Eocene-Oligocene climatic and biotic evolution. Princeton University Press, New Jersey, pp 1–28

Blondel J, Mourer-Chauviré C (1998) Evolution and history of the western Palaearctic avifauna. Trends Ecol Evol 13:488–492

Buffetaut E, Angst D (2014) Stratigraphic distribution of large flightless birds in the Palaeogene of Europe and its palaeobiological and palaeogeographical implications. Earth-Sci Rev 138:394–408

Campbell KE Jr, Tonni EP (1981) Preliminary observations on the paleobiology and evolution of teratorns (Aves: Teratornithidae). J Vertebr Paleontol 1:265–272

Claramunt S, Cracraft J (2015) A new time tree reveals Earth history's imprint on the evolution of modern birds. Sci Adv 1(11):e1501005

Clarke JA, Ksepka DT, Stucchi M, Urbina M, Giannini N, Bertelli S, Narváez Y, Boyd CA (2007) Paleogene equatorial penguins challenge the proposed relationship between penguin biogeography, diversity, and Cenozoic climate change. Proc Natl Acad Sci U S A 104:11545–11550

Corlett RT, Primack RB (2006) Tropical rainforests and the need for cross-continental comparisons. Trends Ecol Evol 21:104–110

Cox CB (2000) Plate tectonics, seaways and climate in the historical biogeography of mammals. Mem Inst Oswaldo Cruz 95:509–516

Cracraft J (1973) Continental drift, paleoclimatology, and the evolution and biogeography of birds. J Zool 169:455–545

Cracraft J (2001) Avian evolution, Gondwana biogeography and the Cretaceous-Tertiary mass extinction event. Proc R Soc Lond Ser B 268:459–469

Cristoffer C, Peres CA (2003) Elephants versus butterflies: the ecological role of large herbivores in the evolutionary history of two tropical worlds. J Biogeogr 30:1357–1380

Degrange FJ, Pol D, Puerta P, Wilf P (2021) Unexpected larger distribution of Paleogene stem-rollers (Aves, Coracii): new evidence from the Eocene of Patagonia, Argentina. Sci Rep 11:1363

Ericson PGP, Christidis L, Cooper A, Irestedt M, Jackson J, Johansson US, Norman JA (2002) A Gondwanan origin of passerine birds supported by DNA sequences of the endemic New Zealand wrens. Proc R Soc Lond Ser B 269:235–241

Ericson PGP, Irestedt M, Johansson US (2003) Evolution, biogeography, and patterns of diversification in passerine birds. J Avian Biol 34:3–15

Feduccia A (1999) The origin and evolution of birds, 2nd edn. Yale University Press, New Haven

Figueirido B, Janis CM, Pérez-Claros JA, De Renzi M, Palmqvist P (2012) Cenozoic climate change influences mammalian evolutionary dynamics. Proc Natl Acad Sci U S A 109:722–727

Flynn JF, Finarelli JA, Spaulding M (2010) Phylogeny of the Carnivora and Carnivoramorpha, and the use of the fossil record to enhance understanding of evolutionary transformations. In: Goswami A, Friscia A (eds) Carnivoran evolution. New views on phylogeny, form and function. Cambridge University Press, Cambridge, pp 25–63

Fukami T, Wardle DA, Bellingham PJ, Mulder CP, Towns DR, Yeates GW, Bonner KI, Durrett MS, Grant-Hoffman MN, Williamson WM (2006) Above- and below-ground impacts of introduced predators in seabird-dominated island ecosystems. Ecol Lett 9:1299–1307

Goedert JL (1988) A new late Eocene species of Plotopteridae (Aves: Pelecaniformes) from northwestern Oregon. Proc Calif Acad Sci 45:97–102

Goedert JL, Cornish J (2002) A preliminary report on the diversity and stratigraphic distribution of the Plotopteridae (Pelecaniformes) in Paleogene rocks of Washington State, USA. In: Zhou Z, Zhang F (eds) Proceedings of the 5th symposium of the Society of Avian Paleontology and Evolution, Beijing, 1–4 June 2000. Science Press, Beijing, pp 63–76

Goin FJ, Case JA, Woodburne MO, Vizcaíno SF, Reguero MA (1999) New discoveries of "opposum-like" marsupials from Antarctica (Seymour Island, medial Eocene). J Mamm Evol 6:335–365

Hackett SJ, Kimball RT, Reddy S, Bowie RCK, Braun EL, Braun MJ, Chojnowski JL, Cox WA, Han K-L, Harshman J, Huddleston CJ, Marks BD, Miglia KJ, Moore WS, Sheldon FH, Steadman DW, Witt CC, Yuri T (2008) A phylogenomic study of birds reveals their evolutionary history. Science 320:1763–1767

Haddrath O, Baker AJ (2001) Complete mitochondrial DNA genome sequences of extinct birds: ratite phylogenetics and the vicariance biogeography hypothesis. Proc R Soc Lond Ser B 268:939–945

Harrison CJO (1979) Small non-passerine birds of the Lower Tertiary as exploiters of ecological niches now occupied by passerines. Nature 281:562–563

Harshman J, Braun EL, Braun MJ, Huddleston CJ, Bowie RCK, Chojnowski JL, Hackett SJ, Han K-L, Kimball RT, Marks BD, Miglia KJ, Moore WS, Reddy S, Sheldon FH, Steadman DW, Steppan SJ, Witt CC, Yuri T (2008) Phylogenomic evidence for multiple losses of flight in ratite birds. Proc Natl Acad Sci U S A 36:13462–13467

Hassanin A, Veron G, Ropiquet A, Jansen van Vuuren B, Lécu A, Goodman SM, Haider J, Nguyen TT (2021) Evolutionary history of Carnivora (Mammalia, Laurasiatheria) inferred from mitochondrial genomes. PLoS One 16:e0240770

Hooker JJ, Collinson ME, Sille NP (2004) Eocene-Oligocene mammalian faunal turnover in the Hampshire Basin, UK: calibration to the global time scale and the major cooling event. J Geol Soc 161:161–172

Houde P, Olson SL (1989 [1988]) Small arboreal nonpasserine birds from the Early Tertiary of western North America. In: Ouellet H (ed) Acta XIX Congressus Internationalis Ornithologici. University of Ottawa Press, Ottawa, pp 2030–2036

Houde P, Olson SL (1992) A radiation of coly-like birds from the early Eocene of North America (Aves: Sandcoleiformes new order). In: Campbell KE (ed) Papers in avian paleontology honoring Pierce Brodkorb. Nat Hist Mus Los Angeles Cty, Sci Ser 36:137–160

James HF (2005) Paleogene fossils and the radiation of modern birds. Auk 122:1049–1054

Janis CM (1993) Tertiary mammal evolution in the context of changing climates, vegetation, and tectonic events. Annu Rev Ecol Syst 24:467–500

Kuhl H, Frankl-Vilches C, Bakker A, Mayr G, Nikolaus G, Boerno ST, Klages S, Timmermann B, Gahr M (2021) An unbiased molecular approach using 3'UTRs resolves the avian family-level tree of life. Mol Biol Evol 38:108–121

Lindow BEK, Dyke GJ (2006) Bird evolution in the Eocene: climate change in Europe and a Danish fossil fauna. Biol Rev 81:483–499

Mayr G (1998) „Coraciiforme" und „piciforme" Kleinvögel aus dem Mittel-Eozän der Grube Messel (Hessen, Deutschland). Cour Forsch-Inst Senckenberg 205:1–101

Mayr G (2001) New specimens of the Middle Eocene fossil mousebird *Selmes absurdipes* Peters 1999. Ibis 143:427–434

References

Mayr G (2005) The Paleogene fossil record of birds in Europe. Biol Rev 80:515–542

Mayr G (2006) First fossil skull of a Paleogene representative of the Pici (woodpeckers and allies) and its evolutionary implications. Ibis 148: 824–827

Mayr G (2009) Paleogene fossil birds, 1st edn. Springer, Heidelberg

Mayr G (2011) Two-phase extinction of "Southern Hemispheric" birds in the Cenozoic of Europe and the origin of the Neotropic avifauna. Palaeobiodiv Palaeoenv 91:325–333

Mayr G (2015) A reassessment of Eocene parrotlike fossils indicates a previously undetected radiation of zygodactyl stem group representatives of passerines (Passeriformes). Zool Scr 44:587–602

Mayr G (2017a) Avian higher-level biogeography: Southern Hemispheric origins or Southern Hemispheric relicts? J Biogeogr 44:956–957

Mayr G (2017b) Avian Evolution: The fossil record of birds and its paleobiological significance. Wiley-Blackwell, Chichester

Mayr G (in press) A partial skeleton of *Septencoracias* from the early Eocene London Clay reveals derived features of bee-eaters (Meropidae) in a putative stem group roller (Aves, Coracii). Palaeobiodiv Palaeoenv

Mayr G, Peters DS (1998) The mousebirds (Aves: Coliiformes) from the Middle Eocene of Grube Messel (Hessen, Germany). Senck leth 78: 179–197

Mayr G, Weidig I (2004) The early Eocene bird *Gallinuloides wyomingensis*—a stem group representative of Galliformes. Acta Palaeontol Pol 49:211–217

Mayr G, Rana RS, Rose KD, Sahni A, Kumar K, Singh L, Smith T (2010) *Quercypsitta*-like birds from the early Eocene of India (Aves, ?Psittaciformes). J Vertebr Paleontol 30:467–478

Miyaki CY, Matioli SR, Burke T, Wajntal A (1998) Parrot evolution and paleogeographical events: mitochondrial DNA evidence. Mol Biol Evol 15:544–551

Mourer-Chauviré C (1980) The Archaeotrogonidae from the Eocene and Oligocene deposits of "Phosphorites du Quercy", France. In: Campbell KE (ed) Papers in avian paleontology honoring Hildegarde Howard. Nat Hist Mus Los Angeles Cty Contrib Sci 330:17–31

Mourer-Chauviré C (1982) Les oiseaux fossiles des Phosphorites du Quercy (Eocène supérieur à Oligocène supérieur): implications paléobiogéographiques. Geobios, mém spéc 6:413–426

Mourer-Chauviré C (1983) Les Gruiformes (Aves) des Phosphorites du Quercy (France). 1. Sous-ordre Cariamae (Cariamidae et Phorusrhacidae). Systématique et biostratigraphie. Palaeovertebr 13:83–143

Mourer-Chauviré C (1988) Les Aegialornithidae (Aves: Apodiformes) des Phosphorites du Quercy. Comparaison avec la forme de Messel. Cour Forsch-Inst Senckenberg 107:369–381

Mourer-Chauviré C (1999) Les relations entre les avifaunes du Tertiaire inférieur d'Europe et d'Amérique du Sud. Bull Soc Géol France 170: 85–90

Mourer-Chauviré C (2003) Birds (Aves) from the Middle Miocene of Arrisdrift (Namibia). Preliminary study with description of two new genera: *Amanuensis* (Accipitridae, Sagittariidae) and *Namibiavis* (Gruiformes, Idiornithidae). Mem Geol Surv Namibia 19:103–113

Mourer-Chauviré C, Pickford M, Senut B (2015) Stem group galliform and stem group psittaciform birds (Aves, Galliformes, Paraortygidae, and Psittaciformes, family incertae sedis) from the Middle Eocene of Namibia. J Ornithol 156:275–286

Mourer-Chauviré C, Pickford M, Senut B (2017) New data on stem group Galliformes, Charadriiformes, and Psittaciformes from the middle Eocene of Namibia. Contrib MACN 7:99–131

Oliveros CH, Field DJ, Ksepka DT, Barker FK, Aleixo A, Andersen MJ, Alström P, Benz BW, Braun EL, Braun MJ, Bravos GA, Brumfield RT, Chesser RT, Claramunt S, Cracraft J, Cuervo AM, Derryberry EP, Glenn TC, Harvey MG, Hosner PA, Joseph L, Kimball RT, Mack AL, Miskelly CM, Peterson AT, Robbins MB, Sheldon FH, Silveira LS, Smith BT, White ND, Moyle RG, Faircloth BC (2019) Earth history and the passerine superradiation. Proc Natl Acad Sci 116:7916–7925

Olson SL (1989 ["1988"]) Aspects of global avifaunal dynamics during the Cenozoic. In: Ouellet H (ed) Acta XIX Congressus Internationalis Ornithologici. University of Ottawa Press, Ottawa, pp 2023–2029

Olson SL, Hasegawa Y (1979) Fossil counterparts of giant penguins from the North Pacific. Science 206:688–689

Paul GS (2002) Dinosaurs of the air: The evolution and loss of flight in dinosaurs and birds. Johns Hopkins University Press, Baltimore

Prothero DR (1994) The late Eocene-Oligocene extinctions. Annu Rev Earth Planet Sci 22:145–165

Prum RO, Berv JS, Dornburg A, Field DJ, Townsend JP, Lemmon EM, Lemmon AR (2015) A comprehensive phylogeny of birds (Aves) using targeted next-generation DNA sequencing. Nature 526:569–573

Rose KD (2006) The beginning of the age of mammals. Johns Hopkins University Press, Baltimore

SKF S, Smith K, Habersetzer J (eds) (2018) Messel - An ancient greenhouse ecosystem. Schweitzerbart, Stuttgart

Simpson GG (1971) A review of the pre-pliocene penguins of New Zealand. Bull Am Mus Nat Hist 144:319–378

Smith AG, Smith DG, Funnell BM (1994) Atlas of Mesozoic and Cenozoic coastlines. Cambridge University Press, Cambridge

Solé F, Smith T, De Bast E, Codrea V, Gheerbrant E (2016) New carnivoraforms from the latest Paleocene of Europe and their bearing on the origin and radiation of Carnivoraformes (Carnivoramorpha, Mammalia). J Vertebr Paleontol 36:e1082480

Stidham TA, Eberle JJ (2016) The palaeobiology of high latitude birds from the early Eocene greenhouse of Ellesmere Island, Arctic Canada. Sci Rep 6:20912

van Tuinen M, Sibley CG, Hedges SB (1998) Phylogeny and biogeography of ratite birds inferred from DNA sequences of the mitochondrial ribosomal genes. Mol Biol Evol 15:370–376

van Valkenburgh B (1999) Major patterns in the history of carnivorous mammals. Annu Rev Earth Planet Sci 27:463–493

Vickers-Rich PV (1991) The Mesozoic and Tertiary history of birds on the Australian plate. In: Vickers-Rich P, Monaghan TM, Baird RF, Rich TH (eds) Vertebrate palaeontology of Australasia. Pioneer Design Studio and Monash University Publications Committee, Melbourne, pp 721–808

Warheit KI (1992) A review of the fossil seabirds from the Tertiary of the North Pacific: plate tectonics, paleoceanography, and faunal change. Paleobiol 18:401–424

Warheit KI (2002) The seabird fossil record and the role of paleontology in understanding seabird community structure. In: Schreiber EA, Burger J (eds) Biology of marine birds. CRC Marine Biology Series, Boca Raton, Florida, pp 17–55

Warheit KI, Lindberg DR (1988) Interactions between seabirds and marine mammals through time: interference competition at breeding sites. In: Burger J (ed) Seabirds & other marine vertebrates: Competition, predation, and other interactions. Columbia University Press, New York, pp 292–328

Webb SD (1977) A history of savanna vertebrates in the New World. Part I: North America and the Great Interchange. Annu Rev Ecol Syst 8:355–380

Woodburne MO, Case JA (1996) Dispersal, vicariance, and the late Cretaceous to early Tertiary land mammal biogeography from South America to Australia. J Mamm Evol 3:121–161

Woodburne MO, Gunnell GF, Stucky RK (2009) Climate directly influences Eocene mammal faunal dynamics in North America. Proc Natl Acad Sci U S A 106:13399–13403

Zachos J, Pagani M, Sloan L, Thomas E, Billups K (2001) Trends, rhythms, and aberrations in global climate 65 Ma to Present. Science 292:686–693

Printed by Books on Demand, Germany